Activated Carbon Adsorption

Roop Chand Bansal
Meenakshi Goyal

CRC Press
Taylor & Francis Group
Boca Raton London New York

CRC Press is an imprint of the
Taylor & Francis Group, an **informa** business
A TAYLOR & FRANCIS BOOK

Published in 2005 by
CRC Press
Taylor & Francis Group
6000 Broken Sound Parkway NW, Suite 300
Boca Raton, FL 33487-2742

First issued in paperback 2020

ISBN 13: 978-0-367-57807-7 (pbk)
ISBN 13: 978-0-8247-5344-3 (hbk)

Visit the Taylor & Francis Web site at
http://www.taylorandfrancis.com

and the CRC Press Web site at
http://www.crcpress.com

Library of Congress Cataloging-in-Publication Data

Bansal, Roop Chand, 1937-
  Activated carbon adsorption / Roop Chand Bansal and Meenakshi Goyal.
     p. cm.
  Includes bibliographical references and indexes.
  ISBN 0-8247-5344-5
  1. Carbon, Activated. 2. Carbon--Asorption and adsorption. I. Goyal, Meenakshi. II. Title.

TP245.C4B36 2005
662'.93--dc22   2004024878

Preface

Activated carbons are versatile adsorbents. Their adsorptive properties are due to their high surface area, a microporous structure, and a high degree of surface reactivity. They are, used, therefore, to purify, decolorize, deodorize, dechlorinate, separate, and concentrate in order to permit recovery and to filter, remove, or modify the harmful constituents from gases and liquid solutions. Consequently, activated carbon adsorption is of interest to many economic sectors and concern areas as diverse as food, pharmaceutical, chemical, petroleum, nuclear, automobile, and vacuum industries as well as for the treatment of drinking water, industrial and urban waste water, and industrial flue gases.

Interest in activated carbon adsorption of gases and vapors received a big boost during and after the first World War, while an increasing attention to the activated carbon adsorption from aqueous solutions was initiated by the pollution of the environment, which includes air and water, due to rapid industrialization and ever-increasing use of the amount and the variety of chemicals in almost every facet of human endeavor. Life has initiated increasing attention to the activated carbon adsorption from aqueous solutions. It was, therefore, thought worthwhile and opportune to prepare a text that describes the surface structure of activated carbons, the adsorption phenomenon, and the activated carbon adsorption of organics and inorganics from gaseous and aqueous phases.

A vast amount of research has been carried out in the area of activated carbon adsorption during the past four or five decades, and research data are scattered in different journals published in different countries and in the proceedings and abstracts of the International Conferences and Symposia on the science and technology of activated carbon adsorbents. This book critically reviews the available literature and tries to offer suitable interpretations of the surface-related interactions of the activated carbons. The book also contains consistent explanations for surface interactions applicable to the adsorption of a wide variety of adsorbates that could be strong or weak electrolytes.

The book has been written with a view to equip the surface scientists (chemists, physicists, and technologists) with the surface processes, their energetics, and with the adsorption isotherm equations, their applicability to and deviations from the adsorption data for both gases and solutions. To carbon scientists and technologists, the book should help understand the parameters and the mechanisms involved in the activated carbon adsorption of organic and inorganic compounds. The book thus combines in one volume the surface physical and chemical structure of activated carbons, the surface phenomenon at solid-gas and solid-liquid interfaces, and the activated carbon adsorption of gaseous adsorbates and solutes from solutions.

This unified approach will provide the reader access to the relevant literature and promote further research toward improving and developing newer activated carbon adsorbents and develop processes for the efficient removal of pollutants from drinking water and industrial effluents. The book can also serve as a text for studies relating to adsorption and adsorption processes occurring on solid surfaces.

The authors are grateful to Elsevier, Ann Arbor Science publishers, South African Institute of Mining and Metallurgy, Marcel Dekker Multi-Science Publishing Co., Society of Chemistry and Industry, and various authors for permission to reproduce certain figures and tables. Professor Bansal also acknowledges the understanding, the cooperation, and the encouragement of his wife Rajesh Bansal. Dr. Meenakshi Goyal is grateful to her husband Er. Arvinder Goyal for his patience and help, and to her son Nikhil and daughter Mehak, who accepted her extreme busyness and continued to attain excellence in their schools during the preparation of the manuscript. We also thank Tulsi Ram and Ruby Singh for typing the manuscript and preparing figures and tables.

Roop Chand Bansal
Meenakshi Goyal

Introduction

ACTIVATED CARBONS

Activated carbon in its broadest sense includes a wide range of processed amorphous carbon-based materials. It is not truly an amorphous material but has a microcrystalline structure. Activated carbons have a highly developed porosity and an extended interparticulate surface area. Their preparation involves two main steps: the carbonization of the carbonaceous raw material at temperatures below 800°C in an inert atmosphere and the activation of the carbonized product. Thus, all carbonaceous materials can be converted into activated carbon, although the properties of the final product will be different, depending on the nature of the raw material used, the nature of the activating agent, and the conditions of the carbonization and activation processes.

During the carbonization process, most of the noncarbon elements such as oxygen, hydrogen, and nitrogen are eliminated as volatile gaseous species by the pyrolytic decomposition of the starting material. The residual elementary carbon atoms group themselves into stacks of flat, aromatic sheets cross-linked in a random manner. These aromatic sheets are irregularly arranged, which leaves free interstices. These interstices give rise to pores, which make activated carbons excellent adsorbents. During carbonization these pores are filled with the tarry matter or the products of decomposition or at least blocked partially by disorganized carbon. This pore structure in carbonized char is further developed and enhanced during the activation process, which converts the carbonized raw material into a form that contains the greatest possible number of randomly distributed pores of various sizes and shapes, giving rise to an extended and extremely high surface area of the product. The activation of the char is usually carried out in an atmosphere of air, CO_2, or steam in the temperature range of 800°C to 900°C. This results in the oxidation of some of the regions within the char in preference to others, so that as combustion proceeds, a preferential etching takes place. This results in the development of a large internal surface, which in some cases may be as high as 2500 m²/g.

Activated carbons have a microcrystalline structure. But this microcrystalline structure differs from that of graphite with respect to interlayer spacing, which is 0.335 nm in the case of graphite and ranges between 0.34 and 0.35 nm in activated carbons. The orientation of the stacks of aromatic sheets is also different, being less ordered in activated carbons. ESR studies have shown that the aromatic sheets in activated carbons contain free radical structure or structure with unpaired electrons. These unpaired electrons are resonance stabilized and trapped during the carbonization process, due to the breaking of bonds at the edges of the aromatic sheets, and thus, they create edge carbon atoms. These edge carbon atoms have unsaturated valencies and can, therefore, interact with heteroatoms such as oxygen, hydrogen, nitrogen, and sulfur, giving rise to different types of surface groups. The elemental composition of a typical activated carbon has been found to be 88% C, 0.5% H, 0.5% N, 1.0% S, and 6 to 7% O, with the balance representing inorganic ash constituents. The oxygen content of an activated carbon can vary, however, depending on the type of the source raw material and the conditions of the activation process.

The activated carbons in general have a strongly developed internal surface and are usually characterized by a polydisperse porous structure consisting of pores of different sizes and shapes. Several different methods used to determine the shapes of the pores have indicated ink-bottle shaped, regular slit shaped, V-shaped, capillaries open at both ends, or with one end closed, and many more. However, it has been difficult to obtain accurate information on the actual shape of the pores. It is now well accepted that activated carbons contain pores from less than a nanometer to several thousand nanometers. The classification of pores suggested by Dubinin and accepted by the International Union of Pure and Applied Chemistry (IUPAC) is based on their width, which represents the distance between the walls of a slit-shaped pore or the radius of a cylindrical pore. The pores in activated carbons are divided into three groups: the micropores with diameters less than 2 nm, mesopores with diameters between 2 and 50 nm, and macropores with diameters greater than 50 nm. The micropores constitute a large surface area (about 95% of the total surface area of the activated carbon) and micropore volume and, therefore, determine to a considerable extent the adsorption capacity of a given activated carbon, provided however that the molecular dimensions of the adsorbate are not too large to enter the micropores. The micropores are filled at low relative vapor pressure before the commencement of capillary condensation. The mesopores contribute to about 5% of the total surface area of the carbon and are filled at higher relative pressure with the occurrence of capillary condensation. Attempts, however, are now on to prepare mesoporous carbons. The macropores are not of considerable importance to the process of adsorption in activated carbons, as their contribution to surface area does not exceed 0.5 m^2/g. They act as conduits for the passage of adsorbate molecules into the micro- and mesopores.

Because all the pores have walls, they will comprise two types of surfaces: the internal or microporous surface and the external surface. The former represents the walls of the pores and has a high surface area that may be several thousands in many activated carbons, and the latter constitutes the walls of the meso- and macropores as well as the edges of the outward facing aromatic sheets and is comparatively much smaller and may vary between 10 and 200 m^2/g for many of the activated carbons.

Besides the crystalline and porous structure, an activated carbon surface has a chemical structure. The adsorption capacity of an activated carbon is determined by the physical or porous structure but strongly influenced by the chemical structure of the carbon surface. In graphites that have a highly ordered crystalline structure, the adsorption capacity is determined mainly by the dispersion component of the van der Walls forces. But the random ordering of the aromatic sheets in activated carbons causes a variation in the arrangement of electron clouds in the carbon skeleton and results in the creation of unpaired electrons and incompletely saturated valencies, which would undoubtedly influence the adsorption properties of activated carbons. Activated carbons are invariably associated with certain amounts of oxygen and hydrogen. In addition, they may contain small amounts of nitrogen. X-ray diffraction studies have shown that these heteroatoms are bonded at the edges and corners of the aromatic sheets, or to carbon atoms at defect positions, giving rise to carbon-oxygen, carbon-hydrogen, and carbon-nitrogen surface compounds. As the edges constitute the main adsorbing surface, the presence of these surface compounds modifies the surface characteristics and surface properties of activated carbons.

Carbon-oxygen surface groups are by far the most important surface groups that influence the surface characteristics such as the wettability, polarity, and acidity, and the physico-chemical properties such as catalytic, electrical, and chemical reactivity of these materials. In fact, the combined oxygen has often been found to be the source of the property by which a carbon becomes useful and effective in certain respects. For example, the presence of oxygen on the activated carbon surface has an important effect on the adsorption capacity of water and other polar gases and vapors on their aging during storage, on the adsorption of electrolytes, on the properties of carbon blacks as fillers in rubber and plastics, and on the lubricating properties of graphite as well as on its properties as a moderator in nuclear reactors. In the case of carbon fibers, these surface oxygen groups determine their adhesion to plastic matrices and consequently improve their composite properties.

Although the identification and estimation of the carbon-oxygen surface groups have been carried out using several physical, chemical, and physio-chemical techniques that include their desorption, neutralization with alkalies, potential, thermometric, and radiometric titrations, and spectroscopic methods such as IR spectroscopy and x-ray photoelectron spectroscopy, the precise nature of the chemical groups is not entirely established. The estimations obtained by different workers using varied techniques differ considerably because the activated carbon surface is very complex and difficult to reproduce. The surface groups can not be treated as ordinary organic compounds because they interact differently in different environments. They behave as complex structures presenting numerous mesomeric forms depending upon their location on the same polyaromatic frame.

The aromatic sheets constituting the activated carbon structure have limited dimensions and therefore have edges. In addition these sheets are associated with defects, dislocations, and discontinuities. The carbon atoms at these places have unpaired electrons and residual valencies, and are richer in potential energy. These carbon atoms are highly reactive and are called active sites or active centers and determine the surface reactivity, surface reactions, and catalytic reactions of carbons. The impregnation of activated carbons with metals and their oxides, dispersed as fine particles, makes them extremely good catalysts for certain industrial processes. The impregnation of metals also modifies the gasification characteristics and varies the porous structure of the final product. Several inorganic and organic reagents when present on the carbon surface also modify the surface behavior and adsorption characteristics of activated carbons and make them useful for the removal of hazardous gases and vapors by chemisorption and catalytic decomposition.

ADSORPTION

Adsorption arises as a result of the unsaturated and unbalanced molecular forces that are present on every solid surface. Thus, when a solid surface is brought into contact with a liquid or gas, there is an interaction between the fields of forces of the surface and that of the liquid or the gas. The solid surface tends to satisfy these residual forces by attracting and retaining on its surface the molecules, atoms, or ions of the gas or liquid. This results in a greater concentration of the gas or liquid in the near vicinity of the solid surface than in the bulk gas or vapor phase, despite the nature of the gas or vapor. The process by which this surface excess is caused is called adsorption. The adsorption involves two types of forces: physical forces that may be dipole moments, polarization forces, dispersive forces, or short-range repulsive interactions and chemical forces that are valency forces arising out of the redistribution of electrons between the solid surface and the adsorbed atoms.

Depending upon the nature of the forces involved, the adsorption is of two types: physical adsorption and chemisorption. In the case of physical adsorption, the adsorbate is bound to the surface by relatively weak van der Walls forces, which are similar to the molecular forces of cohesion and are involved in the condensation of vapors into liquids. Chemisorption, on the other hand, involves exchange or sharing of electrons between the adsorbate molecules and the surface of the adsorbent resulting in a chemical reaction. The bond formed between the adsorbate and the adsorbent is essentially a chemical bond and is thus much stronger than in the physisorption.

Two types of adsorptions differ in several ways. The most important difference between the two kinds of adsorption is the magnitude of the enthalpy of adsorption. In physical adsorption the enthalpy of adsorption is of the same order as the heat of liquefaction and does not usually exceed 10 to 20 KJ per mol, whereas in chemisorption the enthalpy change is generally of the order of 40 to 400 KJ per mol. Physical adsorption is nonspecific and occurs between any adsorbate-adsorbent systems, but chemisorption is specific. Another important point of difference between physisorption and chemisorption is the thickness of the adsorbed phase. Although it is multimolecular in physisorption, the thickness is unimolecular in chemisorption. The type of adsorption that takes place in a given adsorbate-adsorbent system depends on the nature of the adsorbate, the nature of the adsorbent, the reactivity of the surface, the surface area of the adsorbate, and the temperature and pressure of adsorption.

When a solid surface is exposed to a gas, the molecules of the gas strike the surface of the solid when some of these striking molecules stick to the solid surface and become adsorbed, while some others rebound back. Initially the rate of adsorption is large because the whole surface is bare, but the rate of adsorption continues to decrease as more and more of the solid surface becomes covered by the adsorbate molecules. However, the rate of desorption, which is the rate at which the adsorbed molecules rebound from the surface, increases because desorption takes place from the covered surface. With the passage of time, the rate of adsorption continues to decrease, while the rate of desorption continues to increase, until an equilibrium is reached, where the rate of adsorption is equal to the rate of desorption. At this point the solid is in adsorption equilibrium with the gas. It is a dynamic equilibrium because the number of molecules sticking to the surface is equal to the number of molecules rebounding from the surface.

As the amount adsorbed at the equilibrium for a given adsorbate-adsorbent system depends upon the pressure of the gas and the temperature of adsorption, the adsorption equilibrium can be represented as an adsorption isotherm at constant temperature, the adsorption bar at constant pressure, and the adsorption isostere for a constant equilibrium adsorption. In actual practice the determination of adsorption at constant temperature is most convenient and, therefore, the adsorption isotherm is the most extensively employed method for representing the equilibrium states of an adsorption system. The adsorption isotherm gives useful information regarding the adsorbate, the adsorbent, and the adsorption process. It helps in the determination of the surface area of the adsorbent, the volume of the pores, and their size distribution. It also provides important information regarding the magnitude of the enthalpy of adsorption and the relative adsorbility of a gas or a vapor on a given adsorbent with respect to chosen standards. The adsorption data can be represented by several isotherm equations, the most important being the Langmuir, the Freundlich, the Brunauer-Emmett-Teller (BET), and Dubinin equations. The first two isotherm equations apply equally to physisorption as well as to chemisorption. The BET and Dubinin equations are most important for the analysis of physical adsorption of gases and vapors on porous carbons.

The Langmuir isotherm equation is the first theoretically developed adsorption isotherm that was derived using thermodynamic and statistical approaches. The applicability of the equation to the experimental data was carried out by a large number of investigators, but deviations were often noticed. According to this isotherm equation, the plot of p/v against p should be linear from $\theta = 0$ to $\theta = \infty$, and it should give a reasonable value of Vm (the monolayer capacity), which should be temperature independent. However, few data conform to this criterion. Similarly, several chemisorption results are known where the Langmuir equation is valid only within a small restricted range. Thus, although the Langmuir isotherm equation is of limited significance for the interpretation of the adsorption data because of its idealized character, the equation remains of basic importance for expressing dynamic adsorption equilibrium. Furthermore, it has provided a good basis for the derivation of other, more complex, models. The assumptions that the adsorption sites on solid surfaces are energetically homogeneous and that there are no lateral interactions between the adsorbed molecules are the weak points of this model.

Brunauer, Emmet, and Teller derived the BET equation for multimolecular adsorption by a method that is the generalization of the Langmuir treatment of unimolecular adsorption. These workers proposed that the forces acting in multimolecular adsorption are the same as those acting in the condensation of vapors. Only the first layer of adsorbed molecules, which is in direct contact with the adsorbent surface, is bound by adsorption forces originating from the interaction between the adsorbate and the adsorbent. Thus, the molecules in the second and subsequent layers have the same properties as in the liquid or gaseous phase. The BET equation has played a significant role in studies of adsorption because it represents the shapes of the actual isotherms. It also gives reasonable values for the average enthalpy of adsorption in the first layer and satisfactory values for Vm, the monolayer capacity of the adsorbate which can be used to calculate the specific surface area of the solid adsorbent.

The BET equation is applicable within the relative pressure range of 0.05 to 0.35. The failure of the equation above and below this range of relative pressures has been attributed to the faulty and simplifying assumptions of the theory. The failure below a relative pressure of 0.05 is due to the heterogeneity of the adsorbent surface. Activated carbon and inorganic gel surfaces that are important adsorbents are generally energetically heterogeneous (i.e., the enthalpy of adsorption varies from one part of the surface to another). At higher relative pressures, the BET equation loses its validity because adsorption by capillary condensation along with physical adsorption also takes place. The assumption that the adsorbate has liquid-like properties after the first layer is difficult to reconcile because both porous and nonporous adsorbents exposed to a saturated vapor sometimes adsorb strictly a limited amount and not the infinitely large quantity as postulated by the BET model. Thus, the limited validity of the BET equation is due to the shortcomings in the model itself rather than to our lack of knowledge of the various parameters, such as the number of layers, the heat of adsorption, or the evaporation constant in the higher layers.

The potential theory of adsorption and the Dubinin equation, which is based on it, have been developed primarily for microporous adsorbents, for which they have proved to be better than all other theories. Dubinin and coworkers, while investigating the effect of surface structure of activated carbons on the adsorbability of different vapors and of different solutes from solutions on active carbons, observed that over a wide range of values of adsorption, the characteristic curves of different vapors on the same adsorbent were related to each other. In fact, it was observed that if the adsorption potential corresponding to a certain volume of adsorption space on the characteristic curve for one vapor was multiplied by a constant, called the affinity coefficient, the adsorption potential corresponding to the same value of adsorption space on the characteristic curve of another vapor was obtained. Based on these observations, the characteristic curves for microporous activated carbons were expressed analytically by a Gaussian distribution equation between the total limiting volume of the adsorption space and the adsorption potential. This further made it possible to obtain an equation of the adsorption isotherm and to calculate the appropriate micropore volume. The Dubinin equation is valid over the range of relative pressures from 1×10^{-5} to 0.2 or 0.4, which corresponds to about 85 to 95% filling of the micropores. At relative pressures below 10^{-5}, extremely ultra-fine micropores that are not accessible to larger molecules are filled. Thus, the potential theory of adsorption together with the Dubinin equation represent the temperature dependence of adsorption and enable calculation of important thermodynamic functions, such as the heat and entropy of adsorption. The Dubinin equation has been further modified by Kaganer to yield a method for calculating the specific surface area from these isotherms. He confined his attention to monolayer region and assumed that adsorption at very low relative pressures results in the formation of a unimolecular layer on the walls of all the pores. This method thus yields monolayer capacity rather than the micropore volume. The method is applicable in the low pressure region of the isotherm (below relative pressure of 10^{-4}). The surface areas calculated by Kaganer method for activated carbons were within few percent of those calculated from the BET equation.

The Freundlich isotherm equation is a limiting form of the Langmuir isotherm and is applicable only in the middle ranges of vapor pressure. The equation is of greater significance for chemisorption, although some physical adsorption data have also been found to fit this equation.

Adsorption from solutions on activated carbons has wide applications in food, pharmaceutical, and other process industries to remove unwanted components from the solution. However, a theoretical analysis of adsorption from solution and the derivation of a suitable adsorption equation have been comparatively difficult because both the components of a solution compete with each other for the available surface. Furthermore, the thermal motion of the molecules in the liquid phase and their mutual interactions are much less well understood. It is, therefore, difficult to correctly assess the nature of the adsorbed phase, whether unimolecular or multimolecular. The adsorption of a solute from a solution is usually determined by the porosity and the chemical nature of the adsorbent, the nature of the components of the solution, the concentration of the solution, its pH, and the mutual solubility of the components in the solution. The adsorption of a nonpolar solute will be higher on a nonpolar adsorbent. But since there is competition between the solute and the solvent, the solvent should be polar in nature for the solute to be adsorbed preferentially. The other factor that also determines the adsorption from solutions is the steric arrangement or the chemical structure of the adsorbate molecule. As the activated carbons have a highly microporous structure, some of the pores may be inaccessible to larger molecules of the adsorbate. Thus, the experimentally simple technique of adsorption from solution can be developed into a method to determine surface area, microporosity, oxygen content, and the hydrophobicity of the carbon surface. The adsorption from solutions is also receiving further attention because of the growing importance of environmental control involving purification of waste water using activated carbons.

Adsorption from solutions can be classified into adsorption of solutes that have a limited solubility (i.e., from dilute solutions) and adsorption of solutes that are completely miscible with the solvent in all proportions. In the former case, the adsorption of the solvent is of little consequence and is generally neglected. In the latter case, the adsorption of both components of the solution plays its part and has to be considered. The adsorption in such a system is the resultant of the adsorption of both the components of the solution. The adsorption from such solutions is represented in the form of a composite isotherm, which is a combination of the isotherms for the individual components.

ACTIVATED CARBON ADSORPTION

Carbon surface has a unique character. It has a porous structure which determines its adsorption capacity, it has a chemical structure which influences its interaction with polar and nonpolar adsorbates, it has active sites in the form of edges, dislocations and discontinuities which determine its chemical reactions with other atoms. Thus, the adsorption behavior of an activated carbon can not be interpreted on the basis of surface area and pore size distribution alone. Activated carbons having equal surface area but prepared by different methods or given different activation treatments show markedly different adsorption properties. The determination of a correct model for adsorption on activated carbon adsorbents with complex chemical structure is therefore, a complicated problem. A proper model must take into consideration both the chemical and the porous structure of the carbon, which includes the nature and concentration of the surface chemical groups, the polarity of the surface, the surface area, and the pore size distribution, as well as the physical and chemical characteristics of the adsorbate, such as its chemical structure, polarity, and molecular dimensions. In the case of adsorption from solutions, the concentration of the solution and its pH are also important additional factors.

Thus, activated carbons are excellent and versatile adsorbents. Their important applications are the adsorptive removal of color, odor, and taste, and other undesirable organic and inorganic pollutants from drinking water, in the treatment of industrial waste water; air purification in inhabited spaces, such as in restaurants, food processing, and chemical industries; for the purification of many chemical, food, and pharmaceutical products; in respirators for work under hostile environments; and in a variety of gas-phase applications. Their use in medicine and health applications to combat certain types of bacterial ailments and for the adsorptive removal of certain toxins and poisons, and for the purifications of blood, is being fast developed. Activated carbons can be used in various forms: the powdered form, the granulated form, and now the fibrous form. Powdered activated carbons (PAC) generally have a finer particle size of about 44 μm, which permits faster adsorption, but they are difficult to handle when used in fixed adsorption beds. They also cause a high pressure drop in fixed beds, which are difficult to regenerate. The granulated activated carbon (GAC) have granules 0.6 to 4.0 mm in size and are hard, abrasion resistant, and relatively dense to withstand operating conditions. Although more expensive than PAC, they cause low hydrodynamic resistance and can be conveniently regenerated. GAC can be formulated into a module that can be removed after saturation, regenerated by heat treatment in steam, and used again. The fibrous activated carbon fibers (ACF) are expensive materials for waste water treatment, but they have the advantage of the capability to be molded easily into the shape of the adsorption system and produce low hydrodynamic resistance to flow.

The most important application of activated carbon adsorption where large amounts of activated carbons are being consumed and where the consumption is ever increasing is the purification of air and water. There are two types of adsorption systems for the purification of air. One is the purification of air for immediate use in inhabited spaces, where free and clean air is a requirement. The other system prevents air pollution of the atmosphere from industrial exhaust streams. The former operates at pollutant concentrations below 10 ppm, generally about 2 to 3 ppm. As the concentration of the pollutant is low, the adsorption filters can work for a long time and the spent carbon can be discarded, because regeneration may be expensive. Air pollution control requires a different adsorption setup to deal with larger concentrations of the pollutants. The saturated carbon needs to be regenerated by steam, air, or nontoxic gaseous treatments. These two applications require activated carbons with different porous structures. The carbons required for the purification of air in inhabited spaces should be highly microporous to affect greater adsorption at lower concentrations. In the case of activated carbons for air pollution control, the pores should have higher adsorption capacity in the concentration range 10 to 500 ppm. It is difficult to specify the pore diameters exactly, but generally in the micro- and meso- range are preferred because they fill in this concentration range.

The effluent gases from industry and processing units contain a large number of pollutants, such as oxides of nitrogen and sulfur, H_2S, and vapors of CS_2, styrene, and several solvents, such as ethanol or toluene. Many of these compounds can be economically recovered when present in large amounts. However, when present in low concentrations, these volatile organic compounds need to be removed from the flue gases before they are mixed with air. Activated carbon is one of the important adsorbents that are used for the recovery of useful compounds when economically viable and for adsorptive removal of the pollutant gases and vapors when present in small amounts. In addition, many of these VOCs are released from the exhaust of automobiles on the roads. In order to reduce this VOC release, catalyst converters are being used to convert VOC into CO_2 and water vapors. The release of these VOCs can be further decreased by fitting the automobiles with activated carbon canisters. However, in addition to the porous structure of activated carbons, their surface chemistry is also of considerable interest.

For personal protection when working in a hostile environment, the activated carbons used in respirators are also different. When working in the chemical industry, the respirators can use ordinary activated carbons because the pollutants are generally of low toxicity. However, for protection against warfare gases such as chloropicrin, cynogen chloride, hydrocynic acid, and nerve gases, special types of impregnated activated carbons are used in respirators and body garments. These activated carbons can protect by physical adsorption, chemisorption, and catalytic decomposition of the hazardous gases.

More than 800 specific organic and inorganic chemical compounds have been identified in drinking water. These compounds are derived from industrial and municipal discharge, urban and rural runoff, natural decomposition of vegetable and animal matter, and from water and waste water chlorination practices. Liquid effluents from industry also discharge varying amounts of a variety of chemicals into surface and ground water. Many of these chemicals are carcinogenic and cause many other ailments of varying intensity and character. Several methods such as coagulation, oxidation, aeration, ion exchange, and activated carbon adsorption have been used for the removal of these chemical compounds. Many studies including laboratory tests and field operations have indicated that the activated carbon adsorption is perhaps the best broad spectrum control technology available at the present moment.

An activated carbon in contact with a salt solution is a two-phase system consisting of a solid phase that is the activated carbon surface and a liquid phase that is the salt solution containing varying amounts of different ionic and molecular species and their complexes. The interface between the two phases acts as an electrical double layer and determines the adsorption processes. The adsorption capacity of an activated carbon for metal cations from the aqueous solutions generally depends on the physico-chemical characteristics of the carbon surface, which include surface area, pore size distribution, electro-kinetic properties, the chemistry of the carbon surface, and the nature of the metal ions in the solution. Activated carbons are invariably associated with acidic and basic carbon-oxygen surface groups.

The acidic groups that have been postulated as carboxyls, lactones, and phenols render the carbon surface polar and hydrophilic, and the basic groups have been postulated as pyrones and chromenes structures.

A perusal of the literature indicates that the more important parameters that influence and determine the adsorption of metal ions from aqueous solutions are the carbon-oxygen functional groups present on the carbon surface and the pH of the solution. These two parameters determine the nature and concentration of the ionic and molecular species in the solution. Electrokinetic studies have shown that the nature and concentration of the carbon surface charge can be modified by changing the pH of the carbon-solution system. The activated carbon surface has a positive charge below pH_{zpc} (zero point charge) and a negative charge above ZPC up to a certain range of pH values. The origin of the positive charge on the activated carbon surface has been attributed to the presence of basic surface groups, the excessive protonation of the surface at low pH values and to graphene layers that act as Lewis bases resulting in the formation of acceptor-donor complexes important for the adsorption of many organic compounds from aqueous solutions. At higher pH values, the carbon surface has a negative charge, due to the ionization of acidic carbon-oxygen surface groups. Thus, the adsorption of metal ions mainly involves electro-static attractive and repulsive interactions between metal ionic species in the solution and the negative sites on the carbon surface produced by the ionization of acidic groups. The dispersive interactions between the ionic species in the solution and the graphene layers and the surface area of the carbon surface play a smaller role in the adsorption of inorganics.

In the adsorption of organics, however, the situation is quite different. The organic compounds present in water can be polar or nonpolar, so that not only electrostatic interactions but also dispersive interactions will play an important role. In addition, the hydrogen bonding is also an important consideration in the adsorption of certain polar organic molecules. The molecular dimensions of the organic molecules also have a wide variation. Thus, the porous structure of the activated carbon, which includes the existence of mesopores, shall also have an important consideration for the adsorption of essentially nonpolar organic molecules, because a certain proportion of the microporosity may not be accessible to very large organic molecules.

This book has been written in eight chapters, which cover activated carbons; their surface structure; the adsorption on solid surfaces and the models of adsorption; adsorption from solution phase; the preparation, characterization of, and adsorption by carbon molecular sieves; important applications of activated carbons with special emphasis on medicinal and health applications; and the use of activated carbons in environmental clean up.

The crystalline, microporous, and chemical structures of the activated carbon surface are discussed in Chapter 1. This chapter discusses classification of pores and their contribution to surface area and adsorption capacity; the nature and characteristics of carbon-oxygen surface groups; the methods of their identification and estimation using physical, chemical, and physico-chemical methods, which include XPS and the latest innovations in infrared spectroscopy. Chapter 1 also delineates the influence of these surface groups on the adsorption characteristics and adsorption properties.

The adsorption on a solid surface, the types of adsorption, the energetics of adsorption, the theories of adsorption, and the adsorption isotherm equations (e.g., the Langmuir equation, BET equation, Dubinin equation, Temkin equation, and the Freundlich equation) are the subject matter of Chapter 2. The validity of each adsorption isotherm equation to the adsorption data has been examined. The theory of capillary condensation, the adsorption-desorption hysteresis, and the Dubinin theory of volume filling of micropores (TVFM) for microporous activated carbons are also discussed in this chapter.

The adsorption from binary solutions on solid adsorbents in general and on activated carbons in particular is discussed in Chapter 3. The nature and types of adsorption and adsorption isotherms from dilute solutions and from completely miscible binary solutions are described. The composite isotherm equation is derived. The shapes and classification of composite isotherms and the influence of adsorbate-adsorbent interactions, the heterogeneity of the carbon surface, and the size and orientation of the adsorbed molecules on the shapes are examined. The thickness of the adsorbed layer and the determination of individual adsorption isotherms from a composite isotherm are also described.

Chapter 4 briefly describes the preparation of carbon molecular sieves by pore blocking of activated carbons by decomposition of H_2S or CS_2, and depositing sulfur, by decomposition of benzene or other hydrocarbons and deposition of carbon, and by impregnation of PVC followed by its decomposition. The characterization of carbon molecular sieves by molecular probe methods using adsorption of inorganic gases and organic vapors varying in size and shape and by immersional heats of wetting in liquids of varying sizes is discussed. The applications of CMS for the separation of different gaseous mixtures are also discussed.

Chapters 5 to 8 are devoted to important applications of activated carbon adsorption. The most general liquid phase and gas phase applications of activated carbons with special reference to the nature of the carbon surface and the form of the activated carbon are discussed in Chapter 5, with special emphasis on medicinal and health applications. Different types of carbons prepared from different source raw materials and using different activation treatments are examined for the control of drug overdose, control of antibacterial activities against certain bacteria to remove toxins and poisons from the human body, and for the purification of blood by hemoperfusion. The next two chapters are concerned with the adsorptive removal of inorganic (Chapter 6) and organic (Chapter 7) pollutants from drinking and waste waters. The various parameters that are involved in the removal of hazardous organics and inorganics are reviewed and the mechanisms involved are suggested. The subject matter of Chapter 8 is the adsorptive removal of hazardous gases and vapors from industrial flue gases and automobile exhaust. The use of activated carbon in respirators for work under hostile environments is also discussed.

Contents

1 Activated Carbon and Its Surface Structure

Active carbon in its broadest sense is a term that includes a wide range of amorphous carbonaceous materials that exhibit a high degree of porosity and an extended inter-particulate surface area. They are obtained by combustion, partial combustion, or thermal decomposition of a variety of carbonaceous substances. Active carbons have been obtained in granular and powdered forms. They are now also being prepared in spherical, fibrous, and cloth forms for some special applications. The granular form has a large internal surface area and small pores, and the finely divided powdered form is associated with larger pore diameters and a smaller internal surface area. Carbon cloth and fibrous activated carbons (activated carbon fibers) have a large surface area and contain a comparatively higher percentage of larger pores.

Active carbons in the form of carbonized wood charcoal have been used for many centuries. The Egyptians used this charcoal about 1500 BC as an adsorbent for medicinal purposes and also as a purifying agent. The ancient Hindus in India purified their drinking water by filtration through charcoal. The first industrial production of active carbon started about 1900 for use in sugar refining industries. This active carbon was prepared by the carbonization of a mixture of materials of vegetable origin in the presence of metal chlorides or by activation of the charred material by CO_2 or steam. Better quality gas-adsorbent carbons received attention during World War I, when they were used in gas masks for protection against hazardous gases and vapors.

Active carbons are unique and versatile adsorbents, and they are used extensively for the removal of undesirable odor, color, taste, and other organic and inorganic impurities from domestic and industrial waste water, solvent recovery, air purification in inhabited places, restaurants, food processing, and chemical industries; in the removal of color from various syrups and pharmaceutical products; in air pollution control from industrial and automobile exhausts; in the purification of many chemical, pharmaceutical, and food products; and in a variety of gas-phase applications. They are being increasingly used in the field of hydrometallurgy for the recovery of gold, silver, and other metals, and as catalysts and catalyst supports. They are also well known for their applications in medicine for the removal of toxins and bacterial infections in certain ailments. Nearly 80% (~300,000 tons/yr) of the total active carbon is consumed for liquid-phase applications, and the gas-phase applications consume about 20% of the total production.

Because the active carbon application for the treatment of waste water is picking up, the production of active carbons is always increasing. The consumption of active carbon is the highest in the U.S. and Japan, which together consume two to four times more active carbons than European and other Asian countries. The per capita consumption of active carbons per year is 0.5 kg in Japan, 0.4 kg in the U.S., 0.2 kg in Europe, and 0.03 kg in the rest of the world. This is due to the fact that Asian countries by and large have not started using active carbons for water and air pollution control purposes in large quantities.

Carbon is the major constituent of active carbons and is present to the extent of 85 to 95%. In addition, active carbons contain other elements such as hydrogen, nitrogen, sulfur, and oxygen. These heteroatoms are derived from the source raw material or become associated with the carbon during activation and other preparation procedures. The elemental composition of a typical active carbon is found to be 88% C, 0.5% H, 0.5% N, 1% S, and 6 to 7% O, with the balance representing inorganic ash constituents. The oxygen content of the active carbon, however, may vary between 1 and 20%, depending upon the source raw material and the history of preparation, which includes activation and subsequent treatments. The most widely used activated carbon adsorbents have a specific surface area on the order of 800 to 1500 m^2/g and a pore volume on the order of 0.20 to 0.60 cm^3g^{-1}. The pore volume, however, has been found to be as large as 1 cm^3/g in many cases. The surface area in active carbons is predominantly contained in micropores that have effective diameters smaller than 2 nm.

Active carbons are mainly and almost exclusively prepared by the pyrolysis of carbonaceous raw material at temperatures lower than 1000°C. The preparation involves two main steps: carbonization of the raw material at temperatures below 800°C in an inert atmosphere, and activation of the carbonized product between 950 and 1000°C. Thus, all carbonaceous materials can be converted into active carbons, although the properties of the final product will be different, depending upon the nature of the raw material used, the nature of the activating agent, and the conditions of the activation process. During carbonization most of the noncarbon elements such as oxygen, hydrogen, nitrogen, and sulfur are eliminated as volatile gaseous products by the pyrolytic decomposition of the source raw material. The residual elementary carbon atoms group themselves into stacks of aromatic sheets cross-linked in a random manner. The mutual arrangement of these aromatic sheets is irregular and, therefore, leaves free interstices between the sheets, which may become filled with the tarry matter or the products of decomposition or at least blocked partially by disorganized carbon. These interstices give rise to pores that make active carbons excellent adsorbents. The char produced after carbonization does not have a high adsorption capacity because of its less developed pore structure. This pore structure is further enhanced during the activation process when the spaces between the aromatic sheets are cleared of various carbonaceous compounds and disorganized carbon. The activation process converts the carbonized char into a form that contains the largest possible number of randomly distributed pores of various shapes and sizes, giving rise to a product with an extended and extremely high surface area.

The preparation of active carbons from different source raw materials and using different techniques, their porous and surface chemical structures, have been discussed in details in the book *Active Carbon*.[1] Because this book is concerned more with active carbon adsorption, a brief discussion about the more important aspects of active carbon surface chemistry are covered in this book.

1.1 CRYSTALLINE STRUCTURE OF ACTIVATED CARBONS

Active carbons have a microcrystalline structure that starts to build up during the carbonization process. However, the active carbon microcrystalline structure differs from that of graphite with respect to the interlayer spacing, which is 0.335 nm in the case of graphite and ranges between 0.34 and 0.35 nm in active carbons. The orientation of the microcrystallite layers is also different, being less ordered in active carbons. Biscoe and Warren[2] proposed the term *turbostratic* for such a structure. This disorder in microcrystallite layers is caused by the presence of heteroatoms such as oxygen and hydrogen, and by the defects such as vacant lattice sites in active carbons. The three-dimensional structure of graphite and the turbostratic structure of active carbon[3] are compared in Figure 1.1.

$$\text{>}\!\!\overset{C}{\underset{C}{\text{C}\!\!<}} + O \longrightarrow \text{>}\!\!C\!\!<\overset{C-O}{\underset{C-O}{|}}$$

Franklin,[4] on the basis of his x-ray studies, classified active carbons into two types, based on their graphitizing ability. The nongraphitizing carbons, during carbonization, develop strong cross-linking between the neighboring randomly oriented elementary crystallites, resulting in the formation of a rigid immobile mass. The charcoals obtained are hard and show a well-developed microporous structure that is preserved even during the subsequent high-temperature treatment. In the case of PVDC (polyvinylidene chloride) charcoal, which is an example of a nongraphitizing carbon, about 65% of the carbon was arranged in graphitic layers of a mean diameter of 16Å.[4] The remaining carbon was highly disordered, 55% of the graphitic layers being grouped in pairs of parallel planes 0.37 nm apart. The average distance between elementary crystallites is approximately 2.5 nm. The PVDC charcoal does not graphitize even at temperatures higher than 3000°C. The formation of the nongraphitizing

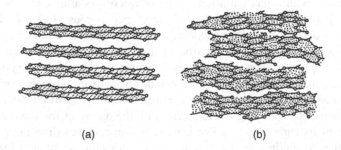

(a)   (b)

FIGURE 1.1 Comparison of three-dimensional crystal lattice of graphite (a) and the turbostratic structure (b). (After Bokros, J.C. in *Chemistry and Physics of Carbon*, Vol. 5, Marcel Dekker, New York, 1969. With permission.)

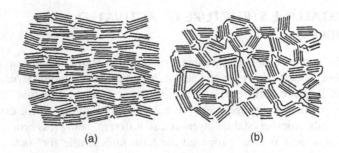

(a)   (b)

FIGURE 1.2 Schematic illustration of the structure of active carbon: (a) easily undergoing graphitization and (b) undergoing graphitization to a small degree. (After Franklin, R.E., *Proc. Roy. Soc.*, A209, 196, 1951. With permission.)

structure with strong cross-links is promoted by the presence of associated oxygen or by an insufficiency of hydrogen in the original raw material.

In the case of PVC (polyvinyl chloride) charcoal, which is an example of a graphitizing carbon, Franklin observed that the elementary crystallites were mobile and had weak cross-linking from the beginning of the carbonization process. The charcoal obtained was weak and had a less-developed porous structure, but the crystallites had a large number of graphitic layers oriented parallel to each other. Franklin observed that, after the elimination of the nonorganized carbon, the growth of the crystallites continued, probably by the addition of layers or even groups of layers. The schematic representating the structures of graphitizing and nongraphitizing active carbons are shown in Figure 1.2.

The difference in abilities to undergo graphitization results from the difference in the orientation of the crystallites in the two types of carbons.

1.2 POROUS STRUCTURE OF THE ACTIVE CARBON SURFACE

Active carbons with a random arrangement of microcrystallites and with a strong cross-linking between them have a well-developed porous structure. They have relatively low density (less than 2 gm/cm^3) and a low degree of graphitization. This porous structure formed during the carbonization process is developed further during the activation process, when the spaces between the elementary crystallites are cleared of tar and other carbonaceous material. The activation process enhances the volume and enlarges the diameters of the pores. The structure of the pores and their pore size distribution are largely determined by the nature of the raw material and the history of its carbonization. The activation also removes disorganized carbon, exposing the crystallites to the action of the activating agent and leads to the development of a microporous structure. In the latter phase of the reaction, the widening of existing pores and the formation of large pores by burnout of the walls between the adjacent pores also takes place. This causes an increase in the transitional porosity and macroporosity, resulting in a decrease in the micropore volume. According to

Dubinin and Zaverina,[5] a microporous active carbon is produced when the degree of burn-off is less than 50% and a macroporous active carbon when the extent of burn-off is greater than 75%. When the degree of burn-off is between 50 and 75%, the product has a mixed porous structure and contains all types of pores.

Active carbons, in general, have a strongly developed internal surface and they are usually characterized by a polydisperse capillary structure comprising pores of different sizes and shapes. It is difficult to obtain accurate information on the shape of the pores. Several different methods used to determine the shapes of the pores have indicated ink-bottle shape, capillaries open at both ends or with one end closed, regular slit-shaped, V-shaped, and many other shapes.[6,7] It may, however, be mentioned that for all practical purposes, the actual shape of the pores is of no consequence. Generally, the calculations of the pore radii are made by considering the pores to be ink-bottle shaped or straight and nonintersecting cylindrical capillaries.

Active carbons are associated with pores starting from less than a nanometer to several thousand nanometers. Dubinin[8] proposed a classification of the pores that has now been adopted by the International Union of Pure and Applied Chemistry (IUPAC).[9] This classification is based on their width (w), which represents the distance between the walls of a slit-shaped pore or the radius of a cylindrical pore. The pores are divided into three groups: the micropores, the mesopores (transitional pores), and the macropores.

Micropores have molecular dimensions, the effective radii being less than 2 nm. The adsorption in these pores occurs through volume filling, and there is no capillary condensation taking place. The adsorption energy in these pores is much larger compared to larger mesopores or to the nonporous surface because of the overlapping of adsorption forces from the opposite walls of the micropores. They generally have a pore volume of 0.15 to 0.70 cm^3/g. Their specific surface area constitutes about 95% of the total surface area of the active carbon. Dubinin[10] further suggested that for some active carbons, the microporous structure can be subdivided into two overlapping microporous structures involving specific micropores with effective pore radii smaller than 0.6 to 0.7 nm and the super micropores showing radii of 0.7 to 1.6 nm. The micropore structure of active carbons is characterized largely by the adsorption of gases and vapors and, to a smaller extent, by small-angle x-ray scattering technique.

Mesopores, also called *transitional pores*, have effective dimensions in the 2 to 50 nm range, and their volume usually varies between 0.1 and 0.2 cm^3/g. The surface area of these pores does not exceed 5% of the total surface area of the carbon. However, by using special methods, it is possible to prepare activated carbons that have an enhanced mesoporosity, the volume of mesopores attaining a volume of 0.2 to 0.65 cm^3/g and their surface area reaching as high as 200 m^2/g. These pores are characterized by capillary condensation of the adsorbent with the formation of a meniscus of the liquefied adsorbate. The adsorption isotherms show adsorption desorption hysteresis is which stops at a relative vapor pressure of 0.4. Besides contributing significantly to the adsorption of the adsorbate, these pores act as conduits leading the adsorbate molecules to the micropore cavity. These pores are generally characterized by adsorption-desorption isotherms of gases, by mercury porosimetry, and by electron microscopy.

Macropores are not of considerable importance to the process of adsorption in active carbons because their contribution to the surface area of the adsorbate is very small and does not exceed 0.5 m²/g. They have effective radii larger than 50 nm, and frequently in the 500 to 2000 nm range, with a pore volume between 0.2 and 0.4 cm³/g. They act as transport channels for the adsorbate into the micro- and mesopores. Macropores are not filled by capillary condensation and are characterized by mercury porosimetry.

Thus, the porous structure of active carbons is tridisperse, consisting of micro-, meso-, and macropores. Each of these groups of pores plays a specific role in the adsorption process. The micropores constitute a large surface area and micropore volume and, therefore, determine to a considerable extent the adsorption capacity of a given active carbon, provided that the molecular dimensions of the adsorbate are not too large to enter the micropores. Micropores are filled at low relative vapor pressure before the commencement of capillary condensation. The mesopores, on the other hand, are filled at high relative pressures with the occurrence of capillary condensation. The macropores enable adsorbate molecules to pass rapidly to smaller pores situated deeper within the particles of active carbons. Thus, according to Dubinin, the pattern of porous structure in active carbons constitutes macropores opening up directly to the external surface, the transitional pores branching off from the macropores, and the micropores in turn branching off from the transitional pores.

It is worthwhile to mention that Dubinin classification of pores in active carbons is not entirely arbitrary because it takes into account differences in the behavior of molecules adsorbed in micro- and mesopores. Although adsorption-desorption hysteresis is characteristic of mesopores, it has also been observed in the case of micropores at low relative pressures.[11,12] This has been attributed to inelastic distortion of some micropores, resulting in trapping of the adsorbate molecules. Consequently, the accessibility of the micropore system has been found to be increased after a number of adsorption-desorption cycles.[13]

All pores have walls and, therefore, will show two types of surfaces: the internal or microporous surface denoted by S_{mi} and the external surface, S_e. The former represents the walls of the pores and has an area of several hundred square meters per gram of the carbon. It is given by the relationship

$$S_{mi} = \frac{2 \times 10^3 W}{L}$$

where S_{mi} is the surface area in m²/g, W is the volume in cm³/g, and L is the accessible pore width in nanometers. Because the pore width L is very small, the area of the micropores is much larger than the area of mesopores or macropores. The second surface, S_e, which constitutes the walls of the meso- and macropores as well as the edges and the outer facing aromatic sheets, is small and varies between 10 and 200 m²/g for many of the active carbons. The difference between S_{mi} and S_e lies in the volume of the adsorption energy, which can be twice as high on the walls of a micropore as on the open surface.[13,14] This energetic effect decreases rapidly as the pore width increases. As the adsorption in micropores takes place at low relative pressures and as the BET approach is unable to interpret the early stages of the adsorption isotherm at low relative vapor pressures, the surface areas of highly micropores carbons obtained using the BET equation are many times unrealistic.

1.3 CHEMICAL STRUCTURE OF THE CARBON SURFACE

The crystalline structure of a carbon has a considerable influence on its chemical reactivity. However, the chemical reactivity at the basal plane sites is considerably lower than at the edge sites or at defect positions. Consequently, highly graphitized carbons with a homogenous surface consisting predominantly of basal planes are less reactive than amorphous carbons. Grisdale[15] and Hennig[16] observed that the oxidation rates of carbon atoms at the edge sites were 17 to 20 times greater than at the basal plane surface. Similarly, intercalation reactions that involve dimensional changes to the carbon structure are possible only with highly graphitized carbons because of their high degree of order.

Besides the crystalline and porous structure, an active carbon surface has a chemical structure as well. The adsorption capacity of active carbons is determined by their physical or porous structure but is strongly influenced by the chemical structure. The decisive component of adsorption forces on a highly ordered carbon surface is the dispersive component of the van der Walls forces. In graphites that have a highly ordered crystalline surface, the adsorption is determined mainly by the dispersion component due to London forces. In the case of active carbons, however, the disturbances in the elementary microcrystalline structure, due to the presence of imperfect or partially burnt graphitic layers in the crystallites, causes a variation in the arrangement of electron clouds in the carbon skeleton and results in the creation of unpaired electrons and incompletely saturated valences, and this influences the adsorption properties of active carbons, especially for polar and polarizable compounds.

Active carbons are almost invariably associated with appreciable amounts of oxygen and hydrogen. In addition, they may be associated with atoms of sulfur, nitrogen, and halogens. These heteroatoms are derived from the starting material and become a part of the chemical structure as a result of imperfect carbonization, or they become chemically bonded to the surface during activation or during subsequent treatments. There is also evidence that the carbon can adsorb certain molecular species such as amines, nitrobenzene, phenols, and several other cationic species.

X-ray diffraction studies have shown that these heteroatoms or molecular species are bonded to the edges and corners of the aromatic sheets or to carbon atoms at defect positions and give rise to carbon-oxygen, carbon-hydrogen, carbon-nitrogen, carbon-sulfur, and carbon-halogen surface compounds, also known as *surface groups* or *surface complexes*. These heteroatoms can also be incorporated within the carbon layers forming heterocyclic ring systems. Because these edges constitute the main adsorbing surface, the presence of these surface compounds or molecular species modifies the surface characteristics and surface properties of active carbons.

1.3.1 CARBON-OXYGEN SURFACE GROUPS

Carbon-oxygen surface groups are by far the most important surface groups that influence the surface characteristics such as wettability, polarity, and acidity, and physico-chemical properties such as catalytic, electrical, and chemical reactivity of these materials. In fact, the combined oxygen has often been found to be the source of the property by which a carbon becomes useful or effective in certain respects. For example, the oxygen has an important effect on the adsorption capacity of carbons for water and other polar gases and vapors, on their ageing during storage, on the adsorption of electrolytes, on the properties of carbon blacks used as fillers in rubber and plastics, on the lubricating properties of graphite as well as on its properties as a moderator in nuclear reactors. In the case of carbon fibers, these surface groups determine their adhesion to plastic matrices and consequently their composite properties. According to Kipling,[17] the atoms of oxygen and hydrogen are essential components of an active carbon with good adsorptive properties, and the surface of such materials is to be considered as a hydrocarbon surface modified at some points by oxygen atoms.

Although the determination of the number and nature of these surface chemical groups began more than 50 years ago, the precise nature of the functional groups is not entirely established. The estimations obtained by different workers using varied techniques differ considerably because the carbon surface is very complex and difficult to reproduce. The surface groups cannot be treated as ordinary organic compounds because they interact differently in different environments. They behave as complex structures presenting numerous mesomeric forms, depending upon their location on the same polyaromatic frame. Recent electron spectroscopy for chemical analysis (ESCA) studies have shown that irreversible transformation of surface groups occurred when classical organic chemistry methods were used to identify and estimate them. It is Thus, expected that the application of more sophisticated techniques such as FTIR, XPS, NMR spectroscopy, and radiotracer studies will contribute significantly to a more precise knowledge about these surface chemical groups.

Carbons have great tendency to extend this layer of chemisorbed oxygen, and many of their reactions arise because of this tendency. For example, carbons are capable of decomposing oxidizing gases such as ozone[18–21] and oxides of nitrogen,[22,23] chemisorbing oxygen. They also decompose aqueous solutions of silver salts,[24] halogens,[25–27] ferric chloride,[28] potassium and ammonium persulphate,[29–33] sodium hypochlorite,[34] potassium permanganate,[35,36] potassium dichromate,[35] sodium thiosulphate,[37] hydrogen peroxide,[38,39] and nitric acid.[31–33,40,41] In each case, there is chemisorption of oxygen and the formation of carbon-oxygen surface compounds. Carbons can also be oxidized by heat treatment in air, CO_2, or oxygen. The nature and amount of surface oxygen groups formed by different oxidative treatments depend upon the nature of the carbon surface and the history of its formation, its surface area, the nature of the oxidative treatment, and its temperature.

The reaction of activated carbons with oxygen gas at temperatures below 400°C predominantly results in the chemisorptions of oxygen and the formation of carbon-oxygen surface compounds, whereas at temperatures above 400°C the decomposition of the surface compounds and the gasification of the carbon are the predominating reactions.

$$C + O_2 \xrightarrow{\;<400°C\;} C(O) \qquad\qquad \text{Formation of surface compound.}$$

$$C + O_2 \xrightarrow{\;>400°C\;} CO + CO_2 \qquad \text{gasification}$$

$$C(O) \xrightarrow{\;>400°C\;} CO + CO_2 \qquad \text{decomposition of surface compound.}$$

In the case of oxidations in the solution phase, the major reaction is the formation of the surface compound, although some gasification may also take place depending upon the strength of the oxidative treatment and the severity of the experimental conditions. The formation of carbon-oxygen surface compounds using different active carbons and carbon black, and using various oxidative treatment in gaseous and solution phase, has been studied by a large number of investigators and has been very well reviewed.[1,41–45] Thus, we merely point out that carbons have a tendency to pick upon oxygen, at least to some extent under all conditions.

Carbons have an acid-base character. This fact has encouraged many investigators to devote their research effort to understand the cause and mechanism by which a carbon acquires an acid or a base character. Several theories (e.g., the electrochemical theory of Burstein and Frumkin,[46,47] the oxide theory of Shilov and his school,[48] the chromene theory of Garten and Weiss,[43,49] and the pyrone theory of Voll and Boehm,[50] have been proposed to explain the acid-base character of carbons. These theories and the related work have been elaborately reviewed and critically examined in several review articles.[42–44] It is now well accepted that the acid-base character of carbons is developed as a result of surface oxidation and depends on the history of formation and the temperature of oxidation.

Three types of carbon-oxygen surface groups (acidic, basic, and neutral) have been recognized. The acidic surface groups are very well characterized and are formed when carbon is treated with oxygen at temperatures up to 400°C or by reaction with oxidizing solutions at room temperature. These surface groups are thermally less stable and decompose on heat treatment in vacuum or in an inert atmosphere in the temperature range of 350 to 750°C evolving CO_2. These acidic surface groups render the carbon surface hydrophilic and polar in character and have been postulated to be carboxylic, lactone, and phenolic groups.

The basic surface oxygen groups are much less characterized and are obtained when a carbon surface, freed of all surface oxygen groups by heat treatment in vacuum or in inert atmosphere at 1000°C, and after cooling to room temperature, is contacted with oxygen gas. Garten and Weiss[43,49] proposed a pyrone-type structure for basic surface groups, which has also been referred to as a *chromene structure*. This structure has a heterocyclic oxygen-containing ring with an activated $= CH_2$ or $= CHR$ (R is an alkyl group) group. According to Voll and Boehm,[50] the oxygen atoms in the pyrone-like structure are located in two different rings of a graphitic layer. Out of the two differently bonded oxygen atoms on the basic surface sites, one decomposes into CO_2 and CO at 900°C and the other at 1200°C (Figure 1.3).

However, the structure of the basic surface oxygen groups is still a matter of dispute. Morterra et al.[51] are emphatic that the basic properties of carbons cannot

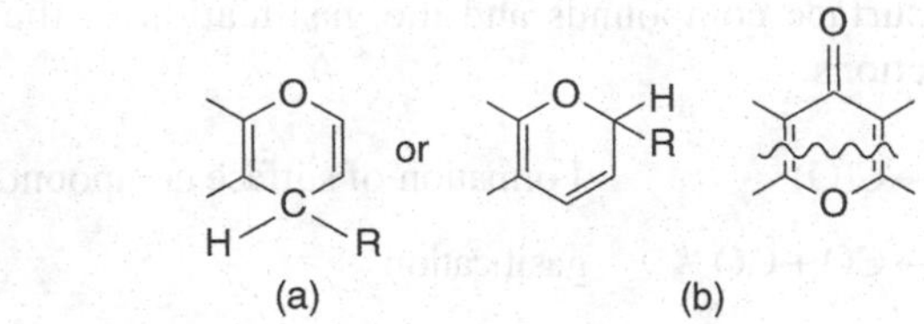

FIGURE 1.3 Functional groups of basic character: (a) chromene (After Garten, V.A. and Weiss, D.E., *Rev. Pure Appl. Chem.*, 7, 69, 1957. With permission.), (b) pyrone-like. (After Boehm, H.P., in *Advances in Catalysis*, Vol. XVI, Academic Press, New York, 1966, p. 179. With permission.)

be assigned to well-defined oxygen structures confirming the earlier view of Puri[44] that the basic character of carbons cannot be attributed to the existence of chromene or any other oxygen-containing surface groups. It appears that there is need for further work before the existence or the structure of basic groups can be accepted.

The neutral surface oxygen groups are formed by the irreversible chemisorption of oxygen at the ethylene type unsaturated sites present on the carbon surface.[44]

The surface compound decomposes into CO_2 on heat treatment. The neutral surface groups are more stable than the acidic surface groups and start decomposing in the temperature range 500 to 600°C and are removed completely only at 950°C. A model of the fragment of an oxidized active carbon surface proposed by Tarkovskya[52] is shown in Figure 1.4.

1.3.2 CHARACTERIZATION OF CARBON-OXYGEN SURFACE GROUPS

A considerable amount of effort has been directed to identify and estimate carbon-oxygen surface groups (or functional groups) using several physical, chemical, and physicochemical techniques that include desorption of the oxide layer, neutralization with alkalies, potentiometric, thermometric, and radiometric titrations, and spectroscopic methods such as IR spectroscopy and x-ray photoelectron spectroscopy. These studies have shown the existence of several groups, the more important being the carboxyls, lactones, phenols, quinones, and hydroquinones. However, these methods have not yielded comparable results and many times the entire amount of the associated oxygen has not been accounted for. Puri[44] suggested caution in the interpretation of

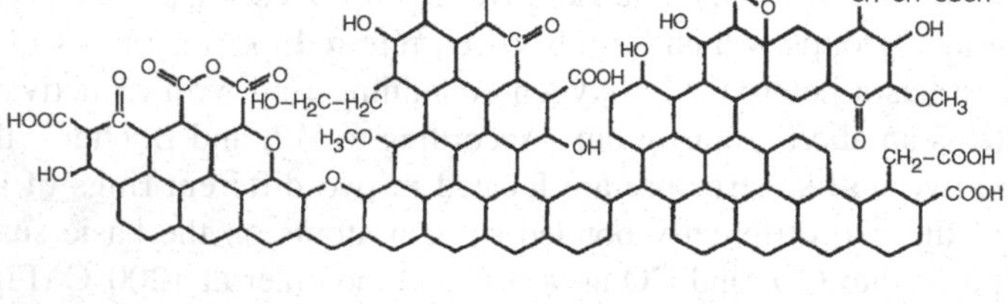

FIGURE 1.4 Model of a fragment of an oxidized active carbon surface. (After Tarkovskya, I.A., Strazhesko, D.N., and Goba, W.E., Adsorbtsiya, *Adsorbenty*, 5, 3, 1977. With permission.)

the results because the surface groups on carbons are unlikely to behave exactly in the same way as those in simple organic compounds. Thus, a brief discussion of the results obtained by these methods in the identification and estimation of the surface oxygen group present on different carbons is appropriate in this chapter.

1.3.2.1 Thermal Desorption Studies

The surface oxygen groups found on as-received carbons, or formed as a result of interaction with oxygen or with oxidizing gases or solutions, have different thermal stabilities because they are formed at different sites associated with varying energies. For example, carboxyl groups decompose at lower temperatures than phenolic or quinone groups. Thus, when a carbon sample is heat treated in vacuum or in an inert atmosphere, different surface groups decompose in different temperature ranges. In general, it has been observed that these surface groups are thermally stable at temperatures below 200°C, independent of the temperature at which they are formed.

The technique generally involves heating the carbon sample in vacuum or in an inert flowing carrier gas at a programmed heating rate. The oxygen-containing surface groups decompose into volatile gaseous products, which are then analyzed by conventional methods such as gravimetry, mass spectroscopy, gas chromatography, and IR spectroscopy. Because carbon is highly reactive with oxygen, the carbon-oxygen surface groups are generally evolved as CO_2, CO, and water vapor, the amount of each gaseous species depending upon the nature of the carbon, its pretreatment, and the thermal desorption temperature. For example, CO_2 is evolved by the decomposition of carboxylic and lactomic groups in the temperature range 350 to 750°C; CO by the decomposition of quinone and phenolic groups in the temperature range 500 to 950°C; water vapor from the decomposition of carboxyls, phenols in the temperature range 200 to 600°C. At lower temperatures some physisorbed and chemisorbed water is also desorbed. Some elementary hydrogen gas formed by the recombination of evolved hydrogen atoms as a result of splitting of C–H bonds is desorbed in the temperature range 500 to 1000°C. It may be pointed out that about 25 to 30% of the elementary hydrogen remains bonded to the interior of the carbon atoms, even after degassing at 1000°C.

Numerous studies on the thermal desorption of different carbons have been reported. Puri and Bansal[53] and Bansal et al.[54] carried out vacuum pyrolysis of a number of carbon blacks, charcoals, and activated carbons, and measured the amount of oxygen evolved as CO_2, CO, and water vapor as a function of heat treatment temperature. The total of the three oxygens (evolved as CO_2, CO, and water vapor) agreed fairly with the total oxygen obtained by ultimate analysis (Table 1.1 to Table 1.3). The desorption of oxygen as CO_2 and CO on evacuation at gradually increasing temperatures (Figure 1.5 and Figure 1.6) shows that these gases are evolved in different temperature ranges, which indicates that the chemisorbed oxygen constitutes different surface groups that involve different sites associated with varying energies. The composition of the evolved gas in a particular temperature range appears to depend upon the nature of the surface group or groups decomposing in that range.

Bansal et al.[55] also studied the decomposition of carbon-oxygen surface groups formed on low temperature oxidation of ultra clean surfaces of activated graphon using a mass spectrometer and observed that both CO_2 and CO were primary

TABLE 1.1
Gases Evolved on Outgassing Various Polymer Carbons at 1200°C

Sample Identification	O_2 Evolved on Outgassing at 1200°C (g/100 g)				Oxygen by Ultimate Analysis (g/100 g)	Hydrogen Evolved on Outgassing at 1200°C (g/100 g)			Hydrogen by Ultimate Analysis (g/100 g)
	CO_2	CO	H_2O	Total		H_2O	H_2	Total	
PF-140°	7.00	13.05	2.05	22.10	22.80	0.26	3.60	3.86	4.90
PF-400°	4.90	5.10	1.30	11.30	10.40	0.16	2.85	3.01	4.20
PF-600°	1.40	4.60	Tr	6.00	3.90	Tr	2.00	2.00	2.90
PF-900°	0.50	1.60	Tr	2.10	1.40	Tr	1.05	1.05	1.30
PVDC-600°	3.50	2.00	0.20	5.70	3.60	0.03	1.49	1.52	0.80
PVC-850° (vac)	Tr	Tr	Tr	Tr	Tr	Tr	Tr	Tr	Tr
PVC-850° (N₂)	Tr	Tr	Tr	Tr	Tr	Tr	Tr	Tr	Tr
PVC-850° (CO₂)	1.01	0.48	0.18	1.67	1.81	0.02	0.30	0.32	0.65
UF-400°	2.80	4.80	2.40	10.00	10.62	0.30	2.95	3.25	4.27
UF-650°	2.59	4.08	Tr	6.67	6.05	Tr	2.25	2.25	3.50
UF-850°	0.25	1.71	Tr	1.96	2.10	Tr	0.30	0.30	0.50

Key: PF = polyfuryl alcohol carbon; PVDC = polyvinylidene chloride carbon; PVC = polyvinyl chloride carbon; UF = urea formaldehyde resin carbon. The number represents the temperature of carbonization.

Source: Bansal, R.C., Dhami, T.L., and Prakash, S., *Carbon*, 15, 157, 1977. Reproduced with permission from Elsevier.

TABLE 1.2
Surface Area and Gases Evolved on Outgassing Different Carbon Blacks

Trade Name	Type	Nitrogen Surface Area (m²/g)	Oxygen Evolved on Outgassing at 1200°C, (g/100 g)				Oxygen by Ultimate Analysis, %	Hydrogen Evolved on Outgassing at 1200°C, (g/100 g)			Hydrogen by Ultimate Analysis %
			CO_2	CO	H_2O	Total		H_2O	H_2	Total	
Pelletex	Furnace	27.1	0.051	0.331	0.133	0.52	0.22	0.016	0.238	0.25	0.38
Kosmos-40	Furnace	31.2	0.088	0.320	0.017	0.48	0.23	0.009	0.209	0.22	0.35
Statex-B	Furnace	48.3	0.115	0.400	0.089	0.60	0.43	0.011	0.248	0.26	0.36
Philblack-A	Furnace	45.8	0.187	0.343	0.000	0.53	0.58	0.000	0.209	0.21	0.35
Philblack-O	Furnace	79.6	0.209	0.428	0.107	0.74	0.79	0.013	0.207	0.22	0.30
Philblack-I	Furnace	116.8	0.348	0.628	0.355	1.33	1.17	0.042	0.139	0.18	0.29
Philblack-E	Furnace	135.1	0.401	0.411	0.435	1.26	1.01	0.054	0.137	0.19	0.31
Vulcan-SC	Channel	194.4	0.428	0.800	0.133	1.36	1.18	0.016	0.094	0.11	0.17
Spheron-9	Channel	115.8	0.536	1.928	0.710	3.17	3.49	0.089	0.321	0.41	0.62
Spheron-6	Channel	120.0	0.500	2.122	0.462	3.08	3.10	0.058	0.284	0.34	0.55
Spheron-4	Channel	152.7	0.547	2.829	0.689	4.06	3.28	0.086	0.232	0.32	0.47
Spheron-C	Channel	253.7	0.575	2.000	0.600	3.17	3.14	0.075	0.152	0.23	0.33
ELF-O	Color	171.0	1.176	2.171	0.710	4.05	4.89	0.089	0.247	0.34	0.47
Mogul-A	Color	228.4	1.894	4.228	0.979	7.10	7.63	0.122	0.236	0.36	0.51
Mogul	Color	308.0	2.205	4.180	1.440	7.82	8.22	0.180	0.132	0.31	0.48

Source: Puri, B.R. and Bansal, R.C., *Carbon*, 1, 451, 1964. Reproduced with permission from Elsevier.

TABLE 1.3
Gases Evolved on Evacuating Mogul and Mogul-A at Different Temperatures

Temperature,°C	Weight Percent Oxygen Evolved As:				Weight Percent Hydrogen Evolved As:		
	CO_2	CO	H_2O	Total	H_2O	H_2	Total
			Mogul				
30–200	—	—	—	—	—	—	—
200–300	0.220	—	0.178	0.398	0.022	—	0.022
300–400	0.661	—	0.417	1.078	0.052	—	0.052
400–500	0.655	—	0.791	1.446	0.099	—	0.099
500–700	0.371	1.314	0.009	1.694	0.001	0.030	0.031
700–800	0.269	2.000	0.019	2.288	0.002	0.085	0.087
800–1000	0.008	0.612	0.026	0.646	0.003	0.082	0.085
1000–1200	0.014	0.251	0.009	0.274	0.001	0.007	0.008
			Mogul-A				
30–200	—	—	—	—	—	—	—
200–300	0.356	—	0.196	0.552	0.024	—	0.024
300–400	0.579	—	0.160	0.739	0.020	—	0.020
400–500	0.519	—	0.533	1.052	0.067	—	0.067
500–700	0.417	1.071	0.044	0.532	0.005	0.035	0.040
700–800	0.004	0.739	0.045	1.788	0.006	0.092	0.098
800–1000	0.010	0.830	0.000	0.840	0.000	0.084	0.084
1000–1200	0.000	0.590	—	0.590	—	0.025	0.025

Source: Puri, B.R. and Bansal, R.C., *Carbon*, 1, 451, 1964. Reproduced with permission from Elsevier.

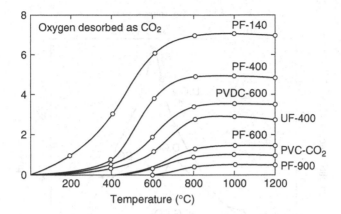

FIGURE 1.5 Oxygen evolved as CO_2 on outgassing polymer charcoals at different temperatures. (After Bansal, R.C., Dhami, T.L., and Prakash, S., *Carbon*, 15, 157, 1977. With permission.)

products obtained by the decomposition of different oxygen functional groups from different sites on the carbon surface.

The nature of the gaseous species evolved on thermal desorption of carbon-oxygen surface groups and the mechanism of their evolution was also studied by Van Driel[56] using gas chromatography, by Lang and Magnier[57] using IR and gas chromatography, by Bonnetain et al.[58,59] using chemical separation techniques, by Tucker and Mulcahy[60] using thermogravimetric technique, and by Dollimore et al.[61] using mass spectrometry. These workers observed that the major part of the surface

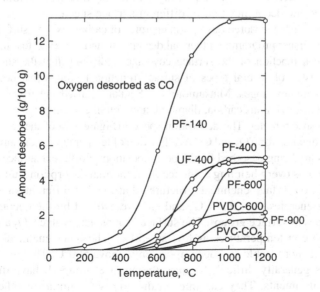

FIGURE 1.6 Oxygen evolved as CO on outgassing polymer charcoals at different temperatures. (From Bansal, R.C., Dhami, T.L., and Prakash, S., *Carbon*, 15, 157, 1977. With permission.)

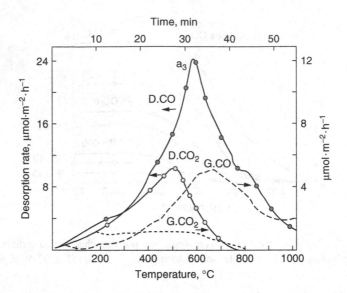

FIGURE 1.7 Desorption of chemisorbed oxygen from oxidized diamond (D) and graphite (G) as a function of temperature. (From Matsumoto, S. and Setaka, N., *Carbon,* 17, 303, 1979. With permission.)

groups decomposed in the temperature range 600 to 800°C and almost completely at 1000°C. The amount of oxygen evolved could be almost completely accounted for by the evolution of CO_2 and CO. The activation energy for desorption was found to increase with decreasing surface coverage, indicating that the evolution of different gases involved the decomposition of different surface species.

Trembley et al.[62] measured desorption energies of carbon oxygen surface groups on graphon, using linear programmed thermal desorption, and observed that the desorption energies were a function of the surface coverage, indicating that the surface oxygen complex consisted of several types of surface functional groups that decomposed in different temperature ranges. Matsumoto and Setaka[63] carried out thermal desorption studies of oxidized vitreous carbon, diamond, and graphite at temperatures up to 950°C using a mass spectrometer. The desorption spectra (Figure 1.7) of the samples showed two different maxima for CO_2 and CO as a function of temperature, indicating once again that the gases are being desorbed by the decomposition of different surface compounds.

Thus, there is overwhelming evidence from thermal desorption studies that there are two types of surface chemical structures that involve different sites associated with varying energies, and that CO_2 and CO are evolved by the decomposition of these two types of surface groups. The surface groups that evolve CO_2 are less stable and decompose at temperatures as low as 350°C. The other chemical groups that evolve CO are more stable and decompose only above 500°C. The interpretation of the results is generally difficult because the surface groups behave differently in different environments. They can interact directly with similar or other groups in the neighborhood. In general, the thermal desorption studies yield valuable information that supplements the results obtained by other independent methods.

1.3.2.2 Neutralization of Alkalies

Titration with alkalies is one of the earliest and simplest methods used to determine the nature and amount of surface acidic groups on carbons. However, the standard conditions under which comparable results can be obtained have been realized during the last few decades. It is now recognized that the base neutralization capacity of a carbon should be determined after degassing the sample at ~150°C so as to free it from any physically adsorbed gases and vapors. The carbon sample is then placed in contact with a 0.1 to 0.2 N alkali solution for 24 to 72 hr. The contact time can be reduced to a few hours if the carbon and the alkali solution are heated under reflux. These conditions are now being followed by many of the investigators.

Puri and coworkers[64] examined a large number of charcoals before and after outgassing, and extensive oxidation treatments in oxygen as well as in oxidizing solutions, and tried to correlate the base neutralization capacity of the charcoal with the oxygen evolved as CO_2 on evacuation. It was found (Table 1.4) that in each case the amount of alkali neutralized was close to the amount of CO_2 evolved on evacuation (termed CO_2 *complex*). As the amount of the CO_2 complex decreased on outgassing or increased on oxidation, the base neutralization capacity of the charcoal decreased or increased correspondingly. When the entire amount of the complex was removed on outgassing around 750°C, the carbon lost almost completely its capacity to neutralize alkalies, even though it still contained appreciable amounts of associated oxygen (*cf.* Table 1.4). This work was later extended to commercial-grade carbon blacks by Puri and Bansal.[66] Their surface acidity, as determined by neutralization of sodium and barium hydroxides, was found to be close to each other as well as to the amount of CO_2 complex contained in each sample. Puri and Mahajan,[67] Anderson and Emmett,[68] and Puri et al.[69] studied the adsorption of ammonia and several amines on several charcoals and carbon blacks and found that the amount adsorbed was close to the amount of CO_2 complex present on the carbon surface.

Thus, Puri and coworkers are of the view that in charcoals as well as in carbon blacks, the same surface group that is involved in the liberation of CO_2 on evacuation is also involved in the neutralization of alkalies. This cannot be a carboxylic group because there is no significant correlation between CO_2 evolved and active hydrogen. This cannot be a lactone group as suggested by Garten and Weiss[43] because it did not show equivalence between CO_2 evolved and alkali neutralized. However, these workers did not rule out the possibility of the existence of certain types of lactone structure that would hydrolyze to give a carboxylic group and a phenolic group, each capable of stoichiometric ionic adsorption.

Boehm[42] differentiated the acidic surface groups on oxidized charcoal and carbon black by selective neutralization technique, using bases of different strengths, namely $NaHCO_3$, Na_2CO_3, NaOH, and C_2H_5ONa (Table 1.5). The strongly acidic groups neutralized by $NaHCO_3$ but not by Na_2CO_3 were postulated as lactones. The weakly acidic group neutralized by NaOH but not by Na_2CO_3 was postulated as a group of phenols. The reaction with sodium ethoxide was not considered a true neutralization reaction because it did not involve an exchange of H⁺ ions by Na⁺ ions. The groups reacting with sodium ethoxide but not with sodium hydroxide were suggested to be carboxyls, which were created by the

TABLE 1.4
Relationship Between Base Neutralization Capacity and CO_2 Evolved on Evacuation of Various Charcoals at 1200°C

Carbon Sample	CO_2 Evolved On Evacuation At 1200°C (meq/100 g)	Sodium Hydroxide Neutralized (meq/100 g)	Barium Hydroxide Neutralized (meq/100 g)
Sugar charcoal			
Original	669	662	664
300°-outgassed	388	381	377
400°-outgassed	231	222	228
500°-outgassed	150	153	147
600°-outgassed	75	74	72
750°-outgassed	0	0	0
1000°-outgassed	0	0	0
Oxidized with H_2O_2	819	810	815
Oxidized with $K_2S_2O_8$	975	970	965
Coconut charcoal			
Original	394	395	400
300°-outgassed	181	165	170
500°-outgassed	38	35	20
600°-outgassed	15	12	16
750°-outgassed	0	0	0
1000°-outgassed	0	0	0
Oxidized with H_2O_2	591	584	582
Oxidized with $K_2S_2O_8$	623	627	619
Wood charcoal			
Original	531	532	535
300°-outgassed	400	399	405
500°-outgassed	50	50	49
600°-outgassed	22	17	25
750°-outgassed	0	0	0
1000°-outgassed	0	0	0
Cotton stalk charcoal			
Original	325	335	342
300°-outgassed	250	245	251
500°-outgassed	62	65	63
600°-outgassed	15	21	16
750°-outgassed	0	0	0
1000°-outgassed	0	0	0

Source: Puri, B.R., Singh, D.D., Nath, J., and Sharma, L.R., *Ind. Engg. Chem.*, 50, 1071, 1958. With permission.

TABLE 1.5
Selective Neutralization of Acidic Surface Structures
on Microcrystalline Carbons

Carbon Sample	Neutralization (meq/g) by			
	Sodium Bicarbonate	Sodium Carbonate	Sodium Hydroxide	Sodium Ethoxide
Sugar charcoal heat-treated in nitrogen at 1200°C	0.16	0.32	0.69	0.85
Sugar charcoal (activated)	0.21	0.43	0.72	0.89
Sugar charcoal activated and then heat-treated	0.35	0.73	1.02	1.38
Eponite	0.16	0.34	0.63	1.06
CK-3	0.76	1.52	2.37	3.15
Philblack-O	0.57	1.09	1.64	2.34
Spheron-6	0.59	1.18	1.96	2.95
Spheron-C	0.64	1.28	1.88	2.56
Sugar charcoal oxidized with $KMnO_4$	0.39	0.64	0.88	0.96
Eponite oxidized with $KMnO_4$	0.78	1.15	1.61	2.21
Eponite oxidized with $(NH_4)_2S_2O_8$	0.88	1.34	1.74	2.35
Eponite oxidized with NaOCl	1.08	1.64	2.15	2.61
CK-3 oxidized with $(NH_4)_2S_2O_8$	0.20	0.31	0.35	0.58

Source: Boehm, H.P., in *Advances in Catalysis*, Vol. XVI, Academic Press, New York, 1966, p. 179.
With permission.

oxidation of the disorganized aliphatic carbon. Puri,[44] however, questioned the validity of the selective neutralization technique.

According to Puri, the same acid group will neutralize different amounts of alkalies of varying strengths. For example, a weak acid-like acetic acid can be neutralized only partially when titrated against Na_2CO_3 or $NaHCO_3$, but the same acid can be neutralized completely by NaOH. Barton et al.[70-73] while studying the surface oxygen structures on a sample of graphite and a carbon black by degassing at different temperatures, using a mass spectrometer and by measuring base neutralization capacities of the degassed samples, suggested that the acidic group present on the surface of graphite was monobasic and that the carbon black contained both a monobasic and a dibasic surface acidic group. Combining these studies with reaction with methyl magnesium iodide[70] and diazomethane,[71] these workers suggested that both the groups on carbon black were lactones, but only one of them had active hydrogen associated with it.

Bansal et al.[74] combined desorption and base neutralization techniques for investigating the acidic surface groups on several polymer carbons. The base neutralization capacity using sodium hydroxide was found to be almost exactly equivalent to the amount of CO_2 evolved on evacuation in the case of polyvinylidene (PVDC), polyvinyl chloride

TABLE 1.6
Alkalis Neutralized by Various Polymer Carbons in Relation to CO_2 Evolved on Evacuation

Sample	Sodium Hydroxide Neutralized (meq/100 g)	CO_2 Evolved on Evacuation (meq/100 g)	Sodium Ethoxide Neutralized (meq/100 g)
PVDC-600°	212	218	256
PVDC-850° (CO_2)	62	63	71
Saran-600°	150	146	171
PF-600°	42	87	56
PF-900°	16	31	21
UF-650°	78	162	109
UF-850°	7	16	11

Source: Bansal, R.C., Bhatia, N., and Dhami, T.L., *Carbon,* 16, 65, 1978. Reproduced with permission from Elsevier.

(PVC), and Saran (a copolymer of PVDC and PVC) charcoals but was almost half of the amount of CO_2 evolved in the case of polyfurfurylalocbol (PF) and urea formoledeloyde (UF) charcoals (Table 1.6). The base neutralization capacity decreased on evacuation, and the decrease at any temperature corresponded to the amount of CO_2 evolved at that temperature. Furthermore, the temperature interval over which the drop in base neutralization occurred appears to be the same (Figure 1.8) as the temperature interval over which CO_2 was evolved from the carbon sample.

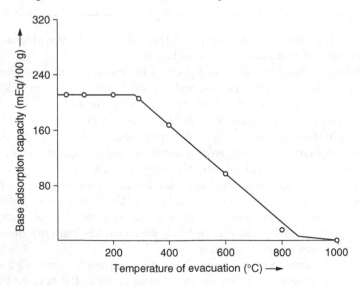

FIGURE 1.8 Base adsorption capacity in relation to evacuation temperature. (From Bansal, R.C., Bhatia, N., and Dhami, T.L., *Carbon,* 16, 65, 1978. With permission.)

The possible structure for a lactone that could explain most of the data obtained by them and by others could be an f-lactone first suggested by Garten and Weiss[43] and later restated by Barton et al.[70] Bansal et al.[74] suggested that the lactone exists in two tautomeric forms. The keto form explains the results obtained for monobasic carbons, and the enol form fits well with the dibasic carbons such as PVDC, PVC, and Saran. The existence of these two forms in different carbons is quite reasonable, because the presence of individual oxygen surface groups and their proportions are very much dependent on the history of their preparation.

The base neutralization technique to measure surface acidic groups, although simple, requires a large sample size. Consequently, several workers[66,75–81] carried out electrometric measurements (potentiometric titrations) using a very small sample size because these measurements are very sensitive and accurate. The procedure essentially consists of preparing a suspension of the carbon in CO_2-free distilled water, adding a standard alkali solution in small amounts, and measuring the current using a precision instrument. The analysis of the shape of the titration curve and the inflection points determine the contribution of functional groups of different acidic strength to the surface acidity of the carbons.

1.3.2.3 Specific Chemical Reactions

Several workers have used more direct analysis of the surface oxygen groups by studying specific chemical reactions of organic chemistry. Most of these reactions have been carried out in conjunction with other methods. The most important surface groups for which chemical analysis has been used are carboxyls, phenols, quinones and lactones. The more commonly used organic chemistry reaction is methylation of the carbon with diazomethane followed by hydrolysis of the methylated product with a mineral acid. The fraction that is hydrolyzed has been attributed to carboxylic groups[82] and to lactones.[43] The unhydrolyzable portion of the methylated product is considered to be due to phenolic groups. The quinone groups have generally been determined by reduction with sodium borohydride. The results of some of these studies on charcoals, carbon blacks, and activated carbons are presented in Table 1.7 and Table 1.8. The readers are directed to consult our earlier books on active carbon[1] and carbon black[83] for more detailed studies.

Surface group estimations using direct chemical analysis have not been able to account for the entire amount of associated oxygen. Thus, several workers have suggested caution in using these methods. Papirer and Guyon,[80] for example, on the basis of their surface group determination using organic methods and spectroscopic analysis, observed that the surface groups on carbons cannot be considered classical organic functional groups, but rather combined structures that may present numerous mesmeric forms largely favored by their locations on the same polyaromatic frame.

1.3.2.4 Spectroscopic Methods

Several spectroscopic techniques have been used to investigate the presence of carbon-oxygen surface groups on carbons. Infrared (IR) and electron spin resonance (ESR) are the two most commonly used spectroscopic techniques, although in more recent years, x-ray photoelectron spectroscopy (XPS) has been increasingly applied for the

TABLE 1.7
Surface Group Analysis by Reaction with Diazomethane

Carbon Sample	Sodium Hydroxide Neutralization Capacity (meq/100 g)	$-OCH_3$ Introduced		Nonhydrolyzable = Phenolic Groups (mEq/100 g)	Sodium Hydroxide Neutralized Minus Total $-OCH_3$,n-lactone Groups (meq/100 g)
		Total	Hydrolyzable = COOH or Lactone Groups (meq/100 g)		
Sugar charcoal	81	82	50	32	—
Sugar charcoal heat-treated in nitrogen	69	72	34	38	—
Sugar charcoal activated in CO_2	71	72	42	30	—
Eponite	62	107	46	61	—
Sugar charcoal heated in air at					
400°	59	11	05	05	48
600°	14	2	1	1	12
800°	13	8	6	6	5
Sugar charcoal oxidized in air after evacuation at 800°					
400°	32	15	10	5	17
600°	22	15	11	4	17

700°	18	15	12	3	3
Carbolac-1	183	146	90	56	37
Mogul-A	118	54	33	21	64
ELF-O	58	24	14	10	34
Spheron-6	37	14	9	6	23
Spheron-9	46	10	5	5	36

Sources: Bansal, R.C., Bhatia, N., and Dhami, T.L., Carbon, 16, 65, 1978; Studebaker, M.L., Huffman, E.W.D., Wolfe, A.C., and Nabors, L.G., Ind. Engg. Chem., 48, 162, 1956; and Studebaker, M.L. and Rinehart, R.W., Rubber Chem. Technol., 45, 106, 1972. With permission.

TABLE 1.8
Surface Group Analysis by Specific Organic Reactions

Carbon Sample	Oxygen as Phenol Group by Grignard Reagent (%)	Oxygen as Phenol Group by Neutralization of Barium Hydroxide (%)	Oxygen as Quinone Group by Reduction With Sodium Borohydride (%)	Oxygen Ether Group by Difference (%)
Pelletex	0.02	0.08	0.16	0.35
Stirling-V	0.03	0.07	0.22	0.23
Kosmos-40	0.08	0.14	0.15	0
Statex-B	0.06	0.17	0.16	0.44
Philblack-A	0.08	0.16	0.24	0.24
Philblack-O	0.11	0.14	0.41	1.16
Philblack-E	0.19	0.33	0.66	1.15
Spheron-9	0.20	0.60	0.92	2.63
Spheron-6	0.18	0.45	0.67	1.82
Spheron-C	0.39	0.76	0.89	1.31
ELF	0.54	0.78	1.10	2.34
Mogul-A	0.52	1.98	1.71	2.84
Mogul	0.71	2.52	2.03	3.18
Carbolac-2	0.96	4.25	2.62	1.84

Source: Studebaker, M.L. and Rinehart, R.W., *Rubber Chem. Technol.*, 45, 106, 1972. With permission.

examination of carbon surface structures. Although a detailed description of the instruments and procedures used in these spectroscopic techniques is beyond the scope of this chapter, the discussion will focus on results obtained on carbons by various spectroscopic techniques listed in Table 1.9.

TABLE 1.9
Spectroscopic Techniques used in Surface Group Analysis of Carbon Materials

Technique	Reference
Infrared spectroscopy	84,92,93
Transmission-absorption (IR-T/A)	96,97,93
Fourier transform (FTIR)	95,96
Photoacroustic (PAS)	85,86,96
Photothermal beam deflection (PDS)	51,89–91,95,96
X-ray photoelectron spectroscopy (XPS)	98–106

Infrared Spectroscopy

Infrared spectroscopy (IR) in its various forms is an important and forceful tool that can provide useful information about surface functional groups on carbons. Special IR studies can also provide information regarding the molecular forces involved in the adsorption processes. Carbon is black in color and so has a tendency to absorb most of the radiation, at least in the visible region. Even its thin sections are opaque. But it can transmit some radiations in the IR region when examined using thin sections. The developments leading to the preparation of halide pellets, in which finely divided carbonaceous samples were uniformly distributed, has made possible the application of IR to the study of carbon surfaces. Carbon blacks present no difficulties because they can be obtained in a very fine state of subdivision. Active carbons are hard and difficult to grind. However, this problem was solved by Friedel and Hofer,[84] who devised a technique by which active carbons and charcoals could be converted into a finely divided state for making halide pellets for IR studies.

The other difficulty in the IR studies of carbons using halide pellets is the exposure of the carbon material to atmospheric gases and vapors that tend to vitriate the results. The development of elaborate techniques for obtaining carbonaceous films and preparation of charcoals by carbonization under vacuum conditions[51] broadened the scope of applications of IR spectroscopy to the study of carbons and their surface groups. Furthermore, the sensitivity of IR measurements has been largely enhanced by using Fourier-Transform (FT), Photoacoustic (PAS), and Photothermal Beam Deflection (PDS) IR spectroscopy.

Transmission and Absorption Infrared Spectroscopy

In IR transmission and absorption (IR-T/A) studies, small amounts of carbon are mixed with KBr or Nujol. The typical composition of the carbon in the KBr pellet ranges from 0.01 to 0.5 weight percent and approximately 0.2 mg carbon/cm^2 of the pellet area. The carbon-Nujol mixture is made into a paste, which is spread into thin layers (less than 0.4 mm thick) for these measurements. As the carbon materials show strong absorption of radiations in the IR region, the spectra show low IR signal-to-noise ratio (s/n) for carbons. Thus, investigators are now turning to alternative and more sensitive techniques to enhance the signal and precision of the measurements.

Attenuated Total Reflectance IR Spectroscopy

In attenuated total reflectance IR spectroscopy (ATR-IRS), the spectra on the sample surface are measured by placing the carbon surface in close contact with a suitable reflecting element. Thus, there is restriction on the thickness of the carbon sample, because it involves measurement of the IR signal, which is reflected from the carbon surface and not the transmitted signal. The reflecting element used in such investigations is generally a trapezoidal prism, and the sample is placed on one or both of the reflecting surfaces. The beam generally enters at a right angle to one of the end faces and is reflected about 25 times before passing out of the reflecting element. The differences in the refractive indices of the sample and the reflecting surface determine the penetration of the IR beam into the sample surface. The reflecting elements most commonly used are the single crystals of AgCl, TiBr, and Ge, which have refractive indices of 2.0, 2.4, and 4.0, respectively. This technique has been successfully used in the case of carbon blacks.

Fourier-Transform Infrared Spectroscopy

The development of computerized Fourier-transform infrared spectroscopy (FTIR) has several advantages over conventional, dispersive IR spectroscopy. FTIR uses an interferometer in place of a grating or slits. This results in the availability of higher energy, of the order of 100 to 200 times over the dispersive system throughout the detector, enabling spectral information to be obtained for all frequencies at the same time. Coupled with an internally calibrated computer system to add a large number of interferograms, FTIR produces markedly superior spectra and can provide more precise information concerning the oxidation of carbons and the formation of carbon-oxygen surface groups. This technique can also allow measurements of lower concentrations of surface functional groups. Thus, the three major advantages of the FTIR technique over the conventional dispersive techniques are the availability of higher energy throughout, the multiplex capability, and the greater accuracy of the frequency scale. The multiplex capability of FTIR improves the s/n ratio by about 150 times over the dispersive IR technique. It may, however, be mentioned that longer sampling times are needed to achieve higher resolutions or lower noise in FTIR spectroscopy, because the s/n is proportional to the number of scans, and the scan time varies inversely with resolution.

Photoacoustic Spectroscopy

In the photoacoustic spectroscopy (PAS-IR) method, the modulated IR radiations are observed by the carbon sample and converted into sound, which is detected by a microphone.[85,86] The carbon sample is placed in a specially designed cell, which contains a gas and a sensitive microphone. Modulated IR radiations from an interferometer illuminate the carbon sample, producing vibrational-thermal relaxation processes that result in modulated heating of the carbon sample. The periodic flow of heat through the gas in the cell produces pressure fluctuations, which are detected by the microphone. The acoustic signal Thus, obtained is of the same general form as the original IR signal. The photoacoustic interferogram is subsequently Fourier transformed to obtain a PAS single-beam spectrum.

The major advantage of the PAS technique is that it can help examine highly scattering and optically opaque materials. The scattered light does not interfere in PAS because only the light that is observed is converted into heat and then into acoustic waves. Furthermore, the carbon sample does not need any extensive preparation for examination by PAS. However, the PAS technique has low sensitivity, particularly in the mid-IR region (400 to 4000 cm^{-1}). More procedural details and theoretical discussion are given by Rosenewaig and others.[87,88]

Infrared Photothermal Beam Deflection Spectroscopy

Infrared photothermal beam deflection spectroscopy (IR-PDS) is better than PAS because it does not have a microphone near the sample. It involves two light sources. One is an interferometer that produces modulated radiations to illuminate the sample, and the other is a laser source that is placed so that its beam grazes the surface of the carbon sample. The absorption of the incident-modulated radiation beam by the sample produces heat, causing thermal gradients that deflect the laser beam. The deflected laser beam is detected by the detector, and the signal reproduced is a measure of the photothermal effect induced on the sample surface. The resulting modulated photothermal interferogram is processed by an FT spectrometer. In these studies the thickness of the sample is of no consequence, but the sample should have a flat area about 3 mm in diameter, and this area should be accessible to IR and laser beam probes.[45] Low and Morterra[51,89,91] have used PDS extensively in their studies on the formation of acidic and basic surface groups by oxidation of carbons prepared by charring cellulose filter paper under well-defined and controlled conditions.

Results of Various IR Studies

The IR spectra of several carbon materials using different IR techniques are presented in Figure 1.9 to Figure 1.15. It will be seen that although FTIR spectra enhance the sensitivity of the measurement, the conventional transmission and absorption measurements also give some meaningful information that can be verified by other methods.

Friedel and coworkers[84,92] obtained IR absorption spectra of a Pittsburgh-activated carbon using a finely ground sample. The spectra (Figure 1.9) show definite bands at 1735, 1590, and 1215 cm⁻¹, which were attributed to carbonyl aromatic surface structures or to conjugated chelated carbonyl and C-0 structures, respectively. Ishazaki and Marti,[93] using direct transmission IR spectroscopy, examined activated carbon Filtrasorb before as well as after neutralization with sodium hydroxide and hydrochloric acid solutions. The spectra of the original and neutralized Filtrasorb (Figure 1.10) show absorption bonds at 1760 to 1710 cm⁻¹, 1670 to 1520 cm⁻¹, 1480 to 1340 cm⁻¹, 1300 to 1230 cm⁻¹, and 1100 to 1000 cm⁻¹ regions. Comparison of the spectra for original and alkali-treated samples showed the presence of acidic surface groups. The 1760 to 1710 cm⁻¹ absorption band was attributed to carboxylic tautomeric structures, and the 1670 to 1500 cm⁻¹ region spectra, which showed considerable overlapping of different absorption bands, was attributed to quinonic and carboxylate groups. In the 1480 to 1340 cm⁻¹ region, which is characteristic of –0–H bonding vibrations, the sample showed a peak at 1465 cm⁻¹, which suffered a strong reduction after neutralization with alkali was assigned to phenolic groups.

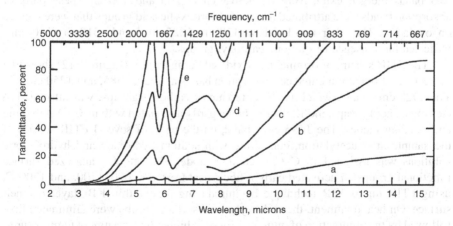

FIGURE 1.9 Infrared spectra of an activated carbon (KBr pellet method). The *b, c, d,* and *e* are scale expansion spectra. (From Friedel, R.A. and Carlson, G.L., *Fuel,* 51, 194, 1972. With permission.)

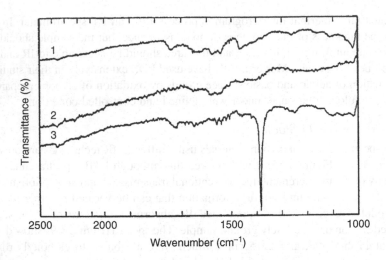

FIGURE 1.10 Direct transmission IR spectra of different carbons. (1) Filtrasorb-200; (2) Filtrasorb-200 neutralized with NaOH; (3) Filtrasorb-200 neutralized with HCl. (From Ishazaki, C. and Marti I., *Carbon*, 19, 409, 1981. With permission.)

The absorption in the 1300 to 1280 cm^{-1} region, which is associated with –C–O stretching vibration, was attributed to lactonic and phenolic, but the 1180 to 1000 cm^{-1} region peak probably arose from phenolic structures.

The internal reflectance spectroscopy (IRS) has also been used[94] to characterize surface oxygen groups on carbons because this technique is easy and does not involve large scattering losses common to transmission IR spectroscopy. The IRS spectra for sugar carbons showed a pair of bands at 1710 to 1750 cm^{-1} and 1750 to 1770 cm^{-1} regions when the carbons were activated between 300 and 700°C (Figure 1.11). The two bands merged extensively on activation at 500 and 700°C. These pairs of absorption bands were attributed to a pair of carboxylic acid groups that were created on oxidation. The other two bands at 1590 to 1625 cm^{-1} and 1510 to 1560 cm^{-1} indicated the presence of quinone carbonyl groups.

The FTIR spectra of original and oxidized Saran carbon (Figure 1.12) show that the degree of oxidation enhances absorption bands at 1729, 1585, and 1250 cm^{-1}.[217] The 1720 cm^{-1} band, which is characteristic of carbonyl groups, was attributed to carboxylic acid groups, the 1250 cm^{-1} being partly associated with its C–O stretching and bending modes. The 1585 cm^{-1} band, on the basis of several FTIR studies of the sample after acetylation, contacting with sodium hydroxide and hydrochloric solutions was assigned to C–C stretching mode. Shin et al.[95] analyzed surface functional groups on activated carbon fibers heat treated at 600, 1100, and 1200°C, using FTIR micro ATR technique by introducing a very thin KBr layer on their surface. On heat treatment, the carboxylic acid surface groups were eliminated first, followed by the elimination of quinone groups at higher heat treatment temperatures.

The IR spectra for a carbon black and a soot obtained by FTIR, ATR-IRS, and PDS have also been used for surface oxygen group analysis. The FTIR spectra obtained by O' Reilly and Mosher (Figure 1.13) for a low surface area thermal black

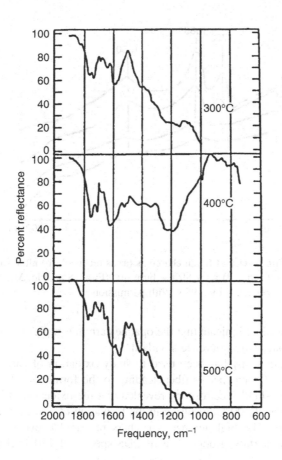

FIGURE 1.11 (a) IRS spectra of sugar charcoals activated (in 1% O_2 and 99% N_2) at 300, 400, and 500°C. (b) IRS spectra of sugar charcoals activated (in 1% O_2 and 99% N_2) at 600° and 700°C. (From Mattson, J.S., Lee, L., Mark, H.B., Jr., and Webber, W.J., Jr., *J. Colloid Interface Sci.*, 33, 284, 1970. With permission.)

with a volatile content of 0.6 shows little evidence for the presence of surface groups (89). However, the spectrum for the high surface area (530 m²/g) and high volatile content (10.8%) of carbon black, Monarch 1300, clearly shows evidence for strong absorption peaks and bands, which could be assigned to surface functional groups. The absorption band at 3430 cm⁻¹ was attributed to absorbed water as its intensity decreased on drying of the sample. The absorption bands at 1720 and 1600 cm⁻¹ were assigned to carbonyl stretching frequencies for carboxylic and quinone groups. The intensities of the two peaks agreed reasonably well with the concentrations of carboxyl and phenol groups determined by other methods.

Rockley and coworkers[96] compared the results obtained from different IR spectroscopic techniques for a Cabot soot prepared by the combustion of spectra-quality hexane in an open flame. The FTIR and ATR-IRS spectra (Figure 1.14) are similar, but the PAS spectra showed different features. The negative absorption bands with

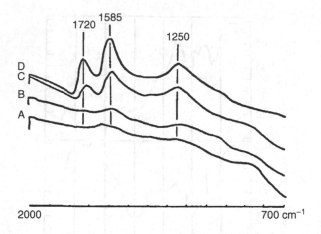

FIGURE 1.12 FTIR spectra of Saran charcoals (a) as received, (b) after 16.5% burn-off, (c) after 50.4% burn-off, and (d) after 89.1% burn-off. (From Starsinic, M., Taylor, R.L., and Walker, P.L., Jr., *Carbon,* 21, 69, 1983. With permission.)

respect to the base line indicate that the opaque carbon samples should be used with caution as photoacoustic reference standard.[97]

IR-PDS spectroscopy has been used to study oxidation of carbons prepared by charring cellulose filter paper or fiber, leading to the formation of acidic and basic surface groups. IR-PDS spectroscopy revealed that the oxidation of low-temperature chars started at about 200°C, resulting in the formation of a large number of surface functional groups. The high temperature chars prepared above 600°C showed no evidence of these surface groups. The IR-PDS spectra of 730°C char (Figure 1.15)

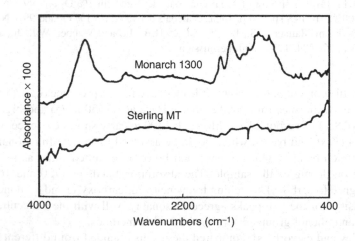

FIGURE 1.13 FTIR spectra of carbon blacks (0.01 to 0.1 weight percent mixed with KBr) after baseline subtraction. Monarch 1300 (furnace black) and Sterling MT (thermal black) are products of Cabot Co. (From O'Reilly, J.M. and Mosher, R.A., *Carbon,* 21, 47, 1983. With permission.)

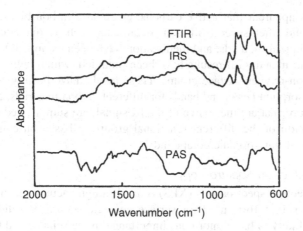

FIGURE 1.14 Absorbance spectra of hexane soot obtained by FTIR, IRS, and PAS (Rockley, M.G., Ratcliffe, A.E., Davies, D.M., and Woodard, M.K., *Appl. Spectroscopy*, 38, 554, 1984. With permission.)

are almost featureless before oxidation at temperatures lower than 300°C Some features started to appear after oxidation at 310°C and showed considerable changes after oxidation at 420°C. The spectra show a weaker shoulder at 1650 cm^{-1}, a sharp band at 1770 cm^{-1}, a symmetrical and dominant band at 1600 cm^{-1}, and a broad band between 1500 and 950 cm^{-1}. The 1760 cm^{-1} band was attributed to lactone-like structures, and the 1600 cm^{-1} band to a C=C mode of the polyaromatic framework.

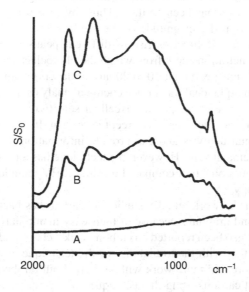

FIGURE 1.15 IR-PDS spectra of cellulose carbon prepared at 730°C. Carbon oxidized (a) at temperatures below 300°C, (b) at 330°C, and (c) at 420°C. (From Morterra, C., Low, M.J.D., and Severdia A.G., *Carbon,* 22, 5, 1984. With permission.)

Thus, it is apparent that active carbons, charcoals, carbon blacks, and carbon films before and after several different treatments such as oxidation, degassing, treatment with alkalies, acids, and methylation have been examined by different IR techniques, and meaningful results have been obtained, which help in the identification of carbon-oxygen surface groups. These results have been well reviewed.[1,83]

The IR absorption peaks and bands for different carbon materials, along with the functional groups that are the source of the IR signal, are summarized in Table 1.10. The identification of the different functional groups is based on comparison with the IR spectra of pure organic compounds.

X-Ray Photoelectron Spectroscopy

X-ray photoelectron spectroscopy (XPS) is an ultrahigh vacuum technique (vacuum of the order of 10^{-9} Torr) used for surface characterization of solid and powder samples. It is likely to be of enormous importance in the science and technology of carbons and graphites. Essentially, the technique measures the kinetic energy of electrons emitted from atoms under the influence of irradiation of the sample with x-rays. The kinetic energy of the emitted electron E_{kin} is related to the binding energy of the electron E_b and is given by

$$E_{kin} = h\nu - E_b - x$$

where x refers to the work function of the surface.

The kinetic energy of the electron emitted from the sample surface is measured by the spectrometer. The binding energy of the electron is dependent upon the chemical environment (the state of bonding to neighboring) of the atom or atoms from which the electron has been emitted. Thus, by measuring the kinetic energy of the emitted electron, the magnitude of the binding energy of the electron with the environment of the atom gives rise to different peaks in the XPS spectrum. Furthermore, the penetration depth from which the photoelectron emerges is seldom more than 10 to 15 nm (i.e., about 10 to 20 atomic layers) from the surface, which makes the XPS technique ideal for surface chemical analysis as well as for the study of adsorbed species. The technique has excellent sensitivity to submonolayer coverage and ability to detect all elements except hydrogen. In addition, it is useful for quantitative elemental analysis and can provide information on bonding from the measurement of chemical shift. However, the technique has a poor lateral resolution and slow rate of data collection compared to other UHV techniques such as augur electron spectroscopy.

A major part of the work on XPS studies on carbon has been carried out using carbon fibers[98–106] and graphites because of their easy manipulation, although a few investigations have also been reported on carbon blacks, chars, and activated carbons. It is felt, however, that with the availability of the technique at different centers of carbon research, more and more work will be carried out on the analysis of surface groups on activated carbons using the technique.

The nitrogen 1s, oxygen 1s and carbon 1s peak in the XPS spectra of oxidized type II PAN-based carbon fiber[98] shows a shoulder on the higher energy side of the carbon 1s peak (Figure 1.16) indicating the presence of two separate kinds of carbon atoms.

TABLE 1.10
Infrared Adsorption Peaks and Signals for Different Surface Groups on Carbons

Carbon	Adsorption Bans and Peaks (cm^{-1})	Bond or Group Source of Infrared Signal	Reference
Cellulose char	3030	Aromatic C—H	179
Coal	2920, 2850	Aliphatic CH$_2$ & CH$_3$ stretch	78.79
Carbon black	1760	Lactone	43
Activated carbon	1735	Caronyl	83
Pyrolyzed polymer	1724–1754	Carbonyl	177
Activated carbon	1710–1760	C=O stretch, carboxyl and lactone	92
Cellulose char	1704	C=O	179
Carbon black	1675–1775	Lactone	180
Cellulose char	1616	C=O	179
Channel black	1600	Carboxyl	43
Channel black	1590	Carbonyl or aromatic group	106
Activated carbon	1590	Aromatic structures and unconjugated carbonyl	83
Pyrolyzed polymer	1575–1550	Lactone	173
Carbon black	1550–1675	Quinone	180
Activated carbon	1500–1670	C=O stretch, carboxyl	92
Cellulose char	1449	CH$_2$	179
Activated carbon	1340–1480	Phenol	92
Activated carbon	1260	C—O—C vibration, lactone	92
Cellulose char	1250	Aromatic C=O	179
Channel black	1230	C—O stretch	106
Activated carbon	1215	C—O absorption or phenoxy adsorption	83
Channel black	1205	Condensed aromatic ring or H-bonded, conjugated carbonyl	181
Carbon black	1195–1205	Phenol	178
Activated carbon	1000–1180	C—O stretch and vibration	

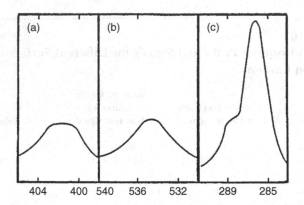

FIGURE 1.16 High-energy photoelectron spectra of oxidized Courtelle Type II carbon fiber of (a) nitrogen 1s,(b) oxygen 1s, (c) carbon 1s. (From Thomas, J.M., Evans, E.L., Barker, M., and Swift, P., in *3rd Conf. Ind. Carbon Graphite,* London Soc. Chem. Ind., 1971, p. 411. With permission.)

The spectrum changed completely when the carbon fiber was heated at 1300°K in nitrogen (Figure 1.17). The nitrogen and oxygen peaks disappeared almost completely and the carbon peak sharpened considerably. On subsequent oxidation of the heated carbon fiber, a significant oxygen 1s peak reappeared. The analysis of the XPS peaks showed that about 5 oxygen atoms per 100 carbon atoms were present at a penetration depth of 10 to 15nm on the surface of the oxidized carbon fiber.

The changes in surface functionality induced by gas-phase oxidation and heat treatment on two samples of carbon (CF) and graphitic (GF) fibers, as studied by Ishatani and Takahagi[99,100] show that the intensity of O(1s) peak increases with oxidation and decreases with heat treatment. The digital difference spectrum technique[100]

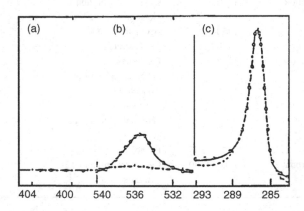

FIGURE 1.17 High-energy photoelectron spectra of oxidized Courtelle Type II carbon fiber after heat treatment at 1300K of (a) nitrogen 1s, (b) oxygen 1s, and (c) carbon 1s. (From Thomas, J.M., Evans, E.L., Barker, M., and Swift, P., in *3rd Conf. Ind. Carbon Graphite,* London Soc. Chem. Ind., 1971, p. 411. With permission.)

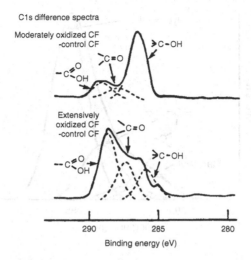

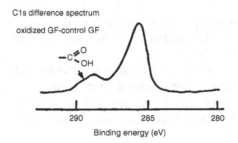

FIGURE 1.18 Carbon 1s digital difference spectra of Type II (CF) and Type I (GF) carbon fiber before and after oxidation. (From Takahagi, T. and Ishitani, A., *Carbon*, 22, 43, 1984. With permission.)

allows the analysis of surface groups on carbons. The difference spectra obtained by subtracting the O(1s) spectra of control fibers from those of oxidized fibers, using the appropriate weight factors, showed three components with different chemical shifts that were assigned to hydroxyl groups (Figure 1.18) (>C—OH, 286 eV), carbonyl groups (>C=O, 287 eV) and carboxyl groups (–COOH 288.6 eV). The curve resolving the C(1s) difference spectra of CF gave the concentrations of the three functions. The higher binding energy component that disappeared on heat treatment (289 eV) was assigned to carboxylic groups and the one with the lower binding energy that stayed after the heat treatment (286 eV) was attributed to disordering of the graphite crystal lattice caused by surface oxidation.

The XPS technique has also been successfully used in the study of chemical groups formed on oxidation and eliminated on heat treatment on carbon foil,[102] on carbons obtained by the electrochemical reduction of poly (tetra fluoro ethylene, PTFE),[103] on bituminous coals,[105] and on carbon blacks.[107] The carbon 1s synthesis of the XPS spectra for coal heated at 325°C obtained by Grint and Perry[105] is shown

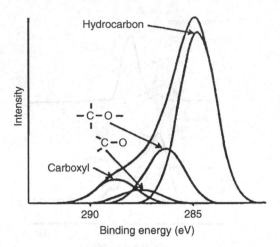

FIGURE 1.19 Carbon 1s peak synthesis for coal heated to 325°C. (From Grint, A. and Perry, D.L., in 15th Bienn. Conf. on Carbon Ext. Abstracts, 1981, p. 462. With permission.)

in Figure 1.19. The groups postulated by these workers were phenolic, ether, and alkyl structures corresponding to 286.20 eV, ketones and aldehydes at 288 eV, and carboxylic groups at 289.2 eV.

Thus, the presence of surface oxygen groups and chemisorbed species on carbon surface can be identified by the chemical shift of the C(Is) peak. Some typical examples from XPS investigations are summed up in Table 1.11. Depth profile studies in XPS spectra can also help obtain concentration profile of surface groups.

1.4 INFLUENCE OF CARBON-OXYGEN SURFACE GROUPS ON ADSORPTION PROPERTIES

It is well known that active carbons contain appreciable amounts of chemisorbed oxygen and hydrogen, which are present in the form of carbon-oxygen surface complexes. Several hypothetical structures have been assigned to these surface complexes. Based on several physical, chemical, and physiochemical techniques used for their analysis, the existence of such functional groups as carbonyls, carboxyls, lactones, quinones, hydroquinones, and phenols has been suggested. However, the various methods of estimation have not yielded comparable results and, even more significant, the entire amount of associated oxygen has not been accounted for. Thus, the problem has not yet been properly elucidated, and the various structures and mechanisms that have been proposed are not outside the realm of speculation. However, the work reported from our laboratories and elsewhere indicates that there are probably definite surface groups or complexes that evolve CO_2 and, similarly, there are distinct surface entities that evolve CO on heat treatment in vacuum or an inert atmosphere.

Whatever the exact nature and structure of these surface oxides, there is little doubt that the chemisorbed oxygen is present mainly at the edges and corners of the giant aromatic sheets. Because these edges and corners constitute the main

TABLE 1.11
XPS Bond Energies and Functional Groups on Different Carbon Materials

Carbon	Binding Energy for C**	Interpretation of Spectra	Reference
Channel black	285.0	C—H bond	106
	288.5	C—O—bond	106
Oxidized carbon fiber	286.0	Hydroxyl group	99
	287.0	Carbonyl group	99
	288.6	Carboxyl group	99
Carbon from PTFE reduced by Li	285.4–285.9	C atoms in the basic C skeleton	102
	290.2–290.8	C atoms in surface COOH groups	102
Carbon fiber	285.0	Hydrocarbon group	100
	287.0	C—O—bond	100
	289.0	C=O bond	100
Graphite oxide	285.0	Carboxyl group	100
	287.2	Phenolic hydroxyl group	100
	289.0	Carbonyl and ether group	100
Air-oxidized carbon fiber	1.5[a]	C—O—group	98
	2.5[a]	C=O group	98
	4.0[a]	Carboxyl group	98
Carbon fiber	1.6[a]	C—O—group	98
	3.0[a]	C=O group	98
	4.5[a]	Carboxyl group	98
Electrochemically oxidized carbon fiber	11.6[a]	C—O—group	100
	3.0[a]	C=O group	100
	4.5[a]	Carboxyl or ester-type group	100
Electrochemically oxidized carbon fiber	−2.1[a]	C=O and/or quinone-type group	95
	−4.0[a]	Ester-type group	95
	>6.0[a]	CO^{2-}_3 type group	95
Fluorinated graphite	4.7[a]	CF group	184
	6.7[a]	CF_2 group	184
	9.0[a]	CF_3 group	184

[a] Chemical shift for C^{15} peak (eV)

adsorbing surface, these oxygen groups exercise a profound influence on the surface characteristics and surface properties of active carbons.

1.4.1 SURFACE ACIDITY OF CARBONS

Surface acidity of active carbons and carbon blacks has been the subject mater of a large number of investigations because of its importance in determining several decomposition reactions, catalytic reactions, and absorbent properties of these materials. In the case of carbon blacks, Wiegand[108] used surface acidity to classify the family of carbon blacks: ink blacks being strongly acidic, furnace blacks being alkaline or feebly acidic, and channel blacks occupying an intermediate position. The surface acidity of a carbon is measured by its base adsorption (or base neutralization) capacity, which represents the amount of alkali cation exchanged for hydrogen ions furnished by the acidic oxides on the carbon surface. The surface acidity is due to the presence of carbon-oxygen surface chemical structures that have been postulated as carboxyls and lactones, as discussed earlier in this chapter. These surface chemical structures are evolved as CO_2 on heat treatment in vacuum or in an inert atmosphere at temperatures of 300 to 750°C. The base neutralization capacity of a carbon decreases on evacuation or degassing at gradually increasing temperatures, and the decrease at any temperature, corresponds to the amount of CO_2 evolved at that temperature. Furthermore, the temperature interval over which the drop in base neutralization capacity occurs appears to be the same as the temperature interval over which CO_2 was eliminated from the carbon sample. Thus, the surface acidity of carbons depends upon the presence of carbon-oxygen surface chemical groups. This aspect has been discussed in detail earlier in this chapter.

1.4.2 HYDROPHOBICITY

Pure carbon is hydrophobic in character. The hydrophobicity decreases, and the carbon becomes more and more hydrophilic, as the amount of oxygen associated with the carbon surface increases. It has long been known that carbon blacks with high oxygen content can be easily wetted by water. Accordingly, low temperature oxidation of ink, color, and lamp blacks is often used to improve their hydrophilic character. Similarly, high performance carbon fibers that are usually hydrophobic are given a propriety oxidation treatment to produce carbon-oxygen surface functionality to improve their wetting characteristics, which in turn improves their adhesion with the matrix material in high strength composites. Fitzer et al.[109] reported a direct relationship between the amount of surface acidic groups on the carbon fiber surface created by oxidation and the strength of the composite in the case of phenolic as well as epoxy resins as matrix materials. The effect of surface oxygen groups introduced on the carbon fiber surface by oxidation in air at 700°C and composite properties is shown in Figure 1.20. In the case of active carbons, the presence of acidic carbon-oxygen surface groups makes the carbon surface hydrophilic and polar in character, which in turn improves the adsorption properties toward polar gases and vapors.

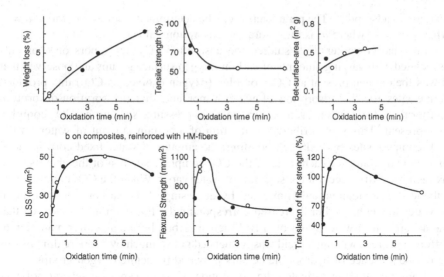

FIGURE 1.20 Effect of surface oxidation in air at 700°C on different carbon fiber and composite properties. (Courtesy of E. Fitzer. With permission.)

1.4.3 ADSORPTION OF POLAR VAPORS

A good deal of work has been reported on the influence of carbon-oxygen surface groups on the adsorption of water vapors. Lawson[110] and King and Lawson[111] observed that the presence of associated oxygen in carbons increases the low pressure adsorption of water vapor and shifts the isotherm to lower pressures. Pierce et al.[112] found that the adsorption isotherm of water on graphon changed significantly after exposure of graphon to water vapor at 80°C. This difference was attributed to interaction of water vapor with graphon, producing a carbon-oxygen surface complex that adsorbed water more readily than the clean surface. Pierce and coworkers[112,113] and Dubinin[114,115] observed that certain active sites on active carbons made available by carbon-oxygen surface groups act as primary adsorption centers at which sorption of water proceeds in the form of isolated clusters through hydrogen bonds. These clusters then grow in size as more adsorption takes place on adsorbed water molecules by hydrogen bonding, and ultimately they merge at higher relative vapor pressures to form two-dimensional islands of the condensed phase on the carbon surface. Dubinin, Zaverini and Serpinski[114,116] were able to derive an equation to calculate the number of these primary adsorption centers. The equation is written as

$$K_1(a_0 - a)(1 - Ka)h = K_2a$$

where a_o is the number of primary adsorption centers and a (both expressed in millimoles per gram) is the amount adsorbed at relative pressure $hx\ p/p_o$. In the equation, $(a - a_o)$ represents the total number of adsorption centers, and $(1 - Ka)$ takes into account the decrease in the number of acting adsorption centers with increase in adsorption. The parameters k can be determined from the condition $a = a_s$ when $p/p_o = 1$, where a_s is the saturation adsorption value.

Puri and coworkers[117,118] studied the adsorption of water vapors on charcoals associated with varying amounts of carbon-oxygen surface groups and observed that it was the oxygen present as CO_2-complex (oxygen evolved as CO_2) that provided active sites for the adsorption of water vapor and that the sorption-desorption isotherms did not meet, even at zero relative pressure, so long as CO_2 complex was present. This was attributed to fixation of a certain amount of water on the CO_2-complex sites by hydrogen bonding; the amount of water fixed corresponded to 1 mole of water for each mole of the CO_2 complex. The rest of the chemisorbed oxygen had little effect on the sorption of water vapor. When the CO_2 complex was enhanced by oxidation and eliminated by degassing of the carbon, the adsorption of water was enhanced or decreased correspondingly. Barton et al.[119] observed that the adsorption of water vapor on a BPL active carbon before and after oxidation to different degrees with nitric acid was influenced not so much by the pore dimensions as by the presence of hydrophilic centers provided by acidic oxygen groups.

Bansal et al.[120] investigated the adsorption of water vapor on several polymer charcoals having different porosities and associated with varying amounts of combined oxygen. The charcoals associated with similar amounts of combined oxygen but possessing different porosities showed similar adsorptions at lower relative pressures but differed appreciably in their adsorption values at higher relative vapor pressures, indicating the influence of associated oxygen and porosity in the two ranges of relative pressures. When the combined oxygen content of the charcoals was decreased, the isotherms showed a continuous decrease in the adsorption of water vapors as more and more of the associated oxygen was removed from the sample. This decrease, however, was restricted only to adsorption at lower relative pressures ($p/p_o < 0.5$). When the amount of combined oxygen on a sample of charcoal was increased by oxidation, an increase in adsorption of water vapor was observed (Figure 1.21).

The presence of carbon-oxygen surface groups also affects the adsorption of other polar or polarizable compounds such as methanol, ethanol, ammonia and amines, dyes, and surfactants. Puri and coworkers[121,122] and Bansal et al.[123,124] studied the adsorption isotherms of methanol on charcoals associated with different amounts of carbon-oxygen surface groups and observed that the adsorption increased or decreased with the amount of the surface oxygen groups. The sorption-desorption isotherms did not meet so long as CO_2-complex was present. The amount of methanol retained by specific interactions on the surface of charcoals corresponded roughly to one half-mole for each mole of the CO_2 complex. The smaller value in the case of methanol as compared to water was probably due to the larger radius of the nonpolar group in methanol. A methyl group of radius 0.2 nm adsorbed on an oxygen atom of radius 0.13 nm, lying flat, may block access of the methyl group to the neighboring oxygen atom. Similarly, the adsorption of amines by active carbons is determined by the amount of the acidic carbon-oxygen surface groups.[125] When these surface groups are removed from the surface, the adsorption of the amine also goes down.

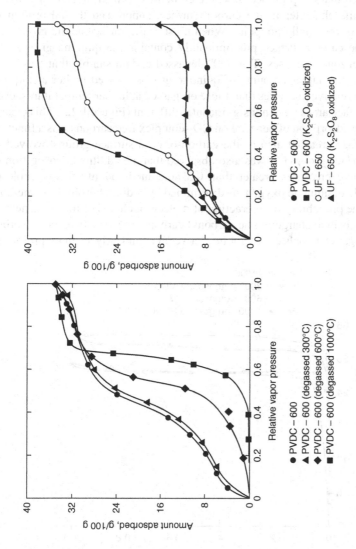

FIGURE 1.21 Water vapor sorption isotherms on polyvinyledene chloride (PVDC and urea formaldehyde resin (UF) carbons before and after outgassing and oxidation. (From Bansal, R.C., Dhami, T.L., and Prakash, S., *Carbon*, 16, 389, 1978. With permission.)

1.4.4 ADSORPTION OF BENZENE VAPORS

Dubinin and coworkers used benzene as a standard vapor for the derivation of the theories of physical adsorption. These workers are of the view that the adsorption of benzene involves purely dispersion interactions. However, Puri et al.[126] in their studies on sugar charcoals and carbon blacks, and Bansal and Dhami[127] on polymer carbons, observed that the presence of acidic carbon-oxygen surface groups, which impart hydrophilic character to the carbon surface, suppressed the adsorption of benzene, which is essentially nonpolar. When most of the surface acidic groups are removed and the carbon surface predominantly contains the quinone groups, the adsorption of benzene increases. The 600°-degassed carbon sample that was associated with relatively larger amounts of quinone groups showed higher adsorption of benzene at all relative pressures than those outgassed at higher temperatures, even through their surface areas were not significantly different (Figure 1.22). This clearly indicates that the adsorption of benzene on CO_2-complex free carbons was a function not only of surface area but also of the carbon-oxygen surface groups evolved as CO (quinone surface groups). It was also observed that the additional adsorption of benzene at relative pressures greater than 0.3 amounted roughly to one mole of benzene per mole of quinone oxygen, as determined by sodium borohydride method. This indicates the probability of interaction of π electron clouds of the benzene ring with the partial positive charge on the carbonyl carbon atom. The benzene adsorbed in the case of sugar charcoals could not be recovered completely, even on prolonged

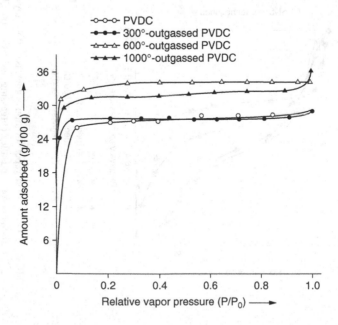

FIGURE 1.22 Adsorption isotherms of benzene on (a) PVDC carbon, (b) Spheron-6, and (c) Mogul before and after outgassing at different temperatures. (From Bansal, R.C. and Dhami T.L., *Carbon,* 15, 153, 1977. With permission.)

evacuation, indicating complexing of benzene at the quinone sites within the micropores.

1.4.5 IMMERSIONAL HEATS OF WETTING

Heats of immersion of carbon blacks in water and methanol have been formed[129] to be a linear function of oxygen content. In charcoals, however, Puri and coworkers,[122,128] observed that the heats of immersion varied linearly with the carbon-oxygen surface groups evolved is CO_2 on evacuation and not so much with respect to total oxygen (Figure 1.23). When these surface groups were enhanced by oxidation, the heat of immersion increased correspondingly. Barton and coworkers[129] observed that the heat of immersion of graphite in n.hexane differed only slightly from that for a completely oxygen-free sample, indicating that the oxygen complexes, which covered about 30% of the graphite surface, did not markedly affect the immersional energetics. However, the heats of immersion values were appreciably different in butyl derivatives for the two graphite samples. This was attributed to the interaction of polar oxygen groups with the dipole moment of the molecules of the immersion liquid.

The heats of immersion in both water and methanol[130,131] decreased continuously on evacuation of the carbon samples at temperatures above 300°C. The continuous decrease in the case of water resulted from the removal of surface oxides that provide hydrophilic centers for hydrogen bonding. The final heat of immersion value, when all the surface oxygen has been removed, agreed fairly well with the value obtained by Wade[132] on graphon, which was completely free of any combined oxygen and

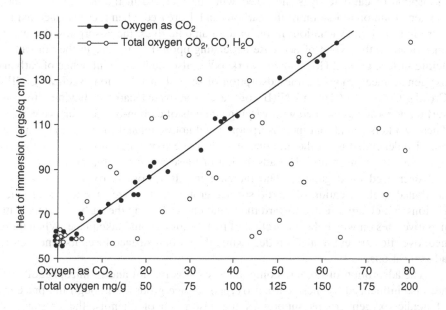

FIGURE 1.23 Heat of immersion of charcoals in relation to chemisorbed oxygen. (After Puri, B.R., Singh, D.D., and Sharma, L.R., *J. Phys. Chem.*, 62, 756, 1958. With permission.)

had a predominantly hydrophobic surface. The decrease in the heat of immersion in water on evacuation was related linearly to the total oxygen content of the carbon, and in case of methanol it varied linearly with the oxygen evolved as CO_2. In the case of a BPL active carbon oxidized to different degrees with HNO_3, the heat of immersion in water was also related to hydrophilic oxide sites,[118] although there was no linear relationship between the heat of immersion and the number of sites. (Figure 1.23)

The heat of immersion in the nonpolar benzene remained more or less unaffected by the removal of CO_2 evolving oxygen groups in the evacuation temperature range up to 550°C and then decreased at higher temperatures. This decrease was attributed to the decrease in the interaction between π electron (pi electrons) clouds of benzene molecule with the partial positive charge on the carbonyl carbon atom which was evolved as CO at temperatures above 550°C.[43] Hagiwara et al.[133,134] observed that the heats of immersion of carbon blacks in ethanol and n.butane increased linearly with an increase in the carbon-oxygen surface groups having active hydrogen.

1.4.6 ADSORPTION FROM SOLUTIONS

The adsorption of organic and inorganic compounds from their aqueous solutions has also been found to be influenced by the presence of carbon-oxygen surface groups. Graham,[135] Puri et al.[136] Goyal and Bansal,[137,138] and Aggarwal[139] studied the adsorption of several cationic and anionic dyes on different charcoals and carbon blacks and observed that the adsorption, although determined by surface area of the carbon, was strongly influenced by the presence of acidic surface oxides. The adsorption of cationic dyes increased with the increase in the amount of surface oxygen group on oxidation of the carbons and decreased when these surface oxides were removed on evacuation. In the case of anionic dyes, the adsorption decreased on oxidation, the extent of decrease depending upon the increase in the amount of acidic surface groups. Goyal and coworkers[31-33] also studied the influence of carbon-oxygen surface groups on the adsorption of several metal cations such as Cr(III), Cr(VI), Co(II), Cu(II), and Ni(II) on several activated carbons having different surface areas and associated with varying amounts of carbon-oxygen surface groups. These workers found that the adsorption could not be related to surface area alone, but also depended upon the amount of the surface oxygen groups and more so on the acidic oxygen groups. The adsorption of these cations increased on oxidation and decreased on degassing. The increase in adsorption on oxidation has been attributed to the creation of acidic surface groups that ionize in water to produce H^+ ions, which are directed toward the liquid phase, leaving the carbon surface with negative sites on which the adsorption of metal cations could take place. When these negative sites are eliminated on degassing, the carbon surface loses its tendency to adsorb cations.

The adsorption of organic compounds such as phenol and p. nitrophenol[140-146] are also influenced by these carbon-oxygen surface groups. Although the presence of acidic oxygen groups suppresses the adsorption of phenols, the presence of quinone groups enhances the adsorption. The adsorption of metal cations and organic compounds will be discussed in detail later in this book.

1.4.7 PREFERENTIAL ADSORPTION

Kipling and coworkers,[147–149] while studying adsorption from binary liquid mixtures, observed that the presence of chemisorbed oxygen imparts polar character to carbons so that they exercise preferential adsorption for a more polar component of the binary mixture. Thus, Spheron-6, which contained combined oxygen, preferred methanol from its solution in benzene and graphon, which is essentially free of oxygen, preferred benzene. Similarly, Spheron-6 preferred benzene from its solutions in n beptane, and graphon preferred n. heptane. This behavior was attributed to the presence of carbon-oxygen surface groups, which interacted with the more polar component of the binary solution. Puri and coworkers,[150] and Bansal and Dhami,[123] while studying the composite adsorption isotherm from benzene methanol and benzene-ethanol solutions on different charcoals associated with varying amounts of oxygen, showed that the nature of the oxygen groups is even more important than the total oxygen in determining surface interaction of carbons in such cases. The composite adsorption isotherms of methanol-benzene mixtures on the various sugar and coconut charcoals (Figure 1.24) showed that the original charcoals that contained larger amounts of CO_2 evolving surface groups showed larger preference for the polar methanol or ethanol over benzenes. Because part of these surface groups was removed by evacuation at 400°C, there was an appreciable fall in the preferential adsorption of alcohols. When these surface groups were removed completely on outgassing the charcoals at 750°C, there was almost a reversal of the preference because the benzene was preferred, though slightly, to ethanol or methanol. (Figure 1.24)

The charcoals outgassed at 750°C retained appreciable amounts of combined oxygen capable of evolving CO. These charcoals show less preference for ethanol

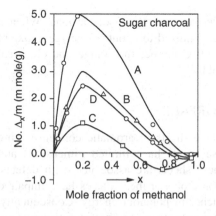
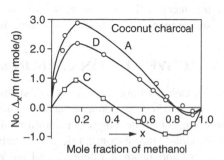

FIGURE 1.24 Composite adsorption isotherms for various samples of charcoal from sugar and coconut shell from methanol-benzene mixtures. Curves: (a) original charcoals, (b) outgassed at 400°C, (c) outgassed at 750°C, (d) outgassed at 1100°C. (After Puri, B.R., in *Chemistry and Physics of Carbon* , Vol. 6, P.L. Walker, Jr., ed., Marcel Dekker, New York, p. 191, 1970. With permission.)

and methanol, and more for benzene, than is shown by the samples outgassed at 100°C, which were almost free of oxygen. It appears that quinone groups that form a part of the CO evolving groups promote preference for benzene. Thus, one part of combined oxygen (present as CO_2-complex) promotes preferential adsorption of alcohols, and another part present as CO-complex promotes preferential adsorption of benzene. Thus, there is need to determine not only the total oxygen but also the form in which it is present to assess correctly the performance and surface behavior of carbons.

1.4.8 CATALYTIC REACTIONS OF CARBONS

Besides the extent of surface area and the availability of active sites, the presence of carbon-oxygen surface groups also plays a significant role in many reactions catalyzed by carbons. Thus, in the case of carbon catalyzed oxidation of [Fe (CN)6]$^{4+}$, NO_2, AsO_3, Fe^{2+} the maximum activity was observed with carbons that were associated with maximum amounts of carbon-oxygen surface groups.[49,127,153] The catalytic activity of carbons for the oxidation of ferrous ions in acid solution by molecular oxygen was found to be about 200 times more in the presence of oxygen-containing groups. The rate of auto-oxidation of stannous chloride in an acid solution was maximum in the presence of carbons activated at 550°C, which contained oxidized carbon surface.[154] The catalytic activity for the oxidation of hydroquinone to quinone, on the other hand, was maximum in the presence of carbons activated at 875°C which contained rather small amounts of combined oxygen.[155] The catalytic oxidation of n-butyl mercapton by carbons has been attributed to the presence of quinone surface.[156] The quinone group results in the formation of the disulphide ion through redox reaction involving the formation of thiyl radical and semiquinone ions.

1.4.9 RESISTIVITY

Chemisorption of oxygen increases the electrical resistance of carbons. When a thoroughly outgassed powdered carbon was oxidized to about 4% oxygen content on treatment with oxygen in the 500 to 700°C temperature range, the electrical resistance increased almost by a hundredfold.[157]

1.5 ACTIVE SITES ON CARBON SURFACES

Active carbons have a structure consisting of sheets of aromatic condensed ring systems stacked in nonpolar layers. These sheets have limited dimensions and therefore constitute edges. In addition, these sheets are associated with defects, dislocations, and discontinuities. The carbon atoms at these places have impaired electrons and residual valences, and are richer in potential energy. Consequently, these carbon atoms are highly reactive and constitute active sites or active centers. A considerable amount of research has been aimed at understanding the number and nature of these active sites because these active centers determine the surface reactivity, surface reactions, and catalytic reactions of active carbons. Because the tendency of carbons to chemisorb oxygen is greater than their tendency to chemisorb any other species, much of the understanding of these active sites comes from the chemisorption of oxygen.

The first suggestion that the carbons are associated with different types of active sites, also called *active centers*, came from the work of Rideal and Wright[158] on the oxidation of carbon surfaces with oxygen gas. They observed three different types of active centers that behaved differently at different oxygen pressures and showed three different rates of oxygen chemisorptions at 200°C. Allardice,[159] while studying the kinetics of chemisorption of oxygen on brown charcoal at temperatures between 25 and 300°C in the pressure range 100 to 700 torr, observed a two-step adsorption that he attributed to the presence of two types of sites. Dietz and McFarlane,[160] while studying the chemisorption of oxygen on evaporated carbon film of high surface area at temperatures between 100 and 300°C and at oxygen pressures of the order of 100 millitorr, observed a rapid initial adsorption followed by a much slower adsorption. Carpenter and coworkers,[161,162] during the initial stages of oxidation of different varieties of coals at different temperatures, found that the chemisorption of oxygen obeyed the Elovich equation only in the first five-minute period. As the time of oxidation increased, the amount of oxygen chemisorbed exceeded the amounts predicted by the Elovich equation. This was attributed to the creation of fresh adsorption sites by the desorption of oxygen as CO_2, CO, and water vapor. Puri and Bansal,[163,164] while investigating the chlorination of sugar and coconut-shell charcoals at different temperatures between 30 and 1200°C, observed that the chemisorbed hydrogen was eliminated in a number of steps, depending on the temperature of the treatment (Figure 1.25). These workers suggested that hydrogen

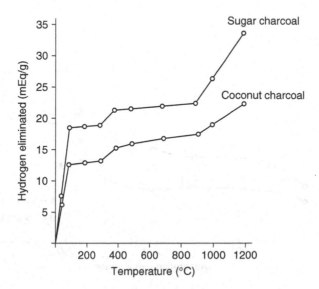

FIGURE 1.25 Elimination of chemisorbed hydrogen from charcoals by chloridation at different temperatures. (From Puri, B.R., Malhotra, S.L., and Bansal, R.C., *J. Indian Chem. Soc.*, 40, 179, 1963. With permission.)

in charcoal was bonded at different types of sites associated with varying energies of activation. Lussow et al.[165] studied the kinetics of chemisorption of oxygen on clean surfaces of graphon activated to different degrees of burn-off between 0 and 35% in the temperature range of 450 to 675°C. The amount of oxygen chemisorbed increased sharply at temperatures above 400°C, the amounts being almost two to three times larger. It was also observed that when the presence of oxygen was increased from 0.5 to 700 torr, the amount of oxygen chemisorbed was almost doubled. These results indicated the existence of more than one type of site. In their later work, Walker and coworkers,[166,167] while studying the rates of oxygen chemisorption on activated graphon, with 14.4% burn-off in the temperature range 300 to 625°C at an oxygen pressure of 0.5 torr, observed two different rates of adsorption, one above and the other below 250°C. Furthermore, there was a sharp increase in the saturation amount of oxygen at temperatures above 250°C or at longer intervals of time. These results indicated the presence of two types of sites that differed in their activation energies (Figure 1.26). The activation energy of adsorption at relatively more active sites was found to be 30 kJ/mol.

Bansal and coworkers,[168–170] characterized ultraclean surfaces of graphon by studying the kinetics of chemisorption of oxygen at low temperature in the range 70 to 160°C and at oxygen pressures between 7.7×10^{-4} and 760 torr. The graphon sample (16.6% burn-off) was cleaned by heat treatment at 1000°C in a vacuum of the order of 10^{-9} torr. The apparatus essentially consisted of a gas inlet system, a vacuum assembly that contained a vacion pump, and an adsorption unit that included a Barotron differential manometer, a residual gas analyzer, and a Cahn microsorption balance. The Elovich plots of these kinetic studies showed linear regions. The number

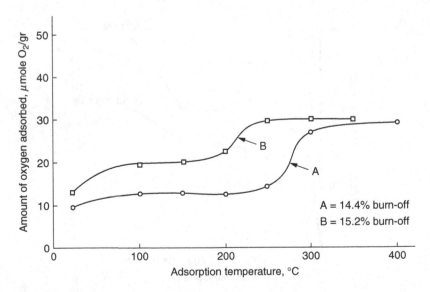

FIGURE 1.26 Saturation amounts of chemisorbed oxygen on activated graphon at different temperatures. (From Hart, P.J., Vastola, F.J., and Walker, P.L., Jr., *Carbon*, 5, 363, 1967. With permission.)

TABLE 1.12

Instantaneous Adsorption Rates Calculated Midway in Each Linear Region; Rates Normalized with Respect to Pressure

Pressure (mtorr)	Rate (atoms/g $\lrcorner$ min $\lrcorner$ mtorr)				
	I (Q = 0.6 \times 10^{18})	II (Q = 1.8 \cdot 10^{18})	III (Q = 4.5 \cdot 10^{18})	IV (Q = 9.6 \cdot 10^{18})	V (Q = 18.0 \cdot 10^{18})
0.77	2.1×10^{15}	—	—	—	—
5.76	2.5	—	—	—	—
11.6	2.7	—	—	—	—
22.9	2.0	2.6×10^{14}	—	—	—
50.6	2.0	2.9	—	—	—
99.2	—	2.9	—	—	—
164	—	3.0	—	—	—
199	—	4.0	2.2×10^{14}	—	—
302	—	3.9	2.0	5.2×10^{13}	—
537	—	—	2.2	6.5	—
693	—	—	2.3	6.9	8.7×10^{12}
5438	—	—	2.0	4.5	7.6
9930	—	—	2.0	4.9	8.9
760×10^3	—	—	—	1.0	3.1

Source: Walker, P.L., Jr., Bansal, R.C., and Vastola, F.J., in *The Structure and Chemistry of Solid Surfaces*, Wiley, 1969, pp. 81-1 to 81-16. With permission.

of linear regions and the time of appearance of a linear region depended on the initial oxygen pressure or the temperature of chemisorption. The number of these linear regions decreased at higher pressures or at higher temperatures. The instantaneous rates of adsorption calculated midway in each linear region when normalized with respect to oxygen pressure were found essentially proportional to the first power of oxygen pressure, which was varied widely. Furthermore, the rate of oxygen chemisorption decreased sharply in advancing from stage I to stage V (Table 1.12) where it was 250-fold less than that for state I. These linear regions are thus postulated to represent different kinetic stages of the same chemisorption process involving adsorption at different types of sites. The activation energies of adsorption calculated from the Arrhenius plots of instantaneous rates at different coverages were found to be independent of surface coverage on any one group of sites, although the activation energies were different for adsorption on different groups of sites (Figure 1.27). In all, five different groups of sites were observed in these studies. The activation energies varied between 3.1 and 12.4 Kcal/mole as the chemisorption proceeded from the most active to the least active sites. Bansal et al.[170] also studied the kinetics of chemisorption of hydrogen on the same sample of graphon as a function of oxygen pressure and adsorption temperature. The rates of hydrogen chemisorption were very low as compared to those of oxygen. Only four types of active sites could be observed

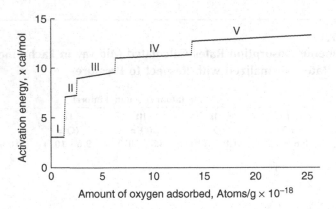

FIGURE 1.27 Variation of activation energy as a function of oxygen chemisorption on activated graphon. (From Bansal, R.C., Vastola, F.J., and Walker, P.L., Jr., *Colloid Interface Sci.*, 32, 187, 1970. With permission.)

with an adsorption temperature of 600°C. Hydrogen chemisorption experiments could not be carried out at higher temperatures because the graphon surface showed burning, producing gaseous species which vitiated the kinetic measurements.

Puri and coworkers[171,172] and Bansal et al.[173] obtained fairly convincing evidence for the existence of entirely different types of active sites, which they called *unsaturated sites*. These active sites could be determined by interacting carbons with aqueous solutions of bromine in potassium bromide. These sites are produced when CO_2-evolving surface-oxygen groups are removed from the carbon surface by degassing at temperatures between 350 and 750°C. One mol of unsaturated sites is produced by the elimination of two mols of chemisorbed oxygen as CO_2.

The concentration of active sites on a carbon surface is measured in terms of the active surface area (ASA).[174] The ASA, according to these workers, is an index of the chemical reactivity of a carbon surface that determines the catalytic and surface reactions of carbons. ASA is determined from the amount of oxygen chemisorbed at 300°C in 24 hr at an initial oxygen pressure of 0.5 torr. Assuming that one oxygen atom is chemisorbed at each active carbon atom, and that each carbon atom occupies an area of 0.083 nm², the amount of oxygen chemisorbed can be converted into ASA. These workers activated graphon to different burn-offs in order to create varying amounts of ASA and observed that ASA increased with the degree of burn-off and could be related to the reactivity of graphon toward oxygen. However, it has now been found that carbons are associated with several types of groups of active sites, each group being associated with varying amounts of activation energies. Consequently, it has been suggested that there is need to obtain reactivity data under proper conditions to ensure that the observed reactivity is not being unduly influenced by the reaction on less activated sites. Hoffman et al.[175] while studying the chemisorption of several hydrocarbons such as propylene, ethylene, propane, and methane on activated graphon, observed that the adsorption of each hydrocarbon increased with burn-off, due to the increase in ASA of the graphon. However, the ASA covered by these hydrocarbons was much less compared to the ASA

TABLE 1.13

Active Surface Area Occupied by the Chemisorption of Different Gases on Graphon at 300°C

Sample	Burn-off (%)			Oxygen ASA (m²/g)	Propylene ASA (m²/g)	Ethylene ASA (m²/g)	n-Butane ASA (m²/g)
	723 K	1223 K	Total				
1	0	0	0	0.264	0.056	0.033	0.012
2	0	0.61	0.61	0.94	0.44	0.32	0.18
3	4.9	0.85	5.75	2.25	1.06	0.72	0.46
4	24.1	0.40	24.5	5.00	1.89	0.19	0.87

Source: Hoffman, W.P., Vastola, F.J., and Walker, P.L., Jr., *Carbon*, 22, 585, 1984. Reproduced with permission from Elsevier.

covered by oxygen at all degrees of burn-off (Table 1.13). Furthermore, propylene covered a large ASA compared to ethylene, methane, or n. butane. This has been attributed to the larger size of these hydrocarbon molecules, which when adsorbed on an active site are likely to shield some of the neighboring sites to make them unavailable for chemisorption. Thus, ASA has a meaning only with respect to the chemisorption of a particular species. Whereas the Walker method can measure reactivity towards oxygen, it fails to measure reactivity of the graphon surface toward these hydrocarbons.

Dentzer et al.[176] examined the adsorption and decomposition of silver diamine complexes from ammonical solution on a graphitic carbon black activated to different degrees of burn-off and observed that the amount of silver adsorbed increased with increase in the degree of burn-off. A linear relationship was observed between the amount of silver adsorbed and the ASA as determined by the Walker method. This was attributed to specific reductive interaction of silver diamine with the carbon active sites producing metallic silver, which was chemisorbed on the active sites.

Bansal and coworkers[168–170] attribute the existence of these active sites to the difference in the geometrical arrangement of the surface carbon atoms. As the carbon surface presents several carbon-carbon distances to the incoming gas molecules, the activated complex formed between the gas atoms and the two surface carbon atoms would be expected to have different potential energy configurations, depending upon the spacing between the carbon atoms, causing a variation of the activation energy of chemisorption. Sherman and Eyring[177] made theoretical calculations of the energy of activation for dissociative chemisorption of hydrogen on a carbon surface and found the values to vary with the carbon-carbon spacing. A comparison of the theoretical values of Sherman and Eyring and the experimental values obtained by Bansal et al.[168,169] are given in Table 1.14, where the carbon-carbon distances selected are those that the hydrogen molecule would most likely encounter when approaching the carbon surface. The spacing 0.246 nm represents the configuration terminating in [101L] face, the spacing of 0.142 and 0.284 nm for termination [112L) face, and the spacings of 0.335 and 0.362 nm for distances between edge carbon atoms in adjacent basal planes. Agreement between the experimental and theoretical values

TABLE 1.14
Comparison of Experimental and Theoretical Values
of Activation Energy for Chemisorption of Hydrogen
on Activated Graphon

Experimental		Theoretical	
Site	E (kcal/mol)	C–C distance (nm)	E (kcal/mol)
I	5.7	0.362	9
II	8.3	0.335	11
III	18.4	0.284	19
IV	30.4	0.246	28
V	—	0.142	50

Source: Bansal, R.C., Vastola, F.J., and Walker, P.L., Jr., *Colloid Interface Sci.*, 32, 187, 1970. Reproduced with permission from Elsevier.

supports the concept that carbon spacing acts like discrete types of sites on which the chemisorption of gases generally can occur.

1.6 MODIFICATION OF ACTIVATED CARBON SURFACE

The most interesting and important property of an activated carbon is that its surface can be suitably modified to change its adsorption characteristics and to tailor-make carbons for a particular application. The modification of a carbon surface can be carried out by the formation of different types of surface groups. These groups include carbon-oxygen surface groups that are formed by oxidation of the carbon surface with oxidizing gases or solutions; carbon-hydrogen surface groups by treatment with hydrogen gas at elevated temperatures; carbon-nitrogen surface groups on treatment with ammonia; carbon-sulfur surface groups on treatment with elementary sulfur; carbon disulphide, H_2S, SO_2, and carbon-halogen surface groups formed by treatment with the halogen in gaseous or solution phase. As these surface groups are bonded or retained at the edge or corners of the aromatic sheets, and because these edges or corners constitute the main adsorbing surface, they are expected to modify the adsorption characteristics and adsorption behavior of these carbons. In addition, the modification of the carbon surface can also be carried out by degassing and by impregnation of the surface with metals, and a detailed discussion about the formation of these surface groups on different chars, activated carbons, and carbon blacks has been given elsewhere[1] and is out of the scope of this book. The influence of carbon-oxygen surface groups on the surface characteristics and surface properties of activated carbon is discussed in section 1.4 of this chapter. A brief discussion about the influence of other modifications on the adsorbent properties of carbons is given below.

1.6.1 MODIFICATION OF ACTIVATED CARBON SURFACE BY NITROGENATION

Activated carbons do not contain significant amounts of nitrogen surface groups. However, gas-phase reactions with dimethylamine[178] at 150°C for about one hour, or with dry ammonia[44,179–182] at 300°C or above, introduce significant amounts of carbon-nitrogen surface groups. Boehm et al.[181] and Rivera-Utrilla et al.[182] found that when an oxidized carbon was heated with dry ammonia, nitrogen groups are formed on the carbon surface. At low temperature the fixation of nitrogen was equivalent to the number of acid oxygen groups and was attributed to the formation of ammonium salts. However, at high-temperature treatment a substitution of the hydroxyl groups by amino groups was postulated. The carbon became hydrophobic and markedly decreased the adsorption capacity of the carbon for methylene blue, which is a basic dye (Table 1.15).

Puri and Mahajan[67] observed that the interaction of a sugar charcoal with dry ammonia gas involved neutralization of the surface acidic groups and fixation of some additional amounts of ammonia in nonhydrolyzable form. But these workers did not attribute this fixation of ammonia to any particular group on the carbon surface. However, Puri and Bansal[183] found that the treatment of chlorinated sugar charcoal with ammonia at 300°C resulted in the substitution of a part of the chlorine by amino groups. The resulting carbon was basic in character and showed enhanced adsorption for acids. The increase in the acid adsorption corresponded to the amount of nitrogen fixed, indicating an exchange of C–Cl bond by C–NH$_2$ bond. The interaction of ammonia gas with carbon fibers[183] before and after oxidation showed that ammonia reacted with cyclic anhydrides and lactone groups, resulting in the formation of imide structures. Zwadski[184] treated a carbon film with ammonia gas before and after oxidation with nitric acid and, using IR spectroscopy, observed no absorption corresponding to imide groups. The IR spectra also indicated reaction of ammonia with the lactone structures on the carbon film.

TABLE 1.15
Reaction of Acid Surface Oxides with Ammonia (Activated Charcoal Carboraffin Oxidized with HNO$_3$)

Reaction Temperature (°C)	Nitrogen Uptake (mg atom/100 g)	Neutralization of: KOH (alcoholic) (meq/100 g)	Neutralization of: NaOH (in H$_2$O) (meq/100 g)	Adsorption of Methylene Blue (mg/g)
(untreated)	6	750	400	535
110	390	400	—	252
255	540	230	—	158
330	620	170	—	136
410	710	120	—	—

Source: Boehm, H.P., Hoffman, U., and Clauss, A., in Proc. 3rd Conf. on Carbon, Pergamon Press, New York, 1959, p. 241. With permission.

1.6.2 MODIFICATION OF CARBON SURFACE BY HALOGENATION

The surface character of charcoals, activated carbons, and carbon blacks has been modified by several investigators by treatment with halogens. The adsorption of the halogens is both physical[185,186] and chemical, and proceeds through several mechanisms (Table 1.16) that include addition at the unsaturated sites,[183–190] exchange with chemisorbed hydrogen,[48–55] and surface oxidation of the carbon.[182,190] The mechanisms involved depend upon the nature of the carbon surface, the oxygen and hydrogen contents of the carbon (Table 1.17), the experimental conditions, and the nature of the halogenating species. The halogen fixed on the carbon surface in the form of carbon-halogen surface compounds is thermally highly stable and cannot be removed on heat treatment in vacuum up to 1000°C if the carbon has no residual hydrogen. However, a part of the halogen could be exchanged with –OH groups on treatment with alkali hydroxides and with –NH$_2$ groups on heat treatment with ammonia gas. Consequently, these chemisorbed or physisorbed species may completely modify the surface properties and surface reactions of carbons. For example, the fixation of chlorine or bromine may produce a polar but nonhydrogen bonding adsorbent similar to the abundant polar

TABLE 1.16
Contributions of Various Reactions Responsible for Cl$_2$ Interaction with Carbons (Data Derived from Table 1.10)

Step	Addition (meq/g)	Exchange (meq/g)	Dehydrogenation (meq/g)
Continex N-110 as received			
1	0.177	0.048	—
2	0.022	0.203	—
3	0.005	0.147	—
4	—	0.136	0.004
5	0.002	0.122	—
6	—	0.100	0.003
7	0.009	0.126	—
8	—	0.173	0.010
9	0.007	0.134	—
Total	0.222	1.153	0.017
TCM 128 as received			
1	0.16	0.32	—
2	0.13	0.17	—
3	—	0.18	0.03
4	0.01	0.05	—
5	—	0.02	0.02
Total	0.30	0.74	0.05
Carbopack-B			
	0.09	0.01	

Source: Tobias, H. and Soffer, A., *Carbon*, 23, 291, 1985. Reproduced with permission from Elsevier.

TABLE 1.17
Amount of Chlorine Fixed by Different Carbons

Carbon	Oxygen Content (%)	Hydrogen Content (%)	Chlorine Fixed (meq/g)
Sugar charcoal original	32.52	3.38	6.70
400°-outgassed	16.86	3.37	5.35
500°-outgassed	12.76	3.07	4.21
700°-outgassed	7.48	2.58	2.87
1000°-outgassed	1.42	1.15	2.70
1200°-outgassed	0	1.02	2.57
Coconut shell charcoal original	15.45	1.24	11.35
700°-outgassed	7.25	1.05	5.63
1000°-outgassed	0.52	0.62	3.50
1200°-outgassed	0	0.35	2.31

Source: Puri, B.R., Malhotra, S.L., and Bansal, R.C., *J. Indian Chem. Soc.*, 40, 179, 1963. With permission.

carbons that are associated with oxygen surface groups. The carbon-chlorine or carbon-bromine bond may be exchanged with other functional groups to obtain new kinds of surface-modified carbon adsorbents and electrodes. The microporosity and the adsorption stereoselectivity of microporous carbons may be modified, giving rise to newer carbons with absolutely new adsorption characteristics and behavior.

Puri and Bansal[183] studied the surface characteristics and surface behavior of sugar and coconut-shell charcoals modified by treatment with chlorine gas. The true and bulk densities of the charcoals increased linearly with increase in the amount of chlorine fixed by the carbon. However, the status of surface acidity of the charcoals remained more or less unchanged. The adsorption isotherms of water vapor on sugar charcoals outgassed at (a) 500°, (b) 700°, and (c) 1000°C in Figure 1.28, containing varying amounts of chemisorbed chlorine show that the chlorinated samples adsorbed more water vapor at lower relative pressures and less at higher relative pressures, and that the effect increased with the fixation of chlorine, so long as the temperature of chlorination was 300°C or less. By chlorination at these temperatures, the sorption of water vapor decreased at all relative vapor pressures. This indicated that the fixation of chlorine resulted more in conditioning of pore structure and size distribution of capillary pores, or in changing the location or frequency of active sites, involved in the adsorption of water vapor than in altering the chemical nature of the carbon surface.

The availability of chemisorbed chlorine for substitution by other groups was examined by Puri et al.[191] and Boehm.[42] by refluxing the chlorinated carbons with 2.5 M sodium hydroxide and by treatment with ammonia, which indicated the substitution of chlorine by amino groups. The presence of amino groups imparted a basic character to the carbon surface, and there was a noticeable increase in the acid adsorption capacity (Table 1.18). Equivalence in chlorine eliminated, nitrogen

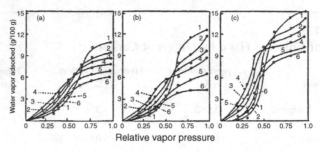

Curve	Sample	Chlorine fixed (meq/g)		
		a	b	c
1	Untreated	0	0	0
2	Chlorinated at 100°	1.09	0.97	0.58
3	Chlorinated at 200°	1.67	1.32	1.70
4	Chlorinated at 300°	2.12	1.69	2.02
5	Chlorinated at 400°	3.50	3.35	3.07
6	Chlorinated at 500°	3.56	3.28	2.99

FIGURE 1.28 Water vapor adsorption isotherm on sugar charcoal. (From Puri, B.R. and Bansal, R.C., *Indian J. Chem.*, 5, 566, 1967. With permission.)

fixed, and increase in basicity (i.e., increase in adsorption of HCl) on ammonia treatment of chlorinated carbons clearly showed that a part of the combined chlorine was available for exchange reactions with other groups. Tobias and Soffer[192] observed that chlorination of carbons followed by degassing at the same temperature as the chlorination or at higher temperature creates double bonds, which were different from the graphite surface π bonds. These additional double bonds were available for adsorption of hydrogen and HCl. Thus, carbons modified by chlorine treatment may be of importance in producing basic activated carbons and in catalysis.

1.6.3 MODIFICATION OF CARBON SURFACE BY SULFURIZATION

The carbon-sulfur surface compounds have been reported on a wide variety of charcoals, activated carbons, carbon blacks, and coals. They are formed either during or subsequent to the formation of the carbon. In the case of activated carbons, they are generally formed by heating the carbon in the presence of elementary sulfur[193–195] or sulfurous gases such as CS_2,[195–200] H_2S,[203–205] and SO_2.[200,201,206] These sulfur surface compounds are nonstoichiometric falling within a wide range of composition depending on the nature of the carbon, the experimental conditions, and the magnitude of the carbon surface. They frequently contain appreciable amounts of sulfur, which may be as high as 40 to 50%. These surface sulfur compounds can be neither extracted with solvents nor decomposed completely on heat treatment in inert atmosphere, but they can be removed almost completely as H_2S on treatment with H_2 between 500 and 700°C. The chemisorption of sulfur on the carbon surface involves bonding to the peripheral carbon atoms, addition at the double bond sites, penetration into the porous structure, and exchange for hydrogen, as well as oxygen associated with the carbon surface. As the peripheral carbon atoms, due to their unsatisfied valences, determine the adsorption

TABLE 1.18
Interaction of Sugar and Coconut Charcoals with Ammonia at 300°C

Sample	Chlorine Fixed (meq/g)	Chlorine Eliminated as NH_4Cl on Treatment with NH_3 (meq/g)	Nitrogen Fixed on Treatment with NH_3 (meq/g)	Increase in the Amount of Acid Adsorbed after Treating Chlorinated Product with NH_3 (meq/g)	Chlorine Hydrolyzed on Refluxing with 2.5 N NaOH (meq/g)
1000°C-degassed sugar charcoal treated with chlorine at 400°C	3.09	0.65	0.68	0.71	0.40
1200°C-degassed sugar charcoal treated with chlorine at 400°C	3.07	0.49	0.52	0.53	0.45
1000°C-degassed coconut charcoal treated with chlorine at 400°C	3.39	0.64	0.62	0.59	0.64
1200°C-degassed coconut charcoal treated with chlorine at 400°C	2.31	0.38	0.36	0.40	0.40

Source: Puri, B.R., Malhotra, S.L., and Bansal, R.C., *J. Indian Chem. Soc.*, 40, 179, 1963. With permission.

characteristics of activated carbons, it is reasonable to believe that the presence of surface sulfur compounds will influence the surface properties of these materials.

The influence of sulfur surface compounds on the adsorption of polar and nonpolar vapors of varying molecular dimensions was examined by Puri and Hazra.[195] The adsorption of water vapors increased appreciably at relative pressures lower than 0.4 and decreased at higher relative pressures. The effect increased with increase in the amount of sulfur fixed and was attributed to the variation of the pore-size distribution caused by the fixation of sulfur along the pore walls. The adsorption isotherms of methanol and benzene vapors indicated that these larger molecules found smaller and smaller areas as more and more sulfur was being incorporated into the pores. Bansal et al.[201,202] prepared carbon molecular sieves by blocking pores of PVDC charcoals by depositing sulfur in the micropores.

Sulfurized Saran carbon carbonized at 900°C and loaded with varying amounts of sulfur between 1 and 12% was used by Sinha and Walker[203] for the removal of mercury vapors from the air or steam. When the contaminated stream was passed through the carbon bed at 150°C, the breakthrough time increased and the mercury buildup in the effluent stream was also very low compared to the unsulfurized carbon. This was attributed to the reaction of mercury with sulfur on the carbon surface, forming mercuric sulfide. Lopez-Gonzalev et al.[204] found that sulfurized activated carbons were better adsorbents for the removal of $HgCl_2$ from aqueous solutions.

1.6.4 ACTIVATED CARBON MODIFICATION BY IMPREGNATION

Activated carbons impregnated with metals and their oxides dispersed as small particles have been and are being widely used in several gas-phase reactions both in industry and for human protection against hazardous gases and vapors. These carbons were used for the first time in World War I for protection of the upper respiratory tracts of army personnel against warfare gases. In addition, the impregnation of metals in carbonaceous materials modify the gasification characteristics and alter the porous structure of the final carbon product. Thus, such an impregnation has also been used to obtain active carbons (Table 1.19) with a given microporous structure.[205]

TABLE 1.19
Micropore Volume of Carbons Obtained by the Carbonization of Coal in the Presence of Alkali Hydroxide

Hydroxide Content (wt. %)	Micropore Volume (cm³/g)	
	Potassium Hydroxide	Sodium Hydroxide
0	0.169	0.169
10	0.078	0.049
20	0.177	0.067
30	0.386	0.142
70	0.627	—

Source: Ehrburger et al. (205). With permission.

Activated carbons impregnated with potassium iodide and similar compounds[206] and with amines,[207,208] including several pyridines, have been widely used in the nuclear industry for the retention of radioactive iodine compounds from coolant release and ventilation systems. The impregnated potassium iodide reacts with the oxygen groups on the carbon surface and modifies their adsorption behavior, thereby improving the efficiency of the activated carbon to retain radioactive methyl iodide. The adsorptive properties of coal-based active carbons were modified by impregnating with five different pyridines.[207] The reactivity of the impregnated carbon with cynogen chloride increased with an increase in the amount of a given impregnant. The reactivity, however, varied from one pyridine to another but not necessarily in the order of their expected nucleophilicities or basicities.

Barnir and Aharoni[209] compared the adsorption of cynogen chloride on activated carbons before and after impregnation with Cu(II), Cr(VI), Ag(I), and ammonia in a given ratio. The adsorption of cynogen chloride, which was reversible in the case of active carbon, became irreversible after impregnation although the adsorption capacity remained unchanged. Thus, the impregnated carbon evolved CO_2 on heating while the active carbon evolved cynogen chloride. Reucroft and Chion[210,211] also compared the adsorption behavior of a BPL-activated carbon before and after impregnation with Cu(II), CrO_4^{2-}, and Ag(I) for chloroform, cynogen chloride, phosgene, and hydrogen cyanide. The impregnated carbons showed both chemisorption and physisorption, the chemisorption being more pronounced in the case of phosgene, cynogen chloride, and hydrogen cyanide. All the impregnated carbons retained appreciable amounts of the three adsorbates after evacuation at 150°C. The adsorption of water vapor and its binary mixtures with hydrogen cyanide increased by tenfold on the impregnated carbons compared to BPL activated carbon at lower relative vapor pressures.

An activated carbon cloth prepared by carbonization of viscose rayon in the presence of $CuCl_2$, when impregnated with oxidizing agents such as $KMnO_4$, $Na_2Cr_2O_7$, and ClO_2, with an organic tertiary amine triethylene diammine (TEDA) and AgNO3 from aqueous solutions showed enhanced adsorption capacity toward low-boiling pollutants such as SO_2, NO_2, H_2S, HCN, and CNCl.[212] The exact nature of the reaction depended upon the nature of the impregnant and the oxidizing agent (Table 1.20). The presence of $CuCl_2$ as the impregnant enhanced the adsorption capacity for all the gases, while the presence of organic amine (TEDA) enhanced considerably the adsorption of CNCl. Hall et al.[213] investigated the adsorption of phosphine (PH_3) and water vapor on an activated carbon cloth before and after impregnation with $AgNO_3$ and $Cu(NO_3)_2$ from aqueous solutions. The impregnated carbon cloth samples showed enhanced adsorption of PH_3 above a certain pressure limit that depended on the impregnant content of the carbon cloth. The amount of phosphine adsorbed was five times greater when the impregnant was $AgNO_3$ and three times greater when $Cu(NO_3)_2$ was the impregnant. The adsorption of water vapor was enhanced by about 25% at relative pressures lower than 0.3. Kaistha and Bansal[214] found that impregnation of activated carbons with sodium ions (Na^+) on treatment with solutions of NaOH and $NaHCO_3$ resulted in appreciable increase in the adsorption of oxalic acid. The extent of increase in adsorption was found to depend on the degree of impregnation of the sodium ion. The increase in the amount

TABLE 1.20
Adsorption Capacity of Impregnated Carbons for H_2S, SO_2, and HCN

Impregnant	Weight Uptake (vg/cm²)		
	H_2S	SO_2	HCN
None	50	220	120
AgNO₃	210	—	—
Na₂Cr₂O₇	310	480	120
Cu (5%)	360	470	320
Cu (5%) 1 Na₂Cr₂O₇	760	960	890

Source: Capon, A., Alves, V.R., Smith, M.E., and White, M.P., 15th Bienn. Conf. on Carbon, 1981. Preprints p. 232. With permission.

of oxalic acid adsorbed on the impregnated carbons was very close to the amount of sodium ions present on the carbon surface, indicating that the adsorption of oxalic acid involved an exchange mechanism.

Rong et al.[215] studied the removal of formaldehyde using a rayon-based activated carbon fiber impregnated with p. aminobenzoic acid. The impregnated carbon fiber showed enhanced adsorption capacity both in static and dynamic adsorption processes. In the dynamic process, the breakthrough time was also considerably enhanced. The adsorption was both physical and chemical in nature.

REFERENCES

1. Bansal. R.C., Donnet, J.B., and Stoeckli, H.F., *Active Carbon,* Marcel Dekker, New York, 1988.
2. Biscoe, J. and Warren, G.E., *J. Appl. Phys.,* 13, 364, 1942.
3. Bokros, J.C. in *Chemistry and Physics of Carbon,* P.L. Walker, Jr., ed., Vol. 5, Marcel Dekker, New York, 1969.
4. Franklin, R.E., *Proc. Roy. Soc.,* A209, 196, 1951.
5. Dubinin, M.M. and Zaverina, E.D., *Dokl. Akad. Nauk SSSR,* 65, 295, 1949.
6. de Boer, J.H., in *The Structure and Properties of Porous Materials,* D.H. Everett, F.S. Stone, ed., Butterworth, London, p. 68, 1958.
7. Barrer, R.M., Mckenzie, N., and Reay, J.S.S., *J. Colloid Sci.,* 11, 479, 1956.
8. Dubinin, M.M., *Zh. Fiz. Khim.,* 34, 959, 1960.
9. IUPAC Manual of Symbols and Terminology, Appendix 2, Part I, Colloid and surface chemistry, *Pure and Appl. Chem.,* 31, 578, 1972.
10. Dubinin, M.M., *Izv. Akad. Nauk SSSR, Ser. Khim.,* p. 1961, 1979.
11. Bailey, A., Cadenhead, D.H., Davies, D.A., Everett, D.H., and Miles, A.J., *Trans. Faraday Soc.,* 67, 231, 1971.
12. Linares-Solane, Rodriguez-Reinoso F., Martin-Maritnez, J.M., Lopez-Gonzalez, J.D., *Adsorp. Sci. Technol.,* 1, 317, 1984.
13. Stoeckli, H.F., Perret, A., and Mona, P., *Carbon,* 18, 443, 1980.
14. Stoeckli, H.F. and Perret, A., *Helv. Chim. Acta,* 58, 2318, 1975.
15. Grisdale, R.O., *J. Appl. Phys.,* 24, 1288, 1953.

16. Hennig, G.R., *Proc. 5th Conf. on Carbon,* Vol. I, p. 143, 1962, Pergamon Press.
17. Kipling, J.J., *Quart. Rev.,* 10, 1, 1956.
18. Puri, B.R., Aggarwal, V.K., Bhardwaj, S.S., and Bansal, R.C., *Indian J. Chem.,* 11, 1020, 1973.
19. Donnet, J.B. and Papirer, E., *Rev. Gen. Caoutchouc Plastiques,* 42, 889, 1965.
20. Donnet, J.B. and Papirer, E., *Bull. Soc. Chim.,* France, p. 1912, 1965.
21. Dietz, V.R. and Bitner, J.L., *Carbon,* 11, 393, 1973.
22. Smith, R.N., Swinehard J., and Lessini D., *J. Phys. Chem.,* 63, 544, 1959.
23. Puri, B.R., Bansal, R.C., and Bhardwaj, S.S., *Indian J. Chem.,* 11, 1168, 1973.
24. Puri, B.R., Singh, S., and Mahajan, O.P., *J. Indian Chem. Soc.,* 42, 427, 1965.
25. Puri, B.R., Mahajan, O.P., and Singh, D.D., *J. Indian Chem. Soc.,* 37, 171, 1960.
26. Behrman, A.S. and Gustafson, H., *Ind. Engg. Chem.,* 27, 426, 1935.
27. Puri, B.R., Singh, D.D., Sharma, L.R., and Chander, J., *J. Indian Chem. Soc.,* 35, 130, 1958.
28. Puri, B.R. and Mahajan, O.P., *J. Indian Chem. Soc.,* 39, 292, 1962.
29. Puri, B.R. and Bansal, R.C., *Carbon,* 1, 457, 1964.
30. Puri, B.R. and Sharma, S.K., *Chem. Ind.,* London, p. 160, 1966.
31. Goyal, M., Rattan, V.K., Aggarwal, D., and Bansal, R.C., Colloids and surfaces, *A Physico-Chem. and Engg. Aspects,* 190, 229, 2001.
32. Aggarwal, D., Goyal, M., and Bansal, R.C., *Carbon,* 37, 1989, 1999.
33. Goyal, M., Rattan, V.K., and Bansal, R.C., *Indian J. Chem. Technol.,* 6, 305, 1999.
34. Donnet, J.B., Bouland, J.C., and Jalger, J., *Compt. Rendu,* 256, 5340, 1963.
35. Puri, B.R., Sharma, A.K., and Mahajan, O.P., *Res. Bull. Panjab Univ.,* Chandigarh, 15, 285, 1964.
36. Donnet, J.B., Beuber, F., Reitzer, C., Odoux, J., and Reiss G., *Bull. Soc. Chim.,* France, p. 1927, 1962.
37. Gandhi, D.L., Ph.D. dissertation, Panjab Univ., Chandigarh, 1976.
38. Puri, B.R. and Kalra, K.C., *Carbon,* 9, 313, 1971.
39. Puri, B.R., Sharma, L.R., and Singh, D.D., *J. Indian Chem. Soc.,* 35, 770, 1958.
40. Puri, B.R. and Mahajan, O.P., *J. Indian Chem. Soc.,* 39, 292, 1962.
41. Donnet, J.B. and Henrich G., *Bull. Soc. Chim.,* France, p. 1609, 1960.
42. Boehm, H.P., in *Advances in Catalysis,* Vol. XVI, Academic Press, New York, 1966, p. 179.
43. Garten, V.A. and Weiss, D.E., *Rev. Pure Appl. Chem.,* 7, 69, 1957.
44. Puri, B.R., in *Chemistry and Physics of Carbon,* Vol. 6, P.L. Walker, Jr., ed., Marcel Dekker, New York, p. 191, 1970.
45. Kinoshita, K., in *Carbon, Electrochemical, and Physicochemical Properties,* John Wiley and Sons, New York, p. 174, 1988.
46. Burstein, R. and Frumkin, A., *Z. Phys. Chem.,* A141, 219, 1929.
47. Frumkin, A., *Kolloid Z.,* 52, 107, 1930.
48. Shilov, N., Shatunovska, H., and Layrowskaja, D., *Z. Phys. Chem.,* A150, 421, 1930.
49. Garten, V.A. and Weiss, D.E., *Aust. J. Chem.,* 16, 309, 1957.
50. Voll, M. and Boehm, H.P., *Carbon,* 9, 481, 1971.
51. Morterra, C., Low, M.J.D., and Severdia A.G., *Carbon,* 22, 5, 1984.
52. Tarkovskya, I.A., Strazhesko, D.N., and Goba, W.E., Adsorbtsiya, *Adsorbenty,* 5, 3, 1977.
53. Puri, B.R. and Bansal, R.C., *Carbon,* 1, 451, 1964.
54. Bansal, R.C., Dhami, T.L., and Prakash, S., *Carbon,* 15, 157, 1977.
55. Bansal, R.C., Vastola, F.J., and Walker, P.L., Jr., *Carbon,* 8, 443, 1970.
56. Van Driel, J., in *Activated Carbon: A Fascinating Material,* A. Capelle and F.D. Vooys, eds., Norit N.V., Netherlands, 1983, p. 40.

57. Lang, F.M. and Magnier, P., *Carbon*, 2, 7, 1964.
58. Bonnetain L., Duval, X., and Letort M., *Proc. 4th Conf. on Carbon*, Pergamon Press, Oxford, 1962, p. 107.
59. Bonnetain, L., *J. Chem. Phys.*, 58, 34, 1961.
60. Tucker, B.G. and Mulcahy, M.F.R., *Trans. Faraday Soc.*, 65, 247, 1969.
61. Dollimore, J., Freedman, C.M., Harrison, B.H., and Quiros, D.F., *Carbon*, 8, 587, 1970.
62. Trembley, G., Vastola, F.J., and Walker, P.L., Jr., *Carbon*, 16, 35, 1978.
63. Matsumoto, S. and Setaka, N., *Carbon*, 17, 303, 1979.
64. Puri, B.R., Meyer Y.P., and Sharma, L.R., *J. Indian Chem. Soc.*, 33, 781, 1956.
65. Puri, B.R., Singh, D.D., Nath, J., and Sharma, L.R., *Ind. Engg. Chem .*, 50, 1071, 1958.
66. Puri, B.R. and Bansal, R.C., *Carbon*, 1, 457, 1964.
67. Puri, B.R. and Mahajan, O.P., *J. Indian Chem. Soc.*, 41, 586, 1964.
68. Anderson, R.B. and Emett, P.H., *J. Phys.*, 56, 753, 1952.
69. Puri, B.R., Talwar, C., and Sandle, N.K., *J. Indian, Chem. Soc.*, 41, 581, 1964.
70. Barton, S.S. and Harrison, B.H., *Carbon*, 13, 283, 1975.
71. Barton, S.S., Gillespie, D.J., Harrison, B.H., and Kemp, W., *Carbon*, 16, 363, 1978.
72. Barton, S.S., Boulton, G.L., and Harrison, B.H., *Carbon*, 10, 395, 1972.
73. Barton, S.S., Gillespie, D.J., and Harrison, B.H., *Carbon*, 11, 649, 1973.
74. Bansal, R.C., Bhatia, N., and Dhami, T.L., *Carbon*, 16, 65, 1978.
75. Studebaker, M.L., in *Proc. 5th Conf. on Carbon*, Vol. II, Pergamon Press, New York, 1963, p. 189.
76. Epstein, B.D., Dalle-Molle, E., and Mattson, J.H., *Carbon*, 9, 609, 1971.
77. Kinoshita, K. and Bett, J.A.S., *Carbon*, 12, 525, 1974.
78. Matsumura, Y., Hagiwara, S., and Tokahashi, H., *Carbon*, 14, 247, 1976.
79. Matsumura, Y., Hagiwara, S., and Takahashi, H., *Carbon*, 14, 163, 1976.
80. Papirer, E. and Guyon, E., *Carbon*, 16, 127, 1978.
81. Matsumura, Y. and Tokahashi, H., *Carbon*, 17, 109, 1979.
82. Studebaker, M.L., Huffman, E.W.D., Wolfe, A.C., and Nabors, L.G., *Ind. Engg. Chem.*, 48, 162, 1956.
83. Donnet, J.B., Bansal, R.C., and Wang, M., in *Carbon Black*, Second Edition, Marcel Dekker, New York, 1993, p. 175.
84. Friedel, R.A. and Hofer, L.J.E., *J. Phys. Chem.*, 74, 2921, 1970.
85. Rosenewaig. A., *Physics Today*, p. 23, Sept. 1975.
86. Gerson, D.J., Wong, J.S., and Cosper, J.M., *American Laboratory*, 63, Nov. 1984.
87. Rosenewaig, A., in *Advances in Electronics and Electron. Physics*, L. Marton, ed., Academic Press, New York, 1978, p. 207.
88. Rosenewaig, A. and Gersho, A., *J., Appl. Phys.*, 47, 64, 1976.
89. Morterra, C. and Low, M.J.D., *Carbon*, 23, 301, 1985.
90. O'Reilly, J.M. and Mosher, R.A., *Carbon*, 21, 47, 1983.
91. Morterra, C. and Low, M.J.D., *Carbon*, 23, 335, 1985.
92. Friedel, R.A. and Carlson, G.L., *Fuel*, 51, 194, 1972.
93. Ishazaki, C. and Marti I., *Carbon*, 19, 409, 1981.
94. Mattson, J.S., Lee, L., Mark, H.B., Jr., and Webber, W.J., Jr., *J. Colloid Interface Sci.*, 33, 284, 1970.
95. Shin, S., Jang, J., Yoon, S.H., and Mochida, L., *Carbon*, 35 1739 1997.
96. Rockley, M.G., Ratcliffe, A.E., Davies, D.M., and Woodard, M.K., *Appl. Spectroscopy*, 38, 554, 1984.
97. Bannet, C.A. and Patty, R.R., *J. Photoacoustics*, 1, 237, 1982.

98. Thomas, J.M., Evans, E.L., Barker, M., and Swift, P., in *3rd Conf. Ind. Carbon Graphite,* London Soc. Chem. Ind., 1971, p. 411.
99. Ishitani, A., *Carbon,* 19, 269, 1981.
100. Takahagi, T. and Ishitani, A., *Carbon,* 22, 43, 1984.
101. Proctor, A. and Sherwood, P.M.A., *Carbon,* 21, 53, 1983.
102. Young, V., *Carbon,* 20, 35, 1982.
103. Kavan, L., Bostl, Z., Dausek, F.F., and Jansta J., *Carbon,* 22, 77, 1984.
104. Fischer, F.G. and Feichtinger in *15th Bunin Conf. on Carbon Ext. Abstracts,* 1981, p. 472.
105. Grint, A. and Perry, D.L., in *15th Bienn. Conf. on Carbon Ext. Abstracts,* 1981, p. 462.
106. Gardner, S.D., He, G., and Pittman, C.U., Jr., *Carbon,* 34, 1221 1996.
107. Papirer, E., Guyon, E., and Perol, N., *Carbon,* 16, 133, 1978.
108. Wiegand, W.B., *Can. Chem. and Process Ind.,* 28, 157, 1944.
109. Fitzer, E., and Weiss, R., *Synergium,* 83, Intern. Conf. Interface-Interphase Composite Mater., *Soc. Plastic Engg.,* University of Leige, 1983.
110. Lawson, C.G., *Trans. Faraday Soc.,* 32, 473, 1936.
111. King, A. and Lawson, C.Q., *Trans. Faraday Soc.,* 30, 1094, 1934.
112. Pierce, C. and Smith, R.N., *J. Phys. Colloid Chem.,* 54, 784, 1950.
113. Pierce, C., Smith, R.N., Wiley, J.K., and Corder, H.J., *Am. Chem. Soc .,* 73, 4551, 1951.
114. Dubinin, M.M., Zaverina, E.D., and Serpinski V.V., *J. Chem. Soc.,* 1760, 1955.
115. Dubinin, M.M., *Carbon,* 18, 355, 1980.
116. Dubinin, M.M. and Serpinski, V.V., *Carbon,* 19, 402, 1981.
117. Puri, B.R., Murari, K., and Singh, D.D., *J. Phys. Chem.,* 65, 37, 1961.
118. Puri, B.R., *Carbon,* 4, 391, 1966.
119. Barton, S.S., Evans, M.J.B., Holland, J., and Koresh, J.E., *Carbon,* 22, 265, 1984.
120. Bansal, R.C., Dhami, T.L., and Prakash, S., *Carbon,* 16, 389, 1978.
121. Puri, B.R. and Murari, K., *Res. Bull. Panjab Univ.,* 13, 9, 1962.
122. Puri, B.R., *Carbon,* 4, 391, 1966.
123. Bansal, R.C., Singh, A., and Dhami, T.L., *Indian J. Chem.,* 13, 1321, 1975.
124. Bansal, R.C. and Dhami T.L., *Carbon,* 15, 153, 1977.
125. Puri, B.R., Talwar, C., and Sandle, N.K., *J. Indian Chem. Soc.,* 41, 581, 1964.
126. Puri, B.R., Kaistha, B.C., Vardhan, Y., and Mahajan, O.P., *Carbon,* 11, 329, 1973.
127. Bansal, R.C. and Dhami T.L., *Carbon,* 18, 297, 1980.
128. Puri, B.R., Singh, D.D., and Sharma, L.R., *J. Phys. Chem.,* 62, 756, 1958.
129. Barton, S.S., Boulton, G.L., and Harrison, B.H., *Carbon,* 10, 391, 1972.
130. Barton, S.S. and Harrison, B.H., *Carbon,* 10, 745, 1972.
131. Barton, S.S. and Harrison, B.H., *Carbon,* 13, 47, 1975.
132. Wade, W.H., *J. Colloid Interface Sci.,* 31, 111, 1969.
133. Hagiwara, S., Tsutsumi, K., and Takahashi H., *Carbon,* 19, 107, 1981.
134. Hagiwara, S., Tsutsumi, K., and Takahashi, H., *Carbon,* 9, 693, 1971.
135. Graham, D., *J. Phys. Chem.,* 59, 896, 1955.
136. Puri, B.R., in *Activated Carbon Adsorption,* I.H. Suffet and M.J. McGuire, eds., Ann Arbor Science Publishers, Ann Arbor MI, 1981, p. 353.
137. Goyal, M., Ph.D. dissertation, Panjab University, Chandigarh, 1997.
138. Goyal, M. and Bansal, R.C., Symposium on Porous Carbons, National Physical Lab., New Delhi, India, 1997.
139. Aggarwal, D., Ph.D. dissertation, Panjab University, 1998.
140. Goyal, M. and Bansal, R.C., personal communication, 2003.
141. Bansal, R.C., Aggarwal, D., Goyal, M., and Kaistha, B.C., *Indian J. Chem. Technol.,* 9, 290, 2002.

142. Goyal, M., *Carbon Science,* under publication, 2004.
143. Puri, B.R., Bhardwaj, S.S., and Gupta, U., *J. Indian Chem. Soc.,* 53, 1095, 1978.
144. Mahajan, O.P., Castilla, M.C., and Walker, P.L., *Sep. Sci. Technol.,* 15, 1733, 1980.
145. Graham, D., *J. Phys. Chem.,* 59, 896, 1995.
146. Coughlin, R.W., Ezra, F.S., and Tan, R.N., *J. Colloid Interface Sci.,* 28, 386, 1968.
147. Kipling, J.J., in *Adsorption from Solutions of Electrolytes,* Academic Press, New York, 1965.
148. Kipling, J.J. and Tester, D.A., *J. Chem. Soc.,* London, 1952, p. 4123.
149. Gasser, C.G. and Kipling J.J., in *Proc. 4th Conf. on Carbon,* Univ. of Buffalo, Pergamon Press, New York, 1960, p. 55.
150. Puri, B.R., Kumar, S., and Sandle N.K., *Indian J. Chem.,* 1, 418, 1963.
151. Bansal. R.C. and Dhami, T.L., *Carbon,* 15, 153, 1977.
152. King, A., *J. Chem. Soc.,* 1936, p. 1688.
153. Puri, B.R. and Kalra, K.C., *Indian J. Chem.,* 10, 72, 1972.
154. Bente, P.F. and Walton, J.H., *J. Phys. Chem.,* 47, 133, 1943.
155. Firth, J.B. and Watson, F.H., *J. Chem. Soc.,* 1923, p. 1750.
156. Oswald, A.A. and Wallace, T.J., in *The Chemistry of Organic Sulfur Compounds,* Vol. 2, N. Kharasch and C.Y. Meyers, eds., Ann Arbor Science Publishers, Ann Arbor MI, 1981, p. 379.
157. Hirabayashi, H. and Toyode, H., *Tanso,* 4, 2, 1954.
158. Rideal, E.K. and Wright, M.W., *J. Am. Chem. Soc.,* 1347, 1925.
159. Allardice, D.J., *Carbon,* 4, 255, 1966.
160. Dietz, V.R. and Mac Farlane, E.F., in *Proc. 5th Conf. on Carbon,* Vol. 2., Pergamon Press, New York, p. 219.
161. Carpenter, D.L. and Giddings, D.W., *Fuel,* 43, 375, 1964.
162. Carpenter, D.L. and Sergeant, G.D., *Fuel,* 45, 311, 1966.
163. Puri, B.R., Malhotra, S.L., and Bansal, R.C., *J. Indian Chem. Soc.,* 40, 179, 1963.
164. Puri, B.R. and Bansal, R.C., *Chem. Ind.,* London, 1963, p. 574.
165. Lussow, R.D., Vastola, F.J., and Walker, P.L., Jr., *Carbon,* 5, 591, 1967.
166. Hart, P.J., Vastola, F.J., and Walker, P.L., Jr., *Carbon,* 5, 363, 1967.
167. Walker, P.L., Jr., Vastola, F.J., and Hart, P.J., in *Proc. Symp. Fundamentals of Gas Surface Interactions,* San Diego CA, Academic Press, New York, 1967, p. 307.
168. Bansal, R.C., Vastola, F.J., and Walker, P.L., Jr., *Colloid Interface Sci.,* 32, 187, 1970.
169. Walker, P.L., Jr., Bansal, R.C., and Vastola, F.J., in *The Structure and Chemistry of Solid Surfaces,* Wiley, 1969, pp. 81–1 to 81–16.
170. Bansal, R.C., Vastola, F.J., and Walker, P.L., Jr., *Carbon,* 9, 185, 1971.
171. Puri, B.R., Sandle, N.K., and Mahajan, O.P., *Chem. Ind.,* London, 1963, p. 4880.
172. Puri, B.R. and Bansal, C., *Carbon,* 3, 523, 1966.
173. Bansal, R.C., Dhami, T.L., and Prakash, S., *Carbon,* 18, 395, 1980.
174. Laine, N.R., Vastola, F.J., and Walker, P.L., Jr., *J. Phys. Chem.,* 67, 2030, 1963.
175. Hoffman, W.P., Vastola, F.J., and Walker, P.L., Jr., *Carbon,* 22, 585, 1984.
176. Dentzer, J., Ehrburger, P., and Lahaye, J., *J. Colloid Interface Sci.,* 112, 170, 1986.
177. Sherman, A. and Eyring, H.J., *Am. Chem. Soc.,* 54, 2661, 1932.
178. Studebaker, M.L., *Rubber Chem. Technol.,* 30, 1401, 1957.
179. Studebaker, M.L. and Nabors, L.G., *Rubber Age,* 80, 661, 1957.
180. Emmett, P.H., *Chem. Rev.,* 43, 69, 1948.
181. Boehm, H.P., Hoffman, U., and Clauss, A., in *Proc. 3rd Conf. on Carbon,* Pergamon Press, New York, 1959, p. 241.
182. Rivera-Utrila, J., Ferro-Garcia, M.A., Mata-Arjona, A., and Gonzalev-Gomez, C., *J. Chem. Tech. Biotechnol.,* 34A, 243, 1984.

183. Puri, B.R. and Bansal, R.C., *Indian J. Chem.*, 5, 566, 1967.
184. Zawadski, J., *Carbon*, 19, 19, 1981.
185. Gandhi, D.L., Sharma, S.K., Kumar, A., and Puri, B.R., *Indian J. Chem.*, 13, 1317, 1975.
186. Puri, B.R. and Bansal, R.C., *Indian J. Chem.*, 5, 381, 1967.
187. Puri, B.R. and Bansal, R.C., *Carbon*, 3, 533, 1966.
188. Puri, B.R., Malhotra, S.L., and Bansal, R.C., *J. Indian Chem. Soc.*, 40, 179, 1963.
189. Puri, B.R. and Bansal, R.C., *Carbon*, 5, 189, 1967.
190. Puri, B.R., Mahajan, O.P., and Gandhi. D.L., *Indian J. Chem.*, 10, 848, 1972.
191. Puri, B.R., Tulsi, S.S., and Bansal, R.C., *Indian J. Chem.*, 4, 7, 1966.
192. Tobias, H. and Soffer, A., *Carbon*, 23, 291, 1985.
193. Puri, B.R., Kaistha, B., and Hazra, R.S., *J. Indian Chem. Soc.*, 45, 1001, 1968.
194. Puri, B.R., Balwar, A.K., and Hazra, R.S., *J. Indian Chem. Soc.*, 44, 975, 1967.
195. Puri, B.R. and Hazra, R.S., *Carbon*, 9, 123, 1971.
196. Karpinski, K. and Swinarski, A., *Chem. Stosow. Ser.*, A9, 307, 1965.
197. Puri, B.R., Kaistha, B.C., and Hazra, R.S., *Chem. Ind.*, London, 1967, p. 2087.
198. Bansal, R.C. and Gupta, U., *Indian J. Technol.*, 18, 131, 1980.
199. Stacy, W.O., Vastola, F.J., and Walker, P.L., Jr., *Carbon*, 6, 1917, 1968.
200. Puri, B.R., Jain, C.M., and Hazra, R.S., *J. Indian Chem. Soc.*, 43, 67, 1966.
201. Bansal, R.C., Bala, S., and Sharma, N., *Indian J. Technol.*, 27, 206, 1989.
202. Bansal, R.C., Intern. Conf. on Carbon, *CARBON '90*, Paris, July 16–20, 1990.
203. Sinha, R.K. and Walker, P.L., Jr., *Carbon*, 10, 754, 1972.
204. Lopez-Gonzalev, J.D., Morino-Castilla, C., Guererro-Ruiz, A., and Rodriguez-Reinoso, F., *J. Chem. Tech. Biotechnol*, 32, 575, 1982.
205. Ehrburger, P., Addoun, A., Addoun, F., and Donnet, J.B., Presented at *FUNCAT COGAS* Conf., 1985.
206. Billinge, B.H.M., Docherty, J.B., and Borvan, M.J., *Carbon*, 22, 83, 1984.
207. Baker, J.A. and Poziomek, E.J., *Carbon*, 12, 45, 1974.
208. Collins, D.A., Taylor, L.R., and Taylor, R., in *Proc. 4th AEC Air Cleaning Conf.*, CONF. 660904, Vol. I, 1957, p. 59.
209. Barnir, J. and Aharoni, C., *Carbon*, 13, 363, 1975.
210. Reucroft, P.J., and Chion, C.T., *Carbon*, 15, 285, 1977.
211. Chion, C.T. and Reucroft, P.J., *Carbon*, 15, 49, 1977.
212. Capon, A., Alves, V.R., Smith, M.E., and White, M.P., 15th Bienn. Conf. on Carbon, 1981 Preprints p.232.
213. Hall, P.G., Gittius, P.M., Wiun, J.M., and Robertson, J., *Carbon*, 23, 353, 1985.
214. Kaistha, B.C. and Bansal, R.C., *Indian J. Environ. Protection*, 8, 608, 1988.
215. Rong, H., Ryu, Z., and Zhong, J., Intern. Conf. on Carbon, '02, Beijung, 2002, Paper PI 94 D 107.
216. Studebaker, M.L. and Rinehart, R.W., *Rubber Chem. Technol.*, 45, 106, 1972.
217. Studebaker, M., Taylor, R.L., and Walker, P.L., Jr., *Carbon*, 21, 69, 1983.

2 Adsorption Energetics, Models, and Isotherm Equations

Activated carbons are unique and versatile adsorbents. Their adsorbent properties are essentially attributed to their large interparticulate surface area, universal adsorption effect, high adsorption capacity, a high degree of surface reactivity and a favorable pore size which makes the internal surface accessible and enhances the adsorption rate. The most widely used activated carbons have a surface area of about 800 to 1500 m²/g. This surface area is contained predominantly within micropores which have effective diameter less than 2 nm. In fact, a particulate of active carbon is made up of a complex network of pores that have been classified into micropores (diam. < 2 nm), mesopores (diam. between 2 and 50 nm), and macropores (diam. > 50 nm). Most of the adsorption on active carbons takes place in micropores and only small amount in mesopores, the macropores acting only as conduits for the passage of the adsorbate into the interior mesopores and the micropore surface. The pore size distribution in a given carbon depends on the type of the raw material and the method and conditions under which the carbon has been prepared. The large surface area of active carbons is the result of the activation process in which a carbonaceous char with little internal surface is oxidized in an atmosphere of air, carbon dioxide, or steam at a temperature of 800 to 900°C. This results in the oxidation of some of the regions within the char in preference to others so that a preferential etching takes place, as the combustion proceeds, causing an increase in surface area.

2.1 ADSORPTION ON A SOLID SURFACE

Whatever the nature of the forces holding a solid together, it can be regarded as producing a field of force around each ion, atom, or molecule. At the surface of the solid, these forces can not suddenly disappear and thus reach out in space beyond the surface of the solid. Due to these unsaturated and unbalanced forces, the solid has a tendency to attract and retain on its surface molecules and ions of other substances with which it comes into contact. Thus, when a solid surface comes in contact with a gas or a liquid, the concentration of the gas or liquid is always greater on the surface of the solid than in the bulk gas or liquid phase. The process by which this surface excess is created is called *adsorption*. The balance of the forces is partially restored by the adsorption of the gas or the liquid on the surface of the solid. The substance attached to the surface is called *adsorbate*, and the substance to which it is attached is known as the *adsorbent*.

Adsorption of a gas on a solid or liquid surface is a spontaneous process. It is, therefore, accompanied by a decrease in free energy of the system. Furthermore, the gaseous molecules in the adsorbed state have fewer degrees of freedom than in the gaseous state. This results in a decrease in entropy during adsorption. Using the thermodynamic relationship

$$\Delta G = \Delta H - T\Delta S$$

it follows that the term ΔH, which is the heat of adsorption, must be negative indicating that adsorption is always an exothermic process, respective of the nature of the forces involved in the adsorption process. However, a few adsorption cases have been reported to be endothermic. For example, the adsorption of hydrogen on glass, adsorption of oxygen on silver, and adsorption of hydrogen on iron are endothermic. The endothermicity of these adsorption processes is attributed to an increase in the entropy of the adsorbate due to the dissociation of the molecule during the adsorption process or to an increase in the entropy of the adsorbent.

Depending upon the nature of the forces involved, the adsorption is of two types: physical or van der Walls adsorption, and chemisorption or chemical adsorption. In the case of physical adsorption, the adsorbate is bound to the surface by relatively weak van der Walls forces identical with molecular forces of cohesion that are involved in the condensation of vapors into liquids. Chemisorption, on the other hand, involves exchange or sharing of electrons between the adsorbate molecules and the surface of the adsorbent, resulting in a chemical reaction. The bond formed between the adsorbate and the adsorbent is essentially a chemical bond and is thus much stronger than in physical adsorption.

Because the nature of the forces involved in the two types of adsorption are different, the two differ in several ways. The most common difference between the two kinds of adsorption is in the magnitude of the heat of adsorption. In the case of physical adsorption, the heat of adsorption is of the same order as the heat of condensation and does not usually exceed 10 to 20 KJ per mole, whereas in chemisorption the heat of adsorption is usually 40 to 400 KJ per mole. It may, however, be mentioned that in some cases the heat of chemisorption does not differ substantially from the heat of physical adsorption. Physisorption does not require any activation energy so that the rate of adsorption is very high, even at low temperatures. The chemisorption, on the other hand, requires activation energy; the rate of adsorption is low and depends upon the temperature of adsorption. However, when the surface of the adsorbent is very reactive, as in the case of ultraclean carbon surfaces, the rate of chemisorption can be very high, even at low temperatures.[1-3] Similarly, physisorption in microporous adsorbents may be very much retarded by the slow diffusion of the adsorbate into the fine pores and may require activation energy. Physical adsorption is nonspecific and occurs between any adsorbate-adsorbent system, while chemisorption is specific. For example, carbon monoxide is not chemisorbed by an iron catalyst at 450°C, whereas nitrogen can be chemisorbed with a surface coverage of about 50%. Another point of difference between physisorption and chemisorption is the thickness of the adsorbed phase. Whereas it is multimolecular in physisorption, the thickness is monomolecular in chemisorption.

The type of adsorption that takes place in an adsorbate-adsorbent system depends upon the reactivity of the surface, the nature of the adsorbate, the nature of the adsorbent, and the temperature of adsorption. For example, the adsorption of oxygen on active carbon is physical adsorption to a large extent at temperatures below −100°C, and it is chemisorption at room temperature and above. When it is not certain that the process of adsorption is either physisorption or chemisorption, or when both are occurring in appreciable proportions, then it is preferable to use a less committal term: *sorption*.

2.2 ADSORPTION EQUILIBRIUM

When a solid surface is exposed to a gas, the molecules of the gas strike the surface of the solid. Some of the striking molecules stick to the solid surface and become adsorbed while the others rebound. Initially the rate of adsorption is large as the whole surface is bare but as more and more of the surface becomes covered by the molecules of the gas, the available bare surface decreases and so does the rate of adsorption. However, the rate of desorption, which is the rate at which adsorbed molecules rebound from the surface, increases because desorption takes place from the covered surface. As time passes, the rate of adsorption continues to decrease while the rate of desorption increases until an equilibrium is reached between the rate of adsorption and the rate of desorption. At this stage the solid is in adsorption equilibrium with the gas, and the rate of adsorption is equal to the rate of desorption. It is a dynamic equilibrium because the number of molecules sticking to the surface is equal to the number of molecules rebounding from the surface.

For a given adsorbate-adsorbent system, the equilibrium amount adsorbed x/m is a function of pressure and temperature; i.e.,

$$\frac{x}{m} = f(p, T) \tag{2.1}$$

where x/m is the amount adsorbed per unit mass of the adsorbent at the equilibrium pressure p, and T is the temperature of adsorption. The adsorption equilibrium can be approached in three different ways.

2.2.1 ADSORPTION ISOTHERM

If the temperature is kept constant, then for a given adsorbent-adsorbate system x/m depends on the equilibrium pressure, and the equilibrium can be represented as

$$\frac{x}{m} = f(p) \quad [T = \text{constant}] \tag{2.2}$$

Such an equilibrium is called an *adsorption isotherm* (Figure 2.1).

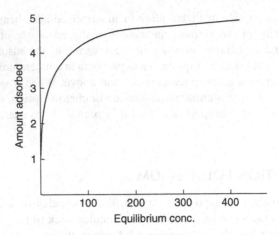

FIGURE 2.1 A typical adsorption isotherm.

2.2.2 ADSORPTION ISOBAR

When pressure is kept constant and T is varied, we obtain an *isobar* (Figure 2.2).

$$\frac{x}{m} = f(T) \quad [\,p = \text{constant}\,]$$

2.2.3 ADSORPTION ISOSTERE

An *isostere* is obtained when, for a constant equilibrium amount adsorbed, the temperature is varied and the pressure essential to keep x/m to be constant is a function of the temperature (Figure 2.3).

$$p = f(T) \quad \left[\frac{x}{m} = \text{constant}\right] \tag{2.3}$$

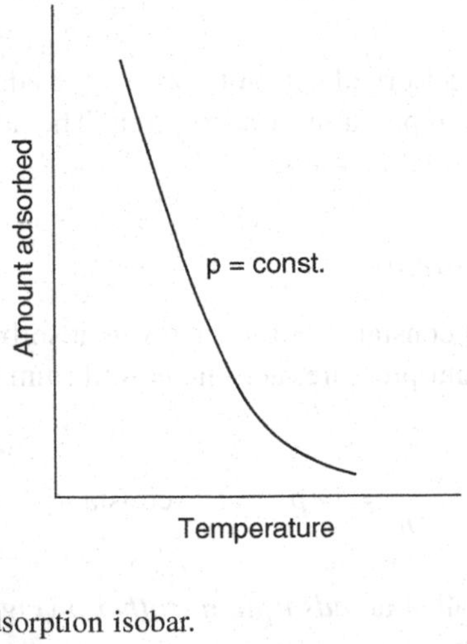

FIGURE 2.2 A typical adsorption isobar.

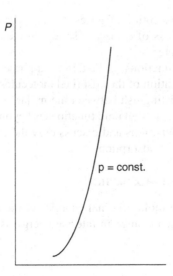

FIGURE 2.3 A typical adsorption isostere.

In practice, the experimental adsorption data are usually represented in the form of adsorption isotherms because investigation of the adsorption process at constant temperature is most convenient. Furthermore, the theoretical analysis of adsorption data for certain assumed models usually arrives at adsorption isotherms and not isobars or isosteres. The adsorption bar is determined less frequently, and the direct measurement of an isostere is rare. Adsorption isobars are sometimes useful in ascertaining the adsorption mechanism in a particular system and to determine whether more than one type of adsorption is involved. Adsorption isosteres, on the other hand, are often used for calculating heats of adsorption from adsorption measurements at two or more temperatures, using the Clasius-Clapeyron equation. In an adsorption isotherm, it is usual to express the amount adsorbed in millimoles, or milliliters, of the gas or vapor at NTP per gram of the adsorbent. Because the three types of equilibria (i.e., the isotherm, the isobar, and the isostere) are equilibrium functions, it is possible to obtain one relationship from the another. For example, from adsorption isotherms for a given system at several temperatures, the isobars and isosteres can be obtained. Similarly, isotherm can be obtained from isobars and isosteres.

2.3 ENERGETICS OF ADSORPTION

The measurement and interpretation of the thermodynamic function of the adsorbed phase, such as the free energy of adsorption, the enthalpy of adsorption, and the entropy of adsorption have been the subject matter of large number of investigations. These functions have been evaluated using adsorption isotherms and are compared with those obtained from theoretical considerations. In all treatments of physical adsorption of gases, the adsorbent has been assumed to be inert so that the gain or loss of energy is solely due to an increase or decrease in adsorption of the adsorbate.

The thermodynamic functions of the adsorbed phase are a function of pressure, temperature, and concentration of the adsorbed molecules on the surface of the solid so that it is essential to distinguish between the molar and partial molar properties. As the derivation of the thermodynamic functions is beyond the scope of this chapter, we shall only define the functions and discuss only the calculations of the enthalpy of adsorption or the heat of adsorption.

2.3.1 MOLAR ENERGY OF ADSORPTION

If u_a and μg represent the molar internal energies of the adsorbed phase and of the adsorbate, respectively, then change in internal energy as a result of adsorption can be represented as

$$\frac{\Delta u_a}{n_a} = u_a - \mu g = \Delta u_a \qquad (2.4)$$

where $\frac{\Delta u_a}{n_a}$ is the change in molar energy of the system caused by the transfer of n_a moles of the adsorbate from the gaseous phase to the adsorbed state. This change in molar internal energy of the system is the *molar integral energy of adsorption* Δu_a, which depends upon all the adsorbate-adsorbate and adsorbate-adsorbent interactions at all coverages so that u_a is the mean molar internal energy for all the adsorbed molecules.

When an infinitesimally small amount, dn_a moles, of the adsorbate is transferred to the surface of the solid from the gas phase at constant volume, then the change in the internal energy of the system is given by the relationship

$$\Delta \bar{u}_a = \bar{u}_a - u_g \qquad (2.5)$$

where $\Delta \bar{u}_a$ is now the *differential molar energy of adsorption* also known as the *derivative energy of adsorption*. The differential molar energy $\bar{u}_a$ of the adsorbed state is given by

$$\bar{u}_a = \left(\frac{\partial u_a}{\partial n_a} \right)_{T,A} \qquad (2.6)$$

2.3.2 MOLAR INTEGRAL ENTHALPY OF ADSORPTION

As in the case of the molar integral energy of adsorption, the molar integral enthalpy of adsorption ΔH_a can be given by

$$\Delta H_a = H_a - H_g$$

where H_a and H_g are the molar enthalpies of the adsorbed state and of the adsorbate in the gaseous phase. Similarly, molar differential enthalpy of adsorption also known as *derivative enthalpy of adsorption* can be written as

$$\Delta \overline{H}_a = \overline{H}_a - H_g \tag{2.7}$$

where

$$\overline{H}_a = \left(\frac{\partial H_a}{\partial n_a} \right)_{T,p}$$

2.3.3 MOLAR INTEGRAL ENTROPY OF ADSORPTION

Molar entropy of adsorption is also an important parameter in understanding the state of the adsorbed phase. The molar integral entropy of adsorption ΔS_a, can be defined as

$$\Delta S_a = S_a - S_g \tag{2.8}$$

where S_a is the mean molar entropy of all the adsorbed molecules over the whole range of surface coverages at a given adsorbed amount and S_g is the molar entropy of the molecules in the gas phase.

The molar differential entropy of adsorption resulting from the transfer of dn_a moles of the adsorbate from gas to the adsorbed phase likewise can be represented as

$$\Delta \overline{S}_a = \overline{S}_a - S_g \tag{2.9}$$

when $\overline{S}_a$ is given by

$$\overline{S}_a = \left(\frac{\partial S_a}{\partial n_a} \right)_{T,A}$$

2.3.4 HEAT OF ADSORPTION

It is an important thermodynamic function that can be used to characterize the surface of a solid. As has been mentioned earlier, adsorption is an exothermic process and takes place with the evolution of heat. The magnitude of the heat of adsorption is considerable, and its knowledge is important both for practical as well as theoretical purposes. There are two ways of expressing the heat of adsorption, each useful in its own context. First of these is the *integral heat of adsorption*, which is defined as the total amount of heat (Q) given out when one gram of an outgassed solid (active carbon, for example) adsorbs x grams of a gas or vapor and is expressed as Joules per gram of the adsorbent. For instance, in a particular case of adsorption on charcoal, the integral heat of adsorption of H_2 gas at 20°C was 64.5 J/G of charcoal when the adsorption was 0.125 g of hydrogen per g of the charcoal.

The second type of heat of adsorption is differential heat of adsorption $(-\Delta H)$, which can be expressed as Joules per g mole of the adsorbate. Let us suppose we have a solid that has already adsorbed $\times$ grams of a gas or vapor. The solid is allowed to adsorb additional Δx grams of the adsorbate and, in so doing, the solid evolves ΔQ Joules. The differential heat of adsorption $(-\Delta H)$ is then equal to

$$-\Delta H = M\,\frac{\Delta Q}{\Delta x} \qquad (2.10)$$

when Δx is infinitesimally small and M is the molecular mass. Thus, the differential heat of adsorption is expressed as Joules per mole of the adsorbate.

2.3.5 ISOSTERIC HEAT OF ADSORPTION

The differential heat of adsorption is also called the isosteric heat of adsorption, DH, and can be measured either by using a calorimeter or by using a family of isotherms measured at two or more temperatures.

When a gas or vapor is in adsorption equilibrium with a solid surface, the chemical potential of the adsorbate in the adsorbed phase can be expressed as a function of internal energy μ_a as

$$\mu_a = \left(\frac{\partial u_a}{\partial n_a}\right) S,\,V,\,m \qquad (2.11)$$

when entropy (S), volume (V) and the mass of the adsorbent (m) are constant. As the system is in equilibrium, the chemical potential of the adsorbate in the adsorbed state is equal to the chemical potential of the adsorbate in the gas phase (μ_g).

$$\mu_g = \mu_g \qquad (2.12)$$

or

$$d\mu_a = d\mu_g \qquad (2.13)$$

Now for a gas

$$d\mu_g = -S_g dT + V_g dp \qquad (2.14)$$

where S is entropy and V the volume of the gas. The corresponding equation for the adsorbed state can be expressed as

$$d\mu_a = -\overline{S}_a dT + \overline{V}_a dp + \left(\frac{\partial \mu_a}{\partial n_a}\right)_{T,p} dn_a \qquad (2.15)$$

where $\overline{S}_a$ is the mean molar entropy, and $\overline{V}_a$ is the mean molar volume for all the molecules adsorbed for a complete range of surface coverage up to an adsorption of a given amount

In view of Equation 2.13.

$$-S_g dT + V_g dp = -\overline{S}_a dT + \overline{V}_a dp + \left(\frac{\partial \mu_a}{\partial n_a} dn_a \right)_{T,p} \qquad (2.16)$$

and at constant n_a

$$S_g dT + V_g dp = -\overline{S}_a dT + \overline{V}_a dp \qquad (2.17)$$

or

$$(S_g - \overline{S}_a)dT = (V_g - \overline{V}_a)dp \qquad (2.18)$$

or

$$\left(\frac{dp}{dT} \right)_{na} = \left(\frac{S_g - \overline{S}_a}{V_g - \overline{V}_a} \right) \qquad (2.19)$$

because $V_g \gg \overline{V}_a$ and substituting $V_g = \dfrac{RT}{p}$ we get

$$\left(\frac{dp}{dT} \right)_{na} = (S_g - \overline{S}_a) \frac{p}{RT} \qquad (2.20)$$

or

$$\left(\frac{\partial l_n p}{\partial T} \right)_{na} = \frac{S_g - \overline{S}_a}{RT} = -\Delta \overline{S}_a / RT \qquad (2.21)$$

where $\Delta \overline{S}_a =$ is the differential molar entropy.

Because $\Delta H_a = T\Delta S_a$ we get

$$\left(\frac{\partial \ln p}{\partial T} \right)_{na} = -\frac{\Delta H_a}{RT^2} \qquad (2.22)$$

The quantity $-\Delta H_a$ is the differential molar enthalpy of adsorption which is also termed as the isosteric enthalpy of adsorption. Thus, the isosteric heat of adsorption ΔH can be written with the opposite sign as

$$\left(\frac{\partial \ln p}{\partial T}\right)_{na} = \frac{\Delta H}{RT^2} \qquad (2.23)$$

ΔH being positive, the heat is evolved when adsorption takes place.

The integrated form of Equation 2.23 can be written as

$$\left(\ln \frac{p_1}{p_2}\right)_{na} = -\frac{\Delta H(T_2 - T_1)}{T_1 T_2} \qquad (2.24)$$

where p_1, and p_2 are the equilibrium pressures at temperatures T_1 and T_2, respectively, when the amount adsorbed is n_a. This equation helps to evaluate the heat of adsorption ΔH from the experimental adsorption data at two temperatures. In other words, at any constant amount adsorbed, the equilibrium adsorption pressure can be determined from adsorption isotherms at two different temperatures. In general the equation can be written as

$$(\ln p)_{na} = -\frac{\Delta H}{RT} + \text{constant} \qquad (2.25)$$

Equation 2.24 and Equation 2.25 represent the relationship between p and T for a given amount absorbed and is called an adsorption isostere. The adsorption isostere can be obtained by measuring adsorption isotherms for the given adsorbate-adsorbent system at several different temperatures. The points corresponding to a particular surface coverage (i.e., for a definite amount adsorbed per unit mass of the adsorbent) are plotted in the $\ln p$ versus $1/T$ coordinate. The adsorption isosteres so obtained are linear. The slopes of the linear plots gives the value of heats of adsorption at a given surface coverage.

It may be mentioned that the heat of adsorption calculated by using Equation 2.24 and Equation 2.25 are applicable to a particular value of amount adsorbed. However, these values of heat of adsorption can be modified by energetically nonuniform nature of the carbon surface or by the interactions between the adsorbed molecules in the adsorption layer. It is, therefore, better that the calculations of ΔH be repeated for a series of values of n_a and then obtain a curve between ΔH and n_a.

Heat of adsorption can also be measured using a calorimeter. In these measurements the heat ΔQ Joules evolved by the adsorption of a certain amount of gas or vapor is determined. If the amount adsorbed Δx grams is small, then the differential heat of adsorption is approximately given by $\Delta Q.M/\Delta x$, where M is the molar mass of the adsorbate.

The integral heat of adsorption can also be obtained using a calorimeter. The integral heat of adsorption corresponding to an adsorption amount x grams is obtained by exposing the outgassed adsorbent (active carbon) to the vapor of the

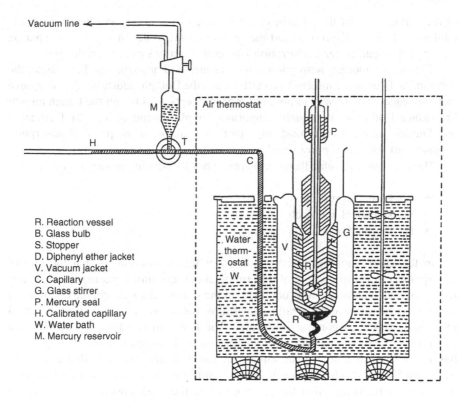

Vacuum line

M

Air thermostat

H

T

P

C

R. Reaction vessel
B. Glass bulb
S. Stopper
D. Diphenyl ether jacket
V. Vacuum jacket
C. Capillary
G. Glass stirrer
P. Mercury seal
H. Calibrated capillary
W. Water bath
M. Mercury reservoir

Water
therm-
ostat
W

V

G

R

B

R

R

FIGURE 2.4 A phase-change calorimeter.

adsorbate so that the adsorption increases from zero to x grams, and measuring the heat evolved.

The more important calorimeters used for measuring heats of adsorption are phase-change calorimeters, such as the Bunsen-type ice calorimeter. More recently, phase-change calorimeters using diphenyl ether as the calorimetric fluid are commonly used. In these calorimeters, the calorimetric fluid exists in a solid-liquid equilibrium state so that the heat evolved as a result of adsorption melts a corresponding amount of the solid fluid and causes a movement of mercury in the capillary tube connected to the calorimeter. This mercury movement in the capillary is first calibrated by standard reactions so that the movement of the thread of mercury in the capillary directly gives the amount of heat evolved.

One such phase-change calorimeter using diphenyl other as the calorimetric fluid is shown in Figure 2.4.

2.4 ADSORPTION ISOTHERM EQUATIONS

The adsorption isotherm is the most extensively employed method for representing the equilibrium states of an adsorption system. It can give useful information regarding the adsorbate, the adsorbent, and the adsorption process. It helps in the determination of the surface area of the adsorbent, the volume of the pores, and their size distribution, the heat of adsorption, and the relative absorbability of a gas or a vapor on a given adsorbent. Several adsorption isotherm equations have been derived.

The more important adsorption isotherms are the Langmuir, the Freundlich, the Temkin, the Brunauer-Emmett-Teller (BET), and the Dubinin equations. The first three isotherm equations are very important for chemisorption, although the Langmuir and Freundlich isotherms are equally important for physisorption. The BET equation and Dubinin equations are most important for analysis of the physical adsorption of gases and vapors on porous carbons.

There are three possible theoretical approaches for deriving adsorption isotherms:

- The kinetic approach
- The statistical approach
- The thermodynamic approach

In the kinetic approach, the condition of the equilibrium is that the rate of adsorption is equal to the rate of desorption at equilibrium. Equating the two rates in an isotherm equation can be obtained. In the statistical approach, the equilibrium constant is represented by a ratio of partition functions of vacant sites, adsorbed molecules, and the gas-phase molecules. The isotherm equation can be obtained by equating this ratio to the corresponding ratio of concentrations; this approach has the advantage that it gives numerical value to the constants that cannot be evaluated by the kinetic approach. The equilibrium can also be approached thermodynamically, using either the conditions that the work done in transferring an infinitesimal amount of gas from the gas phase to the surface at constant temperature is zero, or alternatively the Gibbs adsorption equation.

2.4.1 LANGMUIR ISOTHERM EQUATION

The Langmuir isotherm equation is the first theoretically developed adsorption isotherm. Many of the equations proposed later and which fit the experimental results over a wide range are either based on this equation, or these equations have been developed using the Langmuir concept. Thus, the Langmuir equation still retains an important position in physisorption as well as chemisorption theories. The equation has also been derived using thermodynamic and statistical approaches but we shall discuss the commonly used kinetic approach for its derivation.

The American scientist I. Langmuir[4] derived this equation based on certain assumptions. More important of these assumptions are

- The adsorbed entites (atoms or molecules or ions) are attached to the surface at definite localized sites.
- Each site accommodates one and only one adsorbed entity.
- The energy state of each adsorbed entity is the same at all sites on the surface independent of the presence or absence of other absorbed entities at neighboring sites. Thus, the Langmuir model (also called localized model) assumes that the surface is perfectly smooth and homogenous and that the lateral interactions between the adsorbed entities are negligible.

In the original kinetic approach, the equilibrium is assumed to be dynamic in that the rate at which the molecules from the gas phase strike the solid surface and condense on the bare sites is equal to the rate at which molecules evaporate from the occupied sites.[5-7] In other words, the rate of adsorption is equal to the rate of desorption. If at any pressure p of the gas the fraction of the sites occupied is θ, and the fraction of the bare sites is θ_0 (so that $\theta + \theta_0 = 1$) then the rate of condensation (adsorption) is

$$r_{ads} = akp\,\theta_0 \qquad (2.26)$$

where p is the pressure and k is a constant given by the kinetic theory of gases $k = N/(2\pi kmT)^{1/2}$; a is the condensation coefficient that is the fraction of the striking molecules that actually condense on the surface.

The evaporation (desorption) of an adsorbed molecule from the surface is essential an activated process in which the energy of activation, E, for desorption may be identified with $-\Delta H$, the differential (isosteric) heat of desorption. The rate of evaporation r_{des} can be written as

$$r_{des} = Z_m\theta v e^{-E/RT} \qquad (2.27)$$

where Z_m is the number of molecules adsorbed (occupied sites) per unit area of the surface so that $Z_m\theta$ represents the corresponding number of adsorbed molecules, and v is the frequency of oscillation of the molecules perpendicular to the surface and closely related to the frequency of vibration of the atoms or molecules of the adsorbent. Thus, at equilibrium, $r_{ads} = r_{des}$.

$$akp\theta_0 = Z_m\theta_1\, ve^{-E/RT} \qquad (2.28)$$

$$akp\,(1 - \theta) = Z_m\theta v e^{-E/RT} \qquad (2.29)$$

so that

$$\frac{\theta}{1-\theta} = \frac{akp}{Z_m ve^{-E/RT}} = bp \qquad (2.30)$$

where

$$b = \frac{ak}{Z_m ve^{-E/RT}} = bp \qquad (2.31)$$

or

$$\theta = \frac{bp}{1+bp} \qquad (2.32)$$

The fraction θ of the surface covered can also be written as the ratio of volume v of the gas or vapor adsorbed at pressure p and V_m, the volume of the adsorbate required to form a monomolecular layer. Thus, Equation 2.32 can also written as

$$\frac{V}{V_m} = \frac{bp}{1+bp} \tag{2.33}$$

$$V = \frac{V_m bp}{1+bp} \tag{2.34}$$

This relationship shows that V tends asymptotically to V_m as p tends to infinity. Equation 2.33 is the well known Langmuir isotherm equation. θ can also be written as n/n_m so that Equation 2.33 now becomes

$$\frac{n}{n_m} = \frac{bp}{1+bp} \tag{2.35}$$

where n is the number of moles adsorbed per g of the adsorbent and n_m is the monolayer capacity in moles.

Similarly, θ can also be expressed in terms of the amount adsorbed in grams as x/x_m so that

$$\frac{x}{x_m} = \frac{bp}{1+bp} \tag{2.36}$$

where x in grams is the amount adsorbed per g of the adsorbent, and x_m is the monolayer capacity in grams.

Depending upon the pressure, Equation 2.34 reduces to two limiting expressions. At low pressure the value of bp is less than unity; i.e., at the beginning of the adsorption process it reduces to the approximate form

$$V = V_m bp \tag{2.37}$$

which indicates a proportionality between the amount adsorbed and the equilibrium pressure.

At high pressures $Kp \gg 1$ (i.e., for the advanced stage of adsorption), Equation 2.34 is reduced to

$$V = V_m \tag{2.38}$$

This indicates that at higher pressure the adsorption is independent of pressure because it has attained the highest value equal to V_m, the volume required to cover the surface by a monolayer.

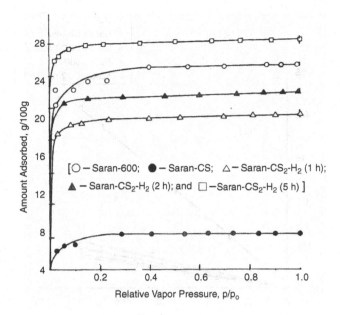

FIGURE 2.5 Adsorption of iso-octane vapours on different Saran charcoal.

Langmuir adsorption isotherms for the adsorption of isooctane vapors are shown in Figure 2.5. This isotherm is generally valid for chemisorption as well as in physisorption of gases on microporous carbons.[7-9]

Rearrangement of Equation 2.34 gives the linear form of the equation

$$p/V = \frac{1}{V_m b} + \frac{p}{V_m} \qquad (2.39)$$

Figure 2.6 shows the linear Langmuir plots. The slope of the linear plot gives the value of V_m, while the value of b can be obtained from the intercept. The value of V_m can be used to calculate the surface area of the adsorbent. It may, however, be worthwhile to mention that the linear Langmuir isotherm does not necessary mean that the adsorption process meets the requirements of the idealized localized monolayer model even when reasonable values of V_m and b are obtained. This is due to the fact that perfectly energetically homogeneous surfaces are rare and difficult to obtain. In some cases the effect of nonhomogeneity of the surface may be compensated by adsorbate-adsorbent interactions and thus give rise to linear Langmuir plots. It is, therefore, suggested that along with the Langmuir adsorption isotherms, differential heats of adsorption should also be measured which, according to the Langmuir concept, should be independent of the degree of surface coverage.

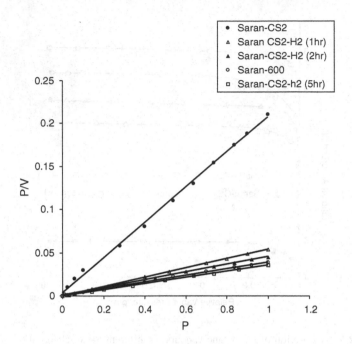

FIGURE 2.6 Linear Langmuir plots for adsorption of iso-octane on different Saran charcoal [own work].

2.4.1.1 Langmuir Isotherm for Dissociative Adsorption

Adsorption of gases such as oxygen and hydrogen generally takes place by dissociation of the gas molecule on the surface of the solid. The adsorbed species are, therefore, atoms. In such cases the adsorption process may be considered as a reaction between the gas molecule and the two surface sites. The rate of adsorption r_{ads} in such cases is given by

$$r_{ads} = kp \, (1 - \theta)^2 \tag{2.40}$$

The process of desorption now involves interaction between two adsorbed atoms on neighboring sites and is, therefore, given by

$$r_{des.} = V\theta^2 \tag{2.41}$$

Following the earlier arguments for deriving the Langmuir adsorption isotherm equation and equating Equation 2.40 and Equation 2.41, we get

$$\frac{\theta}{1-\theta} = (bp)^{1/2} \tag{2.42}$$

or

$$\theta = \frac{(bp)^{1/2}}{1+(bp)^{1/2}} \tag{2.43}$$

In general, when the adsorbed molecule dissociates into n entities, each of which occupies a surface site, the Langmuir equation for dissociative adsorption can be written as

$$\theta = \frac{(bp)^{1/n}}{1+(bp)^{1/n}} \tag{2.44}$$

2.4.1.2 Langmuir Isotherm for Simultaneous Adsorption of Two Gases

Langmuir isotherm for two gases adsorbed simultaneously and without dissociation on the same adsorbent is of considerable use in the interpretation of catalyzed reactions involving two reactants. Markham and Benton[10] first extended the Langmuir isotherm to a binary gaseous mixture, preserving all the assumptions of the original theory. Following the procedure completely analogous to that for a single component, these workers obtained expressions for the fractions covered by the two components θ_A and θ_B as

$$\theta_A = \frac{b_A p_A}{1+b_A p_A + b_B p_B} \tag{2.45}$$

and

$$\theta_B = \frac{b_B p_B}{1+b_A p_A + b_B k_B} \tag{2.46}$$

The total fraction of the surface covered by components A and B is then given by

$$\theta = \theta_A + \theta_B = \frac{b_A p_A + b_B p_B}{1+b_A p_A + b_B p_B} \tag{2.47}$$

When θ is expressed as the ratio of V and V_m, then

$$\theta = V_A + V_B = V_m \frac{b_A p_A + b_B p_B}{1+b_A p_A + b_B p_B}$$

In general, when i substances are being simultaneously adsorbed, we get

$$\theta_A = \frac{b_A p_A}{1+\sum b_i p_i} \tag{2.48}$$

and

$$\theta_B = \frac{b_B p_B}{1 + \sum bipi} \tag{2.49}$$

This step, however, assumes that the value of V_m is the same for both components A and B.

Although the adsorption of gaseous mixtures is, in practice, much more important than the adsorption of a single pure gas in catalysis, the knowledge of the subject is very incomplete and the information available is not always reliable.

2.4.1.3 Applicability of the Langmuir Isotherm

The Langmuir equation

$$\theta = \frac{V}{V_m} = \frac{bp}{1 + bp}$$

can be transformed into the following three linear forms:

$$\frac{p}{V} = \frac{1}{bV_m} + \frac{p}{V_m} \tag{2.50}$$

$$\frac{V}{V_m p} = b - \frac{bV}{V_m} \tag{2.51}$$

$$\frac{V_m}{V} = 1 + \frac{1}{bp} \tag{2.52}$$

Thus, it is apparent that plots of $\frac{p}{V}$ against p, $\frac{V}{P}$ against $-V$, and $\frac{1}{V}$ against $\frac{1}{P}$ should yield straight lines from which the values of V_m and b can be calculated. The last linear plot viz $\frac{1}{V}$ against $\frac{1}{P}$ is generally avoided because it lays too much emphasis on the low pressure region of the isotherm and adsorption at low pressure is rarely representative of the entire surface.

The applicability of the Langmuir isotherm equation to the experimental data was carried out by a large number of workers, but deviations were often noticed. According to the Langmuir isotherm equation, the plot of p/v against p should be linear from $\theta = 0$ to $\theta = $ infinity, and it should give a reasonable value of V_m, and V_m should be temperature independent. When the experimental data was subjected to these criteria, only few data conformed to Langmuir model. For example, when Langmuir's own data on the adsorption of carbon dioxide, oxygen, and nitrogen was plotted, it gave reasonably good linear plots, but the values of V_m obtained form the linear plots differed by 3 to 80% from the values of surface area measured by other methods. Culver and Health,[8] observed that for a Saran charcoal the surface area calculated from Langmuir V_m value was ~3000 m²g⁻¹. This value is larger than the value 2630 m²g⁻¹ which could be possible only if one gram of this carbon could be present as layers of graphite one atom thick and accessible to gas molecules on both sides, an assumption that is difficult to reconcile.

Similarly, several chemisorption results are known for which the Langmuir equation is valid only within a small restricted range. Titoff[11] observed that the value of V_m fell from 231 cm³ (NTP)g⁻¹ at 195.6 K to 113 cm³ (NTP)g⁻¹ at 303 K. Furthermore, the Langmuir theory requires the heat of adsorption to be independent of the surface coverage while direct determinations of heat of adsorption are found to decrease invariably with increasing surface coverage. Thus, although Langmuir isotherm equation is of limited significance for interpretation of the adsorption data because of its idealized character, the equation remains of basic importance for expressing dynamic adsorption equilibrium. Furthermore, it has provided a good basis for the derivation of other more complex models. The assumptions that the adsorption sites on solid surfaces are energetically homogenous and that there are no lateral interactions between the adsorbed molecules are the weak points of the Langmuir model.

2.4.2 BRUNAUER, EMMETT, AND TELLER (BET) ISOTHERM EQUATION

The adsorption isotherms of gases at temperatures not far removed from their condensation points show two regions for most of the adsorbents: at low pressures the isotherms are concave, and at higher pressures convex, to the pressure axis. The higher pressure convex portion has been attributed to capillary condensation by some workers and to the formation of multimolecular layers by others.

The formation of multimolecular layers was explained by assuming that the uppermost layer of the adsorbent induces dipoles in the first layer of adsorbate molecules, which in turn induces dipoles in the second layer, and so on until several layers are built up. Brunauer, Emmett, and Teller[5,12] were of the view that the polarization of the second layer of adsorbed gas molecules by the first layer gas molecules will be much too small to constitute a major portion of the binding energy between the two adsorbed layers, at least in those cases in which the gas molecules do not possess considerable permanent dipole moments. These workers proposed that the forces acting in multimolecular adsorption are the same as those acting in the condensation of vapors. Only the first layer of adsorbed molecules, which is in direct contact with the surface of the adsorbent, is bound by adsorption forces originating from the interaction between the adsorbate and the adsorbent. Thus, the molecules in the second and subsequent layer have the same properties as in the liquid state.

On this basis they derived an equation for multimolecular adsorption by a method that is a generalization of the Langmuir treatment of unimolecular adsorption. This equation is known as the Brunauer, Emmett, Teller equation or, more commonly, the BET equation. This equation has played a significant role in the studies of adsorption because it represents the shapes of the actual isotherms. Furthermore, it also gives reasonable values for the average heat of adsorption in the first layer and satisfactory values for V_m, the volume of the gas or vapor required to form a unimolecular layer on the surface of the adsorbent.

2.4.2.1 Derivation of the BET Equation

The basic assumption of the BET theory is that the Langmuir isotherm can be applied to every adsorption layer. The theory postulates that the first layer of adsorbed molecules acts as a base for the adsorption of the second layer of molecules, which in turn acts as a base for the third layer, and so on, so that the concept of localization is maintained in all layers. Furthermore, the forces of interaction between the adsorbed molecules are also neglected, as in the case of Langmuir concept.

Let us suppose S_0, S_1, S_2 ... S_i represents the surface areas that are covered by 0,1,2 ... i molecular layers of adsorbed molecules. Because at equilibrium the surface area S_o is constant, the rate of condensation on the bare surface is equal to the rate of evaporation from the first layer, so that

$$a_1 p S_o = b_1\, S_1 e^{-E1RT} \tag{2.53}$$

where p is the pressure and E_1 is the heat of adsorption in the first layer. This is essentially Langmuir equation for unimolecular adsorption involving the assumption that a_1, b_1, and E_1 are independent of the number of adsorbed molecules already present in the first layer.

Now, at equilibrium, S_1 should also be constant. S_1 can change by

- Condensation on the bare surface
- Evaporation from the first layer
- Condensation on the first layer
- Evaporation from the second layer

because condensation and evaporation can occur only on the exposed surface. Thus, the equilibrium for the second layer analogous to the first layer can be written as

$$a_2\, p S_1 = b_2 S_2 e^{-E2/RT} \tag{2.54}$$

(i.e., the rate of condensation on top of the first layer is equal to the rate of evaporation from the second layer).

Extending the argument to the second and consecutive layers, the general equation of equilibrium between the $(i-1)$ layer and ith layer can be written as

$$a_i\, p S_{i-1} = b_i S_i e^{-Ei/RT} \tag{2.55}$$

The total surface area S of the adsorbent is equal to the surface area of all the layer and is, therefore, given by

$$S = \sum_{i=0}^{i=\alpha} i S_i \tag{2.56}$$

and the total volume adsorbed as

$$V = V_o \sum_{i=0}^{i=\alpha} i S_i \qquad (2.57)$$

where V_o is the volume of the gas adsorbed per unit surface when it is covered with a complete unimolecular layer of adsorbed molecules.

Dividing Equation 2.57 by Equation 2.56, we get

$$\frac{V}{SV_0} = \frac{V}{V_m} \frac{\sum_{i=0}^{i=\alpha} i S_i}{\sum_{i=0}^{i=\alpha} S_i}$$

$$\frac{V}{V_m} = \frac{o + \sum_{i=1}^{i=\alpha} i S_i}{S_o + \sum_{i=1}^{i=\alpha} S_i} \qquad \text{(when } i = o, iS_i = o, \text{ and } S_i = S_1) \qquad (2.58)$$

where V_m is the volume of the gas required to form a completed unimolecular layer on the adsorbent surface.

In order to carry out the above summation, Brunauer, Emmett, and Teller made the following simplifying assumptions:

a. In all layers except the first, heat of adsorption is equal to the molar heat of condensation; i.e.,

$$E - \nabla E \times \nabla E \div E_\tau \nabla E \Lambda$$

where E_L is the heat of condensation of the gas.

b. In all layers except the first, the evaporation and condensation conditions are identical; i.e.,

$$\frac{b_2}{a_2} = \frac{b_3}{a_3} = \frac{b_4}{a_4} \dots \frac{b_i}{a_i}$$

These two assumptions are tantamount to saying that the condensation and evaporation properties of the molecules in the second and higher adsorbed layers are the same as those of the liquid state.

Thus, $S_1, S_2 \dots S_i$ can be represented in terms of S_o as

$$\sum + \nabla \sum o \frac{a_1}{b_1} \pi > \epsilon \, E/PT \, \nabla v \sum o$$

Where

$$\left(y = \frac{a_1}{b_2}\, p.e^{E1/RT} \right) \qquad (2.59)$$

and

$$\Sigma - \nabla \Sigma + \frac{a_1}{b_2}\, \pi > \epsilon\ E/PT\ \nabla_\chi \Sigma +$$

where

$$\left(x = \frac{a_2}{b_2}\, p.e^{-E2/RT} \right) \qquad (2.60)$$

Similarly

$$S_3 = x\, S_2 = X^2 S_1 \ \dots\ \text{and so on,} \qquad (2.61)$$

so that

$$S_i = x\, S_{i-1} = x^{i-1} S_1 = y x^{i-1}\, S_o = c x^i\, S_o \qquad (2.62)$$

Where

$$c = \frac{y}{x} = \frac{\frac{a_1}{b_1}\, pe^{-E1/RT}}{\frac{a_2}{b_2}\, pe^{-E2/RT}} \qquad (2.63)$$

or

$$c = e^{(E_1 - E_2)/RT} \left(\because \frac{a_1}{b_1} = \frac{a_2}{b_2} \right) \qquad (2.64)$$

or

$$c = e^{(E_1 - E_L)/RT} \ (\because E_2 = E_L) \qquad (2.65)$$

Substituting Equation 2.63 in Equation 2.59, we get

$$\frac{V}{V_m} = \frac{\sum_{i=1}^{i=\alpha} icx_i S_0}{S_o + \sum_{i=1}^{i=\infty} cx_i S_o} \tag{2.66}$$

$$= \frac{CS_o \sum_{i=1}^{i=\infty} ix_i}{S_o \left[1 + C\sum_{i=1}^{i=\infty} x_i\right]} \tag{2.67}$$

The summation in the denominator is the sum of infinite geometrical progression:

$$\therefore \sum_{i-1}^{i=\infty} x^i = \frac{x}{1-x} \tag{2.68}$$

and the summation in the numerator can be transformed to:

$$\sum_{i=0}^{i=\alpha} ix_i = x\frac{d}{dx}\sum_{i=1}^{i=\infty} x_i = \frac{x}{(1-x)^2} \tag{2.69}$$

Substituting Equation 2.68 and Equation 2.69 into Equation 2.67, we obtain

$$\frac{V}{V_m} = \frac{cS_o\left[\frac{x}{(1-x)^2}\right]}{S_o\left[1+c\left(\frac{x}{1-x}\right)\right]} = \frac{cx}{(1-x)(1-x+cx)} \tag{2.70}$$

For adsorption on a free surface, an infinite number of layers of the adsorbed molecules can be built up at saturation vapor pressure so that at $p = po$, x in Equation 2.70 must be unity to make $V =$ infinity; i.e.,

$$x = \left(\frac{a_2}{b_2}\right)pe^{E2/RT} = 1 \tag{2.71}$$

when

$$x = p/p_o \tag{2.72}$$

where p/p_o is the relative vapor pressure.

Substituting Equation 2.72 into Equation 2.70, we get

$$V = \frac{V_m c p}{(p_o - p)[1 + (c-1)p/p_o]}$$ (2.73)

Equation 2.73 is called the *simple* or *infinity form* of the BET equation and gives an S-shaped isotherm. The equation can be tested by writing it in the linear form as

$$\frac{p}{V(p_o - p)} = \frac{1}{V_m c} + \frac{c-1}{V_m c} p/p_o$$ (2.74)

so that a plot of $\frac{p}{V(p_o-p)}$ against p/p_o gives a straight line with slope equal to $\frac{c-1}{V_m c}$ and intercept as $1/(V_m c)$. Thus, from the slope and intercept the two constants V_m and c can be calculated. The former is the volume of the gas required to form a monolayer and is used to calculate the specific surface area of the adsorbent if we know the molecular area a_m, the average area occupied by a molecule of the adsorbate in the completed monolayer. Alternatively the BET equation can be written as

$$\frac{p}{n(p_o - p)} = \frac{1}{n_m c} + \frac{c-1}{n_m c} \frac{p}{p_o}$$ (2.75)

where n is the number of moles adsorbed at pressure p and n_m is the monolayer capacity in moles.

The specific surface area can be calculated by using the relationship

$$S_{\text{BET}} = a_m N \cdot V_m$$ (2.76)

where N is Avogadro's number. If the monolayer capacity is expressed in volume of gas adsorbed (reduced to NTP) per gram of the adsorbent, and a_m is expressed in nm² (nanometer)² per molecule, then

$$S_{\text{BET}} = \frac{V_m}{22400} \cdot N \cdot a_m 10^{-18} \ m^2/g$$ (2.77)

Alternatively the monolayer capacity can be expressed in grams of adsorbate x_m adsorbed per gram of the adsorbent, then the specific surface area is given by

$$S_{\text{BET}} = N \frac{x_m}{M} \cdot N \cdot a_m \cdot 10^{-18} m^2/g = N n_m a_m 10^{-18} \ m^2/g$$ (2.78)

where M is the molecular mass of the adsorbate and n_m number of moles

According to the classical BET method, the specific surface area of the adsorbent is usually measured by adsorption of nitrogen at 77 K using molecular area a_m of nitrogen as 0.162 nm^2.

2.4.2.2 Applicability of the BET Equation to Active Carbons

The BET Equation 2.74 is very useful for explaining adsorption data on nonporous and macroporous surfaces, but the equation sometimes loses its applicability in case of adsorption on microporous adsorbents. In the case of active carbons, which are energetically and structurally inhomogeneous, the assumptions of the BET theory are not fulfilled even approximately. Therefore, it is not surprising that the BET plots deviate from linearity, even in regions of the applicability of the BET equation. Furthermore, active carbons are generally highly microporous, the pores being only a few molecular diameters in width. The adsorption data thus shows a higher adsorption so that the value of V_m obtained is very large compared to the expected values. This is due to the fact that the adsorption in micropores takes place through volume filling of micropores at very low relative vapor pressures. Moreover the formation of adsorption layers does not exceed some finite value n due to inadequate space available in the micropores.

When the thickness of the adsorbed layer cannot exceed a certain finite value n, then the summation of the two terms in Equation 2.67 is carried out to n terms only and not to infinity. This gives rise to the equation.

$$V = \frac{V_m cx[1 - (n+1)x^n + nx^{n+1}]}{(1-x)[1 + (c-1)x + cx^{n+1}]} \tag{2.79}$$

where $x = p/po$, and V_m and c have the same meaning. This equation is called n layer BET equation. A plausible interpretation of n is that the width of the pores puts a limit on the number of layers that can be built up even at saturation vapor pressure. Equation 2.79 is the most general form of the BET isotherm equation, and it reduces to Langmuir equation when $n = 1$, and to the simple form of the BET equation (Equation 2.72) when $n = $ infinity.

It should be noted that when x (i.e., p/p_o) has a small value and n is as large as 4 or 5, then Equation 2.72 becomes a very good approximation of Equation 2.79. Therefore, to use Equation 2.79, one should plot the experimental isotherm in the low pressure region, evaluate c and V_m from the slope and the intercept, and then using these values in Equation 2.79 calculate the value of n.

Some authors[13-15] are of the view that, in the case of microporous carbons, the concept of BET surface area loses its meaning because the adsorption takes place through volume filling of micropores. However, Choma[16] Jankowska et al.[17] are of the view that the values of surface area obtained by BET method, even for typical microporous carbons, are realistic and have physical meaning in characterizing the porous structure of adsorbents consisting of mesopores or having a mixed micro and mesoporous structure. These workers suggest that since Equation 2.74 gives acceptable value of V_m for adsorption in meso and macropores, and Equation 2.79 for adsorption in micropores, the acceptance of the value of V_m or surface area depends on the acceptability of the two values of V_m and the assumed value of the molecular area of the adsorbate molecule (a_m). Thus, in addition to BET surface area, the values of V_m and a_m should also be mentioned.

The BET adsorption isotherms are most frequently linear in the relative pressure range 0.05 to 0.35. At lower relative pressures, the equation is not usually valid because the influence of surface heterogeneity of the surface becomes significant. At higher relative pressures, it loses its validity because adsorption by capillary condensation along with physical adsorption also takes place. In addition the assumption of the BET theory relating to the heat of adsorption of the layers beyond the first layer is not completely fulfilled. Cases, however, are known[12,18] when the range of the validity of the BET isotherm has been displaced to lower relative pressures. This is due to the fact that the equation describes the region of the isotherm in which a statistical monomolecular layer is obtained (i.e., when the surface coverage is approximately 0.8 to 1.2 statistical monolayers) and the relative pressure corresponds to these values of the coverage. The relative pressure in turn depends on the constant c, the net heat of adsorption (Table 2.1).

2.4.2.3 Criticism of the BET Equation

It has been mentioned earlier that the BET equation is applicable within the relative pressure range of 0.05 to 0.35. The failure of the equation above and below this range of relative pressure is attributed to the faulty and simplifying assumption of the theory. The failure below a relative pressure of 0.05 is due to the heterogeneity of the adsorbent surface. Active carbon and inorganic gel surfaces that are important adsorbents are generally energetically heterogeneous; i.e., the heat of adsorption varies from one part of the surface to another. The surface of carbons is composed of groups of surface sites each group being homogeneous in itself but differing from

---

TABLE 2.1
Dependence of the Relative Pressure Corresponding to the Formation of a Statistical Monomolecular Layer on the Constant C; Adsorption of Nitrogen at T = 77.4 K

C	$F_1 - E_L$ $(-RT \ln C)$	Relative Pressure at which the Value of Is	
		1	0.8–1.2
10	350	0.239	0.167–0.315
30	531	0.154	0.084–0.245
100	710	0.092	0.033–0.199
300	886	0.055	0.012–0.179
1,000	1,060	0.030	0.004–0.171

Source: Emmett, P.H., in Catalysis, P.H. Emmett, ed., Reinhold, NY, 1954, p. 31. With permission.

other groups.[18] The heat of adsorption in a given group of sites is the same while it differs with the value from another group of surface sites. The failure of the model above 0.35 relative pressure is due to the existence of narrow micropores that limit the thickness of the adsorbed film.

The assumption that the adsorbate has liquid-like properties after the first layer is difficult to reconcile, because both porous and nonporous (planar) adsorbents exposed to a saturated vapor sometimes adsorb strictly a limited amount and not the infinitely large quantity as postulated by BET. Another drawback of the BET model is essentially the coordination number of the molecules in the higher layers. The BET model assumes that each molecule that is adsorbed in any layer after the first layer, gives out its full latent heat of liquefaction, whether it has horizontal neighbors or not, and shall show a coordination number of 12. But in the absence of horizontal neighbors, the coordination number is much less that 12, and, therefore, the heat evolved (i.e., the heat of adsorption) should be only a fraction of the latent heat of condensation.

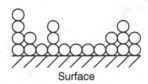
Surface

Thus, the limited validity of the BET equation is due to the shortcomings in the model itself rather than to our lack of knowledge of the various parameters such as the number of layers, the variation of heat of adsorption E, or the evaporation constant b, etc. in the higher layers.

2.4.2.4 Alternative Approach to Linearization of the BET Equation

Application of the *traditional* BET equation (Equation 2.74) presents several difficulties, the more important being the uncertainty in the determination of relative pressure range over which the equation can be applied; and this problem is further complicated in the case of microporous carbons, where the range of applicability is very low. IUPAC has recommended the use of a relative pressure range of 0.05 to 0.30. However, it has been suggested by Sing et al.[19] that the most adequate range for each particular situation occurs in the region where the equation becomes linear. Several other investigators[20–24] have recognized the shortcomings of the classical linear equation and have proposed different treatments for determining the applicability range of the BET equation. However, all of these alternatives have an empirical basis.

Keli et al.[25] suggested an alternative linear form of the BET equation as

$$\frac{1}{n(1-x)} = \frac{1}{n_m} + \frac{1}{n_m c}\left(\frac{1-x}{x}\right) \tag{2.80}$$

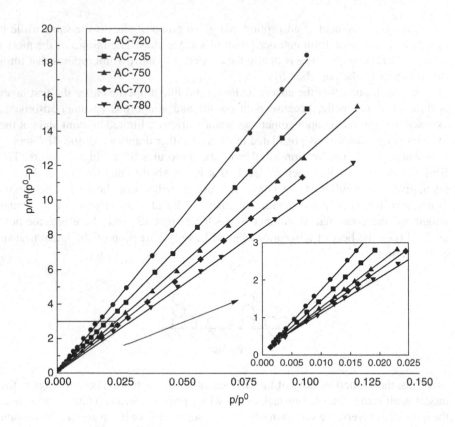

FIGURE 2.7 BET plots on activated carbons (AC Series) using traditional equation. (*Source:* Parra, J.B., de Sousa, J.C., Bansal, R.C., Pis, J.J., and Pajares, J.J., *Adsorption Sci. and Technol.*, 12, 51, 1995. With permission.)

This linear form of the BET equation has been used by Parra et al.[20] to calculate the BET parameters, viz. the surface area and the c constant and also to determine the relative pressure range for the applicability of the BET equation in the case of two series of activated carbons and two samples of activated carbon fibers. The results obtained by the alternative equation have been compared with those obtained from the classical linear form of Equation 2.74.

The BET plots obtained using the two linear forms of the BET equation on two series of active carbons differ considerably (Figure 2.7 and Figure 2.8). Although the plots obtained using the alternative approach appear to be more sensitive to deviations from linearity and show an asymptotic behavior both at high and low relative pressures (Figure 2.9 and Figure 2.10), the plots obtained using the classical approach are almost linear (Figure 2.7 and Figure 2.8) over the same relative pressure range. It appears that a more precise determination of the relative pressure range over which the BET equation is linear can be made using the alternative approach compared with the classical approach. The relative pressure range obtained for the linear regions of the BET plots using the two equations are recorded in Table 2.2. It is seen that the relative pressure range when determined using the classical

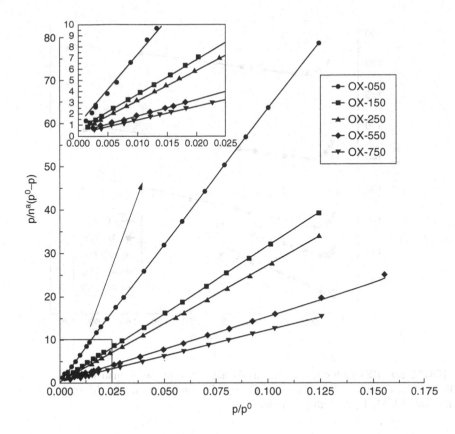

FIGURE 2.8 BET plots on activated carbons (OX Series) using the traditional equation. (*Source*: Parra, J.B., de Sousa, J.C., Bansal, R.C., Pis, J.J., and Pajares, J.J., *Adsorption Sci. and Technol.*, 12, 51, 1995. With permission.)

equation (Figure 2.7 and Figure 2.8) generally does not vary appreciably with variation in the degree of burn-off (AC series of carbons) or in the degree of oxidation (OX series of carbons), although both these treatments result in a considerable change in the textural characteristics of these carbons. On the other hand, when the relative pressure range is calculated using the alternative equation, the relative pressure range for the applicability of the BET equation shifts regularly to higher relative pressure ranges in the case of AC series and to lower pressure ranges in the case of OX series. This variation in relative pressure range may be attributed to a change in the microporous character of these carbons caused by the degree of burn-off or the degree of oxidation. Similar variations in *DR* equation parameters such as $E_o L$ have been observed (Table 2.3).

Surface area and the *c* constant are the two most significant parameters obtained from the BET equation. The values of these two parameters obtained using the two equations are shown in Table 2.4. It is seen that the values of surface area obtained from the two equations are quite similar, the values being slightly larger (about 2 to 4%) as obtained from the alternative equation. However, the *c* parameter values

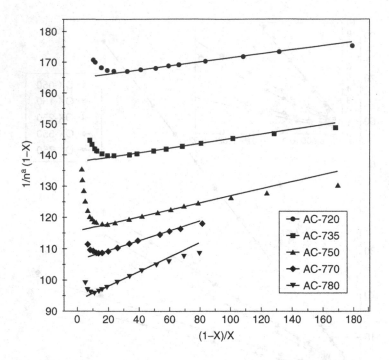

FIGURE 2.9 BET plots on AC activated carbon series using the alternative equation. (*Source:* Parra, J.B., de Sousa, J.C., Bansal, R.C., Pis, J.J., and Pajares, J.J., *Adsorption Sci. and Technol.*, 12, 51, 1995. With permission.)

differ appreciably for the two equations. It is seen that, in general, the values obtained from the classical equation present a certain degree of incoherence as the degree of burn-off or the degree of oxidation is increased. In the case of samples AC-720 and AC-735, the values of c are even negative, which have no meaning. The values of the c parameter obtained from the alternative equation, on the other hand, show a certain degree of order; decreasing regularly with increasing degree of burn-off and increasing regularly with increase in the degree of oxidation. This regular variation in c value appears to be reasonable in view of the changes in the microtextural character of these carbons on these treatments.

Relationship between the c parameter calculated from both the equations and the degree of burn-off and between the c values and the characteristic energy E_o are shown in Figure 2.11 and Figure 2.12. The characteristic energy represents the microporous character of these carbons. The c values calculated by the modified equation show almost a linear relationship with the degree of burn-off or the E_o values, while those calculated from the classical equation show a considerable scatter. It is apparent, therefore, that the c values calculated from the modified approach are more reasonable and consistent with the microporous character of the activated carbons. The variation in the c value with the degree of burn-off or with the degree of oxidation is also reflected in a gradual change in the shape of the knee in the nitrogen adsorption isotherms (Figure 2.13 and Figure 2.14). Thus, the variation in the c parameter of the

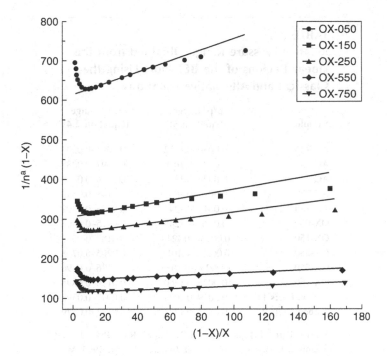

FIGURE 2.10 BET plots on OX series activated carbons using alternative equation. (*Source:* Parra, J.B., de Sousa, J.C., Bansal, R.C., Pis, J.J., and Pajares, J.J., *Adsorption Sci. and Technol.*, 12, 51, 1995. With permission.)

BET equation for microporous carbons appears to be influenced more by the mean pore size of the micropores than by the nature of the carbon, and this is in consonance with the values calculated using the modified equation.

The modified equation was also applied to nitrogen adsorption isotherms at 77 K on porous materials other than activated carbons. The adsorption isotherms and the corresponding BET plots using the two equations determined on a nonporous $\gamma-Al_2O_3$ that is frequently used as a reference material and a mesoporous silica gel (Merck) were also determined. As in the case of activated carbons, the BET plots obtained using the classical equation were quite linear, and those obtained using the modified equation in the same relative pressure range showed greater sensitivity toward deviations both at high and low relative pressures. The range of relative pressure for the applicability of the BET equation could also be found more precisely.

Thus, it appears that although the problems associated with the applicability of the BET equation to adsorption on microporous materials are not completely solved, the modified linear BET equation does represent an improvement over the classical linear equation.

2.4.2.5 Classification of Adsorption Isotherms

The graphic representation between amount adsorbed x and the pressure p at any constant temperatures is referred to as an *adsorption isotherm*. When the adsorbate

TABLE 2.2
Relative Pressure Range Obtained from the
Linear Regions of the BET Plots Using the
Classical and Alternative Equations

Sample	p/p⁰ Range (Equation 3)	p/p⁰ Range (Equation 2.4)
AC-720	0.0030–0.1052	0.0091–0.0293
AC-735	0.0030–0.1050	0.0097–0.0288
AC-750	0.0026–0.1260	0.0126–0.0395
AC-770	0.0014–0.1031	0.0148–0.0595
AC-780	0.0016–0.1242	0.0125–0.0809
OX-050	0.0014–0.1248	0.0215–0.0807
OX-150	0.00232–0.1248	0.0188–0.0705
OX-250	0.00244–0.1019	0.0205–0.0708
OX-550	0.00236–0.1035	0.0126–0.0496
OX-750	0.0026–0.1260	0.0126–0.0395
Carbon fibers 1	0.0019–0.1044	0.0097–0.0393
Carbon fibers 2	0.0022–0.1278	0.0173–0.0579

Source: Parra, J.B., de Sousa, J.C., Bansal, R.C., Pis, J.J., and Pajares, J.J., *Adsorption Sci. and Technol.*, 12, 51, 1995. With permission.

TABLE 2.3
Parameters of the DR Equations for the Nitrogen
Adsorption Isotherms

Sample	E_0 (kJ/mol)	L (nm)	W_0 (cm³/g)	S_{mic} (m²/g)
AC-720	28.43	0.63	0.223	704
AC-735	26.13	0.73	0.272	741
AC-750	23.26	0.91	0.3257	715
AC-770	22.74	0.95	0.333	700
AC-780	19.32	1.36	0.396	581
OX-050	19.65	1.31	0.060	91
OX-150	20.61	1.17	0.117	199
OX-250	20.77	1.15	0.137	238
OX-550	22.87	0.94	0.257	545
OX-750	23.26	0.91	0.326	715
Carbon fibers 1	24.17	0.85	0.363	859
Carbon fibers 2	20.00	1.26	0.509	811

Source: Parra, J.B., de Sousa, J.C., Bansal, R.C., Pis, J.J., and Pajares, J.J., *Adsorption Sci. and Technol.*, 12, 51, 1995. With permission.

TABLE 2.4
BET Parameters Obtained from the Classical
and Alternative Equations

Sample	Equation 2.3		Equation 2.4	
	S_{BET} (m²/g)	C	S_{BET} (m²/g)	C
AC-720	570	−5120	593	2486
AC-735	684	−7824	709	1835
AC-750	827	2316	847	1007
AC-770	905	1228	922	675
AC-780	1027	776	1050	393
OX-050	157	767	160	432
OX-150	313	820	320	467
OX-250	366	861	371	514
OX-550	658	2192	673	974
OX-750	827	2316	847	1007
Carbon fibers 1	910	27378	943	1569
Carbon fibers 2	1284	1147	1322	526

Source: Parra, J.B., de Sousa, J.C., Bansal, R.C., Pis, J.J., and Pajares, J.J., *Adsorption Sci. and Technol.*, 12, 51, 1995. With permission.

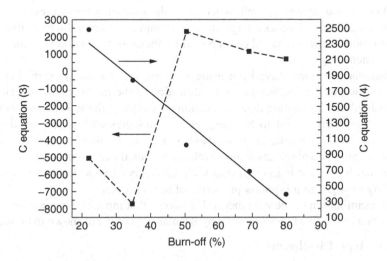

FIGURE 2.11 Relation between the degree of burn-off and the C parameter calculated from the traditional and alternative equations. (*Source:* Parra, J.B., de Sousa, J.C., Bansal, R.C., Pis, J.J., and Pajares, J.J., *Adsorption Sci. and Technol.*, 12, 51, 1995. With permission.)

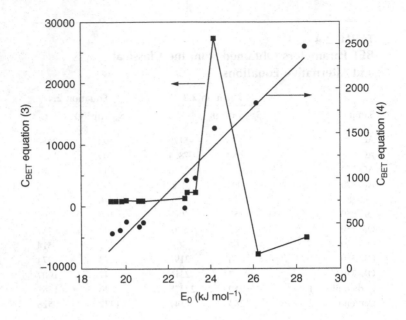

FIGURE 2.12   Relationship between characteristic energy E_0 and the C parameter calculated from the traditional and Alternative equations. (*Source:* Parra, J.B., de Sousa, J.C., Bansal, R.C., Pis, J.J., and Pajares, J.J., *Adsorption Sci. and Technol.*, 12, 51, 1995.)

is a vapor, it is preferable to express the results in terms of relative vapor pressure (p/po) rather than against p itself, where po is the saturation vapor pressure. During the last six or seven decades a large number, running into several tens of thousands, of adsorption isotherms have been reported by thousands of workers in the field of surface chemistry.

Numerous attempts have been made to derive mathematical expressions to fit the adsorption isotherms, but no single adsorption isotherm equation has been found to explain all the adsorption data. The common feature of these isotherms, however, is that all isotherms tend to be linear at low pressures and correspondingly low adsorption values; i.e., the amount adsorbed x is proportional to pressure p. This low-pressure region of the adsorption isotherm is sometimes referred to as the Henrys law region, because it is in consonant with the Henrys law, according to which the solubility of a gas in a liquid is proportional to its pressure.

An examination of these isotherms has shown that most of them conform to one or the other of the five types proposed by Brunauer et al.[5,12] and shown in Figure 2.15.

2.4.2.6   Type I Isotherms

When a solid contains very fine micropores that have pore dimensions only a few molecular diameters, the potential field of force from the neighboring walls of the pores will overlap causing an increase in the interaction energy between the solid surface and the gas molecules. This will result in an increase in adsorption, especially at low relative pressures. There is a possibility and considerable evidence that the

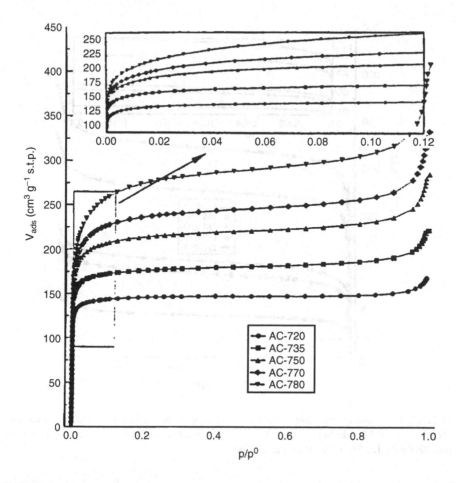

FIGURE 2.13 Adsorption isotherms of nitrogen at 77 K on AC activated carbons. (*Source:* Parra, J.B., de Sousa, J.C., Bansal, R.C., Pis, J.J., and Pajares, J.J., *Adsorption Sci. and Technol.*, 12, 51, 1995.)

interaction energy may be large enough to bring about complete filling of the pores at quite a low relative vapor pressure, giving rise to a Type I isotherm. These isotherms are thus characterized by a plateau that is almost horizontal and parallel to the pressure axis. The adsorption at higher relative pressures being small and tending to level off. Type I isotherms are generally common to chemisorption, although some physical adsorptions such as those in highly microporous active carbons and in carbon molecular sieves also conform to this Type (Figure 2.16 and Figure 2.17).

The fact that the adsorption in Type I isotherms does not increase continuously as in Type II isotherms but attains a limiting value shown by the plateau is due to the pores being so narrow that they cannot accommodate more than a single molecular layer. The shape of the isotherm can be explained by the Langmuir model even though this model was derived for adsorption on an open surface or on a nonporous solid.

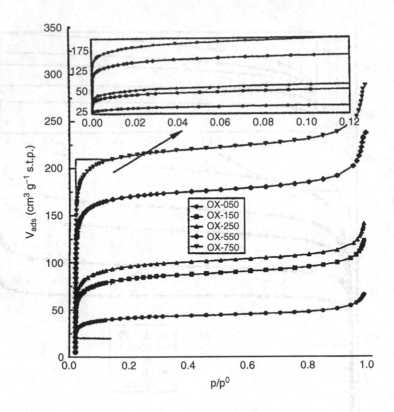

FIGURE 2.14 Adsorption isotherms of nitrogen at 77 K on OX series activated carbons. (*Source:* Parra, J.B., de Sousa, J.C., Bansal, R.C., Pis, J.J., and Pajares, J.J., *Adsorption Sci. and Technol.*, 12, 51, 1995. With permission.)

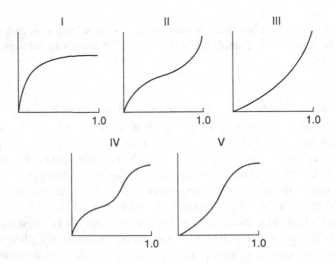

FIGURE 2.15 Five main types of adsorption isotherms.

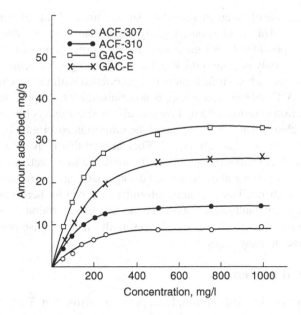

FIGURE 2.16 Adsorption isotherms of Ni(II) ions on different activated carbons. (*Source:* Goyal, M., Rattan, V.K., and Bansal, R.C., *Indian J. Chem. Technol.*, 6, 305, 1999. With permission.)

It may, however, be mentioned that not all Type I isotherms conform to the Langmuir model. While a good linear plot is obtained between p/v against p (or p/po) for many, the plot is distinctly curved for others. Furthermore, the Langmuir equation assumes n_m to be equal to the monolayer capacity that can be converted into surface area.

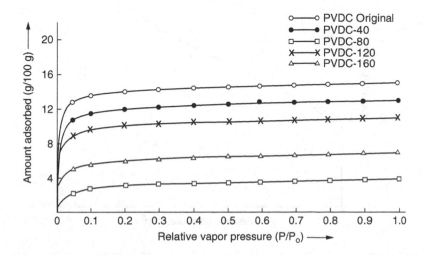

FIGURE 2.17 Adsorption isotherms of heptane vapors on molecular sieve carbons obtained by pore blocking by benzene pyrolyses.

However, in the case of Saran charcoal,[8] the surface area calculated from n_m came out to be ~2970 m²/g. This value is only slightly larger than the area value of 2630 m²/g, which could be provided by all the carbon atoms in one gram of carbon present as layers of graphite only one atom thick and accessible to gas molecules on both sides. Such a structure of carbon surface cannot be reconciled with the mechanical strength of the carbon. A further evidence toward nonconformity of Type I isotherms to the Langmuir equation was provided by Pierce et al.[21] These workers observed that steam activation of a charcoal at 900°C increased the saturation adsorption by threefold, but the adsorption isotherm was still Type I. They argued that if the width of the pores was only two molecular diameters before activation, the steam activation will enhance the pore diameter by removal of the tar and disorganized carbon, resulting in a change in shape of the isotherm. Thus, the nonconformity of Type I isotherms to the Langmuir equation is due to the faulty assumption that the heat of adsorption does not vary with surface converage. In actual practice, however, the heat of adsorption varies with surface coverage in many cases.

2.4.2.7 Type II Isotherms

These isotherms correspond to multiplayer physical adsorption. They generally show a rather long linear portion, a feature that is not strictly compatible with the requirements of the BET equation (Figure 2.18). The point at which this linear portion begins was termed *Point B* by Emmett and Brunauer,[26] and was taken by them as the point at which the monolayer is completed so that the adsorption at Point B

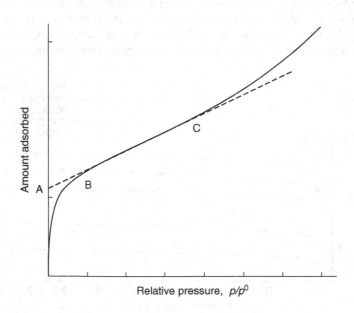

FIGURE 2.18 A typical Type II isotherm showing "Point A" and "Point B." (Gregg, S.J. and Sing, K.S.W., in *Adsorption, Surface Area, and Porosity*, Acad. Press Inc. London, 1982. With permission.)

should be equal to the monolayer capacity X_m. In an earlier paper,[27] these workers had suggested that the point A (Figure 2.18), where the extrapolated linear branch cuts the adsorption axis, might represent the monolayer capacity.

Point A was discarded in favor of Point B, because the value of monolayer capacity X_B calculated from Point B for a variety of systems agreed well with the value of monolayer capacity X_m, calculated using BET equation. Thus, Young and Crowell[28] studied the adsorption of nitrogen at 77 K on 68 different solids and observed that the ratio of values of X_m and X_B ranged between 0.75 and 1.53, with the average being close to unity at 1.03. Brennen et al.,[29] while studying the adsorption of krypton and xenon on a number of evaporated films, found that the values of X_B and X_m differed by about 20% per cent. Sing[30] observed that the two values agreed within about 5% for adsorption isotherms of nitrogen on a number of silica and alumina samples. Drain and Morrison[31,32] also found that in the case of adsorption of nitrogen, oxygen and argon, the low relative pressure adsorption isotherms that included Point B gave a value of X_m, which agreed closely with X_B but in the higher pressure isotherms X_m was significantly larger than X_B. Thus, the adsorption at the Point B (X_B) is still held as equal to monolayer capacity X_m obtained by the application of the BET equation.

However, the case of locating Point B depends upon the shape of the knee of the adsorption isotherm. If the value of c the net heat of adsorption is high, the knee is sharp and the Point B can be accurately located even when the linear branch of the isotherm is short. When c has a small value, the knee is rounded so that the location of Point B becomes difficult, and the estimated value of X_B may differ significantly from the BET monolayer capacity X_m. Thus, it is the value of c that makes the two values agree with each other. Gregg and Sing[33] has suggested that the adsorption isotherms, for which the value of c is below 20, may not be used to obtain values of X_m from the Point B, because in such cases it is difficult to locate Point B.

2.4.2.8 Type III and Type V Isotherms

Type III and Type V isotherms are characterized by being convex to the pressure axis. Although in Type III isotherms the convexity continues throughout the isotherm (Figure 2.19), in Type V the isotherm reaches a plateau at fairly high relative pressures, often at p/po equal to or higher than 0.51 in the multiplayer region (Figure 2.20). The convexity of the isotherms is suggestive of the cooperative adsorption, which means that the already adsorbed molecules tend to enhance the adsorption of other molecules. In other words, it implies that the adsorbate-adsorbent interactions are of less importance than the adsorbate-adsorbate interaction. The weak adsorbate-adsorbent interactions result in small adsorption at lower relative pressures. But once a molecule has been adsorbed, the adsorbate-adsorbate interactions will tend to promote the adsorption of more molecules so that the isotherm becomes convex to the pressure axis.

Type III isotherms are generally obtained in the case of nonporous or highly miacroporous adsorbents, and type V on mesoporous or microporous adsorbents for the adsorption of both polar and nonpolar adsorbates, provided, however, that the adsorbent-adsorbate interactions are weak. Because the heat of liquefaction (or evaporation) is a measure of the adsorbate-adsorbate interaction energy, it appears

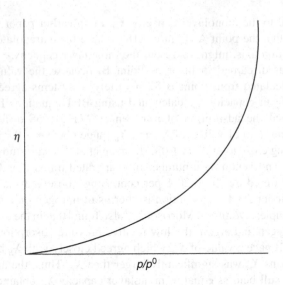

FIGURE 2.19 A typical Type III isotherm.

that the heat of adsorption in the case of Type III isotherms is close to the heat of liquefaction, so the net heat of adsorption $(E_1 - E_L)$ is very small and close to zero.

Adsorption of water vapors on charcoals, active carbons, graphitized carbon blacks, and several types of dehydrated oxide catalysts is generally Type III or Type V. This is due to the fact that the dispersion component of the interaction energy is usually small compared to the polar contribution. Bansal et al.[34] observed that the water vapor adsorption isotherm on a PVDC charcoal was Type IV, but when the carbon

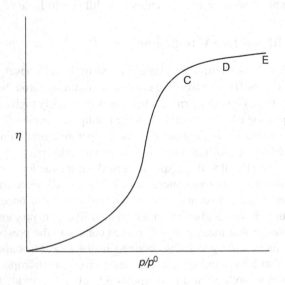

FIGURE 2.20 A typical Type V isotherm.

sample was degassed at increasing temperatures, the shape of the isotherm changed to Type V, which at lower relative pressures was Type III (Figure 2.21). This is due to the fact that the PVDC charcoal is associated with appreciable amounts of polar carbon-oxygen surface groups that result in increasing the interaction of the carbon surface with water molecules, due to the formation of hydrogen bonding. However, when these polar oxygen groups are eliminated on degassing, the interaction of the carbon surface with water continues to decrease as more and more of the oxygen groups are removed. This results in a Type V isotherm. It is evident that the 1000°C-degassed carbon sample that has lost all its oxygen groups (i.e., when the carbon surface is almost completely free of any associated oxygen), the iostherm is clearly Type V. Similarly, for the adsorption of water vapors on clean graphon surfaces or on a silica surface free of hydroxyl groups, the adsorption isotherms are Type III. When the silica surface is hydrated so that it contains hydroxyl groups, these groups can undergo hydrogen bonding with water molecules Thus, increasing the adsorbate-adsorbent interaction. This results in an increase in the c value, and the adsorption isotherm changes from Type III to Type II.

Kiselev et al.[35] modified the surface of a silica aerosol by replacing –OH groups with nonpolar-$Si(CH)_3$ groups on treatment with trimethyl chlorosilane $(CH_3)_3SiCl$. This treatment had the effect of weakening both the dispersion and the polar components of the interaction energy. Kiselev measured the adsorption isotherms of benzene vapors on the aerosol that had gone progressive surface modification and observed that, as the amount of $Si(CH_3)_3$ groups increased on the silica surface, the adsorption isotherm that was Type II on the unmodified silica surface gradually lost its Type II character and was changed to Type III when the surface was completely covered with $Si(CH_3)_3$ groups (Figure 2.22). The heat of adsorption of benzene also decreased progressively until it was less than the heat of liquefaction. Gammage and Gregg[36] studied the adsorption isotherms of butane on ball-milled calcite before and after degassing at 150°C to remove physically adsorbed water. The butane isotherms were Type II on the outgassed sample ($c = 26$) and Type III on the sample before degassing. This is due to the fact that butane, although nonpolar, has a high value of polarizability (82.5×10^{-25} cm^3 per molecule)[32] so that its interaction energy with the ionic solid surface would be comparatively high, producing a Type II isotherm. However, when the surface is covered with a layer of adsorbed water, the adsorbent adsorbate interaction energy is virtually reduced to the weak dispersion energy between water and butane molecules so that a Type III isotherm is obtained.

The question now arises as to whether the BET equation can be used to calculate the monolayer capacity V_m or X_m from Type III isotherms in a manner similar to the one used for Type II isotherms. If the BET model is quantitatively valid for systems yielding Type III isotherms, then the value of V_m should be calculable from Equation 2.74.

$$\frac{p}{V(p_o - p)} = \frac{1}{V_m c} + \frac{c-1}{V_m} p/p_0 \tag{2.81}$$

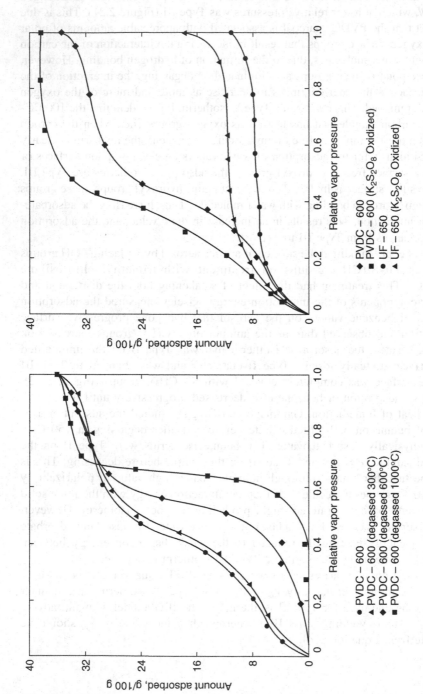

FIGURE 2.21 Water vapor adsorption isotherms on polyvinyledene chloride (PVDC) and urea formaldehyde resin (UF) carbons before and after out gassing. (*Source:* Bansal, R.C., Dhami, T.L., and Prakash, S., *Carbon,* 16, 389, 1978. Reproduced with permission from Elsevier.)

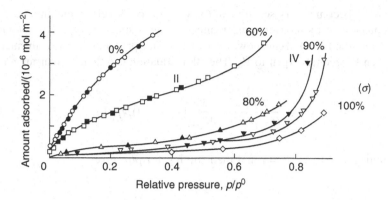

FIGURE 2.22 Benzene vapor adsorption isotherms on silica aerosol before and after covering polar hydroxyl groups with non-polar Si(CH$_3$) groups. (Adapted from Gregg, S.J. and Sing, K.S.W., in *Adsorption, Surface Area, and Porosity*, Acad. Press, London, 1982. Reproduced with permission from Elsevier.)

Because the value of c for Type III isotherms is 1 or 2 (i.e., $c = 1$ or $c = 2$), the substitution of these values in the above equation produces two special cases. Putting $c = 1$ into the above equation, we get the simplified equation:

$$\frac{p}{V(p_o - p)} = \frac{1}{V_m} \qquad (2.82)$$

so that the usual BET plot of $\frac{p}{V(p_o-p)}$ against p/p_o will give a horizontal line parallel to the p/p_o axis at a distance $\frac{1}{V_m}$ from it.

When $c = 2$ is substituted in the above common BET equation, we get.

$$\frac{p}{V(p_0 - p)} = \frac{1}{2V_m} + \frac{1}{2V_m} p/p_0 \qquad (2.83)$$

so that the slope and the intercept of the plot $\frac{p}{V(p_o-p)}$ versus p/p_o are equal and are given by $\frac{1}{2V_m}$.

In practice, however, it is not easy to make an accurate BET plot from Type III isotherm, because the values of V are very small and of lesser accuracy, especially at lower relative pressures. This lower pressure region of the isotherm is the portion over which the BET equation is generally valid.

Furthermore, if we calculate the value of the slope from the first special case plot of $\frac{p}{V(p_o-p)}$ versus p/p_o, the value of the slope

$$\text{Slope} = \frac{c-1}{V_m c} \qquad (2.84)$$

where $c = 1$ becomes very sensitive to the value of c. Therefore, the direct use of the BET equation in the conventional manner is unsuitable for Type III isotherms. A slightly modified method can, however, be used. This procedure consists in calculating p/p_o when V becomes equal to V_m. The BET equation can then be rearranged as

$$\frac{p/p_o}{(1 - p/p_o)} = \frac{V}{V_m c}[1 + (c - 1)p/p_o]$$

(2.85)

Substituting $V = V_m$ or $\frac{V}{V_m} = 1$ and then solving for p/p_o gives

$$\left(\frac{p}{p_o}\right)_m = \frac{-1 \pm \sqrt{c}}{c - 1}$$

(2.86)

Because p/p_o is always positive, the only solution of Equation 2.86 is

$$\left(\frac{p}{p_o}\right)_m = \frac{-1 + \sqrt{c}}{c - 1}$$

(2.87)

Now for the special case $c = 1$, the above equation becomes indeterminate. The value of $p/p_o)_m$ can then be calculated directly from the BET equation.

$$\frac{p}{V(p_o - p)} = \frac{1}{V_m c} + \frac{c - 1}{V_m c}p/p_o$$

by merely substituting $V = V_m$ and $c = 1$ and solving for p/p_o. This gives $(p/p_o)_m = 0.5$ and, for $c = 2$, $(p/p_o)_m = 0.41$. Thus, as the value of c moves from $c = 2$ to $c = 1$, the value of p/p_o changes from 0.41 to 0.50.

It is apparent that for values of c greater than unity, the value of V_m can be derived from the slope and intercept of the BET plot in the usual way. But since there are deviations at lower relative pressures, it is sometimes more convenient to locate the BET monolayer point by the relative pressure $(p/p_o)_m$ at which $V/V_m = 1$. In such cases the value of c is first determined by matching the experimental isotherm with a set of ideal BET isotherms, calculated by substituting different values of c (between 1 and 2, and even nonintegral values if necessary). This value of c can then be inserted into Equation 2.87 and the value of $(p/p_o)_m$ calculated. Then from the usual BET plot, the value of V_m can be read for a suitable value of p/p_o. However, when these values of monolayer capacity obtained from Type III isotherms were compared with V_m values obtained from independent estimates, considerable discrepancies were frequently noticed. Gregg and Sing[33] has attributed these discrepancies to the artificial nature of the BET model. These workers are of the view that to obtain a reasonable value of the monolayer capacity from the adsorption isotherms, it is necessary that the formation of a monolayer shall be virtually complete before the formation of the multiplayer commences so that there is a clearly identifiable point on the isotherm (Point B) corresponding to this condition. In cases where a Type III isotherm is obtained, however, the multilayer is being built up on some parts of the surface while the monolayer is still incomplete on other parts. Thus, the BET procedure for calculating monolayer capacity is not completely applicable for a Type III isotherm.

2.4.2.9 Type IV Isotherm

Type IV isotherms are obtained for solids containing pores in the mesopore range. The shape of the Type IV isotherm follows the same path as the Type II at lower relative pressures until its slope starts decreasing at higher pressures. At the saturation vapor pressure, the isotherm levels off to a constant value of adsorption (Figure 2.15). This portion of the isotherm, which is parallel to the pressure axis, is attributed to filling of the larger pores by capillary condensation. The most significant characteristics of these isotherms is that they show adsorption-desorption hysteresis, which means that the amount adsorbed is always greater at any given relative pressure along the desorption branch than along the adsorption branch (Figure 2.23). This type of isotherm is generally obtained in the case of oxide gels and several porous carbon materials that have larger-sized pores. These adsorption isotherms have led

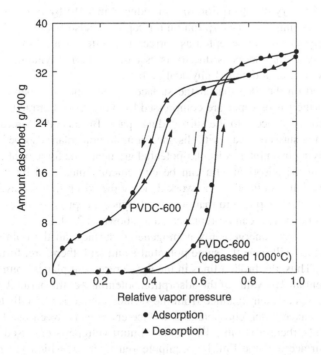

FIGURE 2.23 Adsorption-desorption isotherms of water vapor on PVDC charcoal before and after degassing at 1000°C. (*Source:* Bansal, R.C., Dhami, T.L., and Prakash, S., *Carbon*, 16, 389, 1978. With permission.)

to the development of the theory of capillary condensation, which was first pro-
pounded by Zsigmondy[37] on the principles earlier established by Kelvin (earlier
name Thomson).[38] Broadly speaking, the Zsigmondy model assumes that along the
initial part of the isotherm, the adsorption is restricted to a thin layer on the walls
of the pores until capillary condensation begins in the smallest pores. As the pressure
is gradually enhanced, larger and larger pores are filled until at saturation vapor
pressure the entire pore system is full of the condensate (Figure 2.23).

The connection between mesoporosity and Type IV isotherms has been examined
and demonstrated by several workers,[39–42] using different adsorbate-adsorbent systems and studying adsorption on loose powders and after compacting the powders.
A well-defined Type II isotherm was obtained in each case when the adsorption was
studied, using loose powders and a well-defined Type IV isotherm when compacted
materials were used. The process of compaction produces pores in the form of
interstices between the particles of the original powder. The compacted powders are
expected to have dimensions in the range of mesopores causing an increase in
adsorption involving capillary condensation. These results clearly indicate that the
presence of mesopores results in Type IV isotherms. The capillary condensation and
the subject of adsorption desorption hesteresis shall be discussed shortly.

2.4.3 POTENTIAL THEORY OF ADSORPTION

The potential theory of adsorption first introduced in 1914 by Polanyi[43,44] and later
modified by Dubinin[45] for adsorption on microporous adsorbents is still regarded as
fundamentally sound and accepted as correct and better than all the other theories.
This longevity of the theory is due to its essentially thermodynamic character and
lack of insistence on a detailed physical picture.

It is based on the idea that at the surface of the solid adsorbent, the adsorbed
molecules of the gas or vapor are compressed by the forces of attraction acting from
the surface to a distance into the surrounding space. Because the forces anchoring a
molecule to the surface decay with distance, a multimolecular adsorbed film may be
regarded as lying in an intermolecular potential gradient. The force of attraction at any
given point in the adsorbed film can be conveniently measured by the adsorption
potential ε, which is defined as the work done by the adsorption forces in bringing a
molecule from the gas phase to that point. Polanyis concept of the cross-section for a
typical gas-solid system can be represented as shown in 2.24.

For an idealized, energetically homogeneous surface, all the points equidistant
from the surface will have the same potential ε and will, therefore, form an equipotential plane. Thus, the broken lines in the figure represent planes connecting points
of equal potential. The value of the adsorption potential ε of the parallel equipotential
planes decreases as their distance from the surface increases and falls to zero at the
maximum distance. Each equipotential surface encloses between itself and the surface of the adsorbent a volume W. The maximum volume is enclosed between the
adsorbent surface and the limiting equipotential plane at which the potential has
decreased to zero. Thus, the volumes enclosed between the adsorbent and the
equipotential surfaces $\varepsilon = \varepsilon_1, \varepsilon_2, \varepsilon_3 \ldots \varepsilon_o$ are $W_1\ W_2\ W_3 \ldots W_{max}$. The quantity W_{max}
represents the volume of the entire adsorption space. As W increases from zero to

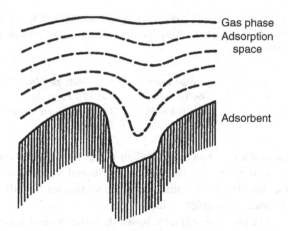

FIGURE 2.24 Schematic of the surface adsorbed layer according to potential theory. (*Source:* Young, D.M. and Crowell, A.D., in *Physical Adsorption of Gases*, Butterworths, London, 1962, p. 198. With permission.)

maximum, ε decreases from its maximum value at the adsorbent surface to zero at the outermost adsorbed layer. The process of formation of the adsorbed film may, therefore, be represented by a relationship between ε and W as

$$\varepsilon = f(W) \tag{2.88}$$

which in reality is a distribution function. The fundamental postulate of the potential theory is the assumption that the adsorption potential is independent of the temperature:

$$\left(\frac{\partial_\varepsilon}{\partial_T}\right)_w - O \tag{2.89}$$

Thus, the curve $\varepsilon = f(W)$ should be temperature-independent for a given gas. Since this curve characterizes the given adsorbate-adsorbent system, it is called the *characteristic adsorption curve* for a given adsorbent. The characteristic curve is the same for all temperatures and, hence, a unique characteristic of a given adsorbent-adsorbate system.

As the adsorbate is compressed in the field of the adsorption potential, its density cannot be the same throughout the adsorption space but increases from the limiting equipotential surface of zero potential to the surface of the adsorbent where the adsorption potential has the highest value. Therefore, the adsorption potential is nothing but the isothermal work required to compress one mole of a vapor from its equilibrium vapor pressure p_1 (gas phase) to the vapor pressure p_2 of the compressed adsorbate on the given equipotential surface. Thus, at constant temperature T, the adsorption potential, can be expressed as

$$\varepsilon_i = -\Delta G = RT \ln \frac{p_2}{p_1} \tag{2.90}$$

The state of the compressed adsorbate in the adsorption space depends on the temperature. Thus, when determining characteristic curve Polanyi distinguished three different cases:

Case	Temperature	State of Film
I	$\ll T_c$	Liquid
II	Just below T_c	Liquid + compressed gas
III	$> T_c$	Compressed gas

When the temperature is much lower than the critical temperature T_c of the adsorbate, the density of almost all the adsorbed vapor may be taken as equal to the density of the liquid adsorbate so that all the adsorbate in the adsorption space is completely liquefied.

When the temperature is marginally lower than the critical temperature, the density of that part of the adsorbate that is enclosed between the equipotential surfaces of the highest values of ε and which is the most compressed can be assumed to be the same as that of the liquid adsorbate at that temperature. Beyond the equipmental surface where the adsorption potential ε is less than that necessary for compressing the vapors to the equlibrium vapor pressure at that temperature, the adsorbate is present as compressed vapor. The density vapor of the adsorbate present as compressed vapor decreases with increasing distance from the surface of the adsorbent until, at the boundary of the adsorption space, the density falls to the density of the adsorbate in the surrounding space (gaseous phase).

When the temperature is above the critical temperature, the adsorbate cannot be converted into liquefied state so that the adsorption space contains only compressed gas, the density of which decreases from the adsorbent surface to the boundary of the adsorption space.

Among the above three cases, the first situation is of greatest practical significance and can be easily evaluated because at temperatures sufficiently below the critical temperature, the adsorbed gas or vapor can be considered as completely liquefied. If it is further assumed that the liquid is incompressible and that the vapor in the gaseous phase behaves ideally, then Equation 2.90 for the adsorption potential can be written as

$$\varepsilon = RT \ln \frac{p_s}{p} \tag{2.91}$$

where p_s is the saturation vapor pressure of the liquid adsorbate and p is the equilibrium vapor pressure so that the adsorption potential below the critical temperature represents the isothermal work required for compressing the vapor from their equilibrium pressure in the gaseous phase to the saturation vapor pressure of the liquid adsorbate at that temperature.

As the liquefied adsorbed film has been assumed to be incompressible, the density of the condensate that is taken as constant throughout its volume can be taken as equal

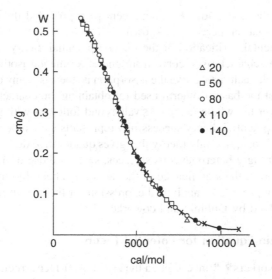

FIGURE 2.25 Characteristic curve for benzene adsorption on active carbon.

to the normal density of the liquid adsorbate at that temperature. Thus, we can write

$$W = \frac{x}{\rho} \tag{2.92}$$

where W is the volume of the adsorbate filling the adsorption space or the volume adsorbed, x is the mass of the adsorbed film in grams, and ρ is the density of the liquid adsorbate at temperature T. Thus, knowing W and ε, a plot can be made between W and ε, which is called a *characteristic curve*. A typical characteristic curve is shown in Figure 2.25. Equation 2.92 can also be written as

$$W = \frac{X_m \cdot M}{\rho} = X_m \cdot V_m \tag{2.93}$$

where X_m is the mass of the adsorbed film in moles and V_m is the molar volume of the adsorbate at that temperature.

From Equation 2.91 and Equation 2.92, it is clear that for every pair of values for x and p (i.e., for every point on the adsorption isotherm), there is a corresponding pair of values of ε and W (i.e., a point on the characteristic curve). Because ε, the adsorption potential, is independent of temperature, the characteristic curve will correspond to points on adsorption isotherms for a given vapor on a given adsorbent for all temperatures. This means that the equivalence of adsorption isotherms for a given vapor on a given adsorbent at different temperatures can all be transformed, point by point, to a single characteristic curve. Alternatively we can say that from a known characteristic curve, which can be obtained from the isotherm measured for a single temperature, it is possible to obtain isotherms at any other arbitrary temperature for the given adsorbate-adsorbent pair, provided the molar volume of the adsorbate at that temperature is known.

The experimental verification of the Polanyi potential theory can be carried out by calculating the characteristic curve from one experimental isotherm and in determining from the characteristic curve the adsorption isotherms at any other temperature. It is essential that the basic isotherm used for obtaining the characteristic adsorption curve should cover the whole range of ε values and followed with great accuracy.

The Polanyi potential theory successfully represents the temperature dependence of adsorption. It is also the only theory that gives quantitative description of physical adsorption on strongly heterogeneous surfaces, such as those of active carbons and oxide gels. However, the significance of the theory had been limited for a long time because it did not provide an analytical expression for the adsorption isotherm. This problem was solved by Dubinin and coworkers.[45–48]

2.4.3.1 Dubinin Equation for Potential Theory

Dubinin and coworkers[45–48] investigated the effect of surface structure on the absorbability of different gases and adsorption of different solutes from solutions on active carbons, and observed that over a wide ranges of values of adsorption, the characteristic curves of different vapors on the same adsorbent were related to each other. In fact, it was observed that if the adsorption potential ε, corresponding to a certain volume of adsorption space W, on the characteristic curve for one vapor was multiplied by a constant, the adsorption potential corresponding to the same value of W on the characteristic curve of another vapor was obtained. Thus, we obtain

$$\varepsilon = \varepsilon_0 \beta \tag{2.94}$$

where β is a constant called the *affinity coefficient* for a given pair of vapor, and its value is independent of the temperature and the nature of the porosity of the given adsorbent (active carbons). It (β) is thus a measure of the absorbability of a given vapor on a given carbon with respect to the vapor selected as a standard. The standard vapor used is benzene for which $\beta = 1$

Equation 2.94 can also be written as

$$\frac{\varepsilon}{\varepsilon_o} = \beta \tag{2.95}$$

where ε_o is the value of the potential for the standard vapor. For a standard vapor the characteristic curve is given by

$$w = f(\varepsilon_o)$$

or

$$w = f\frac{\varepsilon}{\beta} \tag{2.96}$$

Dubinin and Radushkevich[50] suggested that for microporous active carbons (or porous materials), volume of adsorption space can be expressed as a Gaussian function of the corresponding adsorption potential $W = W(\varepsilon)$, so that the characteristic curve for microporous active carbons can be written as

$$W = W_o \exp. \left(-K\,\varepsilon_o^2\right) \qquad (2.97)$$

where W_o is the total volume of all the micropores, and K is a constant characterizing the pore-size distribution. The above equation can also be written as

$$W = W_o \exp. \left[-K\left(\frac{\varepsilon}{\beta}\right)^2\right] \qquad (2.98)$$

or

$$\left(\because\ \varepsilon_o = \frac{\varepsilon}{\beta}\right)$$

or

$$W = W_o \exp. \left[-\frac{K}{\beta^2}\left(RT\,\ln\frac{p_s}{p}\right)^2\right] \left(\varepsilon = RT\,\ln\frac{p_s}{p}\right)$$

or

$$W = W_o \exp. \left[-\frac{K}{\beta^2}(RT)^2\left(\ln\frac{p_s}{p}\right)^2\right]$$

$$\therefore \ln W = \ln W_o - \frac{K}{\beta^2}(RT)^2\left(\ln\frac{p_s}{p}\right)^2$$

or

$$\ln W = \ln W_o - D\left(\ln\frac{p_s}{p}\right)^2 \qquad (2.99)$$

where $D = \frac{K}{\beta^2}(RT)^2$ and W_o is the micropore volume. This equation (Equation 2.99) is known as the Dubinin equation and can be used to obtain from one measured isotherm, the isotherms for the same vapor at any other temperature over the range

TABLE 2.5
Affinity Coefficient β for Some Vapors and Gases
on Active Carbon (From 47)

Benzene	1.00	Dichloromethane	0.66
Cyclohexane	1.04	Ethyl chloride	0.76
Toluene	1.25	Tetrafluoroethylene	0.59
Propane	0.78	Hexafluoropropylene	0.76
n-Butane	0.90	Chloropicrin	1.28
n-Pentane	1.12	Ethyl ether	1.09
n-Hexane	1.35	Acetone	0.88
n-Heptane	1.59	Formic acid	0.61
Methanol	0.40	Acetic acid	0.97
Ethanol	0.61	Carbon disulphide	0.70
Methyl chloride	0.56	Ammonia	0.28
Methyl bromide	0.57	Nitrogen	0.33
Chloroform	0.86	Krypton	0.37
Tetrachloromethane	1.05	Xenon	0.50

Source: Dubinin, M.M. and Timofeev. D.P., *Dokl. Akad. Nauk., SSSR*, 54, 1946. With permission.

well below the critical temperature provided the molar volume and the saturation pressure are known. The equation can also be used to calculate adsorption isotherms for other vapors at any temperature in the range mentioned above, provided in addition to the two quantities mentioned above, we also know the affinity coefficient β. For approximation, it is possible to replace β by the ratio of the molar volumes of the given and standard vapor at the particular temperature, or by the ratio of the parachors or roughly by the ratio of the van der Walls constant a.[51] Affinity coefficients (β) for some vapors and gases on active carbons are given in Table 2.5.[52-54]

Equation 2.99 shows that a plot of ln W against $(\ln \frac{p_s}{p})^2$ should give a straight line with Slope D and intercept W_o. Dubinin plotted results of a large number of adsorbates such as nitrogen, and benzene, saturated hydrocarbons on active carbons and found that the equation was valid over the range 1×10^{-5} to 0.2 relative vapor pressure in those cases where the solid was truly microporous.

The important quantity of interest in the Dubinin equation is W_o, the micropore volume that is evaluated from the intercept of the linear plot on the adsorption axis where $\ln \frac{p_o}{p} = 1$. The method thus makes it possible to calculate the micropore volume from the low-pressure part of the isotherm and offers the possibility of using different adsorbates as molecular probes. For example, using benzene molecule, which is much larger than the nitrogen molecule, it is possible to calculate the volume of the micropores that are large enough to be accessible to nitrogen molecules but small enough not to be accessible to larger benzene molecules. It may, however, be mentioned that when an active carbon is associated with mesoporosity (i.e., transitional porosity, when the specific surface area of the transitional pores is equal or exceeds a value of about 50 m²/g), setting W_o as equal to the volume of the micropores is sufficiently inaccurate. The results need to be corrected for adsorption in the mesopores.[52-55]

Kaganer[56] has modified Dubinin's treatment to yield a method for the calculation of specific surface area from these isotherms. He confined attention to the monolayer region and assumed that during adsorption at very low relative pressures, a monomolecular layer is formed on the walls of all the pores. This is contradictory to the concept of volume filling of micropores, on the basis of which the Dubinin equation was formulated. Thus, Kaganer assumed that it is the distribution of the adsorption potential over the sites on the surface, which is Gaussian in type and, therefore, can be represented as

$$\theta = \frac{x}{x_m} = \left[-K_1 \varepsilon^2 \right] \tag{2.100}$$

where θ is the fraction of the monolayer occupied, x is the number of moles adsorbed at a given relative pressure, x_m is the number of moles adsorbed at the monolayer, and K_1 is a constant that characterizes the Gaussian distribution, and ε as before is the adsorption potential.

Substituting the value of ε in Equation 2.100, we get

$$\frac{x}{x_m} = \exp\left[-K_1 \left(RT \ \ln \frac{p_s}{p} \right)^2 \right] \tag{2.101}$$

or

$$\log \frac{x}{x_m} = -2.303 K_1 (RT)^2 \left(\log \frac{p_s}{p} \right)^2$$

$$= -D_1 \left(\log \frac{p_s}{p} \right)^2 \left(D_1 = 2.303 K_1 (RT)^2 \right) \tag{2.102}$$

or

$$\log x = \log x_m - D_1 \left(\log \frac{p_s}{p} \right)^2$$

This equation is identical with the Dubinin equation. When $\log x$ is plotted against $(\log p_s/p)^2$, it should yield a straight line which, on extrapolation, intercepts the ordinate axis. The intercept on the ordinate axis gives the value of monolayer capacity rather than the micropore volume. The method thus yields a value of specific surface area and can apply in the low-pressure region (i.e., below relative pressure of 10^{-4}). The surface areas calculated by the Kaganer method for active carbons were within a few percent of those calculated by the BET equation.

Several active carbons show exceptionally high values of surface area when nitrogen at 77 K is used as the adsorbate. This high value is attributed to the fact that nitrogen is not suitable for these measurements because it is not completely inert. Lamond and Marsh[57] compared the surface areas determined by Kaganer method using isotherms of nitrogen and carbon dioxide and suggested that nitrogen at 77 K is not suitable for surface area measurements because it condenses in the micropores at low relative pressures, while carbon dioxide first forms the monolayer and, therefore, yields correct values of surface area.

2.4.4 FREUNDLICH ADSORPTION ISOTHERM

The Freundlich isotherm is a limiting form of the Langmuir isotherm, and is applicable only in the middle ranges of vapor pressures. The general form of the Langmuir isotherm is written as

$$V = \frac{V_m bp}{1 + bp} \qquad (2.103)$$

At low pressures, bp is much smaller than unity and, therefore, can be neglected in the denominator so that the equation becomes. $V = V_m bp$, indicating that the amount adsorbed is proportional to the first power of the pressure. At high pressures, Equation 2.104 becomes $V = V_m$ so that the amount adsorbed becomes independent of the pressure. It is thus apparent that in the middle ranges of pressure, the amount adsorbed can be represented by a fractional exponent $1/n$, which will tend to vary between zero and unity, depending upon whether the pressure increases or decreases. This can be expressed by a general form of the adsorption equation.

$$\frac{V}{V_m} = bp^{1/n} \qquad (2.104)$$

This is known as the *Freundlich equation*, which is followed only at medium pressures. The equation is of greater significance for chemisorption although some physical adsorption have also been explained using this equation.

For adsorption from solution phase, the equation can be written as

$$\frac{V}{V_m} = \frac{x}{m} = Kc^{1/n} \qquad (2.105)$$

where c is the equilibrium concentration and x/m is the amount adsorbed per unit mass of the adsorbent. The constant n is the Freundlich equation constant that represents the parameter characterizing quasi-Gaussian energetic heterogeneity of the adsorption surface. The equation can be written in the linear form as

$$\ln \frac{x}{m} = \ln k + \frac{1}{n} \ln c \qquad (2.106)$$

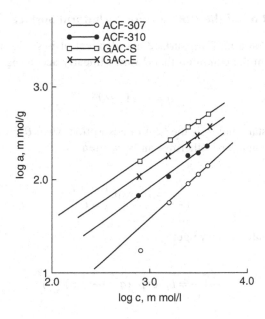

FIGURE 2.26 Linear Freundlich isotherms of Ni(II)ions on different activated. (Source: Goyal, M., Rattan, V.K., and Bansal, R.C., *Indian J. Chem. Technol.*, 6, 305, 1999. With permission.)

so that a plot of ln (x/m) a against ln c gives a straight line with an intercept on the ordinate axis. The values of n and K can be obtained from the slope and the intercept of the linear plot. The value of n is always greater than unity. The linear plots of log x/m versus ln c for several activated carbons are shown in Figure 2.26, where $a = x/m$ is the amount adsorbed per gram of the carbon.

2.4.5 TEMKIN ADSORPTION ISOTHERM

Of the two isotherms considered so far (Langmuir and Freundlich), the Freundlich isotherm is likely to be more widely obeyed than the Langmuir isotherm because the heat of adsorption (q) normally falls with increasing surface converage. This effect is only allowed in the Freundlich isotherm, because it involves a term n that allows for heterogeneity of the solid surface. However, few $q - \theta$ plots are logarithmic in form and, therefore, the Freundlich isotherm shall, with many adsorptions be an approximation to the truth.

Very often the fall in heat of adsorption is more nearly linear than logarithmic, and it is this type of behavior that led to the derivation of the Temkin adsorption isotherm. The isotherm is, in fact, derived from the Langmuir adsorption isotherm by inserting the condition that the heat of adsorption decreases linearly with surface coverage. Such an effect can arise from repulsive forces on a uniform surface or from surface heterogeneity of the surface.

2.4.5.1 Derivation of the Isotherm for a Uniform Surface

Because all sites on a uniform surface are identical, it is sufficient to insert in the Langmuir isotherm the condition that θ varies linearly according to an equation

$$q = q_o (1 - \alpha\theta) \tag{2.107}$$

where α is a constant and q_o is the heat of adsorption when $\theta = 0$. For a single-site adsorption, the Langmuir isotherms can be written as

$$\frac{\theta}{1-\theta} = bp \exp\left(\frac{q}{RT}\right) \tag{2.108}$$

Substituting the value of q we get

$$\frac{\theta}{1-\theta} = bp \exp\left(q_o (1 + \alpha\theta)/_{RT}\right) \tag{2.109}$$

Taking the logarithm and rearranging gives

$$\ln p = \ln A_o + \frac{q_o \alpha\theta}{RT} + \ln \frac{\theta}{1-\theta} \tag{2.110}$$

where $A_o = K \exp\left(\frac{q_o}{RT}\right)$ and is independent of θ.

Now for chemisorption $q_o\alpha \gg RT$ and in the middle ranges of surface coverage, where $\theta/(1 - \theta)$ varies very slowly with θ, the main variation of $\ln p$ is due to the variation of the term $\frac{q_o \alpha\theta}{RT}$. However, this will not apply when θ approaches zero or unity, because then $\frac{\theta}{1-\theta}$ changes very rapidly with θ. But in the middle ranges of surface coverage, it is possible without serious error to neglect the variation of $\ln \theta/(1-\theta)$ and to equate it to zero. Taking these aspects into consideration equation can be written as

$$\ln p = \ln A_o + \frac{q_o \alpha\theta}{RT}$$

or

$$\theta = \frac{RT}{q_0\alpha} \ln A_0 p \tag{2.111}$$

This isotherm characterized by linear variation of $\ln p$ with θ is known as the *Temkin isotherm*. The Temkin isotherm for nonuniform surfaces can be derived by dividing the surface into a number of uniform elements, on each of which the heat of these elements obeys the Langmuir isotherm. The isotherm equation obtained after certain assumptions is identical with the Temkin isotherm derived for a uniform surface. The Temkin isotherm equation is applicable only to the middle ranges of surface coverage and deviation are observed at high or low surface coverages even though the adsorption has the necessary linear fall in heat of adsorption with surface coverage.

The isotherm Equation 2.110 appears to show that θ may he increased indefinitely by increasing p, suggesting that there is no saturation state. This is, of course, because the equation does not apply at higher pressures and at high surface coverages.

2.4.6 CAPILLARY CONDENSATION THEORY

The capillary condensation theory was first put forward by Zsigmondy[37] in 1911 to explain the adsorption of gases and vapors by porous solids such as charcoals and silica gel, and to explain the adsorption-desorption hystersic in Type IV isotherms. The theory postulates that, in addition to the formation of layers, the adsorbate gas or vapor condenses in the small capillary pores of the adsorbent as a result of the lowering of vapor pressure brought about by surface tension effects. The cause of this vapor pressure lowering lies in a decrease in free energy of the adsorbate molecules in fine capillary pores.

Let us first consider an adsorbent with a flat surface in contact with a gas, the pressure of which is gradually increased. At lower relative pressure, the monomolecular adsorption will take place until the surface of the adsorbent becomes covered by a single layer of adsorbate molecules. When the first layer of adsorbed molecules is complete (in the statistical sense) any further increase in pressure causes the formation of second and subsequent layers, resulting in multimolecular layer adsorption. If the adsorbent contains pores with widths equal to several diameters of the adsorbate molecules (i.e., transitional pores), then at higher relative pressures (usually greater than 0.3), the amount adsorbed increases and at a certain stage becomes larger than what would correspond to the formation of multimolecular layers that can be accommodated in the pores. This is because, along with the formation of multimolecular layers, condensation of the gaseous molecules takes place. This adsorption by condensation is called *capillary condensation.*

As the pressure of the gas or the vapor is increased, the thickness of the multimolecular layers in the transitional pores increases until the layer on the opposite walls combine in the narrowest cross-section of the pore and form a meniscus of condensed adsorbate. This meniscus is concave when the adsorbate wets the surface. The molecules of the adsorbate then condense on the meniscus at a pressure lower than the saturation vapor pressure. The lowering of the equilibrium vapor pressure over a concave meniscus, as compared with that over a flat surface at the same temperature, is due to the molecules in a concave surface being held by a large number of neighboring molecules, rather than if they were held on a flat surface. The quantitative relationship between the lowering of vapor pressure and the radius of the capillary, known as the *Kelvin equation*, was given by Thomson (later Lord Kelvin).[38] The Kelvin equation can be written as

$$\ln \frac{p}{p_s} = -\frac{2\gamma V}{rRT} \cos \theta \qquad (2.112)$$

where γ is the surface tension, V is the molar volume of the liquid adsorbate, θ is the angle of contact between the condensate (adsorbate) and the surface of the adsorbent, r is the radius of curvature of the meniscus of the condensed adsorbate in the pore, and p/p_s is the relative vapor pressure. The negative sign implies that for $\theta < 90°$, p is less than p_s, the saturation vapor pressure.

Thus, if a vapor or a gas at pressure p is brought into contact with a porous adsorbent, it should condense as a liquid in all pores having a radius less than that calculated from the Kelvin equation for that particular value of p, assuming that $\theta < 90°$. It is also evident that with increasing relative pressure of the gas or the vapor, wider and wider pores become filled by the condensed adsorbate. Simultaneously, with this process of condensation, building up of multimolecular layers continues in pores whose width is such that the pressure attained is as yet too low for the existing multimolecular layers to permit the formation of a meniscus. At the saturation vapor pressure, the entire pore system is full of the condensate.

It is now generally accepted that in the case of porous solids, both multilayer adsorption and capillary condensation can and do take place, although it is not possible to say at present as to how much of the total adsorption is due to the first and how much due to the second process. It is believed that the pore diameter is an important factor in determining which of the two processes is more effective in a particular case. It has been suggested that the point of inception of the hysteresis, which is observed as a lag between adsorption and the desorption curve, represents the point below which adsorption is due to layer formation and above which adsorption is due to capillary condensation. Assuming the pores to be cylindrical and the angle of contact to be zero, so that the meniscus is hemispherical and by applying the Kelvin equation, it is possible to calculate the minimum radius of pores in which capillary condensation can take place from the relative pressure at the lower limit of the hysteresis loop. Based on these calculations it has been observed that the minimum radius for the onset of capillary condensation varies from system to system but is rarely below 1 nm. The upper limit of the applicability of the Kelvin equation $\gamma = 25$ nm is a practical one given by the experimental difficulty of measuring very small lowering of vapor pressure.

2.4.6.1 Evidence in Support of the Capillary Condensation Theory

The capillary condensation theory is in conflict with the multiplayer theory of adsorption but receives support from certain experimental results.

Existence of Adsorption-Desorption Hysteresis
It is now firmly believed that adsorption-desorption hysteresis is a direct consequence of capillary condensation in pores of the adsorbent. In fact, Zsigmondy put forward the capillary condensation theory to explain the phenomenon of hysteresis. It has been found that adsorption and desorption curves in certain cases do not meet each other along the whole course of the isotherm so that under a certain pressure the equilibrium amount adsorbed is higher when this pressure is approached from the side of higher pressure (i.e., during desorptio) than when it is approached from the side of lower pressures (i.e., during adsorption). The irreproducibility of this adsorption curve

TABLE 2.6
Pore Volume (MI/G) of Several Polymer Carbons Calculated
from Adsorption at Relative Vapor Pressure (0.98)
(Gurvitsch Rule)

Sample Identification	Water	Methanol	Ethanol	Benzene
PVDC-600	0.34	0.35	0.34	0.33
Saran-600	0.29	0.29	0.29	0.33
Saran-600 (Steam Act. 850°C)	0.46	0.45	0.48	0.55
PF-600	0.14	0.15	0.14	0.10
UF-650	0.10	0.03	0.06	—
PVC-850 (N_2)	0.03	0.03	0.01	—

in this range of pressures is due to the fact that the filling of the mesopores involves a different mechanism from their emptying. Several theories have been proposed to explain the existence of the adsorption-desorption hysteresis, but all of them, in one way or the other, are based on the capillary condensation theory. The mechanism of hysteresis will be briefly discussed in a later section.

Gurvitsch Rule

The strongest evidence in favor of the theory of capillary condensation is the widespread conformity to the Gurvitsch rule, which states that the total amount adsorbed when expressed as a volume of liquid (assuming normal liquid density) at or near the saturation vapor pressure is usually, though not invariably, nearly the same for all vapors on a given adsorbent (Table 2.6). This is what could be expected when the filling of all the pores by the adsorbate takes place through condensation into the liquid form.

2.4.7 Applicability of Langmuir, Freundlich or Temkin Isotherms to Adsorption Data

2.4.7.1 Linearity of the Plot

The three isotherms can be written in the linear form as

$$\frac{P}{V} = \frac{1}{KV_m} + \frac{P}{V_m} \quad \text{Langmuir}$$

$$\log \frac{V}{V_m} = \log K + \frac{1}{n} \log P \quad \text{Freundlich}$$

$$\frac{V}{V_m} = \frac{RT}{q_o \alpha} \log A_o P \quad \text{Temkin}$$

Thus, the Langmuir isotherm is obeyed when p against p/V gives a linear plot, Freundlich isotherm when a $\log p$ against $\log V$ is linear, whereas Temkin isotherm gives a linear plot with $\log p$ against V plot.

2.4.7.2   Variation of Heat of Adsorption (q) with Surface Coverage (θ)

The variation of q with surface coverage is different for the three isotherms. The q is independent of θ in Langmuir isotherm, varies logarithmically with θ in Freundlich isotherm, and decreases linearly with θ in case of the Temkin isotherm.

2.4.7.3   Appropriate Range θ

The Langmuir and Freundlich isotherms conform to a wide θ range, and the θ range for Temkin isotherm is usually 20 to 50%.

In addition, the Freundlich isotherm gives a family of $\log p$ against $\log V$ curves at different temperatures, which, when extrapolated, converge on one point. The amount adsorbed at this convergent point is the monolayer capacity.

2.4.8   Adsorption Hysteresis

The term *hysteresis* in adsorption refers to the lag between the adsorption and desorption isotherms i.e., under a certain pressure), the equilibrium amount adsorbed is higher when this pressure is approached from the side of higher pressures than when it is approached from the side of lower pressures (*cf.* Figure 2.23). It was first observed in 1896 by Van Bemmelen[58] with silica-gel water system. Since then it has been the subject matter of numerous investigations, with the primary purpose of formulating a mechanism to account for the phenomenon and as an explanation of the multiplayer adsorption theory. These workers explained hysteresis on the assumption that a liquid in a capillary tube has a greater vapor pressure when being filled than when being emptied, because incomplete wetting causes the curvature of the liquid meniscus to be less during the filling process than during the emptying process.

Patrick and McGavack,[59] attributed hysteresis in the case of SO_2-silica gel system at 25°C to some experimental error. According to these workers, the air held in the adsorbent pores is displaced by the adsorbed gas and thus increases the observed pressure in the apparatus. A similar explanation was given by Coolidge,[60] but in his later work on water charcoal systems,[61] he ascribed hysteresis to the presence of inorganic ash constituents that adsorbed water preferentially. Zsigmondy and coworkers[62,63] and Patrick[64] also attributed hysteresis to the presence of impurities, such as air and other permanent gases, in the capillary that resulted in an imperfect wetting of the capillary walls, as a result of which, the liquid meniscus has a greater difficulty in advancing along the capillary spaces and, consequently, the equilibrium vapor pressure required for a given amount of adsorption was higher on the adsorption branch of the isotherm. In desorption, however, the meniscus is already present from which the desorption can take place so that the desorption pressure is lower.

If these views are true, then one would expect to get hysteresis from one end of the isotherm to the other in the presence of air or other impurities and the complete elimination of hysteresis on the removal of such impurities. As a matter of fact, the bulk of the evidence in the literature[65-71] indicates that completely reproducible hysteresis with carbon-water and silica gel-water systems occurs only over a part of the isotherm in the complete absence of these impurities. By careful experiments on the adsorption of water vapor on activated charcoal, Allmand and coworkers[72,73] observed that the hysteresis was real and not merely due to false measured pressures caused by the accumulation in the vapor phase of the gases expelled from the charcoal. This reality of the hysteresis phenomenon was further established by several workers[68,69] from their extensive investigations on several porous adsorbents.

Kraemer[74] and McBain[75] explained hysteresis on the basis of pore geometry of the adsorbent. These workers considered the pores of the adsorbent as *ink-bottle* shaped, so that the vapor pressure during adsorption is determined by condensation in larger diameters of the bottle, while the pressure at which desorption occurs corresponds to the narrow neck. Rao[69] and Katz[76] found this hypothesis to furnish a qualitative explanation for their observations on adsorption-desorption hysteresis with several gels and sorbates. However, this concept could not explain the complete elimination of the hystereus effect without a marked fall in the adsorption capacity.

Foster[65,77] suggested *open pore* or *delayed meniscus* theory to explain the phenomenon of hysteresis. The pores are considered as regular, nonintersecting capillaries open at both ends. In its simplest form this theory holds that unless the pores are narrow to be bridged by a single molecular layer on the opposite walls, a meniscus will not be readily formed and so condensation cannot take place in accordance with the Kelvin equation. The adsorption thus takes place normally in layers on the opposite walls of the capillaries until the films are thick enough to bridge the pore at one point, and capillary condensation may take place. The adsorption branch of the isotherm is thus completed. Desorption must obviously take place from the meniscus already formed at the ends of the filled pores. As the radius of these spherical menisci is only half of the radius of the peripheral cylindrical meniscus, the condensate will desorb from the pores at a lower relative pressure than is essential for the filling of the pores by condensation.

The open-pore theory was elaborated and given quantitative treatment by Cohn,[78] who calculated thermodynamically the pressure required to fill open and cylindrical capillaries of radius r as a consequence of the deposition of successive layers of the liquid adsorbate (not by capillary condensation) as

$$\log \frac{P_a}{P_o} = \frac{rV}{rRT} \tag{2.113}$$

where V is the molar volume of the adsorbate, r is the radius of the annular ring, γ is the surface tension, and R and T are the gas constant and absolute temperature, respectively. When the capillary is wide and the adsorbed film is thin, then r may be considered equal to the radius of the capillary.

In desorption, however, since the meniscus is already present, the pressure at which evaporation (desorption) takes place must be given by the Kelvin equation as

$$\log \frac{P_d}{P_o} = -\frac{2\gamma V}{rRT} \cos \theta \tag{2.114}$$

where r is the radius of curvature of the meniscus and θ is the angle of contact between the liquid adsorbate and the surface of the adsorbent. When wetting is complete, $\cos\theta = 1$ and it is evident from Equation 2.113 and Equation 2.114 that the radius of curvature corresponding to adsorption is twice as large as the desorption radius, and that p_a and p_d are related by the equation

$$p_a^{\,2} = p_o\, p_d \qquad\qquad (2.115)$$

For narrow capillaries, the thickness of the adsorbed film, D, can not be neglected in comparison with r. Therefore, one must use $(r - D)$ in Equation 2.113 instead of r. Then Equation 2.113 becomes

$$\log_e \frac{p_a}{p_o} = -\frac{\gamma V}{(r - D)RT}$$

and the relationship between p_a and p_d becomes

$$\frac{p_d}{p_o} = \left(\frac{p_a}{p_o}\right)^{2a - \frac{2D}{R}} \qquad\qquad (2.116)$$

A comparison of Equation 2.115 and Equation 2.116 shows that p_d becomes equal to p_a when $\gamma = 2D$ (i.e., when the radius of the adsorbed layer is equal to twice the thickness of the adsorbed film). As the smallest possible value of D is the diameter of the adsorbate molecule, it means that hysterens cannot occur in pores narrower than four molecular diameters; i.e., in cylindrical pores, hysteresis starts when r becomes greater than $2D$.

Bansal,[71] studied the adsorption desorption isotherms of benzene, toluene and o-xylene on sugar charcoal associated with varying amounts of the carbon-oxygen surface groups and observed that the area of the hysteresis loop decreases as the molecular dimensions of the adsorbate increase from benzene to o-xylene (Table 2.7). The point of inception of the hysteresis loop was also found to shift to lower relative vapor pressures as we move up the series of hydrocarbons, which is due to an increase in the molecular size of the adsorbate. The point of inception of hysteresis loop was calculated using Cohn postulates and compared with the values read from the experimental curves (Table 2.8). It is seen that the two values agree closely for all the adsorbate-adsorbent systems. Higute,[67] from thermodynamic considerations, also proposed that the critical radius for the inception of capillary condensation is equal to four times the molecular radius of the adsorbate.

Pierce and Smith[79] observed hysteresis to occur in some adsorbents where there is evidence against the existence of capillaries. Consequently, these workers attempted to explain hysteresis loop observed in water isotherms on nonporous carbons like graphon as well as on porous carbons such as charcoal by postulating

TABLE 2.7
Variation of Area of Hysteresis Loop with the Molecular Dimensions of the Adsorbate

Description of the Sample	Area of Hysteresis Loop (Cm^2) in Case of:		
Sugar Charcoal Original	2570	2139	1671
Sugar Charcoal Evacuated at 750°	1450	1270	1042
Sugar Charcoal Evacuated at 1200°C	1200	900	540

Source: Bansal, R.C., M.Sc. Hons. dissertation, Panjab Univ., Chandigarh, 1960. With permission.

that adsorption first occurs as isolated clusters on certain sites. These clusters grow in size as more and more adsorption takes place and finally merge into one another to form a continuous layer. Further adsorption on porous carbons takes place by capillary condensation. The forces that were confined to certain sites, before the clusters merged to form a layer, now extend to a whole layer as a result of which desorption occurs at higher pressures. McDermot and Arnell[80] supported this *cluster theory* in their work on the adsorption of water vapor by charcoal before and after deoxygenating by treatment with hydrogen at 1000°C.

Puri et al.[70] determined the adsorption-desorption isotherms of water vapor on different samples of charcoals associated with different types of carbon-oxygen surface groups and observed that the presence of acidic surface groups, which are evolved as CO_2 on evacuation, make the charcoal surface hydrophilic and increase the size of the hysteresis loop, the increase being proportional to the amount of acidic surface groups present (Figure 2.27). This was attributed to the chemisorption

TABLE 2.8
Relative Vapor Pressure at the Inception of Hysteresis Loop in the Sorption-Desorption of Different Vapors

Adsorption	Molecular Thickness D (A°)	4 D (A°)	Relative Vapor Pressure Corresponding to 4D Cohn's Postulates	Relative Vapor Pressure at the Inception of Hysteresis
Benzene	5.301	21.204	0.3975	0.3980
Tolune	5.581	22.324	0.3554	0.3600
o-xylene	5.892	23.568	0.3027	0.3140

Source: Bansal, R.C., M.Sc. Hons. dissertation, Panjab Univ., Chandigarh, 1960. With permission.

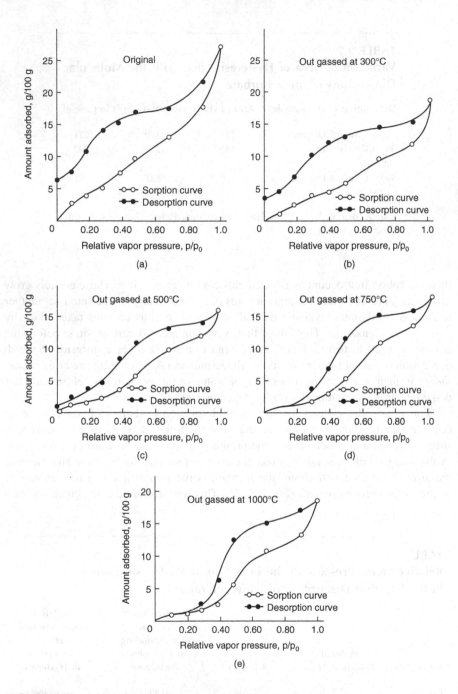

FIGURE 2.27 Adsorption-desorption hysteresis in adsorption of water vapor on sugar charcoal outgassed at different temperatures. (*Source:* Puri, B.R. and Myer, Y.P., *J. Sci. and Ind. Res.*, India, 16B, 52, 1957. With permission.)

of water to the extent of about one mole per mole of the acidic group on the charcoal surface. The chemisorption of water also prevented the desorption branch of the isotherm to meet the adsorption branch even at zero relative pressure. When these groups were removed completely from the charcoal surface, no chemisorption of water could be observed and the two branches in the isotherms met each other at relative pressures well above zero.

2.4.9 THEORY OF VOLUME FILLING OF MICROPORES (TVFM)

It is well established that all activated carbons contain a variety of pores such as micropores, transitional (or meso-) pores, and macropores, and that the adsorption takes place in micropores and on the surfaces of transitional and macropores. Because the specific surface area of macropores is extremely small (usually <2 m^2/g), their contribution to total adsorption value is very small so that the adsorption on the surface of macropores may be neglected. Activated carbons used for the adsorption of gases and vapors generally do not possess developed transitional porosity and the specific surface area of such pores in these carbons does not exceed 50 m^2/g.[55] However, even this value of transitional surface area forms only a small part of the total surface area of the carbon. Dubinin[45] calculated the adsorption values of benzene at 20°C in micropores and on the surface of transitional pores of two active carbons for a range of equilibrium concentrations. A comparison of the adsorption values in micropores and on the surface of transitional pores (Table 2.9) shows that the fraction of vapor adsorption in micropores a_{mie}/atotal, even when the volume of transitional pores in active carbons is highly developed, is atleast 0.85 for the upper limit of equilibrium pressures of practical interest. For active carbons with relatively undeveloped transitional porosity, adsorption on the surface

TABLE 2.9
Composition of Calculated Adsorption Values of Benzene at 20°C Micropores and on the Surface of Transitional Pores of Active Carbons

S_t, m^2/g	$B \cdot 10^4$	p/p,	a_{mi}, mmole/g	a_t, mmole/g	a, mmole/g	a_{mi}/a
50	0.40	1×10^{-4}	2.61	0.02	2.63	0.99
		1×10^{-1}	4.36	0.19	4.55	0.96
	1.00	1×10^{-4}	1.14	0.02	1.16	0.98
		1×10^{-1}	4.14	0.19	4.33	0.96
200	0.40	1×10^{-4}	2.61	0.09	2.70	0.97
		1×10^{-1}	4.36	0.75	5.11	0.85
	1.00	1×10^{-4}	1.14	0.09	1.23	0.93
		1×10^{-1}	4.14	0.75	4.89	0.85

Source: Dubinin, M.M., in *Chemistry and Physics of Carbon*, P.L. Walker, Jr., Ed., Marcel Dekker, New York, 2, 56, 1966. Reproduced with permission from Marcel Dekker.

of transitional pores may be neglected. Thus, it is important in the first instance, to evaluate the relative role of vapor adsorption in micropores in the total value of adsorption.

This has led to the development of the *theory of volume filling of micropores* (TVFM theory). The theory provides a satisfactory description of the shapes of adsorption isotherms where the adsorption takes place largely in micropores. It is based on the assumption that the characteristic adsorption equation expressing the degree of filling of micropores (i.e., the volume filling of micropores) is a function of the differential molar work of adsorption. The adsorption process in this case involves dispersion forces as the main component in the adsorption-adsorbent interactions.

The theory has been developed mainly by Dubinin and coworkers over a period of time[55,83–92] to describe adsorption of gases and vapors by microporus solids in general and active carbons in particular. Dubinin defined the molar work of adsorption as the change in the Gibbs free energy rather than as adsorption potential (Polanyi). Accordingly, the molar work of adsorption A is given by

$$A = -\Delta G = RT \ln \frac{p_o}{p} \tag{2.117}$$

where p_o is the saturation vapor pressure at temperature T and p is the equilibrium vapor pressure. The fundamental TVFM equation in its most general form can be represented.[93]

$$\theta_{mi} = f\left(\frac{A}{E}, n\right) \tag{2.118}$$

where E and n are the distribution function parameters of the system under investigation and θ_{mi} is the degree of filling of the adsorption volumes and is defined as

$$\theta_{mi} = \frac{W}{W_o}$$

where W represents the volume of the micropore filled at temperature T and at relative pressure p/p_o at the equilibrium state, and W_o is the total (limiting) volume of the micropores.

Based on the experimental data and the practical application TVFM equation, Equation 2.117 can be expressed as

$$W = W_o \exp\left[\left(-\frac{A}{E}\right)^n\right] \tag{2.119}$$

If E_o is the characteristic energy of adsorption of a standard vapor (which is usually benzene for active carbons), then the characteristic adsorption energy of another vapor or gas is given by:

$$E = \beta E_o$$

so that the TVFM equation can be expressed by the Dubinin-Astakhov (DA) equation[84,85]

$$W = W_o \exp.\left[\left(-\frac{A}{\beta E_o}\right)^n\right] \qquad (2.120)$$

where β is called the *effinity coefficient* and can be equated to the ratio of Sugdens parachors.[84] It can also be calculated from the molar volumes of the standard and other vapors. It has also been shown by Stoeckli and Morel[94] that for simple adsorbates β is proportional to the minimum of the adsorption energy of the corresponding adsorbate on a graphite surface. This is an indication of the physical meaning of the affinity coefficient, but at the present time the best estimate is still provided by adsorption experiments on a carbon showing equal accessibility to the given adsorbate and the reference vapor. The effinity coefficients and the molar volumes for some adsorbates[94] are given in Table 2.10.

The above DA (Dubinin-Astakhov) equation is applicable to the adsorption of a variety of organic and nonspecific vapors at relative pressures p/p_o less than 0.05 to 0.1, because in this region the influence of the nonmicroporous surface area S_e is negligible.

TABLE 2.10
Experimental Affinity Coefficients χ and Molar Volumes in the Liquid State at 293 K (77 K for N_2) of Typical Adsorptions Used in the Study of Microporous Carbons

Adsorbate	β	V_m (cm³/mol)
Benzene	1.00	88.91
Nitrogen	0.34	34.67
Methylene chloride	0.66	64.02
Cyclohexane	1.04	108.10
Carbon tetrachloride	1.05	96.50
Chlorobenzene	1.19	101.70
n-Hexane	1.29	130.52
n-Heptane	1.62	146.56
n-Hexadecane	4.05	292.57
2,5-Norbonadiene	1.62	101.63
α-Pinene	1.70	158.75
Perchlorocyclopentadiene	1.91	159.30

Source: Bansal, R.C. Donnet, J.B., and Stoeckli, F., in *Active Carbon*, Marcel Dekker, Inc. New York, 1988. Reproduced with permission from Marcel Dekker.

For typical activated carbons, the exponent n is equal to 2, which corresponds to the original empirical equation postulated by Dubinin and Radushkevich in 1947[83] known as the DR equation and represented as

$$W = W_o \exp. \left[-B\left(\frac{T}{\beta}\right)^2 \log^2\left(\frac{p_o}{p}\right) \right] \qquad (2.121)$$

In the modern formulation equation (Equation 2.120) becomes

$$W = W_o \exp. \left[\left(-\frac{A}{\beta E_o} \right)^2 \right] \qquad (2.122)$$

Parameter B in Equation 2.121, which has dimensions of K^{-2}, is called the *structural constant* of the given carbon and is related to the characteristic energy E_o of Equation 2.119 as

$$E_o \text{ (kJ/mol)} = 0.01914/(B)^{1/2}$$

and

$$B = \left[\frac{2.303R}{E_o (\text{kJ/mol})} \right] \qquad (2.123)$$

The DR and DA equations are based on the observation that a plot of W against $(RT)^2 \log^2\left(\frac{p_o}{p}\right)$ or A^n gives rise to a unique curve for a given adsorbate, called the *characteristic curve*.

The DR equation (Equation 2.121) is applicable more accurately to microporous activated carbons containing a narrow distribution of micropores. Strongly activated carbons that have a wider distribution of micropore volume as a function of the micropore size can be described more accurately by the superposition of two distributions[54,95,96] of micropores with effective radius r less than 0.6 to 0.7 nm and of supermicropores with radii of 0.7 to 1.6 nm[98] as

$$W_o = W_{o1} \exp. \left[-\left(\frac{E}{\beta E_{01}}\right)^2 \right] + W_{o2} \left[-\left(\frac{E}{\beta E_{02}}\right)^2 \right] \qquad (2.124)$$

so that $W_o = W_{o1} + W_{o2}$

The parameters W_o, the limiting micropore volume, and the characteristic energy E_{01} and E_{02} of Equation 2.125 and the parameters W_{o1}, and W_{o2}, the limiting micropore volumes of micropores and supermicropores, respectively, can be easily determined graphically from one adsorption isotherm with a wide range of equilibrium pressures.

2.4.9.1 Filling of Micropore Volume in Adsorption

The constants B and W_o of the DR equation (Equation 2.121) are structural parameters and depend upon the microporous structure of activated carbons. When B is sufficiently small, the micropore volume W becomes almost equal to the limiting micropore volume W_o. The constant B expresses that property of the microporous structure that determines the effect of the increase in adsorption energy in micropores. Evidently, it determines the micropore size parameters. The smaller the value of constant B, the steeper the rise of the isotherm in the lower region of equilibrium pressures.

Dubinin[45] wrote Equation 2.122 as

$$\frac{W}{W_o} = F \tag{2.125}$$

where

$$F = -\exp.\left[B\left(\frac{T}{B}\right)^2 \log^2\left(\frac{p_o}{p}\right) \right]$$

is the filling factor expressing the fraction of the filled limiting pore volume under given experimental conditions such as the nature of the vapor, the relative vapor pressure p/p_o and temperature T. Dubinin calculated filling factors for active carbons with different microporous structures for different adsorbates and different relative pressure ranges. These results (Table 2.11) indicate that for well-adsorbed vapors for which β is relatively high, the filling factor F is close to unity at sufficiently

TABLE 2.11
Filling Factors of Micropore Volume

		F		
$B \cdot 10^6$	β	$h = 1 \cdot 10^{-4}$	$h = 5 \cdot 10^{-2}$	$h = 1 \cdot 10^{-1}$
0.40	0.50	0.111	0.483	0.871
0.80		0.012	0.233	0.759
1.20		0.0014	0.113	0.602
0.40	1.00	0.577	0.834	0.966
0.80		0.333	0.695	0.933
1.20		0.193	0.582	0.912
0.40	1.50	0.783	0.923	0.985
0.80		0.614	0.851	0.971
1.20		0.481	0.785	0.955

Source: Dubinin, M.M., in *Chemistry and Physics of Carbon*, P.L. Walker, Jr., Ed., Marcel Dekker, New York, 2, 56, 1966. Reproduced with permission from Marcel Dekker.

high equilibrium relative pressures irrespective of the values of the constant B. In such cases the adsorption properties of active carbons are essentially determined by the development of micropore volume, their dimensions being of secondary importance.

For relatively poorly adsorbed vapors and at small equilibrium relative pressures, the values of the filling factor F are low, differing by orders of magnitude. These values strongly depend on the value of constant B so that in such cases the value of B is the determining factor for adsorption of these vapors, the development of micropore volume playing a secondary role.

Micropores in activated carbons are generally heterogenous. This heterogeneity is introduced by the heterogeneities in the source raw material and the heterogenous nature of the activation process. The exponent n of the DA equation reflects the degree of heterogeneity of the micropore system. For activated carbons n varies practically from 1.5 to 3. According to Dubinin[85] and Finger and Bullow[99] the case $n = 3$ applies to the adsorption in Saran-based charcoals that have relatively a narrow pore size distribution of 0.4 to 0.5 nm in width. More recently, it has been shown that $n = 3$ also applies to molecular sieve carbons with relatively large but homogenous micropores near 0.7 nm.[100] On the other hand, strongly activated and heterogenous carbons may lead to a value of n as low as 1.5.[101]

The micropore distribution must become more homogenous as n increases from 1.5 to 3, the typical range observed in the case of activated carbons. Furthermore, it has been confirmed from experimental data that for $n = 2$, the microporous system presents some degree of heterogeneity and is not as homogenous as assumed earlier.[101] Figure 2.28 shows micropore size distribution in the case of a carbon that was activated to 18% burn-off in CO_2 at 850°C and for which $E_o = 33$ kJ/mole.[95]

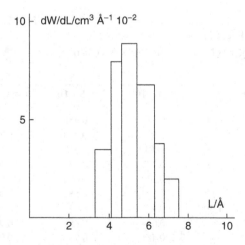

FIGURE 2.28 The micropore distribution of carbon CEP-18 for which $n = 2$, $E = 33$ KJ/mol, and $w = 0.251$ cm/g. (*Source:* Krachenbuchl, F., Stoeckli, H.F., Addoun, A., Ehrburger, F., and Donnet, J.B., *Carbon*, 24, 483, 1986. Reproduced with permission from Elsevier.)

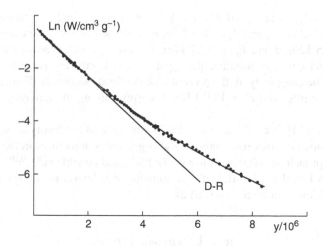

FIGURE 2.29 Adsorption of vapors by a heterogeneous carbon sample F-02 with $n = 1.5$. (Source: Dubinin, M.M; Zaverina, D.E. and Radushkevich, L.V., *Zh. Fiz. Khim.*, 1351, 1947. With permission.)

Although the *DR* plot was linear, about 70% of the pores are found within the 0.4 to 0.6 nm range. For typical activated carbons, as the degree of burn-off increases and the characteristic energy E_o decreases from 35 to 30 kJ/mol to approximately 20 kJ/mol, the classical DR approximation with $n = 2$ holds with a variable degree of approximation.

Deviation in linearity begin to appear at low degrees of micropore filling (Figure 2.29) and their detection depending essentially on the range covered by the adsorption experiment. In such a case, the best fit to the Dubinin-Astakhov equation (Equation 2.121) leads to a value of n smaller than 2, a clear indication of relatively strong heterogeneity in the micropore size distribution. It may be worthwhile to mention that this type of assessment requires accurate adsorption data over a wide range of temperatures and relative vapor pressures. The deviations typically appear in the region of vapor pressure greater than 4 to 5×10^6 that corresponds to low degrees of micropore filling, whereas the classical DR approximation is still valid to a good approximation in the range covered by normal experimental conditions; for benzene at 298 K, the vapor pressure is smaller than 2×10^6). It is equally difficult to get accurate values of $n > 2$ for homogenous carbons and, therefore, the DR equation remains the standard approximation for a variety of applications as well as leading to average values and underlining the importance of techniques based on molecular sieve experiments.[95]

Thus, it is evident that a strong heterogeneity in the micropore distribution is reflected by the deviation from linearity in the classical DR plot and by the value of the exponent n smaller than 2. Such micropore systems may be described either by the DA equation with a suitable exponent n less than 2 or by a combination of DR equations. Izotova and Dubinin,[55] who suggested that the nonlinearity of the graph results from the superposition of two extreme ranges of microporosity and the limiting (total) micropore volume can be obtained using Equation 2.125. Dubinin and coworkers[86,102] calculated W_{o1} and W_{o2} for a carbon F-02 and found the values to be $W_{01} = 0.178$ cm³/g. and $W_{o2} = 9.462$ cm³/g, with the values of characteristic energies as $E_{01} = 25.5$ kJ/mol and $E_{02} = 17.7$ kJ/mol. Although the existence of two distinct classes of micropores is questionable, Equation 2.124 provides useful information on the extent of heterogeneity. In the present case the linear section of (Figure 2.29) leads to $W_o = 0.62$ cm³/g. and $E_o = 18.9$ kJ/mol, corresponding to an average pore width of 1.4 nm.

Stoeckli and Huber[101,103] initiated the development of the theory of volume filling for heterogenous microporous carbons and suggested a refinement in the Izotova and Dubinins approach in collaboration with Dubinin and coworkers.[85–92,104,105] Skoecklis refinement is based on the idea of a continuous distribution of the micropores as a function of B and can be represented as

$$W_t = \int_o^\infty f(B) \, \exp. \, (-By)dB \tag{2.126}$$

As a first approximation and in view of the limitations introduced by the experimental errors on the overall isotherm, these workers postulated a normalized gaussian of half width Δ, for the distribution $dW/dB = f(B)$

$$f(B) = \left(\frac{W_o}{\Delta\sqrt{2\pi}} \right) \exp\left[-\left(\frac{B_o - B}{2\Delta} \right)^2 \right] \tag{2.127}$$

This analytical form is also convenient for solving the integral transform (Equation 2.126), which leads to

$$W_t = W_o \, \exp \, (B_o y) \, \exp\left(\frac{y^2\Delta^2}{2} \right) x0.5[1 - erf(Z)] \tag{2.128}$$

where

$$y = \left(\frac{T}{\beta} \log \frac{P_o}{P} \right)^2 \tag{2.129}$$

and $erf \, (Z)$ is the tabulated error function and is given by

$$Z = \frac{\left(y - \dfrac{B_o}{\Delta} \right)\Delta}{\sqrt{2}} \tag{2.130}$$

TABLE 2.12
Comparison of Parameters Obtained for Various Active Carbons by Fitting the Overall Adsorption Isotherms of Equation 2.121, Equation 2.122, and Equation 2.129

Carbon	Equation 2.27			Equation 2.13	Equation 2.12
	W_0 (cm³/g)	B_0 ($10^{-6} K^2$)	$\Delta(10^{-6}K^2)$	E_0 (kJ/mol)	n
U-02	0.43	0.92	0.21	20.0	1.65
F-02	0.64	1.03	0.29	18.7	1.35
F-6	0.78	1.00	0.36	19.1	1.28
T	0.40	0.61	0	24.5	2.0
CAL	0.44	0.99	0.26	19.2	1.63
CEP-59	0.48	0.66	0.18	25.0	1.67

Source: Dubinin, M.M. and Stoeckli, H.F., *J. Colloid Interface Sci.*, 75, 34, 1980. Reproduced with permission from Elsevier.

Δ in Equation 2.127 expresses heterogeneity, which is a measure of the spread of B around B_o, W_t is the limiting adsorption that corresponds to complete filling of the entire volume of the heterogenuous micropores, and B_o is the structural parameter at which the micropore distribution function $f(B)$ reaches its maximum.

The values of different parameters calculated from DA equation (Equation 2.120), DR equation (Equation 2.121) and Stoeckli-Huber equation (Equation 2.128) obtained from the best fit of the adsorption data are given in Table 2.12.

It is seen that as a general rule any activated carbon with a characteristic energy E_o less than 20 to 22 kJ/mol and with a large micropore volume is potentially heterogeneous and further information require extended adsorption data. Table 2.12 also indicates that an increase in the degree of heterogeneity results in a corresponding decrease in the value of the exponent n of the DA equation. Dubinin and Stoeckli[86] also showed that a linear relationship exists between n and Δ. This correlation, suggested by mathematical modelling based on Equation 2.120 and Equation 2.128, is illustrated in Figure 2.30. The corresponding empirical relationship is

$$N = 2.00 - 1.78 \times 10^6 . \Delta \tag{2.131}$$

which indicates that an approximate micropore distribution can be calculated directly from the DA equation (Equation 2.120).

Dubinin et al and Stoeckli[86,90,91,105] derived another equation that directly relates the micropore distribution to the adsorption data. This equation.

$$\frac{W_o}{2\sqrt{1+m\delta^2 A^2}} \exp . \left(\frac{-mx^2 o A^2}{1+2m\delta^2 A^2} \right) \left\{ 1 + \exp . \left[\frac{x_o}{-\sqrt{1+2\,m\delta^2 A^2}} \right] \right\} \tag{2.132}$$

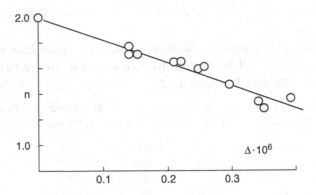

FIGURE 2.30 Empirical relationship between exponent n (Equation 2.120) and parameter D (Equation 2.128). (Source: Dubinin, M.M. and Stoeckli, H.F., *J. Colloid Interface Sci.*, 75, 34, 1980. With permission.)

has been called the *Dubinin-Stoeckli equation*, in which $A = -RT \, l_n \, p_o/p$ and $m = (\frac{1}{\beta k_o})^2$. For k_o, an average value of 12 kJ/mol for typical heterogenous carbons with characteristic energy of about 20 kJ/mol has been suggested. For practical purposes Equation 2.132 can also be expressed in terms of the accessible pore width $L = 2x$ rather than x. Figure 2.31 illustrates the micropore distribution resulting from the two approaches (Equations 2.128 and 2.132).

The agreement between the two approaches is good, but a major advantage of Equation 2.133 over equation Equation 2.128 is the fact that it directly relates the adsorption isotherm to W_o, x_o, and δ, a set of physically relevant quantities.

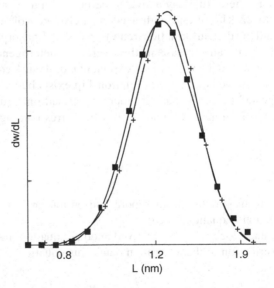

FIGURE 2.31 Micropore distribution of carbon AU-11 obtained from best fit of the adsorption data using Equation 2.128 and Equation 2.132. (Adapted from Dubinin, M.M., Effremov, S.N., Kataeva, L.I., and Ustinov, E.A., *Izv. Akad. Nauk SSSR, Ser. Khim.*, 255, 1, 1985).

REFERENCES

1. Bansal, R.C., Vastola, F.J., and Walker, P.L., Jr., Colloid and Interface, *Colloid and Interface Sci.*, 32, 187, 1, 1970.
2. Bansal, R.C., Vastola, F.J., and Walker, P.L., Jr., *Carbon,* 8, 113, 1, 1970.
3. Bansal, R.C., Vastola, F.J., and Walker, P.L., Jr., *Carbon,* 9, 185, 1, 1971.
4. Langmuir, I., *J. Am. Chem. Soc.,* 40, 1361, 1, 1918.
5. Brunauer, S., in *The Adsorption of Gases and Vapors,* Princeton Univ. Press, Princeton, 1943.
6. Young, D.M. and Crowell, A.D., in *Physical Adsorption of Gases,* Butterworth, London, 1962.
7. Dacey, J.R. and Thomas, D.G., *Trans. Faraday Soc.,* 50, 740, 1, 1954.
8. Culver, R.E. and Heath, N.S., *Trans. Faraday Soc.,* 51, 1569, 1575, 1, 1955.
9. Bansal, R.C., Goyal, M., Aggarwal, D., and Kaistha, B.C., *Indian J. Chem. Technol.,* 9, 290, 2002.
10. Markham, E.C. and Benton, A.F., *J. Am. Chem. Soc.,* 53, 497, 1, 1935.
11. Titoff, A., *Z. Phys. Chem.,* 74, 650, 1, 1910.
12. Brunauer, S., Emmett., P.H., and Teller, A., *J. Am. Chem. Soc.,* 60, 309, 1918.
13. Bering, B.P., Dubinin, N.N., and Serpinski V.V., *J. Colloid Interface Sci.,* 39, 185, 1, 1972.
14. Dubinin, M.M., *Izv. Akad. Nauk SSSR, Ser. Khim.,* 9, 1, 1981.
15. Dubinin, M.M., Efremov, S.N., Kataeva, L.J., and Vlin, V.I., *Izv. Akad. Nauk SSSR, Ser. Khim.,* 1711, 1, 1981.
16. Choma, J., *Polish J. Chem.,* 57, 507, 1983.
17. Jankowska, H., Swiatkowsky. A., and Choma, J., *Active Carbon,* Ellis Horwood, LTD, 1991.
18. Emmett, P.H., in *Catalysis,* P.H. Emmett, ed., Reinhold NY, 1954, p. 31.
19. Sing, K.S.,W., Everett, D.H., Haul, R.A.W., Mosen, L., Pierotti, R.A., Rouquerol, J., and Siemieniewska, T., *Pure Appl. Chem.,* 57, 603, 1, 1985.
20. Parra, J.B., de Sousa, J.C., Bansal, R.C., Pis, J.J., and Pajares, J.J., *Adsorption Sci. and Technol.,* 12, 51, 1, 1995.
21. Pierce, C., Wiley, J.W., and Smith, R.N., *J. Phys. Chem.,* 53, 669, 1, 1949.
22. Emmett, P.H. and McIver, D.S., *J. Am. Chem. Soc.,* 60, 824, 1, 1954.
23. Dollimore, D., *Thermochim. Acta,* 38, 15, 1, 1980.
24. Fernendez-Colinas, J., Grillet, Y., Rouquerol, F., and Rouquerol, J., *Langmuir,* 5, 1205, 1, 1989.
25. Keli, T., Takagi, T., and Kanataka, S., *Anal. Chem.,* 33, 1965, 1, 1961.
26. Emmett, P.H. and Brunauer, S., *J. Am. Chem. Soc.,* 59, 1553, 1, 1937.
27. Brunauer, S. and Emmett, P.H., *J. Am. Chem. Soc.,* 57, 1754, 1, 1935.
28. Young, D.M. and Crowell, A.D., in *Physical Adsorption of Gases,* Butterworths, London, 1962, p. 198.
29. Brennen, D., Graham, M., and Hayes, F.H., *Nature,* 199, 1152, 1, 1963.
30. Sing, K.S.W., *Chemistry and Industry,* p. 321, 1964.
31. Drain, L.E. and Morrison, J.L., *Trans. Faraday Soc.,* 48, 840, 1, 1952.
32. Drain, L.E. and Morrison, J.L., *Trans. Faraday Soc.,* 49, 650, 1, 1953.
33. Gregg, S.J. and Sing, K.S.W., in *Adsorption, Surface Area, and Porosity,* Acad. Press, London, 1982.
34. Bansal, R.C., Dhami, T.L., and Prakash, S., *Carbon,* 16, 389, 1, 1978.
35. Kiselev, A.V., *Quart Rev. Chem. Soc.,* 15, 116, 1, 1961.
36. Gammage, R.B. and Gregg, S., *J. Colloid Interface Sci.,* 38, 118, 1, 1972.

37. Zsigmondy A., Z. Anorg. Chem., 71, 71, 356, 1, 1911.
38. Thomson, W.T., Phil. Mag., 42, 448, 1, 1871.
39. Gregg, S.J. and Langford, J.P., J. Chem. Soc. Faraday Trans., 73, 747, 1, 1977.
40. Kiselev, A.V., in The Structure and Properties of Porous Materials, P.H. Emmett and F.S. Stone, eds., Butterworths, London, 1958, p. 195.
41. Carman, P.C. and Raal, F.A., Proc. Royal Soc., 209A, 59, 1, 1951.
42. Zweitring P., Proc. Intern. Symp., Reactivity of Solids, 1956, p. 111.
43. Polanyi, M., Vehr. Deutsch. Phys. Ges., 16, 16, 1012, 1914.
44. Polanyi M., Vehr. Deutsch. Phys. Ges., 18, 55, 1916.
45. Dubinin, M.M., in Chemistry and Physics of Carbon, P.L. Walker, Jr., ed., Marcel Dekker, New York, 2, 56, 1, 1966.
46. Dubinin, M.M. and Chmutov. K., Fiziko Khimicheskie Protivogazovovo dela, Physico-Chemical Fundamentals of Mark Fittering, Voy. Akad. Khim. Zashch, Moscow, 1939.
47. Zaverina, E.D. and Dubinin, M.M., Zhhur. Fiz. Khim., 13, 151, 1, 1939.
48. Dubinin, M.M. and Timofeev. D.P., Dokl. Akad. Nauk SSSR, 54, 1, 1946.
49. Dubinin, M.M. and Timofeev, D.P., Zhur. Fiz. Khim., 22, 133, 1, 1948.
50. Dubinin, M.M. and Radushkevich, L.V., Dokl. Akad. Nauk SSSR, 53, 331, 1, 1947.
51. Smisek, M. and Cerny, S., in Active Carbon, Elsevier Pub. Co., 1970, p. 119.
52. Dubinin, M.M., Zhur Zh. Fiz. Khim., 39, 1305, 1, 1965.
53. Dubinin, M.M., Zaverina, E.D., and Timofeev, D.P., Izv. Akad., Nauk SSSR, Otd. Khim. Nauk., 1957, p. 670.
54. Dubinin, M.M. and Zhukovaskaya, E.G., Izv. Akad. Nauk SSSR, Otd. Khim. Nauk. 1558, p. 535.
55. Izotova, T.I. and Dubinin, M.M., Zh. Fiz. Khim., 39, 2796, 1, 1965.
56. Kaganer, M.G., Zh. Fiz. Khim., 33, 2202, 1, 1959.
57. Lamond, T.G. and Marsh. H., Carbon, 1, 281, 1, 1964.
58. Bemmelen, V., Z. Anorg. Allgam. Chem., 13, 234, 1, 1985.
59. Patrick, W.A. and Mc Gavack Jr., J. Am. Chem. Soc., 42, 946, 1, 1920.
60. Coolidge, A.S., J. Am. Chem. Soc., 46, 596, 1, 1924.
61. Coolidge, A.S., J. Am. Chem. Soc., 49, 708, 1, 1927.
62. Zsigmondy, R., Z. Anorg. Chem., 71, 356, 1, 1911.
63. Zsigmondy, R., Bachman, W. and Stevenson, E.J., Z. Anorg. Chem., 75, 189, 1, 1912.
64. Patrick, W.A., Colloid Symposium Monograph., 7, 129, 1, 1930.
65. Foster, A.G., Proc. Roy. Soc., A 146, 129, 1, 1934.
66. Foster, A.G., Trans. Faraday Soc., 32, 473, 1, 1936.
67. Higuti, I., Bull. Inst. Phys. Chem. Res., Tokyo, 20, 130, 1, 1941.
68. Lambart, B. and Foster, A.G., Proc. Roy. Soc., A136, 367, 1, 1932.
69. Rao, K.S., J. Phys. Chem., 45, 506, 1, 1941.
70. Puri, B.R. and Myer, Y. P., J. Sci. and Ind. Res., India, 16B, 52, 1957.
71. Bansal, R.C., M.Sc., Hons. dissertation, Panjab Univ., Chandigarh, 1960.
72. Allmand, A.J., Rand, P.G.T., Mauning, J.E., and Shiels, D.C., J. Phys. Chem., 33, 1682, 1, 1929.
73. Allmand, A.J., Rand, P.G.T., and Mauning, J.E., J. Phys. Chem., 33, 1694, 1, 1919.
74. Kraemer, E.O., A Treatise on Physical Chemistry, S.S. Taylor, ed., D. Von Nostrand, New York, Chapter XX, p. 1661, 1, 1931.
75. McBain, J.W. and Britton, G.T., J. Am. Chem. Soc., 52, 2198, 1, 1930.
76. Katz, S.M., J. Phys. Chem., 53, 1166, 1, 1949.
77. Foster, A.G., Trans. Faraday Soc., 28, 645, 1, 1932.
78. Cohn, L.H., J. Am. Chem. Soc., 60, 433, 1, 1938.

79. Pierce, G. and Smith, R.N., *J. Phys. Chem.*, 54, 784, 1, 1950.
80. McDermot, H.L. and Arnell, J.C., *J. Phys. Chem.*, 58, 492, 1, 1954.
81. Bansal, R.C., Aggarwal, D., Goyal, M. and Kaistha, B.C., *Indian J. Chem. Technol.*, 9, 290, 2002.
82. Goyal, M., Rattan, V.K., and Bansal, R.C., *Indian J. Chem. Technol.*, 6, 305, 1, 1999.
83. Dubinin, M.M., Zaverina, D.E. and Radushkevich, L.V., *Zh. Fiz. Khim.*, 1351, 1, 1947.
84. Dubinin, M.M. and Astakov, V.A., *Izv. Akad. Nauk SSSR, Ser. Khim.*, p. 5, 1971.
85. Dubinin, M.M., in *Progress in Surface and Membrane Science*, D.A. Cadenhead, ed., Academic Press, New York, 1975, Vol. 9.
86. Dubinin, M.M. and Stoeckli, H.F., *J. Colloid Interface Sci.*, 75, 341, 1980.
87. Dubinin, M.M., *Carbon*, 19, 321, 1, 1981.
88. Dubinin, M.M., *Carbon*, 21, 359, 1, 1983.
89. Dubinin, M.M., *Izv. Akad. Nauk SSSR, Ser. Khim.*, 1983, 487.
90. Dubinin, M.M., *Carbon*, 23, 373, 1, 1985.
91. Dubinin, M.M. and Polyakov, N.S., *Izv. Akad. Nauk SSSR, Ser. Khim.*, 1943, 1, 1985.
92. Dubinin, M.M., *Carbon*, 20, 195, 1, 1982.
93. Jankowska, H., Swiatkowska, A., and Chema, J., in *Active Carbon*, Ellis Horwood, New York, 1991, p. 141.
94. Stoeckli, H.F. and Moreli, D., *Chimia*, 34, 502, 1, 1980.
95. Bansal, R.C. Donnet, J.B., and Stoeckli, F., in *Active Carbon*, Marcel Dekker, New York, 1988.
96. Dubinin, M.M., Polstyanov, E.F., *Izv. Akad. Nauk SSSR, Ser. Khim.*, 1691, 1, 1979.
97. Dubinin, M.M., Polstyanov, E.F., *Izv. Akad. Nauk SSSR, Ser. Khim.*, 793, 1, 1966.
98. Dubinin, M.M., *Izv. Akad. Nauk SSSR, Ser. Khim.*, 996, 1, 1974.
99. Finger, G. and Bullow, M., *Carbon*, 17, 87, 1, 1979.
100. Krachenbuchl, F., Stoeckli, H.F., Addoun, A., Ehrburger, F., and Donnet, J.B., *Carbon*, 24, 483, 1, 1986.
101. Huber, U., Stoeckli, H.F., and Houriet, J.P., *J. Colloid Interface Sci.*, 67, 195, 1, 1978.
102. Dubinin, M.M., *Izv. Akad. Nauk SSSR, Ser. Khim.*, 18, 1, 1980.
103. Stoeckli, H.F., *J. Colloid Interface Sci.*, 59, 184, 1, 1977.
104. Dubinin, M.M., Polyakev, N.S., and Ustinov, E.A., *Izv. Akad. Nauk SSSR, Ser. Khim.*, 2688, 1, 1985.
105. Dubinin, M.M., Effremov, S.N., Kataeva, L.I., and Ustinov, E.A., *Izv. Akad. Nauk SSSR, Ser. Khim.*, 255, 1, 1985.

3 Activated Carbon Adsorption from Solutions

Adsorption from liquid phase can take place at any of the three interfaces: liquid-solid, liquid-liquid, or liquid-vapor. In practice, however, more attention has been directed and more is known about the liquid-solid interface. This is due to the fact that purification of liquids such as water, wine, and oils, and their decolorization and detoxification have been carried out for centuries using charcoals and active carbons. With the expansion of chemical, pharmaceutical and food industries the range of substances to be purified by carbons has increased enormously.

The activated carbon adsorption from solutions was recognized with the early studies of Freundlich and Heller[1] on lower fatty acids and phenols, and of Kipling[2] on weak electrolytes. Most of these studies are aimed at determining the adsorption isotherms of the solutes, which are of importance for the determination of surface area and the micropore distribution in charcoals and active carbons. The measurements of adsorption can be carried out both simply and rapidly without the necessity for fabricating a special apparatus such as a vacuum system, which is so essential for measuring the adsorption of gases and vapors. It has been thought that the surface area can be determined using an adsorption isotherm equation, and the pore-size distribution could be determined by the molecular probe method, using molecules of different dimensions as solutes (adsorbates). This method has been used with a certain degree of success by Dubinin and Zaverina.[3] The application of the molecular probe method in the characterization of active carbons is discussed elsewhere in this book.

It may be worthwhile mentioning here that the term *specific surface area* has a definite meaning in the case of microporous carbons, because the size of the solute molecule has a profound influence on the measured surface area. Thus, if a carbon is to be used for a particular application, its surface area should be determined with a molecular probe of the appropriate size and shape.

The above procedures for measuring surface area and pore-size distribution appear to be very simple, but recent work has shown that the description of adsorption from solutions is over-simplified. There is need for a more critical approach if reliable conclusions are to be derived from solution adsorption experiments.

The adsorption isotherms from the data obtained from solution adsorption on solids, in general, and on active carbons, in particular, can be obtained by plotting the amount adsorbed as a function of change in concentration of the two components in the solution. In reality, this change in concentration is caused by the removal of

145

any one or both of the components of the solution. The isotherm should, therefore, represent not the absolute adsorption of the solute, although it should strictly represent *surface excess*. For example, the adsorption data can be fitted to the Freundlich equation

$$\frac{x}{m} = KC^{1/n} \tag{3.1}$$

where x is the weight of the adsorbate adsorbed by mg of the solid adsorbent, C is the concentration of the solution at equilibrium, and K and n are constants. The same Freundlich equation has been used for the adsorption of gases on solid surfaces by replacing concentration by pressure. The Freundlich adsorption equation has been extensively used to adsorption isotherms from solutions.

3.1 TYPES OF ISOTHERMS FOR ADSORPTION FROM SOLUTION PHASE

Two types of adsorption can be recognized when considering adsorption from binary solutions on solid surfaces: the *preferential* or *selective adsorption* and the *true adsorption*.

3.1.1 PREFERENTIAL ADSORPTION

This used to be called apparent adsorption and represents the surface excess, which is defined as the excess of solute in moles present in unit area of the solid-liquid interface over that present in a region of the bulk liquid containing the same number of moles of the solution.[4] In other words, it represents the extent by which the bulk liquid is impoverished with respect to one component, because the surface layer is correspondingly enriched.

3.1.2 ABSOLUTE ADSORPTION

Absolute adsorption (or simply *adsorption*) is equivalent to the actual quantity of that component present in the adsorbed phase, as opposed to its surface excess relative to the bulk liquid. Simply, it is the surface concentration.

It may be mentioned that preferential adsorption is directly related to the experimentally measured quantity and can be expressed directly and unambiguously. Thus, isotherms based on preferential adsorption can normally be used in practical applications. The determination of absolute adsorption, however, have led to a much better understanding of the adsorption process and especially of the meaning of preferential adsorption.

3.2 TYPES OF ADSORPTION ISOTHERMS

Adsorption isotherms on microcrystalline carbons, which include charcoals, carbon blacks, and active carbons, from solutions of solids and liquids have been found to be of the forms shown in Figure 3.1. The Freundlich isotherm equation could be applied to isotherm 1(c), which is for solutions of solids of limited solubility. The other three forms of isotherms show a maximum in each case that could not be

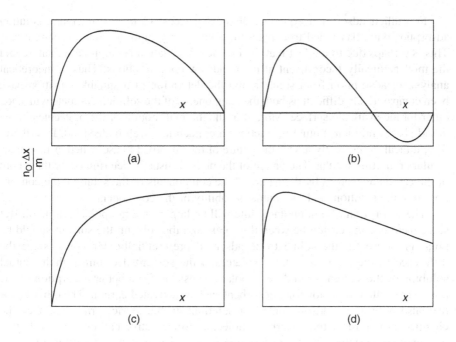

FIGURE 3.1 Types of adsorption isotherms: (a) and (b) for completely miscible liquids, (c) and (d) for solutions of solids.

explained by the Freundlich equation or any modification of the Langmuir equation. This is due to the fact that the experimental measurement in adsorption from solution is the change in concentration that is a measure of the extent of adsorption, determined by multiplying the change in concentration by the weight of the solution used. It is assumed in these calculations that the second component of the solution (i.e., the solvent) is not involved in the adsorption process. This criteria may be approximately valid for solutions of solutes with low solubility (i.e., for dilute solutions). However, for solutions of two components that are completely miscible (e.g., ethyl alcohol and water), neither can be regarded as a solvent or as a solute over the whole range of concentrations, so it is important to consider that each component of the solution can be adsorbed.

Because there are two types of adsorption, it is now possible to understand qualitatively the importance of isotherms of the type shown in Figure 3.1(a), Figure 3.1(b), and Figure 3.1(d). At lower concentrations over the first part of the isotherms, component one is adsorbed preferentially, meaning thereby that, at equilibrium, component one is present is greater proportions in the adsorbed layer than in the bulk liquid phase. Negative adsorption in Figure 3.1(b) indicates preferential adsorption of component two at higher concentrations. For completely miscible liquid components (e.g., ethyl alcohol-benzene or ethyl alcohol-water), the isotherm must fall to zero adsorption at each end of the concentration range because these points represent a pure component, or pure component two, and there is no change in concentration for pure liquids.

Freundlich adsorption equation, although successful to explain many solution adsorption data, has failed to explain the data at very high and low concentrations. This is perhaps due to the fact that the Freundlich equation is empirical in nature and thermodynamically inconsistent at high and low concentrations. Thus, a theoretical analysis of adsorption from solution and the derivation of a suitable equation have been comparatively difficult as both the components of the solution compete with each other for the available surface. Moreover, the thermal motion of the molecules in the liquid phase and their mutual interactions are much less well understood. It is, therefore, difficult to correctly assess the nature of the adsorbed phase, whether monomolecular or multimolecular. The nature of the phase is usually determined by the nature of the carbon as well as by the nature of the components of the solution, the concentration of the solution, and the mutual solubility of the components.

The adsorption of a nonpolar solute will be larger on a nonpolar adsorbent. But since there is competition between the solute and the solvent, the solvent should be polar in nature for the solute to be adsorbed preferentially. Moreover, since the difference in the polarities of the solute and the solvent determines their mutual solubilities, the solvent should be as polar as possible for a better adsorption of the nonpolar solute on the nonpolar adsorbent such as activated carbon. The other factor that also determines adsorption from solutions is the steric arrangement or the chemical structure of the adsorbate molecule. Since active carbons have a highly microporous structure, some of the micropores may not be accessible to larger molecules of the solute. Thus, the experimentally simple technique of adsorption from solutions can be developed into a method to characterize microporous carbons in terms of surface area, microporosity, oxygen content, and hydrophobicity of the carbon surface. The adsorption from solutions is getting further attention because of the growing importance of environmental control, involving purification of water by adsorption on activated carbons.

3.2.1 CLASSIFICATION OF ADSORPTION FROM SOLUTIONS

Adsorption from solutions can be classified into adsorption of solutes that have a limited solubility (i.e., from dilute solutions and adsorption of solutes that are completely miscible with the solvent in all proportions. In the former case, the adsorption of the solvent is of little consequence and is generally neglected. In the latter case, however, the adsorption of both the components of the solution plays its part and has to be considered. The importance of adsorption form solution can be judged by the fact that several monographs have been written on the subject.[4-6] Still, our knowledge of the various factors influencing adsorption from the solution phase is inadequate.

3.2.2 ADSORPTION FROM DILUTE SOLUTIONS

Adsorption of solutes on active carbons from aqueous and nonaqueous solutions can be carried out simply by placing a known weight of an active carbon in contact with different concentrations of the solute in stoppered or sealed Corning-glass test tubes. The contents are shaken in a mechanical shaker for a certain period of time, which may vary between a few minutes and several hours, depending upon the nature of the carbon surface and the nature of the solute. After the equilibrium is attained, the contents are allowed to stand, preferably overnight, for the carbon to settle down. The supernatant liquid is then centrifuged to settle down the last traces of fine particles of carbon and the clear liquid is analyzed for solute concentration using standard analytical procedures. When the solvent is a volatile organic liquid, care is taken that the solvent does not evaporate. It is advisable that a blank run is carried out for each concentration along with the experimental run under exactly similar conditions to obtain accurate results. It is also important to ensure that neither the solvent nor the solute contains any impurities that might be preferentially adsorbed by the carbon and may vitiate the results.

Most adsorption studies of adsorption from solutions lead to adsorption isotherms, which explains the influence of experimental conditions and the properties of the components of the solution on the adsorption process. Adsorption from solutions depends upon the properties of the organic and inorganic solutes such as their molecular mass, molecular size and geometry, their polarity and solubility, and the structure and nature of the carbon surface. The increase in adsorption capacity with increasing molecular mass of the solute, as in the case of benzene and phenol derivatives on activated carbons, results from a greater affinity of larger molecules to the carbon surface.[7-9] Abe et al.[10] carried out extensive studies on the influence of several different parameters, such as the molar volume, number of carbon atoms in the molecule, and solubility of the compound on the adsorbabilities of a large number of organic compounds, and observed a linear relationship between the logarithm of absorbability and many of these parameters. Adsorption from solutions is also influenced by the nature of the carbon surface (i.e., its surface area, porous and chemical structure, and its surface properties). In addition, the experimental conditions such as the pH and the ionic strength of the solution and the temperature of adsorption also determine the adsorption capacity and the nature of the adsorption process. Depending upon these parameters, the adsorption could be physical, chemical, or electrostatic. Puri et al.[11] studied the adsorption of hydroquinone from aqueous solutions on different carbon blacks and observed that the amount of physically adsorbed hydroquinone was a function of the specific surface area and did not depend on the surface properties of the adsorbent. However, the chemically adsorbed amount was found to depend on the nature of the carbon surface.

Lemeiur and Morrison[12] attempted to calculate surface area of carbons from adsorption of fatty acids from aqueous solutions but failed to find values comparable to BET values. In fact the values of surface area increased with increase in the molecular dimensions of the acid. In the case of carbon blacks, which are essentially nonporous, the surface areas agreed with the BET area. Adsorption from dilute solutions has also been found to be greatly influenced by the nature of the carbon surface. Puri[13] and Arora[14] studied the adsorption isotherms of stearic acid from its solutions in benzene and carbon tetrachloride on a number of active carbons, carbon blacks, and a sample of graphon, and found that the maximum amount of stearic acid adsorbed from the two solvents differed appreciably on the active carbons and carbon blacks but was of the similar values on graphon. Further, the amount of stearic acid adsorbed was less in the case of carbon blacks compared to that graphon, although the carbon blacks had much larger surface areas. This was attributed to the

TABLE 3.1
Maximum Surface Coverage of Various Carbons of Different Oxygen
Contents by Stearic Acid Adsorbed from Carbon Tetrachloride Solution

Carbon	Oxygen Content (g/100 g)	Nitrogen Surface Area (m²/g)	Maximum Surface Coverage (%)
Activated carbons			
A	0.92	1096	20
B	0.91	988	17
C	0.72	924	19
D	1.41	528	14
E	1.32	898	16
Carbon blacks			
Graphon	0	86	100
Mogul	7.8	308	23
Spheron-C	3.2	254	25
Spheron-4	4.1	153	30
Spheron-6	3.1	120	28
Spheron-6 outgassed at 600°	1.3	115	40
Spheron-6, outgassed at 1000°	0.5	100	74
Spheron-6, outgassed at 1400°	0	91	100
ELF-0	4.1	171	27
ELF-0, outgassed at 600°	2.1	189	54
ELF-0,outgassed 15 1000°	0.6	178	77
ELF-0, outgassed at 1400°	0	142	100

Source: Arora, V.M., Ph.D. dissertation, Panjab Univ., Chandigarh, India, 1977. With permission.

sensitivity of the adsorption of stearic acid to oxygen surface groups present on carbon blacks. It was suggested that the stearic acid molecules undergo dimerization in organic solvents and the dimer behaves as a nonpolar molecule. Therefore, its adsorption on polar oxygen group sites was inhibited. The adsorption of stearic acid increased on removal of the oxygen surface groups from the carbon black surface by evacuation at 600°C, 1000°C, and 1400°C (Table 3.1).

Adsorption of phenol and its derivatives from aqueous solutions on active carbons and carbon blacks has been the subject matter of a large number of investigations. Jaroniec and coworkers,[15–16] Enrique et al.[17] Worch and Zakke,[18] and Magne and Walker[19] studied the adsorption of several phenols from aqueous solutions and found that the adsorption was partly physical and partly chemical in character. Aytekin,[20] Chaplin,[21] and Kiselev and Krasilinkov[22] observed that the adsorption isotherms of phenol from aqueous solutions were step-wise, suggesting the possibility of rearrangement of phenol molecules in the adsorbed phase and their interaction with active sites on the carbon surface. Morris and Weber,[23] however, found that the adsorption isotherms of phenols on active carbons show two plateaus, even before the completion of the monolayer. The upper plateau was observed at three times the concentration of the lower plateau. Singer and Yen,[24] while studying the adsorption of phenol and alkyl substituted phenols on carbons, observed that the alkyl substituted phenols were more strongly adsorbed than phenol itself, and that the adsorption increased with the length of the chain, as well as with an increase in the number of substitutents on the phenol molecule. This has been attributed to the fact that the substitution of alkyl groups on the phenol molecule decreases its solubility by rendering it less polar.[25] These workers also found that the position of the alkyl group did not affect the extent of adsorption.

Abuzaid and Nakhla,[26] and Vidic et al.[27] found that the adsorption of phenol by activated carbons from aqueous solutions in the presence of molecular oxygen in the test environment resulted in a threefold increase in the adsorption capacity of the carbon. This has been attributed to the oxygen induced polymerization reactions on the surface of the carbon. Juang et al.[28] studied liquid-phase adsorption of eight phenolic compounds on a PAN-based activated carbon Fiber in the concentration range of 40 to 500 mg/L and observed that the chlorinated phenols showed better adsorption than methyl substituted phenols. Moreno-Castilla et al.[29] studied the adsorption of several phenols from aqueous solutions on activated carbons prepared from original and demineralized bituminous coal and found that the adsorption capacity depended upon the surface area and the porosity of the carbon, the solubility of the phenolic compound, and the hydrophobicity of the substituent. The adsorption was attributed to the electron donor-acceptor complexes formed between the basic sites on the surface of the carbon and the aromatic ring of the phenol.

The adsorption of phenol by active carbons has also been found to be influenced by the presence of carbon-oxygen surface groups on carbons. Redke et al.[30] and Hanmin and Yigun,[31] have reported that the adsorption of phenol increased on degassing of the carbon, which reduced the surface acidity without any appreciable change in the physical properties of the carbon. Graham[32] and Coughlin et al.[33] studied the adsorption at lower concentrations of phenol and reported negative influence of surface-oxide groups of carbon on the adsorption of phenol. However, Clauss et al.[34] while working at relatively higher concentrations, did not observe any such effect. The negative effect of the surface oxides was attributed to the depletion of π electron bond of graphite-like layers as a result of lowering of van der Walls forces of attraction. Puri et al.[35] while working at moderate concentrations, observed that the carbon-oxygen surface groups on carbons influence the adsorption of phenol. Mahajan et al.[36] and Keizo et al.[37] also observed that phenol uptake by porous carbons decreased sharply upon surface oxidation, and it increased when the chemisorbed oxygen was eliminated on heat treatment in nitrogen. Rachkovskaya et al.[38] however, observed that phenol adsorption increased considerably on activation of the carbon with CO_2 or water vapor. Bansal and coworkers,[39] Goyal,[40] and Aggarwal[41] also made a systematic study of the adsorption of phenol and p. nitrophenol from aqueous solutions on granulated and fibrous activated carbons having different surface areas and are associated with varying amounts of carbon-oxygen surface groups in the concentration range of 20 to 1000 mg/L. The adsorption of phenol and p. nitrophenol did not depend upon surface area alone but was also influenced by the presence of carbon-oxygen surface groups. The amount of the surface groups was enhanced by

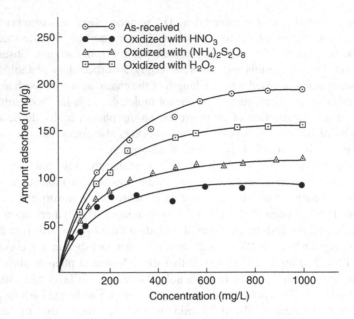

FIGURE 3.2 Adsorption isotherms of phenol on ACF-307 before and after oxidation. (After Bansal, R.C., Aggarwal, D., Goyal, M., and Kaistha, B.C., *Indian J. Chem. Technol.*, 9, 290, 2002. With permission.)

oxidation with nitric acid, ammonium persulphate, and hydrogen peroxide, and was decreased by degassing the activated carbons at gradually increasing temperatures of 400°C, 650°C, and 950°C. The oxidation on the carbons was found to decrease the adsorption of phenols, the extent of decrease depending upon the nature of the oxidative treatment (Figure 3.2). The adsorption of the phenols increased on degassing of the carbon samples, the extent of increase depending upon the nature of the carbon-oxygen surface groups being eliminated at that temperature of degassing (Figure 3.3). The results show[39] that the oxidation of carbons with nitric acid, which preferentially creates carboxylic or lactonic acidic groups (Table 3.2), suppress the adsorption of phenol. However, when these surface groups are removed by degassing, the adsorption of phenol tends to increase. The adsorption of phenol is maximum when the carbon surface is almost completely free of the acidic surface groups and when carbonyl groups evolved as CO on degassing are the dominating surface groups. These electrophillic quinone groups present on the carbon surface are electron donors and may be involved in a strong bonding with the π electrons of the benzene ring.

The adsorption of higher *n*-alkanols from their dilute solutions in *n*. heptane[42] exhibited a pronounced step indicating a strong cooperative adsorption mechanism leading to a closely-packed monolayer of alkanol molecules oriented with their chain axis oriented parallel to the graphite basal planes. This unusual adsorption behavior in dilute nonaqueous solutions was attributed to an order-disorder transition of the alkanols adsorbed in the pores at the liquid-carbon interface. Prakash,[43] while studying the adsorption isotherms of two cationic pesticides, diquat and paraquat from aqueous

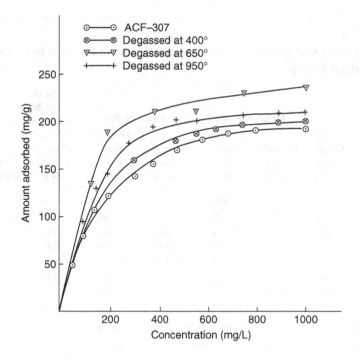

FIGURE 3.3 Adsorption isotherms of phenol on ACF-307 before and after degassing. (After Bansal, R.C., Aggarwal, D., Goyal, M., and Kaistha, B.C., *Indian J. Chem. Technol.*, 9, 290, 2002. With permission.)

TABLE 3.2
Amount of Oxygen Evolved on Degassing Different Oxidized Samples at 950°C

Sample	Oxygen Evolved (g/100 g) As			
	CO_2	CO ACF-307	H_2O	Total
As-received	1.00	5.30	1.30	7.60
Oxidised with-HNO_3	12.90	7.47	2.40	22.77
$(NH_4)_2S_2O_8$	5.40	7.51	4.91	17.82
H_2O_2	2.55	7.42	2.10	12.07
		ACF-310		
As-received	1.90	4.20	1.40	7.50
Oxidised with-HNO_3	11.96	7.20	2.20	21.36
$(NH_4)_2S_2O_8$	4.70	6.30	2.10	13.10
H_2O_2	2.1	6.20	2.10	10.40

Source: Bansal, R.C., Aggarwal, D., Goyal, M., and Kaistha, B.C., *Indian J. Chem. Technol.*, 9, 290, 2002. With permission.

TABLE 3.3
Adsorption of Methylene Blue and Metanil Yellow in Relation to Surface Area and Surface Acidity

Carbon	Total Surface Area (m²/g)	Methylene Blue Accessible Area (m²/g)	Methylene Blue Accessible Area (%) of Total Surface	Metanil Yellow Accessible Area (m²/g)	Metanil Yellow Accessible Area (%) of Total Surface	Ratio of Accessible Areas of Metanil Yellow and Methylene Blue	Acidity of Carbon Surface (mEq/100 m²)
Graphon	83.9	83.9	100	83.9	100	1.00	0
1	1120	757	68	721	64	0.95	0.12
2	1130	836	74	629	56	0.76	0.45
3	1300	912	70	880	68	0.97	0.06
4	1300	602	46	329	25	0.55	0.94
4	1300	950	73	721	55	0.76	0.80
5	600	430	72	323	54	0.75	0.70
6	580	374	64	294	51	0.79	0.60

Source: Graham, D., *J. Phys. Chem.*, 59, 1955, 896. With permission.

solutions on active carbons, found that the adsorption involved on intraparticles transport rate-control mechanism. Competitive adsorption of the two pesticides from their equimolecular mixtures showed that the adsorption involved two different mechanisms.

Adsorption of cationic and anionic dyes from dilute aqueous solutions has been used to characterize carbons for their surface area, microporous structure, and polarity.[13,14,32] Graham[32] studied the adsorption of two dyes, methylene blue and metanil yellow, of opposite character but of approximately the same molecular dimensions from aqueous solutions on a number of active carbons, measuring separately the influence of pore size and the surface acidic groups. Graphon, which essentially is nonporous, has uniform surface, and is free of surface heterogeneities was used as the standard model substance. The surface available for adsorption was only 46% of the total in the case of methylene blue and only 25% in the case of metanil yellow, although the pores with diameters greater than those of these dyes contributed to about 75% of the total surface area (Table 3.3). This has been attributed to the presence of acidic carbon-oxygen surface groups, which influenced the adsorption characteristics towards these dyes. It was found that the available surface increased when these carbon-oxygen surface groups were removed by heat treatment in nitrogen at 900°C. In general, the adsorption of metanil yellow, which is an anionic dye, was less than that of methylene blue, which is a cationic dye. The adsorption of metanil yellow also showed a linear relationship with the amount of acidic surface groups present on the carbon surface (Figure 3.4). Puri[13] and Arora[14] observed the adsorption isotherms of methylene blue and Rhodamine B on a graphon and a number of active carbons to be Type I of the BET classification, indicating a completion of the monolayer. However, when surface areas were calculated, the

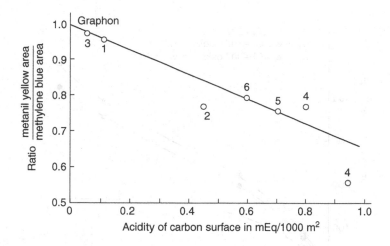

FIGURE 3.4 Influence of surface acidity of a carbon surface on its adsorption for metanil yellow. (After Graham, D., *J. Phys. Chem.*, 59, 1955, 896. With permission.)

values agreed with the BET areas for graphon but differed appreciably for active carbons. (Table 3.4). It was suggested that some of the micropores in active carbons were not accessible to larger dye molecules.

Goyal[40] and Goyal and Bansal[44] studied the adsorption isotherms of methylene blue on two samples of granulated activated carbons and two samples of activated carbon Fibers in the concentration range of 20 to 3000 ppm. The isotherms followed Langmuir isotherm equation. The linear Langmuir plots were used to calculate the monolayer capacity, which was further used to calculate surface areas covered by

TABLE 3.4
Specific Surface Areas as Obtained by Dye Adsorption and Nitrogen Adsorption (BET) Methods and Relative Proportions of Ultramicropores in Various Activated Carbons

Activated Carbon	N_2 Adsorption (77 K)	Surface Area (m²/g) from Dye Adsorption			
		Methylene Blue	Rhodamine B	Average Dye Value	Area Constituted by Ultramicropores
A	109	916	896	906	190
B	988	797	763	780	208
C	924	638	613	625	299
D	528	564	548	346	182
E	898	232	228	623	275

Source: Arora, V.M., Ph.D. dissertation, Panjab Univ., Chandigarh, India, 1977. With permission.

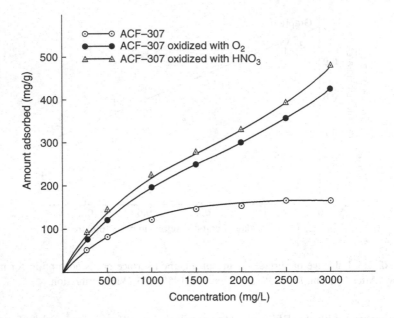

FIGURE 3.5 Influence of surface oxidation on the adsorption of methylene blue on ACF-307. (After Goyal, M., Ph.D. dissertation, Panjab Univ., Chandigarh, India, 1997. With permission.)

methylene blue, using 119 A^{02} as the molecular area of methylene blue. Although the whole of the BET surface area was available for adsorption in granulated carbons, only 50% of the BET area was available in the case of activated carbon Fibers. This indicated that about 50% of the BET area in the case of fibrous activated carbons was present in fine micropores, which were in accessible to larger methylene blue molecules. The adsorption of methylene blue increased on oxidation (Figure 3.5) of the carbon with nitric acid and oxygen gas and decreases on degassing at 400°C, 650°C, and 950°C (Figure 3.6). The increase in adsorption is more on oxidation of the carbons with nitric acid than on oxidation with oxygen. The decrease in adsorption is maximum on the 650°-degassed samples that have lost most of their acidic carbon-oxygen surface groups. The results indicate that the adsorption of methylene blue on carbon depends upon the amount of acidic surface groups, the nonacidic surface groups having no influence. Because methylene blue is a cationic dye, it has large interactions with carbons that have negatively-charged sites when placed in aqueous solutions. As oxidized carbons have carboxylic or lactonic acidic groups which ionize in water producing negative COO⁻ sites, the larger adsorption of methylene blue on oxidized carbons can be visualized. As these carbons are degassed, there are less and less negative sites on the carbon surface, due to the removal of acidic surface groups, the adsorption of methylene blue decreases.

The adsorption isotherms of cationic polymers, such as nonylphenol (NP) diethylamine, NP trietylamine, NP tetraethyl pentamine, octyl phenol (OP) diethylamine, and OP tetraethylene pentamine from dilute aqueous solutions on an activated carbon

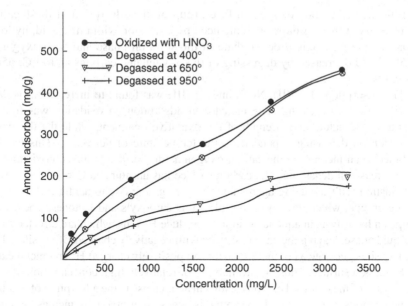

FIGURE 3.6 Influence of carbon-oxygen surface groups on the adsorption of methylene blue by activated carbon. (After Goyal, M., Ph.D. dissertation, Panjab Univ., Chandigarh, India, 1997. With permission.)

showed a two-stage adsorption.[45] At low adsorptions, the molecules were postulated to lie flat on the surface. A reorientation of the molecules occurred in the second stage as the adsorption increased until at saturation the molecules were almost perpendicular to the surface. The variation of the heat of adsorption with surface coverage also showed two inflections corresponding to the two-stage adsorption process. The adsorption isotherms of ionic surfactants, such as dodecyl ammonium chloride, dedecyl pyridenium bromide, and sodium dodecyl sulfate from aqueous solutions on carbons are S-shaped with two maxima.[46,47] The equilibrium concentration for the second maximum coincided with the CMC of the surfactants. The adsorption was found to take place in two steps: monomolecular adsorption of the Langmuir type and multiplayer adsorption of the BET type. The polar sites on the polar adsorbents attracted polar groups of the surfactant molecule by electrostatic attractive forces and formed oriented monomolecular layers and then multilayers between the nonpolar chains of the surfactant molecules. The adsorption clearly showed the influence of adsorbate-adsorbate and adsorbate-adsorbent interaction forces during adsorption from solutions.[48-51]

Bansal and coworkers[52-54] studied the adsorption isotherms of Cu(II), NI (II), and Cr(IV) and Cr(VI) ions on granulated and fibrous activated carbons from aqueous solutions. The adsorption isotherms are Type I of the BET classification, showing initially a rapid adsorption tending to be asymptotic at higher concentrations. The uptake of these cations was generally smaller in case of the fibrous activated carbons (activated carbon fibers) compared with granulated carbons, which could not be explained on the basis of surface area alone. In order to examine the influence of carbon-oxygen surface group on the adsorption of these cations, the amounts of these groups was enhanced by oxidation with nitric acid, hydrogen peroxide, and ammonium persulphate in the solution phase and with oxygen gas at 350°C and decreased by degassing of the oxygenated carbons at 400°C, 650°C, and 950°C.

The adsorption of Cu(II), Ni(II) and Cr (III) was found to increase on oxidation and decrease on degassing. The increase in adsorption on oxidation was found to depend on the nature and strength of the oxidative treatment, while the decrease in adsorption on degassing depended on the temperature of degassing. This has been attributed to an increase in the carbon-oxygen acidic surface groups on oxidation and their decrease on degassing. The carbon surface has unsaturated $C = C$ bonds which on oxidation with oxidizing agents add oxygen, giving rise to acidic carbon-oxygen surface groups, which have been postulated as carboxyls and lactones. These acidic groups on hydrolysis in aqueous solutions produce H^+ ions, which are directed toward the liquid phase, leaving the carbon surface with negatively charged COO^- sites. These negative sites generate a competition between positively charged H^+ and metal cations for the carbon surface. Thus, the availability of relatively high concentration of COO^- sites in case of the oxidized carbons results in an increase in the adsorption of positively charged cations. When the carbon sample is degassed at gradually increasing temperatures, these oxygen surface groups are eliminated in increasing amounts from the carbon surface. This causes a decrease in the concentration of surface COO^- sites, thereby decreasing the adsorption of metal cations. When the oxygen groups are removed almost completely by degassing at 950°C, the concentration of COO^- sites is reduced to almost nil, resulting in a further decrease in the adsorption of cations.

Freundlich and Langmuir isotherm equations have been employed to explain the earlier results of adsorption from dilute solutions on carbon. However, as mentioned earlier, the Freundlich isotherm lacks a theoretical basis, and the Langmuir isotherm assumes a constant energy of adsorption over the surface, which is certainly not true in the case of active carbons. Consequently, modified theoretical approaches were advanced to fit the experimental data over a wide range of concentrations. Radke and Prausnitz[55] proposed the equation

$$\frac{1}{a} = \frac{1}{\alpha c} + \frac{1}{[\beta(c)^n]} \tag{3.2}$$

where a is the amount of solute adsorbed, and α, β, and n are constants. This equation was successfully used for the description of several experimental isotherms.[55] This equation reduces to Henry's law at lower concentrations and to the Freundlich isotherm equation at higher concentrations. When n tends to zero, the equation is reduced to Langmuir isotherm equation.

Brown and Everett[56] proposed the Harkins Jura type equation

$$\ln c = \frac{\alpha - \beta}{a^2} \tag{3.3}$$

where a is the amount adsorbed at concentration c, and α and β are constants. The equation was used by Brown and Everett for the analysis of the data on adsorption from solutions. John et al.[57] used the equation.

$$\log [\log \{(c/c_s) \cdot 10^n\}] = A + B \log a \tag{3.4}$$

where c is the concentration of the solution, c_s is the solubility of the solute, n is an integer greater than unity, and A and B are constants. However, all these equations are empirical and, therefore, have not been commonly used. Freundlich adsorption equation, in spite of its inherent shortcomings, is still considered better for its applications to adsorption on carbons from dilute solutions.

3.2.2.1 Potential Theory of Adsorption from Dilute Solutions

The potential theory introduced by Polanyi for adsorption of gases and vapors has been discussed in detail elsewhere in this book. For a single gas or vapor, the adsorption potential ε has been defined as equal to the work done in bringing one mole of the adsorbate from infinity to a specified distance from the surface. Polanyi modified the theory for adsorption from solutions of sparingly soluble solutes.[58] The significance of this approach is that it is not essential to postulate a thickness for the adsorbed layer. This modified equation can be written as

$$\varepsilon_1 = RT \ \ln \frac{c_o}{c} + \varepsilon_2 \frac{V_1}{V_2} \tag{3.5}$$

or

$$V_2 \, \varepsilon_1 - V_1 \, \varepsilon_2 = V_2 \, RT \ln \frac{c_o}{c} \tag{3.6}$$

where ε_1, is the adsorption potential of the solute, ε_2 that of the solvent for a solution of concentration c, c_o is the limiting solubility of the solute, and V_1 and V_2 are the molar volumes of the solute and the solvent, respectively.

Kipling and Tester,[59] Hansen and Fackler,[60] and Manes and coworkers[61–65] applied the Polanyi potential theory to the adsorption of organic compounds from aqueous and nonaqueous solutions on active carbons and carbon blacks. Manes and Hoffer[61] found a relationship between the gas-phase and liquid-phase adsorptions at low surface coverages. However, the solute adsorption at high surface converages did not fit their gas-phase correlation curve. This was presumably attributed to the packing effects that are different in gas-phase and liquid-phase adsorptions. These workers were of the opinion that a better fit can be expected if the adsorbate is a liquid and the solvent is strongly adsorbed, as in the case of adsorption of partially miscible organice liquids from water. The Polanyi potential model is better than the

Langmuir model because it regards the surface as heterogeneous with respect to adsorption energies, and carbon surfaces are truly heterogeneous in nature. The Polanyi potential theory, however, has its limitations. It assumes that all the pores of the adsorbent are available to the adsorbate molecules, which is not completely true. Some of the microspores in carbons are not accessible to larger solute molecules. Furthermore, the Polanyi potential approach does not cite specific or chemical interactions between the adsorbate molecules and carbon-oxygen surface groups that are invariably present on carbons surfaces. Everett and coworkers[66–71] developed thermodynamic treatments that have also been found to be inadequate to explain all the parameters involved in adsorption from solutions.

Urano et al.[72] combined the generalizations of the Polanyi potential theory and the Freundlich isotherm equation to derive a relationship that these workers claimed could successfully predict the adsorption isotherms for several organic compounds from aqueous solutions on active carbons. Their adsorption equation is represented as

$$Q = \left(\frac{W_o}{bV} \right) \left(\frac{c}{c_o} \right)^{1/m} \tag{3.7}$$

where Q is the equilibrium molar adsorption amount, V is the molar volume of the adsorbate, $1/n$ the exponential parameter of the modified Freundlich equation for a standard adsorbate, and b is a constant that was approximated to 1.0 for liquid adsorbate and 1.5 for solid benzene substituted compounds. These workers claim that the equation parameters W_o, n, and r can be calculated from the properties of the activated carbon and the adsorbate as

$$W_o = 1.9 \, V_3 - 0.03 \tag{3.8}$$

$$n = \frac{1}{0.37 (V_3 / V_{20})} = 0.05 \tag{3.9}$$

and

$$r = \frac{V}{V_s} \tag{3.10}$$

where V and V_s are the molar volumes of the adsorbate and the standard adsorbate, and V_3 and V_{20} represent volumes of the pores with diameters less than 3 and 20 nm. These workers compared the experimental adsorption data, using a number of benzene derivatives, phenols and their derivatives, and aliphate compound on five different commercial activated carbons with the data calculated, using this equation and found the agreement to be very close (Figure 3.7), thus making it possible to predict the adsorption capacities of different organic compounds from their aqueous solutions for any activated carbon whose pore size distribution is known.

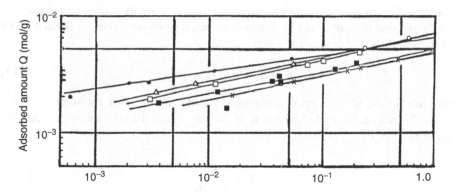

FIGURE 3.7 Comparison of theoretical and experimental adsorption isotherms of benzene derivatives. • = Phenol; Δ = Benzaldehyde; $\square$ = Benzonitrile; $\blacksquare$ = m. toluic acid, $\times$ = m.amino benzoic acid. (After Urano, K., Kochi, Y., and Yamamoto, Y., *J. Colloid Interface Sci.*, 86, 43, 1982. With permission.)

3.2.3 ADSORPTION FROM SOLUTIONS AT HIGHER CONCENTRATIONS (COMPOSITE MISCIBLE SOLUTIONS)

When two components of a solution are miscible with each other in all proportions, then there is no distinction between the solute and the solvent so that neither of the two components can be regarded as a solvent over the whole range of concentrations. In such a system the change in concentration of the solution as a result of adsorption will be the resultant of the adsorption of both the components of the solution. It has been mentioned earlier in this chapter that what is actually measured experimentally is the change in concentration of the solution so that the adsorption isotherm is actually the isotherm of *change in concentration*. More recently, however, this isotherm of change in concentration has been named as the *composite isotherm*, indicating that it combines the results of true (or individual) isotherms for adsorption of the two components of the binary solution.

3.2.3.1 Derivation of Composite Isotherm

Kipling[4] has derived a composite isotherm equation that has been used widely. According to this derivation, when mass m of a solid adsorbent is brought into contact with n_o moles of the solution, there is a decrease in the mole fraction of the solution. Let this decrease in the mole fraction of the solution with respect to component 1 as a result of adsorption be Δx. Let x_o and x be the initial and final mole fractions of the solution with respect to component 1. Then

$$\Delta x = x_o - x \tag{3.11}$$

Let us further suppose that this change is concentration Δx, and it is brought about by the transfer of n_1^s moles of component 1 and n_2^s moles of component 2 to the surface of a unit mass of the solid adsorbent. Let the amounts of the two components in the solution at equilibrium be n_1 and n_2 moles, respectively. Thus, the total number of moles of the solution

$$n_o = n_1 + n_2 + n_1^s m + n_2^s m \qquad (3.12)$$

where $n_1^s\, m$ and $n_2^s\, m$ moles of component 1 and 2 are adsorbed by weight m of the solid. Therefore, the mole fraction x_1 of the component 1 in the solution at equilibrium is given by

$$x_1 = \frac{n_1}{n_1 + n_2} \qquad (3.13)$$

and

$$1 - x_1 = \frac{n_2}{n_1 + n_2} = x_2 \text{ (say)} \qquad (3.14)$$

where $(1 - x)$ is the mole fraction of the solution with respect to component 2. x_o the initial mole fraction of component in the solution is given by

$$\Delta x = \frac{n_1 + n_1^s m}{n_o} = \frac{n_1 + n_1^s m}{n_1 + n_2 + n_1^s m + n_2^s m} \qquad (3.15)$$

$$\therefore \Delta x = x_o - x_1$$

$$\Delta x = \frac{n_1 + n_1^s m}{n_1 + n_2 + n_1^s m + n_2^s m} - \frac{n_1}{n_1 + n_1^s}$$

$$= \frac{n_1^s n_2 m - n_1 n_2^s m}{n_o (n_1 + n_2)} \qquad (3.16)$$

$$= \frac{n_1^s n_2 - n_1 n_2^s}{n_o (n_1 + n_2)}$$

or

$$\frac{n_o \Delta x}{m} = \frac{n_1^s n_2 - n_1 n_2^s}{n_1 + n_2}$$

$$= \frac{n_1^s n_2}{n_1 + n_2} - \frac{n_1 n_2^s}{n_1 + n_2}$$

$$= n_1^s (1 - x) - n_2^s x \qquad (3.17)$$

so that

$$\frac{n_o \Delta x}{m} = n_1^s x_2 = n_2^s x_1 \tag{3.18}$$

where x_1, and x_2 are the mole fractions of components, 1 and 2 in the solution. The function $\frac{n_o \Delta x}{m}$ when moles and mole fraction are used is the apparent adsorption which when plotted against mole fraction gives the Composite isotherm.

The composite isotherm is sometimes expressed in terms of $(x_1^s - x_1)$ when x_1^s is the mole fraction of component 1 in the adsorbed phase. This relationship can be written as

$$x_1^s - x_1 \frac{1}{n^s} \cdot \frac{n_o \Delta x}{m} \tag{3.19}$$

This isotherm equation can be derived from Equation 3.17 as follows:

$$\frac{n_o \Delta x}{m} = n_1^s (1 - x_1) - n_2^s x_1 = n_1^s - x \left(n_1^s - n_2^s \right) \tag{3.20}$$

Substituting

$$n_1^s + n_2^s = n^s \tag{3.21}$$

where n^s is the total number of moles of both components 1 and 2 in the adsorbed phase on unit mass of the solid. Equation 3.20 can be written as

$$\frac{n_o \Delta x}{m} = n_1^s - x_1 n^s \tag{3.22}$$

or

$$\frac{1}{n^s} \cdot \frac{n_o \Delta x}{m} = n_1^s / n^s - x_1$$

$$\frac{1}{n^s} \cdot \frac{n_o \Delta x}{m} = x_1 s - x_1 \tag{3.23}$$

where x_1^s represents the mole fraction of the adsorbed phase with respect to component 1. Equation 3.23 can also be written as

$$n^s \left(x_1^s - x_1 \right) = \frac{n_o \Delta x}{m}$$

or using the value of $\dfrac{n_o \Delta x}{m}$ from Equation 3.17 we get

$$n^s \left(x_1^s - x_1 \right) = n_1^s (1 - x_1) - n_2^s x_1 \qquad (3.24)$$

where $(x_1^s - x)$ represents the preferential adsorption.

Sometimes it is convenient to express the results in terms of weights and weight fractions instead of moles and mole fractions. In that case, Equation 3.17 can be expressed as

$$\frac{W_o \Delta c}{m} = w_1^s (1 - c) - w_2^s c \qquad (3.25)$$

where w_o is the initial weight of the solution brought in contact with mass m of the adsorbent, c is the equilibrium weight fraction of the solution and w_1^s and w_2^s are respectively the weights of the two components adsorbed by a unit mass of the solid adsorbent.

Equation 3.25 is particularly useful when the molecular mass of one component is not known. It is generally felt that concentration of the solution not be expressed in moles per liter. Although it is convenient for dilute solutions but as density of the solution changes with concentration the conversion of adsorption measured in these units into other units is not quite easy

$$\frac{n_o \Delta x}{m} = \frac{w_o \Delta c}{m} \left(\frac{1}{M_2 c + M_1 (1 - c)} \right) \qquad (3.26)$$

when M_1 and M_2 are the molecular masses of the two components 1 and 2, respectively.

3.2.3.2 Classification of Composite Isotherms

Ostwald and de Izaguirre[73] classified composite isotherms broadly into six types (Figure 3.8) and showed a relationship between the size of the composite isotherm and the adsorption of individual components of the solution. By postulating different forms for the individual isotherms, these workers arrived at the resultant

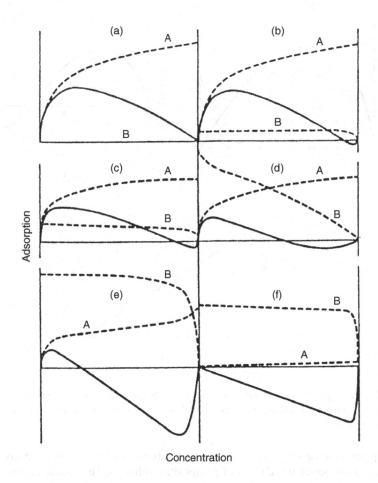

FIGURE 3.8 Composite and individual isotherms for binary systems (solid line, composite; dashed line, individual for components). (After Ostwald, W. and de Izaguirre, R., *Kolloid Z.*, 30, 279, 1922. With permission.)

composite isotherm. The above six examples clearly illustrate that the negative adsorption is possible only if there is adsorption of the solvent also. These isotherms also indicate that a maximum in the composite isotherm can occur even when there is no adsorption of the solvent. The composite isotherm equation (Equation 3.17) then becomes

$$\frac{n_o \Delta x}{m} = n_1^s (1 - x) \tag{3.27}$$

which $\frac{n_o \Delta x}{m}$ indicates that a decrease in the value of x results in a fall in the value of linearly with x.

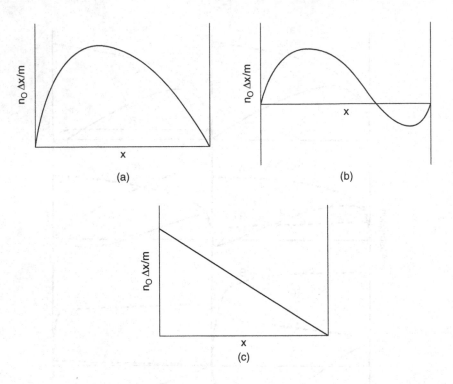

FIGURE 3.9 Types of composite isotherm: (a) U-shaped, (b) S-shaped, (c) linear. (After Kipling, J.J., in *Adsorption from Solutions of Non-Electrolytes*, Acad. Press, London, 1965. With permission.)

Kipling,[4] on the other hand, classified composite isotherms on the basis of their shapes and suggested three types of composite isotherms. The characteristic features of these isotherms shown in Figure 3.9 are

1. A single-branch U-shaped isotherm
2. A two-branch S-shaped isotherm in which the two branches usually have different sizes
3. A linear isotherm

U-shaped or S-shaped isotherms are more common. Although U-shaped isotherms correspond to Type (a) of the Ostwald-de Izaguirre, S-shaped isotherms may replace almost all of the remaining types. Linear isotherm (Type c) is generally obtained for adsorption by molecular sieves when only one component of the solution can enter the pores of the adsorbent while the entry of the second component is inhibited.

Nagy and Schay[74-76] recognize five different types of composite isotherms that are variations of U-shaped and S-shaped isotherms of Kipling. Three isotherm types of Nagy and Schay are U-shaped, and the other two are S-shaped. These five types are presented in Figure 3.10. The S-shape of the isotherms indicates that the surface shows a preference for both the components of the solution but over different

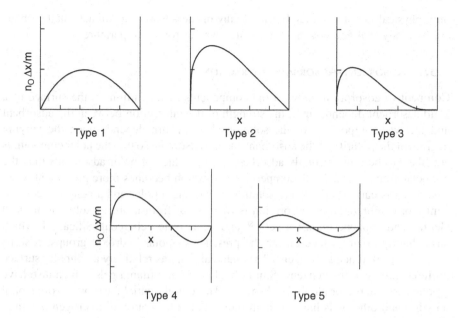

FIGURE 3.10 Classification of composite isotherms. (After Nagy, L.G and Schay, G., *Mag. Kem. Foly*, 66, 31, 1960. With permission.)

concentration ranges. The U-shaped isotherm, on the other hand, indicates a preference for only one component of the solution over the entire concentration range. In the case of active carbons, as for many other adsorbents, all five types of composite isotherms are obtained depending on the properties of the adsorbent surface and the nature of the solvent and the solute in the solution.

3.3 FACTORS INFLUENCING ADSORPTION FROM BINARY SOLUTIONS

It has been mentioned earlier that in the case of the adsorption on carbon adsorbents (or any solid adsorbent) from binary solutions, with completely miscible components in all proportions, both the components of the solution are adsorbed although the surface may show preference for one component depending upon the nature of the solid surface and the nature of the two components of the solution. This indicates that there is competition between the two components of the solution (i.e., between the solvent and solute for the adsorbent surface). Thus, the adsorption from binary solutions is now generally called *competitive adsorption*. Competitive adsorption depends upon several factors, such as the interaction of the solid surface with the two components of the solution (adsorbate-adsorbent interaction), the interaction between the two components of the solution (adsorbate-adsorbate interaction), the porosity of the solid adsorbent, and the heterogeneity of the adsorbent surface. Furthermore, the assumption that adsorption from solutions at room temperature is only physical is not completely true. Many of the adsorptions on activated carbons are both physical and chemical in nature, even at room temperature.

3.3.1 ADSORBENT-ADSORBATE INTERACTION

Competitive adsorption between the components of a solution on the surface of a solid adsorbent depends upon the strength of the interaction between the adsorbent and the two components of the solution. This in turn depends upon the varying degrees of the polarity of the adsorbent and the adsorbate so that the polar compounds are likely to be more strongly adsorbed on the surface of polar adsorbents than the nonpolar compounds. Thus, competitive adsorption becomes more prominent when adsorption is carried out from a solution containing a polar and a nonpolar component. For example, when adsorption is carried out from a binary solution of ethyl alcohol and benzene on silica gel[77-80] and an activated charcoal, Silica gel, which has a highly polar surface due to the presence of polar hydroxyl groups, adsorbs alcohol in preference to benzene. The charcoal that has relatively a nonpolar surface preferentially adsorbs benzene. Similarly alumina and titania gels, which also have a polar surface, prefer alcohol to benzene. These preferential adsorptions of alcohol on silica and other gels have been attributed to the formation of hydrogen bonding. The preferential adsorption of nitro and nitroso[81] derivatives of diphenylamine and aniline on silica gel has also been attributed to the formation of hydrogen bonds.

Activated carbons, charcoals, and carbon blacks have less well-defined surfaces. All these carbons are always associated with a certain amount of chemisorbed oxygen, the amount varying with the source raw material, and the history of formation of the carbon. The associated oxygen on the carbon surface is present in the form of carbon-oxygen surface groups, which have varying thermal stability. The amount of these carbon-oxygen surface groups can be enhanced by oxidation of the carbon surface and decreased by outgassing the carbon sample in nitrogen gas or in vacuum at temperatures up to 1000°C. Some of these oxygen surface groups are acidic in character and make the carbon surface polar. (For details of the formation and estimations of these carbon oxygen surface groups, see Chapter 1).

The influence of these carbon-oxygen surface groups on the competitive adsorption from binary solutions of polar and relatively less polar or nonpolar components has been studied by a large number of workers. Earlier studies using carbon materials were more concerned with the influence of associated oxygen on the competitive adsorption of the two components of the solution, but the later studies made a distinction between the influence of the different types of carbon-oxygen surface groups. Most of these studies have been carried out using alcohols as the polar component of the binary solution. For example, earlier studies of Bartell and Lloyd,[84] using charcoals and binary solutions of ethyl alcohol and benzene, observed that the preference for ethyl alcohol relative to benzene increases with oxidation and is reduced by the removal of oxygen from the charcoal surface by degassing.

Kipling and coworkers,[59,82,83] in a series of publications have pointed out that the presence of associated oxygen imparts polar character to the carbon surface, as a result of which they exercises preferential adsorption for a more polar component of the binary solution. Gasser and Kipling,[83] while studying the adsorption of cyclohexane,

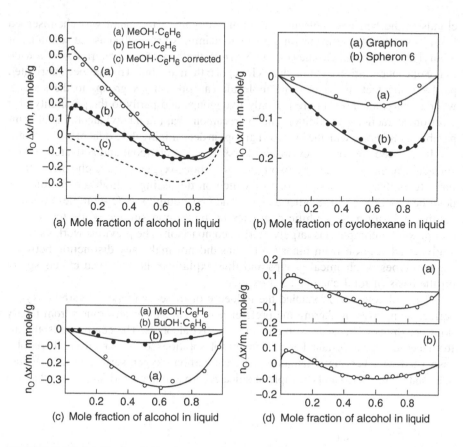

FIGURE 3.11 (a) Composite isotherms on Spheron-6 from binary solutions of alcohols and benzene. (b) Composite isotherms on Spheron-6 and graphon from binary solutions of cyclohexane and benzene. (c) Composite isotherms on graphon from binary solutions of alcohols in benzene. (d) Composite isotherms on Spheron-6 from binary solutions of alcohols in benzene for (1) *n*-propyl alcohol in benzene and (2) *n*-butyl alcohol in benzene. (From Gasser, C.G. and Kipling, J.J., *Proc. 4th Carbon Conf.*, Univ. of Buffalo, Pergamon Press, New York, 1960, p. 55. With permission.)

methanol, ethanol, and n. butanol from their solutions in benzene on Spheron-6 and graphon (Spheron-6 heat treated in inert atmosphere at 2700°C) observed that the composite isotherms were S-shaped in the case of alcohol-benzene solutions on Spheron-6 (Figure 3.11(a), (b)), indicating a preference for alcohols that are the more polar components of the solution. The composite isotherms were, however, U-shaped in the case of graphon, indicating a preference for benzene, the less polar component of the solution. This has been attributed to the presence of associated oxygen on the surface of Spheron-6 and virtual absence of this oxygen on the surface of graphon. The marked difference for preferential adsorption of benzene by Spheron-6 from benzene-cyclohexane solutions (Figure 3.11(b)) has been ascribed to the interaction between polar oxygen groups on the surface with the π electron clouds of the benzene molecule. A certain amount of methanol was chemisorbed from its solutions in benzene on oxygen containing sites. In the case of solutions of n.butyl alcohol from benzene (Figure 3.11(c)), the benzene is preferred over a wide range of concentrations, although it is less polar than alcohol. This has been attributed partly to the fact that a benzene molecule can present six groups to the surface, whereas butyl alcohol can present only five groups and partly to the possibility of π electron interactions. In addition, the hydrocarbon chain in n.butyl alcohol is of comparable significance with the hydroxyl groups in determining the extent of adsorption.[85]

It has been shown by several workers[86–89] that a charcoal and activated carbon surface is usually covered by two types of carbon-oxygen surface chemical structures (complexes or groups), one of which on degassing at high temperatures is desorbed as CO_2 and the other as CO. The first complex (CO_2-complex) imparts acidity, polarity and lyophilic character to the carbon surface while the second complex (CO-complex) is largely inert in comparison. The previous workers while studying adsorption from binary solutions did not make any distinction between the two types of chemical groups and thus explained the observed effects solely on the basis of total chemisorbed oxygen.

Puri and coworkers[90] studied the influence of these two types of carbon-oxygen surface complexes in altering the preference of charcoal for adsorption from binary liquid mixtures of benzene with ethyl and methyl alcohols. The composite isotherms from benzene-ethanol and benzene-methanol solutions on sugar charcoal samples associated with varying amounts of different carbon-oxygen surface groups (Figure 3.12 and Figure 3.13) and a sample of silica gel are S-shaped. Original sugar charcoal

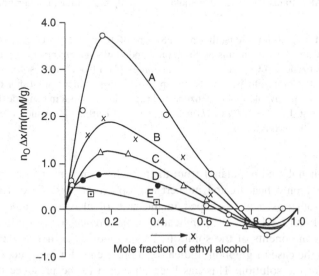

FIGURE 3.12 Composite isotherms on sugar charcoal samples from ethanol-benzene solutions. (Curves A, C, D, and E represent original and degassed samples, and B represents silica gel.) (After Puri, B.R., Kumar, S., and Sandle, N.K., *Indian J. Chem.*, 1, 418, 1963. With permission.)

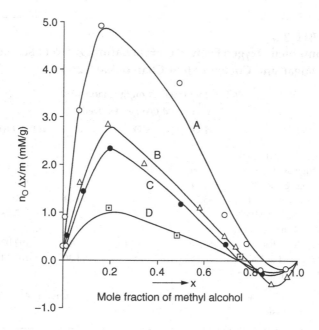

FIGURE 3.13 Composite adsorption isotherms on various sugar charcoal samples from methyl alcohol-benzene mixtures. (Curves A, C, D, and E represent original sugar charcoal and charcoal samples degassed at 400, 1100, and 750°C, respectively.) (After Puri, B.R., Kumar, S., and Sandle, N.K., *Indian J. Chem.*, 1, 418, 1963.)

shows preference for methanol until a mole fraction of less than 0.75 of ethanol and preference for benzene at mole fractions greater than 0.75 of ethanol. The maximum preference for ethanol occurs at a mole fraction of 0.16 of ethanol and for benzene at a mole fraction of 0.12 of benzene. However, the degree of preference for ethanol is much larger and extends over a wider range of concentrations. The composite isotherms on the degassed samples are also S-shaped and show similar trends. However, the magnitude of ethanol preference decreases as the temperature of degassing is enhanced. The results obtained in the case of benzene methanol system show similar trends, except that all the charcoal samples show a greater preference for methanol than for ethanol. This has been attributed partly to the smaller molecular dimensions of methanol so that a larger number of methanol molecules compared to that of ethanol are required to complete the same thickness of the adsorbed layer and partly to fact that the polar-OH group in methanol constitutes a higher proportion of the whole molecule than in ethanol. The results obtained with coconut charcoal samples were also similar, the difference being one of degree. The original charcoal showed the highest, and the 700°-degassed charcoal showed the lowest preference for ethanol or methanol.

The larger preference for alcohols compared to benzene shown by the original charcoals has been attributed to the presence of a larger amount of the CO_2-evolving surface groups (Table 3.5). These surface groups are known to impart polar character to the carbon surface and increase its preference for polar components of the mixture.

TABLE 3.5
Combined Oxygen Evolved on Evacuating Different Samples of Sugar and Coconut-Shell Charcoal at 1200°C

(Values Expressed as mg/g Charcoal)

Sample	Combined Oxygen Evolved As			Total Oxygen
	H_2O	CO_2	CO	
Sugar charcoal				
Original charcoal	73.69	104.80	73.88	252.37
Degassed at 400°	51.11	40.36	68.20	159.67
Degassed at 750°	35.73	—	55.43	91.16
Degassed at 1100°	—	—	2.28	2.28
Coconut shell charcoal				
Original charcoal	68.40	59.34	64.28	192.02
Degassed at 750°	58.84	—	43.30	102.14
Degassed at 1100°	—	—	—	—

After Puri, B.R., Kumar, S., and Sandle, N.K., *Indian J. Chem.*, 1, 418, 1963. With permission.

The results show that it is the nature of the oxygen group, and not necessarily the total combined oxygen, that determines the preference of the surface with a component of the binary solution. The oxygen that is present as CO_2 complex imparts polarity to the carbon surface and increases its capacity to take up the more polar liquids. The oxygen that evolves CO on degassing and is present as quinonic groups[39,91-93] imparts aromaticity to the carbon surface and promotes preference for benzene. These workers have also reported that a part of the methanol or ethanol taken up by the charcoal is held up by chemical or quasi-chemical forces, probably involving hydrogen bonding at the active sites provided by the CO_2-evolving groups.

The Silica gel has polar oxide and hydroxide groups on its surface. The composite isotherms from benzene-methanol solutions, therefore, should resemble those and the original charcoals. This has actually been found to be the case (*cf.* Figure 3.12) and the difference is only one of degree. However, these workers felt that the charcoals, being porous, suffer from inherent steric defects, particularly when adsorbates of different molecular dimensions are involved. Thus, these workers[94] studied the influence of carbon-oxygen surface complexes on selective adsorption of methanol-benzene, methanol-carbon tetrachloride, and ethanol-cyclohexane mixtures by a carbon black Mogul, before as well as after outgassing at 400°C and 700°C to eliminate increasing amounts of the oxygen complexes and also at 1000°C to free it almost completely of the carbon-oxygen surface complexes. The original carbon black as well as the 1000°C-degassed samples were also given oxidative treatments with aqueous hydrogen peroxide and potassium persulphate to obtain carbon samples associated with increasing amounts of the carbon-oxygen surface complexes.

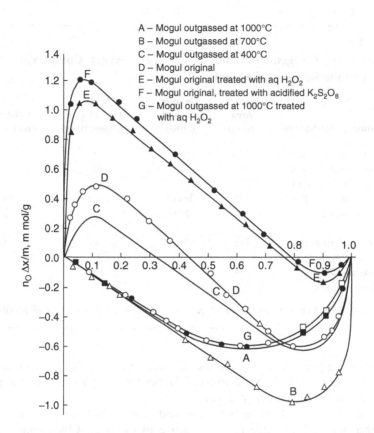

A – Mogul outgassed at 1000°C
B – Mogul outgassed at 700°C
C – Mogul outgassed at 400°C
D – Mogul original
E – Mogul original treated with aq H_2O_2
F – Mogul original, treated with acidified $K_2S_2O_8$
G – Mogul outgassed at 1000°C treated
      with aq H_2O_2

FIGURE 3.14 Composite isotherms from benzene-methanol solutions on Mogul before and after different treatments. (After Puri, B.R., Singh, D.D., and Kaistha, B.C., *Carbon,* 10, 481, 1972. With permission.)

These workers observed that 1000°C-outgassed Mogul sample, which is essentially free of any combined oxygen, shows a strong preference for benzene at all concentrations from benzene-methanol solution,s giving a typical U-shaped isotherm (isotherm A in Figure 3.14).This preference for benzene can be attributed to the interaction of π electrons of the benzene ring with the basal planes of the carbon black as has been suggested by Gasser and Kipling.[83]

When the carbon sample, however, was degassed at a temperature of 700°C where it loses almost all of its oxygen present as CO_2-complex (surface oxygen evolved as CO_2 on degassing) but retains most of its oxygen present as CO-complex (surface oxygen evolved as CO on degassing) the carbon black sample showed even stronger preference for benzene, the less polar component of the solution along the whole range of concentrations. This is not in consonant with the views expressed by the earlier workers that combined oxygen imparts greater preference for the more polar component of the solution. Puri et al., attributed this behavior to the presence of quinone groups that form a part of the CO-complex[95] and tend

TABLE 3.6
**Surface Areas, Oxygen Contents, and Surface Oxygen Complexes
of the Various Samples of Mogul**

Description of the Sample	Surface Area (m₂/g)	Oxygen Content (g/100 g)	Acidic CO₂-Complex (mmoles/100 g)	CO-Complex (mmoles/100 g)
Mogul, original	308	8.1	68	275
Mogul, outgassed at 400°C	306	5.8	42	269
Mogul, outgassed at 700°C	335	2.6	Traces	163
Mogul, outgassed at 1000°C	326	Traces	Traces	Traces
Mogul, original, treated with aq. H_2O_2	310	10.9	145	284
Mogul, original, treated with acidified $K_2S_2O_8$	312	13.2	189	315
Mogul, outgassed at 1000°C treated with aq. H_2O_2	328	4.3	120	28

Source: Puri, B.R., Kumar, S., and Sandle, N.K., *Indian J. Chem.*, 1, 418, 1963. With permission.

to promote the preference for benzene. Bhacca[96] has suggested the possibility of an interaction between the π electrons of the benzene ring with the partial positive charge on the carbonyl carbon atom.

The original as well as the 400°-degassed carbon black samples show a larger preference for methanol due to the presence of the polar CO_2 complex and the composite isotherms as S-shaped. The preference is more on the original sample than on the 400°C-degassed sample because the original sample contains more of the polar oxygen complex (Table 3.6). The preference for methanol was enhanced when the CO_2-complex was enhanced by oxidation with hydrogen peroxide or acidified potassium persulphate. The comparison of isotherms B and F (Figure 3.14) shows clearly how the presence of acidic CO_2-complex has resulted in almost complete reversal in the preference of the carbon surface. This change in the shape of the isotherm from S to U as the temperature of degassing was increased or from U to S when the carbon sample was oxidized with 1 to 12 H_2O_2 or $K_2S_2O_8$ (Figure 3.14) clearly indicates that while a part of the combined oxygen present as acidic CO_2 complex renders the carbon surface polar and enhances its interaction with the more polar component of the solution, the presence of CO-Complex enhances the preference of the surface for benzene, which is due to the interaction of the π electrons of the benzene ring with the partial positive charge on the carbonyl groups.

In order to examine the interaction of π electrons of benzene ring with the quinonic oxygen groups, these coworkers substituted carbon tetrachloride and cyclohexane, two nonpolar liquids, for benzene and studied the composite isotherms from methanol-carbon tetrachloride and ethanol-cyclo-hexane solutions. Ethanol was substituted for methanol because methanol and cyclohexane are not completely soluble in all proportions. In the case of these systems, the interaction of carbon tetrachloride

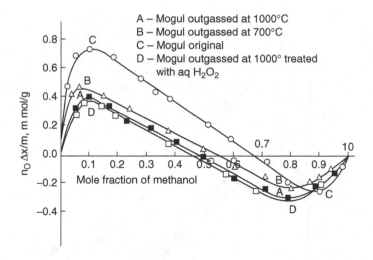

FIGURE 3.15 Composite adsorption isotherms of methanol-carbon tetrachloride mixtures on Mogul before and after various treatments. (After Puri, B.R., Singh, D.D., and Kaistha, B.C., *Carbon,* 10, 481, 1972. With permission.)

or of cyclohexane with the carbon surface was much weaker as compared to the interaction with benzene and the carbon surface preferred methanol or ethanol over a wide range of concentration (Figure 3.15 and Figure 3.16). Furthermore, it is seen that the preference for methanol or ethanol increases with increase in the oxygen content of the carbon surface.

Bansal and Dhami[93] investigated the adsorption from benzene-methanol solutions on a number of charcoals associated with varying amounts of carbon-oxygen surface groups. The composite isotherms (Figure 3.17) are S-shaped (Type IV of the Schay and Nagi classification), except for the charcoal oxidized with hydrogen peroxide. The shape of the isotherms indicates that different portions of the charcoal surface have different preferences depending upon the concentration of the solution. Original sugar charcoal and PVC charcoal prefer methanol up to about 0.82 mole fraction of methanol and benzene at higher concentrations. The magnitude of preference is maximum at 0.20 mole fraction of methanol and 0.91 for benzene in both the cases. However, when original charcoal was degassed at 1000°C, the range of concentration for methanol preference and its magnitude both decrease considerably, and the sample now prefers benzene over a wider concentration range and its magnitude also increases several fold. On the other hand, the magnitude of preference as well as the concentration range change in the opposite manner on oxidation of the original sugar charcoal with hydrogen peroxide. The composite isotherm on the oxidized sample is U-shaped, indicating that the sample prefers methanol at all concentrations. The magnitude of methanol preference also increases considerably.

Because degassing of carbon at 1000°C eliminates the associated oxygen almost completely, and oxidation with H_2O_2 extends this chemisorbed layer of oxygen (Table 3.7), it is apparent that the adsorption of methanol or benzene from their binary solutions is influenced by the associated oxygen. Whereas the formation of carbon-oxygen

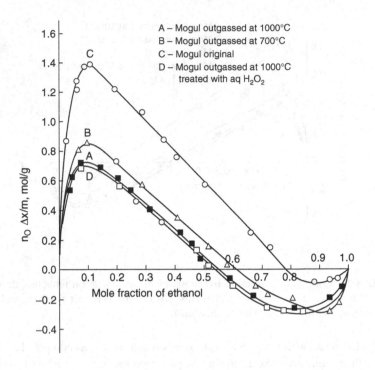

FIGURE 3.16 Composite adsorption isotherms of ethanol-cyclohexane mixtures on Mogul before and after various treatments. (After Puri, B.R., Singh, D.D., and Kaistha, B.C., *Carbon*, 10, 481, 1972.)

surface groups that render the carbon surface more polar enhances its preference for the more polar component methanol, its elimination that renders the carbon surface less polar enhances its preference for benzene, the less polar component of the binary solution. Thus, the competition in adsorption between the two components of a binary solution may be ascribed to differences in the strengths of the adsorbent-adsorbate interactions.

Puri and Murari[98] and Bansal et al.[99] while studying the adsorption isotherms of methanol vapors on different charcoals, found that a certain amount of methanol could not be desorbed and was held irreversibly on those carbons that were associated with chemisorbed oxygen on their surface. In order to see if the chemisorption of methanol takes place from binary solutions, Bansal et al.[93] pretreated the charcoal samples with methanol in order to cover the oxygen groups completely. The methanol treated samples were then evacuated at 120°C to remove any physisorbed methanol, and the composite isotherms were determined on the pretreated samples. The composite isotherms (Figure 3.18) show clearly that the magnitude of preference for methanol decreases considerably in case of the oxygen containing samples. The magnitude of preference in case of the 1000°-degassed sample that is almost completely free of any chemisorbed oxygen remains more or less unchanged, and the two isotherms almost concide with each other at all concentrations. The amounts of chemisorbed and physisorbed methanol was calculated from the linear portions of

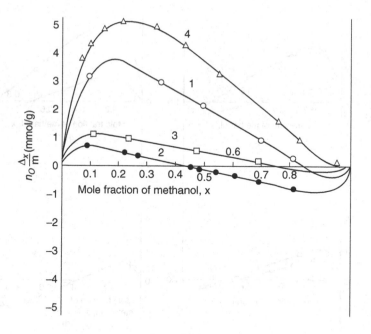

FIGURE 3.17 Composite isotherms from methanol–benzene solutions on various charcoals. ○ 1 = original sugar charcoal; ● 2 = 1000°C outgassed sugar charcoal; □ 3 = PVC charcoal; △ 4 = H_2O_2-treated sugar charcoal. (From Bansal, R.C. and Dhami, T.L., *Carbon,* 15, 153, 1977. With permission.)

the composite isotherms. The results (Table 3.8) show that the oxygenated samples chemisorb appreciable amounts of methanol. It is interesting to note that these values agree fairly well with the values obtained from vapor adsorption isotherms reported by Puri et al.[98] and Bansal et al.[99] which are included in Table 3.8.

TABLE 3.7
Associated Oxygen Obtained on Outgassing Each Sample at 1000°C

| Sample | Associated Oxygen Evolved As (g/100 g) | | | |
	CO_2	CO	H_2O	Total
Original sugar charcoal	11.3	9.2	8.9	29.4
1000°-outgassed charcoal	—	—	—	—
H_2O_2-treated charcoal	13.5	10.1	11.9	35.5
PVC charcoal	1.01	0.18	1.48	2.67

Source: Bansal, R.C. and Dhami, T.L., *Carbon,* 15, 153, 1977. Reproduced with permission from Elsevier.

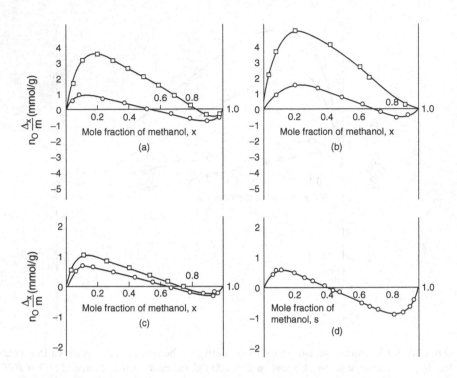

FIGURE 3.18 Composite isotherms from methanol–benzene solutions on carbons before and after methylation: (a) original sugar charcoal, (b) H_2O_2-treated sugar charcoal, (c) PVC charcoal, (d) 1000°C-outgassed sugar charcoal. (From Bansal, R.C. and Dhami, T.L., *Carbon*, 15, 153, 1977. With permission.)

Jankowska et al.[100] studied the adsorption from binary solutions of benzene and alcohol on a wood charcoal modified by heat treatment in vacuum and by oxidation with nitric acid. The presence of acidic oxygen groups enhanced preference for more polar component of the solution and their removal enhanced preference for the less polar component. These workers suggested that in addition to these polar and nonpolar interactions, the adsorption from binary solutions may also be influenced by the nonuniformity and the energetic unhomogeneities of the carbon surface that include unsaturated sites, defects, and free radicals.

Puri and coworkers[90] also studied composite adsorption isotherms on sugar charcoals from binary solutions of methyl alcohol-carbon tetrachloride and methyl alcohol n.amyl alcohol mixtures. The composite isotherms (Figure 3.19 and Figure 3.20) are U-shaped in both systems, indicating that the charcoal surface prefers methyl alcohol at all concentrations. The preference for methanol and its magnitude were found to decrease gradually as the temperature of outgassing was enhanced. These results have been attributed partly to the absence of interaction between the quinone groups on one hand and the carbon tetrachloride or n.amyl alcohol on the other and partly to the fact that the adsorbed molecules of alcohol and acids from solution phase lie with their major axis parallel to the adsorbent surface[82,101] so that the number of molecules of methyl alcohol required per unit surface will be larger than that of

TABLE 3.8
Breakup of Methanol into Physisorption and Chemisorption

| Sample | Methanol | | Benzene | | Methanol Chemisorbed (Vapor Adsorption Isotherm) (mmol/g) |
	Chemisorbed (mmol/g)	Physisorbed (mmol/g)	Chemisorbed (mmol/g)	Physisorbed (mmol/g)	
Original sugar charcoal	3.90	1.20	0	1.10	3.6[a]
1000°-outgassed charcoal	0	0.80	0	1.20	0 [a]
H$_2$O$_2$-treated charcoal	5.0	2.20	0	1.0	—
PVC charcoal	0.64	0.52	0	0.35	0.58

[a] Bansal, R.C. and Dhami, T.L., *Carbon,* 15, 153, 1977.

Source: Puri, B.R. and Murari, K., *Res. Bull.*, Panjab Univ., 13, 9, 1962. With permission.

n.amyl alcohol. Furthermore, some of the fine capillary micropores that contribute more to the internal surface of the carbon are inaccessible to the larger molecules of amyl alcohol. Substituents in the aromatic ring, such as in o and p-xylene, iso-propylbenzene also reduce the preferential adsorption of these compounds from their benzene solutions. These substituent groups increase the distance between the plane of the aromatic ring and the solid surface when the plane is parallel to the surface and between the planes of adjacent rings when the plane is perpendicular to the surface. In each case a decrease in adsorption takes place.

3.3.2 DEPARTURES FROM USUAL COMPOSITE ISOTHERM SHAPES

Several cases of departures of shapes in the composite isotherms have been reported. Puri[13] has reported the composite adsorption isotherms from ethylene glycol-water solutions in which both the components are polar in character, on a carbon black spheron-6 before and after outgassing at 1400°C. Both the isotherms are V shaped. The composite isotherm on the oxygen-free Spheron-6 showed one peak whereas that on the oxygen containing Spheron-6 shows two distinct peaks (Figure 3.21). The composite isotherms with two peaks have been termed *stepped isotherms* by Kipling et al.[4] When the two linear regions were extrapolated to zero concentration, the amounts adsorbed were found to be in the ratio 1:2, indicating the possibility of the formation of a second layer. Similar two-step composite isotherms were obtained

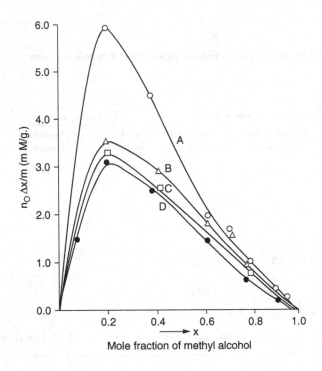

FIGURE 3.19 Composite adsorption isotherms on various sugar charcoal samples from methyl alcohol-carbon tetrachloride mixtures (Curves A, B, C, and D represent original sugar charcoal and charcoal samples degassed at 400, 750, and 1100°C, respectively.) (After Puri, B.R., Kumar, S., and Sandle, N.K., *Indian J. Chem.*, 1, 418, 1963. With permission.)

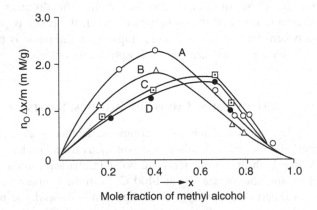

FIGURE 3.20 Composite adsorption isotherms on various sugar charcoal samples from methyl alcohol-*n*-amyl alcohol mixtures (Curves *A*, *B*, *C*, and *D* represent original sugar charcoal and charcoal samples degassed at 400, 750, and 1100°C, respectively.) (After Puri, B.R., Kumar, S., and Sandle, N.K., *Indian J. Chem.*, 1, 418, 1963. With permission.)

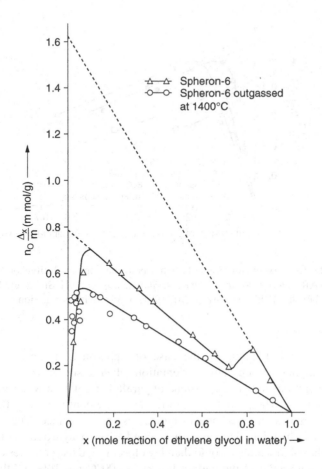

FIGURE 3.21 Composite isotherms on Spheron-6 from glycol-water binary solutions. (After Puri, B.R., in *Activated Carbon Adsorption,* I.H. Suffet and M.J. McGuire, Eds., Ann Arbor Science Publishers, Ann Arbor MI, 1981, p. 353. With permission.)

by Cronford et al.[102] for adsorption from binary solutions of n.butyric acid and cyclohexane on Spheron-6 outgassed at 1000°C. These workers attributed the first peak to a mixed monolayer and the second peak to the formation of a second layer containing a higher proportion of butyric acid.

Puri[13] has also reported composite isotherms for phenol-benzene binary solutions on two samples of carbon blacks namely Spheron-6 and Spheron–C on weight basis and per unit surface and observed the isotherms to be V shaped (Figure 3.22), indicating preference of the carbon surface for phenol. The isotherm reaches a maximum value and then falls linearly downward. The linear portion was extrapolated to zero concentration and the amount of phenol required to form a monolayer was calculated taking molecular areas of phenol as 52 A^{02} and 25 A^{02}, depending on parallel or perpendicular orientations of phenol molecules on the surface. The value of the intercepts were found to be 0.57 mg in the case

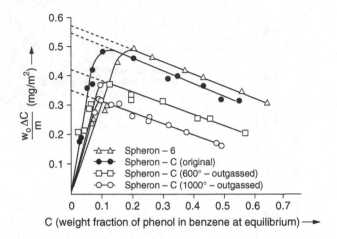

FIGURE 3.22 Composite isotherms of phenol-benzene binary solutions on Spheron-6 and Spheron-C. (After Puri, B.R., in *Activated Carbon Adsorption,* I.H. Suffet and M.J. McGuire, Eds., Ann Arbor Sci. Publ., Ann Arbor MI, 1981, p. 353. With permission.)

of Spheron-6 and 0.55 mg in the case of Spheron-C, indicating that phenol molecules acquire perpendicular orientation when adsorbed on oxygenated surfaces from solutions in benzene instead of parallel orientation when adsorbed from aqueous solutions. The perpendicular orientation evidently permits the accommodation of a larger number of phenol molecules on the surface. The perpendicular orientation of phenol molecules is also conducive to hydrogen bonding as OH groups of phenol molecules would then lay closer to the oxygen sites on the carbon surface. On degassing of the carbon blacks at 600°C and 1000°C, the magnitude of adsorption was found to decrease. The extrapolation of the linear portions to zero concentration showed that with the gradual elimination of oxygen from the carbon surface, the orientation of the phenol molecules in the adsorbed phase tends to change from perpendicular to parallel positions.

The shapes of the composite isotherms on sugar characoals associated with varying amounts of chemisorbed oxygen from binary solutions of benzene and nitrobenzene are also different from the general types given by Schay and Nagi discussed earlier in this chapter. The original charcoal prefers the polar nitrobenzene (Figure 3.23), although the preference is much less than that of methanol or ethanol from their mixtures in benzene. This has been attributed partly to the much larger molecular dimensions of nitrobenzene and partly to its smaller tendency to form hydrogen bonds. The composite isotherm for 750°C-outgassed charcoal is rather unusual as the order of preference varies in a wave-like manner. This may be due to the aromatic character of this charcoal so that it shows almost the same preference for the two components. The wave nature of the isotherm indicates the existence of different sites of varying energies that are occupied at varying concentrations of the solution.

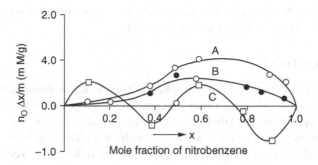

FIGURE 3.23 Composite adsorption isotherms on various sugar charcoal samples from nitrobenzene-benzene mixtures (Curves *A*, *B*, *C*, and *D* represent original sugar charcoal and charcoal samples degassed at 1100 and 750°C, respectively (After Puri, B.R., Kumar, S., and Sandle, N.K., *Indian J. Chem.*, 1, 418, 1963. With permission.)

3.3.3 POROSITY OF THE ADSORBENT

Activated carbons have a spread of pore sizes. Consequently the possibility that they can show a partial molecular sieve effect cannot be overlooked when the components of the binary solution are not of the similar molecular dimensions. This factor would add a degree of preferential adsorption of the components of smaller size molecules irrespective of the competitive adsorption due to other factors. The composite isotherms would, therefore, be of the type obtained on heterogeneous surfaces. This competitive adsorption effect will be more prominent and visible when carbons are produced from the same source raw materials by different procedure or post preparation treatments. For example, carbons that have been produced after varying degrees of activation or carbons that are heat treated at varying temperatures after activation will have different porosities and pore size distribution. The extremely fine micropores get partially blocked as the final heat treatment temperature exceeds 800°C to 900°C, due to the calcinations of the pores. This will produce molecular sieve effect depending upon the heat treatment temperature.

3.3.4 SURFACE HETEROGENEITY

Amorphous carbons that include charcoals, active carbons, and carbon blacks are varied substances and have generally energetically heterogeneous surfaces. In fact the surface can not be considered as elementary carbon but as a *polynuclear hydrocarbon*[103] interspersed with different types of carbon-oxygen surface groups and in some cases with groups containing other elements such as nitrogen, sulfur, and halogens. In addition, the carbon surface has defects and crevices. The presence of these surface groups and defects creates a certain degree of heterogeneity in all these carbon materials, depending on the source raw material and the history of their formation. These heterogenieties on the carbon surface create a certain element of preferential adsorption when binary solutions with different sizes and with varying polarities are used. This will result in different parts of the surface interacting differentially with the two components of the binary solution and will thus cause a change in the shape of the composite isotherm.

3.3.5 STERIC EFFECTS

Wheeler and Lavy[104] studied the adsorption of compounds with substituted aromatic rings such as o-and p-xylene, t-butylbenzene and isopropylbenzene from their benzene solutions and observed that the substituted aromatic compounds were adsorbed much less strongly than benzene at low concentrations. This can be attributed to the influence of the constituent groups in increasing the distance between the plane of the aromatic ring and the adsorbent surface or between the planes of the adjacent rings depending upon whether the plane is parallel or perpendicular to the surface. Similarly, in the case of adsorption of phenols from solutions of cyclohexane, the adsorption of phenols with substitutents in the 2,6 positions was much smaller than that of unsubstituted phenol, indicating that the steric effects reduced the possibility of hydrogen bonding to the surface through the hydrogen group. Puri et al.[90] while studying the adsorption from binary solutions on charcoals, also observed that the preference for nitrobenzene was much smaller than that of methanol or ethanol due to the much larger size of the nitrobenzene molecule.

3.3.6 ORIENTATION OF ADSORBED MOLECULES

Orientation of the molecules in the adsorbed phase also influences the adsorption of a component of the binary solution as it determines the extent of adsorption of a given component. An ideal adsorbate is one that has a spherical molecule as the question of orientation of the molecule at the surface does not then arise. However, for a highly unsymmetrical adsorbate molecule, the orientation that it adopts in the adsorbed phase on the surface of the adsorbent is of considerable consequence because it determines the number of molecules that can occupy a unit area of the surface. This in turn changes the value of the term $\frac{n_o \Delta x}{m}$ in the composite isotherm equation. Although for many small molecules, the areas corresponding to various possible orientations differ by less than a factor of 2, but this may represent a much greater degree of uncertainly than the experimental error in determining the extent of adsorption. For dye molecules, there are three possible orientations, one dimension of which is often much smaller than the other two. Thus, the effect of orientation on the adsorption may be larger. The molecule of stearic acid when adsorbed with its hydrocarbon chain parallel to the surface occupies approximately $112 \, A^{02,13,105}$ which is roughly five times the area that it occupies when the hydrocarbon chain is oriented perpendicular to the surface. The latter orientation is adopted only on very polar surfaces or when chemisorption takes place. In such a case the adsorption of a nonpolar component is almost excluded.

It is, however, worthwhile to mention that the orientation of a molecule in the adsorbed phase cannot be determined directly. It can only be assumed that the orientation adopted by a given adsorbate molecule is the same as in the case of vapor phase adsorption on the concerned solid although this may not be completely true in all cases of adsorption from binary solutions. For example, stearic acid is oriented with the major axis perpendicular to the surface as close packed films on the surface of water. This orientation also occurs in adsorption on some forms of alumina.[106] In the case of carbons, however, the acid is adsorbed with the major axis parallel to the carbon surface,[13,105] an orientation that is probably adopted by normal hydrocarbons and most of their derivations on carbon surfaces.

3.4 DETERMINATION OF INDIVIDUAL ADSORPTION ISOTHERMS FROM COMPOSITE ISOTHERMS

The isotherm of concentration change or the composite isotherm represents the combined adsorption of the two components of the binary solution. It would be of interest to calculate the individual adsorption isotherms of the two components of the solution from the composite isotherm. The composite isotherm equation

$$\frac{n_o \Delta x}{m} = n_1^{s'} - n_2^s x_1$$

contains two unknowns n_1^s and n_2^s each of which must be separately known before the individual isotherms can be determined. This can not be done accurately without additional information relating to n_1^s or n_2^s or both. For determination of n_1^s and n_2^s, it is essential to analyse the adsorbed layer by separating it from the bulk solution phase for which no method exists. Thus, several other methods have been suggested that can be experimental or theoretical. The first experimental approach was suggested by Williams[107] and has been used by several workers for the adsorption by charcoal from binary solutions.[58,108-110] Williams assumed that the solid adsorbent will adsorb the same amount of each of the two components when it is exposed to the vapor in equilibrium with the solution, because it will adsorb when placed in the solution itself so that

$$W = W_1^s + W_2^s \tag{3.28}$$

$$W = n_1^s M_1 + n_2^s M_2 \tag{3.29}$$

where W is the increase in weight per gram of the solid adsorbent when the latter is exposed to the vapor in equilibrium with a solution having the composition represented by x_1 and can be measured directly by weighing in a vapor adsorption experiment, and M_1 and M_2 are molar masses of the two components. It is assumed in this experiment that the equilibrium at the solid-adsorbate interface so established is the same as would be established when the solid is immersed in the solution, although, this may not always be true.

There are now two equations for the two unknowns n_1^s and n_2^s corresponding to the liquid composition x_1. By determining W for a series of liquid compositions covering the whole range from $x_1 = 0$ to $x = 1$, the plots for either n_1^s or n_2^s as functions for x_1, or x_2 i.e., the individual adsorption isotherms for the two components can be determined. The procedure, however, requires numerous measurements of adsorption from the vapor phase. Furthermore, the attainment of equilibrium for individual components that have low saturation vapor pressure may take even weeks.

The Williams method also appears to be inaccurate for adsorbents with wider micropores in which capillary condensation can take place. When such an adsorbent is exposed to the mixture of vapors from the two components of the solution, two types of adsorptions take place. In the immediate neighborhood of the surface, primary adsorption occurs giving an adsorbate of a composition different from that of the solution phase. But the liquid filled in the pores as a result of capillary condensation will have the same composition as that of the solution phase. In other words, the condensate in the pores is equivalent to the bulk solution phase and it is held mechanically in the pores. Thus, it is difficult to separate the primary adsorption from the adsorption by capillary condensation. Jones and Outridge,[111] determined adsorption isotherms on silica gel from vapors of n.butyl alcohol and benzene above their solutions. They calculated the amount of primary adsorption using the Freundlich equation, assuming that each individual isotherm was governed by this equation. However, the individual adsorption isotherms were lower than those obtained by applying William's direct weighing procedure. This difference could be ascribed to adsorption by capillary condensation.

Elton[112] suggested an alternative to Williams equation that was later developed by Kipling and Tester.[59] Elton postulated that the adsorption both from solution and vapor phase is confined to a monolayer that is always complete at all concentrations of the solution so that

$$S = n_1^s S_1 + n_2^s S_2 \tag{3.30}$$

where S is the total area of one gram of the adsorbent, and S_1 and S_2 are the areas that would be occupied by one mole of each respective component (molar areas) if it were adsorbed separately.

If $(n_1^s)_m$ and $(n_2^s)_m$ are the number of moles of the individual components respectively required to completely cover the surface of unit mass of the solid adsorbent (monolayer capacities), then

$$S = S_1 \left(n_1^s\right)_m \tag{3.31}$$

$$S_1 = \frac{S}{\left(n_1^s\right)_m} \tag{3.32}$$

Similarly,

$$S_2 = \frac{S}{\left(n_2^s\right)_m} \tag{3.33}$$

Substituting Equation 3.32 and Equation 3.33 in Equation 3.30, we get

$$\frac{n_1^s}{\left(n_1^s\right)_m} + \frac{n_2^s}{\left(n_2^s\right)_m} = 1 \qquad (3.34)$$

This equation has four unknowns. But the two unknowns $(n_1^s)_m$ and $(n_2^s)_m$ can be determined by direct measurement of adsorption that occurs when the solid is separately exposed to the saturated vapors of each of the two pure component as it has been assumed that the adsorption from the vapor phase over the pure components is the same as from the liquid phase. Thus, combining Equation 3.34 and Equation 3.17, two unknowns n_1^s and n_2^s for a given value of $\times$ can be calculated. Kipling and Tester[59] used this treatment to their composite isotherms from benzene-ethanol, benzene-acetic acid, and benzene-ethylene dichloride solutions on charcoal, and obtained the individual adsorption isotherms (Figure 3.24(a) and Figure 3.24(b))

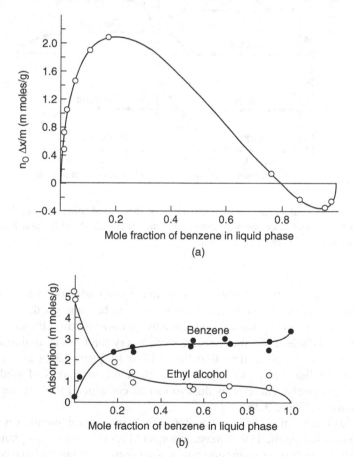

FIGURE 3.24 Composite isotherms (a) and individual isotherms (b) for adsorption from solutions of benzene and ethanol on charcoal. (Kipling, J.J. and Tester, D.A., *J. Am. Chem. Soc.*, 1952, 4123. With permission.)

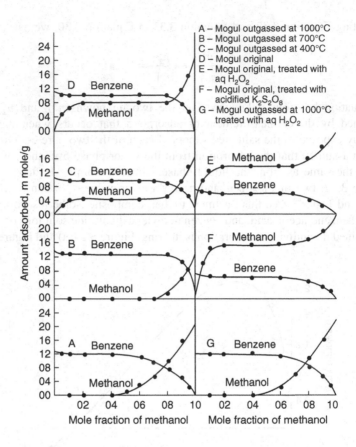

FIGURE 3.25 Individual adsorption isotherms of methanol and benzene from their mixtures on Mogul before and after various treatments. (Puri, B.R., Singh, D.D., and Kaistha, B.C., *Carbon*, 10, 481, 1972. With permission.)

and verified these isotherms with those obtained from adsorption from the vapor phase. The agreement between the two was found to be quite good.

Puri et al.[94] also calculated individual adsorption isotherms (Figure 3.25) from the composite isotherms for methanol-benzene binary solutions assuming adsorption to be monolayer and combining Equation 3.17 and Equation 3.34 as was done by Gasser and Kipling[83] and using 16.7 A^{02} as the molecular areas of methanol and benzene respectively. These individual isotherms correspond to the composite isotherm shown in Figure 3.14.

In addition to the Williams and Gasser-Kipling treatments, several other approaches based on the law of mass action and Freundlich and Langmuir adsorption equations have been suggested but none of these approaches have been found to be practical and, therefore, are seldom applied. The Gasser-Kipling method is still commonly used to calculate the individual isotherms corresponding to a given composite isotherm.

3.5 THICKNESS OF THE ADSORBED LAYER

It has been discussed in the previous section of this chapter that the individual isotherms can be calculated by assuming that the adsorption is monolayer. This, however, does not necessarily prove that adsorption from solutions cannot be multilayer. In fact cases have been reported where adsorption from solution phase is certainly multimolecular.

According to monolayer adsorption, the highest value that $n_o \Delta x/m$ can have corresponds to complete coverage of the adsorbent surface by any one component and no adsorption of the other component so that the composite isotherm equation is reduced to

$$\frac{n_o \Delta x}{m} = \left(n_1^s \right)_m (1-x) \tag{3.35}$$

If, therefore, the value of $\frac{n_o \Delta x}{m}$ for any value of x exceeds that of $(n_1^s)_m (1-x)$, the composite isotherm can not be resolved on the monolayer hypothesis so that adsorption is more than a monolayer. Schay and Nagy pointed out that the evidence supporting the hypothesis that adsorption is confined to a monolayer is to a large extent circumstantial. They provided[74] an analysis of the composite isotherm that was discussed by Cronford and Kipling.[113] According to this viewpoint, some composite isotherms have a substantially linear portion (Figure 3.26) over which the composite isotherm equation

$$\frac{n_o \Delta x}{m} = n_1^s (1-x) - n_2^s x$$

can be written as

$$\frac{n_o \Delta x}{m} = n_1^s - \left(n_1^s + n_2^s \right) x \tag{3.36}$$

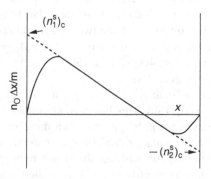

FIGURE 3.26 Composite isotherm showing a linear section.

which defines a straight line. Schay and Nagy consider that over this portion of the isotherm the composition of the adsorbed phase remains constant (i.e., n_1^s and n_2^s), the number of moles of component 1 and 2 in the adsorbed phase remains constant over that range of concentration. Therefore, extrapolation of the linear section of the isotherm gives intercepts on the ordinate through $x = 0$ and $x = 1$, and these correspond to n_1^s and n_2^s, respectively, which define composition of the adsorbed phase over that range of concentration. Knowing the monolayer values $(n_1^s)_m$ and $(n_2^s)_m$ for the two components of the binary solution when adsorbed on the solid adsorbent under investigation, the mean molecular thickness, t, of the adsorbed layer can be calculated using the equation

$$t = \frac{n_1^s}{\left(n_1^s\right)_m} + \frac{n_2^s}{\left(n_2^s\right)_m} \tag{3.37}$$

The monolayer value in each case can be obtained by dividing the surface area of the adsorbent by the molecular area of the adsorbate.

Puri and coworkers[90] calculated thickness of the adsorbed layer in the case of adsorption on sugar charcoals from binary solutions of ethanol-benzene, methanol-benzene, methanol-carbon tetrachloride, and methanol-n.amyl alcohol. The molecular thickness of each component in the adsorbed phase and total molecular thickness of each layer for the various adsorbate-adsorbent systems (Table 3.9) indicate that the thickness of ethanol-benzene as well as methanol-benzene layers is nearly trimolecular on the original sugar charcoal and nearly bimolecular on the original coconut shell charcoal as well as silica gel. The thickness of the adsorbed phase from methanol-carbon tetrachloride solutions on the original sugar charcoal is also trimolecular. This formation of multimolecular layers on charcoal and silica gel samples is indicative of the heterogeneous nature of these materials.[81,82,108] These workers suggested that hetrogeniety in the case of the original charcoals arises from the presence of CO_2 complex and in the case of silica gel due to the presence of oxide and hydroxide surface groups. The ratio of the thickness of the benzene or carbon tetrachloride layer to that of the alcohol layer (methanol or ethanol) indicates that the competitive adsorption is influenced not only by strong interaction between the alcohol molecules and the CO_2 complex or oxide groups (in case of silica gel) on the adsorbing surface but also by interaction or by mutual solubility of the two liquid components. When the CO_2-complex, the cause of surface heterogeneity in charcoals, was removed by degassing at temperatures between 400°C and 750°C, the thickness of the adsorbed layer approaches unity or even less.

In the case of methanol-n.amyl alcohol on the original charcoal, the thickness of the adsorbed layer is only 1.48. This has been attributed to the fact that methanol and amyl alcohol have the same polar group in their molecules, so they both compete for the active CO_2 complex sites present on the charcoal surface. As the alcohol molecules lay flat on the carbon surface, the larger hydrocarbon portion in amyl alcohol occupies more than one active site, renderingthem inaccessible to interaction with the OH group of the alcohols.

TABLE 3.9
Thickness of the Absorbed Layer

	Ethyl Alcohol-Benzene Mixture			Methyl Alcohol-Benzene Mixture			Methyl Alcohol-Carbon Tetrachloride Mixture			Methyl Alcohol-n-Amyl Alcohol Mixture		
Sugar charcoal												
Original charcoal	1.96	0.92	2.88	2.01	0.94	2.95	2.18	0.87	3.05	1.26	0.22	1.48
Degassed at 400°	0.74	0.45	1.19	1.07	0.63	1.70	0.96	0.13	1.09	—	—	—
Degassed at 750°	0.32	0.48	0.80	0.43	0.37	0.80	0.88	0.11	0.99	—	—	—
Degassed at 1100°	0.46	0.40	0.86	0.69	0.36	1.05	0.91	0.11	0.93	—	—	—
Coconut shell charcoal												
Original charcoal	1.31	0.81	2.12	1.33	0.62	1.95	—	—	—	—	—	—
Degassed at 750°	0.41	0.65	1.06	0.42	0.71	1.13	—	—	—	—	—	—
Degassed at 1100°	0.75	0.44	1.19	0.83	0.22	1.05	—	—	—	—	—	—
Silica gel	1.38	0.72	2.10	—								

Source: Puri, B.R., Kumar, S., and Sandle, N.K., *Indian J. Chem.*, 1, 418, 1963. With permission.

TABLE 3.10
Thickness of the Absorbed Layer on Various Charcoals

	Thickness of the Absorbed Layer, t		
Sample	Methanol	Benzene	Total
Original sugar charcoal	1.30	0.68	1.98
1000°-outgassed charcoal	0.18	1.00	1.18
H$_2$O$_2$-treated charcoal	1.62	0.32	1.94
PVC charcoal	0.90	0.70	1.60

Source: Bansal, R.C. and Dhami, T.L., *Carbon*, 15, 153, 1977. Reproduced with permission from Elsevier.

Bansal and Dhami[93] also calculated thickness of the adsorbed phase in the case of adsorption from benzene-methanol binary solutions on sugar charcoals modified by degassing at 1000°C and by oxidation with hydrogen peroxide using the above method. It is seen (Table 3.10) that the thickness is more than unimolecular in case of the charcoal samples associated with carbon-oxygen surface groups and is unimolecular in case of the 1000°-degassed sample that contains little or no associated oxygen. These workers suggested the occurrence of specific adsorption of methanol at oxygen containing sites on the carbon surface, where it could be more strongly held possibly by chemical or quasi-chemical forces involving hydrogen bonding. Benzene probably cannot be specifically held on these sites and occupies the rest of the surface. This phenomenon is more likely to be noticed at low concentrations of methanol, because it then is present to a greater degree as monomeric molecules having free hydroxyl groups, whereas it exists as dimers or higher polymers at higher concentrations in relatively non polar solvents and is probably less able to form strong bonds at the oxygen sites.

3.6 CHEMISORPTION FROM BINARY SOLUTIONS

Adsorption from binary solutions of completely miscible simple organic components as discussed above at room temperature is usually expected to be physical in nature. It has, however, been found that in the case of polar adsorbents such as oxygenated carbon and silica gel, a certain amount of one or both components of the solution is irreversibly held (chemisorbed), the amount of the component chemisorbed depending upon the adsorbate- adsorbent interaction that in turn is determined by the polarity of the adsorbent and that of the adsorbate. Kipling and Peakall,[81] while studying adsorption on silica, titania, and alumina gels from binary solutions of ethanol and benzene, observed that the maximum in the composite isotherm corresponded to more than a monolayer. This could be attributed to multilayer adsorption or to a pore filling process, but the slow establishment of the equilibrium and very little effect of increased temperature on the amount adsorbed indicates that the process involved the occurrence of chemisorption. Chemisorption was also observed in the case of adsorption, on a carbon black Spheron-6 from solutions of acetic acid and alcohols. It was suggested that acetic acid reacts with n lactone groups believed to be present on the surface of this carbon black.

The amount of a component chemisorbed in adsorption from solutions can be determined by exposing the adsorbent to the components, which is likely to be chemisorbed, until the adsorption is complete. The modified adsorbent is then evacuated at about 100°C to remove the amount of the component that is physically adsorbed. The modified adsorbent obtained after evacuation is used for the adsorption from binary solutions. Bansal and Dhami[93] determined the amount of methanol chemisorbed on sugar charcoals from binary solutions of methanol and benzene. These workers treated sugar charcoal samples and a sample of PVC (Polyvinyl Chloride) charcoal with methanol in order to fully cover the oxygenated sites with methanol. The treated samples were then evacuated at 120°C to remove any physisorbed methanol and the composite isotherms determined from methanol-benzene solutions on both the untreated and methanol-treated samples. The composite isotherms in case of oxygen containing samples (Figure 3.18(a), Figure 3.18(b), and Figure 3.18(c)) show clearly that the magnitude of preference for methanol decreases considerably after the pretreatment. However, in the case of the 1000°-degassed sample (Figure 3.18 (d)), the magnitude of preference remains more-or-less the same, and the two isotherm overlap along the entire concentration range.

As discussed earlier, the extrapolation of the linear sections of the composite isotherm to $x = 0$ (Figure 3.20) can give the amount of methanol adsorbed from the binary solution. The amount adsorbed in the case of the untreated sample is the total of physisorbed and chemisorbed methanol while the amount adsorbed on the methanol treated sample is largely the physisorbed value. The breakup of methanol into physisorption and chemisorption obtained using the linear sections of the composite isotherms (Table 3.8) shows that the oxygen containing samples chemisorb appreciable amounts of methanol. Puri and Murari[98] and Bansal et al.[93] while studying the adsorption isotherms of methanol vapors on different charcoals, observed that a certain amount of methanol could not be desorbed and was irreversibly held on those carbons that contained associated oxygen on their surface. These values agree fairly well with the values obtained from the composite isotherms (*cf.* Table 8). The composite isotherms for the adsorption of bromine from solutions in carbon tetrachloride (Figure 3.27) on carbon black Spheron-6 showed a maximum that was too high to correspond to a monolayer adsorption of bromine. It was suggested that both physisorption and chemisorption of bromine occur on Spheron-6. The chemisorption of bromine from solutions in carbon tetrachloride has also been reported by Brooks and Spotswood[115] and Shooter[116] and from aqueous solutions by Puri et al.[117] and Bansal et al.[118]

3.7 TRAUBE'S RULE

Traube in 1891[119] observed that the regularities that are seen in the simple physical properties of a homologous series of organic compounds also extend to their behavior at interfaces. He found that the surface activity of organic solutes in aqueous solutions increased strongly and regularly as in a homologous series. As surface

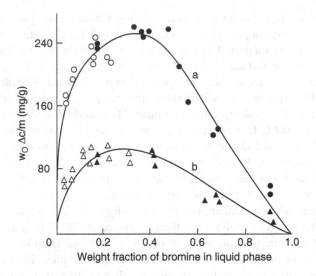

FIGURE 3.27 Composite isotherms from solutions of bromine in carbon tetrachloride on (a) Spheron-6 and (b) Graphon; open symbols represent adsorption from solutions and closed symbols from mixed vapors. (Source: Puri, B.R. and Murari, K., *Res. Bull.*, Panjab Univ., 13, 9, 1962. With permission.)

activity is related to adsorption, Traub's rule has been expressed in terms of the regular increase in the strength of adsorption of successive members of a homologous series as measured by the work done to transfer one mole of the solute from the bulk of solution phase to the solid-phase to the solid-solution interface.

Frendlich,[120] on the basis of adsorption isotherms of homologous series of organic compounds such as fatty acids, aromatic acids, esters and other single-functional group compounds on silica gel and different charcoals and carbon blacks, enunciated the rule that states: " ... the adsorption of organic substances from aqueous solutions increases strongly and regularly as we ascend the homologous series. Figure 3.28 illustrates this rule for the adsorption of a series of lower fatty acids (e.g., formic, acetic, propionic, and butyric) on a carbon surface. The initial slopes and hence *b* values of Langmuir equation increase in the order formic acid, acetic acid, propionic acid, and butyric acid. It was suggested that the molecules of the fatty acids were adsorbed with their major axis parallel to the surface of the carbon so that each $-CH_2$ group contributes an equal amount to the energy of adsorption. These investigations were extended to monocarboxylic acids, dicarboxylic acids, alcohols, and ethyl esters by Claesson.[121] The rule was also found to be applicable to the adsorption of solutions of paraffin in benzene on charcoal.[122]

When the adsorption of same fatty acids from aqueous solutions was studied on a charcoal prepared under strongly oxidizing conditions (Figure 3.29), Traube's rule shows a reversal, which means that the magnitude of adsorption decreases as we ascend the homologous series. This has been attributed to the polar nature of the charcoal surface, due to the presence of polar carbon-oxygen surface groups, which

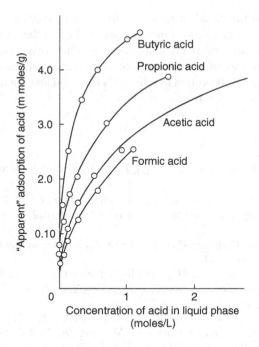

FIGURE 3.28 Adsorption isotherms of lower fatty acids from their dilute aqueous solutions on charcoal. (Adapted from Freundlich, H., *Colloid and Capillary Chemistry*, Methuen, London, 1926. With permission.)

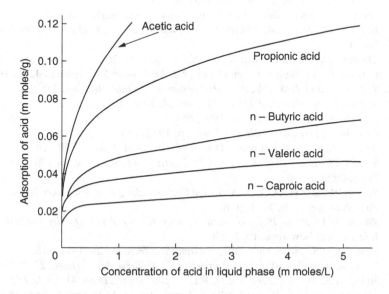

FIGURE 3.29 Adsorption isotherms of fatty acids from aqueous solutions on an oxidized charcoal. (After Kipling, J.J., in *Adsorption from Solutions of Non-Electrolytes*, Acad. Press, London, 1965. With permission.)

results in an increase in the adsorption of polar water molecules on the surface polar sites. Traube's rule in fact had originally been applied to the adsorption of solutes from a polar solvent on a relatively non polar solid. Similar reversal of Traube's rule was observed in the case of adsorption from a relatively nonpolar solvent by a polar solid. Examples are the adsorption of fatty acids from toluene[123] and of nitro paraffins from benzene[124] on silica gel.

REFERENCES

1. Freundlich, H. and Heller, W., *J. Am. Chem. Soc.*, 61, 228, 1939.
2. Kipling, J.J., *J. Chem. Soc.*, 1939, p. 1483.
3. Dubinin, M.M. and Zaverina, E.D., *Acta Phys. Chem., USSR,* 4, 47, 1936.
4. Kipling, J.J., in *Adsorption from Solutions of Non-Electrolytes*, Acad. Press, London, 1965.
5. Ottewill, R.H., Rochester, C.H., and Smith, A.L., in *Adsorption from Solution*, Acad Press, New York, 1983.
6. Mattson, J.S. and Mark, H.B., Jr., in *Activated Carbon: Surface Chemistry and Adsorption from Solution*, Marcel Dekker, New York, 1971.
7. Rovinskaya, T.M., *Koll. Zh.,* 24, 215, 1962.
8. Al-Baharani, K.S. and Martin, R.J., *Water Res.,* 10, 731, 1976.
9. Kogamovsky, A.M. and Rovinskaya, T.M., *Koll. Zh.,* 25, 447, 1963.
10. Abe, I., Hayashi, K., and Kitagaga, M., *Bull. Chem. Soc. Japan,* 53, 1199, 1980.
11. Puri, B.R., Sharma, S.K., and Dosanjh, I.S., *J. Indian Chem. Soc.,* 53, 486, 1976.
12. Lemeiur, R.U. and Morrison, J.L., *Can. J. Res.,* 25B, 440, 1947.
13. Puri, B.R., in *Activated Carbon Adsorption,* I.H. Suffet and M.J. McGuire, Eds., Ann Arbor Sci. Publ., Ann Arbor, MI, 1981, p. 353.
14. Arora, V.M., Ph.D. dissertation, Panjab Univ., Chandigarh, India, 1977.
15. Jaroniec, M., Madey, R., Choma, J., and Piotrouska, J., *J. Colloid Interface Sci.,* 125, 561, 1988.
16. Choma, J. and Jaroniec, M., *Water Sci. Technol.,* 24, 269, 1991.
17. Enrique, C., Guillerino, C., and Louis, M., *Adsorption Sci. Technol.,* 4, 59, 1987.
18. Worch, E., and Zakke, R., *Acta Hydrochim. Hydrobiol.,* 14 1986 305.
19. Magne, P. and Walker, P. L., Jr., *Carbon,* 24, 1986, 101.
20. Aytekin, C., *Spectros Lett.,* 25, 1991, 653.
21. Chaplin, J., *Phys. and Colloid Chem.,* 36, 1932, 653.
22. Kiselev, A.V. and Krasilinikov, D.K., *Doklady Akad. Nauk. USSR,* '86, 1952, 111.
23. Morris, J.C. and Weber, W.J., *Proc. 1st Intern. Conf. Water Pollution Res.,* Pergamon Press, Oxford, 1962.
24. Singer, P.C. and Yen, C.Y., in *Activated Carbon Adsorption,* Vol. I, Ann Arbor Science Pub., Ann Arbor MI, 1981, p. 8.
25. Weber, W.J., Jr., in *Physicochemical Processes for Water Quality Control,* Wiley Interscience, New York, 1972, 230.
26. Abuzaid, N.S. and Nakhla, G.F., *Environ. Sci. Technol.,* 28, 1994, 216.
27. Vidic, R.D., Sudan, M.T., and Brenner, R.C., *Environ. Sci. Technol.,* 27, 1993, 2070.
28. Juang, R.S., Wu, F.C., and Tseng, R.L., *J. Chem. Eng. Data,* 41, 1996, 487.
29. Moreno-Castilla, C, Rivera-Utrilla, J., Lopez-Ramon, M.V., and Carrasco-Martin, F., *Carbon,* 33, 1995, 845.
30. Radeke, K.H., Seidel, A., Spitzer, P., Jung, R., Jankowska, H., and Neffe, S., *Chem. Tech.,* 41, 1989, 32.

31. Hanmin, A. and Yigun, L., *Shichuli Jishu,* 13, 1987, 342.

32. Graham, D., *J. Phys. Chem.,* 59, 1955, 896.

33. Coughlin, R.W., Ezra, F.S., and Tan, R.N., *J. Colloid Interface Sci.,* 28, 1968, 386.

34. Clauss, A., Boehm, H.P., and Hofmann, U., *Z. Anorg. Chem.,* 35, 1957, 35.

35. Puri, B.R., Bhardwaj, S.S., and Gupta, U., *J. Indian Chem. Soc.,* 53, 1976, 1095.

36. Mahajan, O.P., Castilla, M.C., and Walker, P.L., Jr., *J. Sep. Sci. Technol.,* 15, 9, 1980, 1733.

37. Keizo, O., Hiroyuki, T, Kiyoski, Y., and Hiroshi, T., *Nippon Kagaku Kaishi,* 3, 1981, 321.

38. Rackovskaya, L.N, Felenolev, V.B., Levitskii, E.A., Kriksima, T.M., Moroz, E.M., Afanansev, A.D., Efremov, A.I., and Elbert, E.I., *Izv. Akad. Nauk SSSR, Ser. Khim. Nauk.,* 5, 40, 1982.

39. Bansal, R.C., Aggarwal, D., Goyal, M., and Kaistha, B.C., *Indian J. Chem. Technol.,* 9, 290, 2002.

40. Goyal, M., Ph.D. dissertation, Panjab Univ., Chandigarh, India, 1997.

41. Aggarwal, D., Ph.D. dissertation, Panjab Univ., Chandigarh, India, 1998.

42. Furdenegg, G.H., Koch, C., and Liphard, M., in *Adsorption from Solutions*, R.H. Ottewill, C.H. Rochester, and A.L. Smith, Eds., Academic Press, New York, 1983, p. 87.

43. Prakash, S., *Carbon,* 12, 483, 1974.

44. Goyal, M. and Bansal, R.C., *Porous Carbons,* 97, 1998.

45. Capelle, A., in *Activated Carbon: A Fascinating Material,* A Capelle and and F. DeVooys, Eds., Norit. N.V., Netherlands, 1983, p. 191.

46. Bansal, R.C., *13th Bienn. Conf. on Carbon,* Univ. of California at Irvine, CA, 1977, Ext. Abstracts, p. 286.

47. Tamamushi, B. and Tamaki, K., in *Proc. Second Intern. Cong. Surface Activity,* 1957, Vol. 3, Butterworth, London, 1955, p. 449.

48. Tamamushi, B., in *Adsorption from Solutions*, R.H. Ottewill, C.H. Rochester, and A.L. Smith, Eds., Academic Press, New York, 1983, p. 79.

49. Tamamushi, B. and Tamaki, K., *Trans. Faraday Soc.,* 55, 1007, 1959.

50. Eda, K.D., *J. Chem. Soc. Japan,* 80, 343, 1960.

51. Yamada, H., Fukumura, K., and Tamamushi, B., *Bull. Chem. Soc. Japan,* 53, 3054, 1980.

52. Goyal, M., Rattan, V.K., Aggarwal, D., and Bansal, R.C., *Colloids and Surfaces A: Physicochem and Eng. Aspects,* 190, 229, 2001.

53. Goyal, M., Rattan, V.K., and Bansal, R.C., *Indian J. Chem. Technol.,* 6, 305, 1999.

54. Aggarwal, D., Goyal, M., and Bansal, R.C., *Carbon,* 37, 1989, 1999.

55. Radke, C.J. and Prausnitz, J.M., *Ind. Eng. Chem. Fundam.,* 11, 445, 1972.

56. Brown, C.E. and Everett, D.H., in *Colloid Science,* D.H. Everett, Ed., *The Chem. Soc.,* London, Vol. 2, 1975.

57. John, P.T. and Nagpal, K.C., *Indian J. Technol.,* 18, 261, 1980.

58. Polanyi, M., *Z. Physik,* 2, 117, 1920.

59. Kipling, J.J. and Tester, D.A., *J. Am. Chem. Soc.,* 1952, 4123.

60. Hansen, R.S. and Fackler, W.B., Jr., *J. Phys. Chem.,* 57, 6346, 1953.

61. Manes, M. and Hoffer, L.J.E., *J. Phys. Chem.,* 73, 584, 1969.

62. Wohler, D.A. and Manes, M., *J. Phys. Chem.,* 75, 3720, 1971.

63. Chion, C.C.T. and Manes, M., *J. Phys. Chem.,* 77, 811, 1923.

64. Schenz, T.W. and Manes, M., *J. Phys. Chem.,* 79, 604, 1975.

65. Rosen, M.R. and Manes, M., *J. Phys. Chem.,* 80, 953, 1976.

66. Everett, D.H., *Trans. Faraday Soc.,* 60, 1803, 1964.

67. Everett, D.H., *Trans. Faraday Soc.*, 61, 2478, 1965.
68. Ash, S.G., Everett, D.H., and Frindenegg, G.H., *Trans. Faraday Soc.*, 64, 2639, 1988.
69. Ash, S.G., Brown, R., and Everett, D.H., *J. Chem. Thermodyn.*, 5, 239, 1973.
70. Ash, S.G., Brown, R., and Everett, D.H., *Trans. Faraday Soc.*, 70, 123, 1974.
71. Everett, D.H., in *Adsorption from Solutions*, R.H. Ottewill, C.H. Rochester and A.L. Smith, Eds., Academic Press, New York, 1983, p. 1.
72. Urano, K., Kochi, Y., and Yamamoto, Y., *J. Colloid Interface Sci.*, 86, 43, 1982.
73. Ostwald, W. and de Izaguirre, R., *Kolloid Z.*, 30, 279, 1922.
74. Nagy, L.G. and Schay, G., *Mag. Kem. Foly*, 66, 31, 1960.
75. Schay, G., Nagy, L.G., and Szekrenyesy, T., *Periodica Polytech.*, 4, 95, 1960.
76. Schay, G. and Nagy, L.G., *J. Chem. Phys.*, 1961, p. 149.
77. Jones, D.C. and Outridge, L., *J. Chem. Soc.*, 1930, p. 1574.
78. Bartell, F.E., Schaffer, G.H., and Sloan, C.K., *J. Am. Chem. Soc.*, 53, 2501, 1953.
79. Kipling, J.J., *2nd Intern. Cong. Surface Activity*, Butterworths, London, 1957, Vol. III, p. 462.
80. Bartell, F.E., Schaffer, G.H., and Sloan, C.K., *J. Am. Chem. Soc.*, 53, 2507, 1931.
81. Kipling, J.J. and Peakall, D.B., *J. Chem. Soc.*, 1958, 184.
82. Kipling, J.J. and Gasser, C.G., *J. Phys. Chem.*, 64, 710, 1960.
83. Gasser, C.G. and Kipling, J.J., *Proc. 4th Carbon Conf.*, Univ. of Buffalo, Pergamon Press, New York, 1960, p. 55.
84. Bartell, F.S. and Lloyd, L.E., *J. Am. Chem. Soc.*, 60, 2120, 1938.
85. Gasser, C.G. and Kipling, J.J., *13th ACS National Meeting*, Chicago, 1958.
86. Puri, B.R. and Bansal, R.C., *R.G.C. Comptes Rendus*, 41, 215, 1964.
87. Puri, B.R. and Bansal, R.C., *Carbon*, 1, 457, 1964.
88. Bansal, R.C., Bhatia, N., and Dhami, T.L., *Carbon*, 16, 65, 1978.
89. Barton, S.S., Gallispie, D.J., Harrison, B.H., and Kemp. W., *Carbon*, 11, 649, 1973.
90. Puri, B.R., Kumar, S., and Sandle, N.K., *Indian J. Chem.*, 1, 418, 1963.
91. Garten, V.A. and Weiss, D.E., *Rev. Pure Appl. Chem.*, 7, 69, 1957.
92. Puri, B.R., Kumar, B., and Singh D.D., *J. Sci. Ind. Res.*, India, 20, 366, 1961.
93. Bansal, R.C. and Dhami, T.L., *Carbon*, 15, 153, 1977.
94. Puri, B.R., Singh, D.D., and Kaistha, B.C., *Carbon*, 10, 481, 1972.
95. Studebaker, M.L., Hoffman, E.W.D., Wolfe, A.C., and Nabors, L.G., *Ind. Eng. Chem.*, 48, 162, 1956.
96. Bhacca, N.S., *Tetrahedron*, 41, 3124, 1964.
97. Bansal, R.C. and Dhami, T.L., *Carbon*, 15, 163, 1977.
98. Puri, B.R. and Murari, K., *Res. Bull.*, Panjab Univ., 13, 9, 1962.
99. Bansal, R.C., Sethi, A.S., and Dhami, T.L., *Indian J. Chem.*, 13, 321, 1975.
100. Jankowska, H., Swiatowsi, A., Oscik, J., and Kusak, R., *Carbon*, 21, 117, 1983.
101. Gregg, C.G. and Stock, R., *Trans. Faraday Soc.*, 53, 1355, 1937.
102. Cronford, P.V., Kipling, J.J., and Wright, E.H.M., *Trans. Faraday Soc.*, 58, 74, 1962.
103. Wynne-Jones, W.F.K., in *Structure and Properties of Porous Materials*, Everett, D.H. and Stone, Eds., Butterworths, London, 1958, p. 35.
104. Wheeler, O.H. and Lavy, E.M., *Can. J. Chem.*, 37, 1235, 1959.
105. Kipling J.J. and Wright, E.H.M., *J. Chem. Soc.*, 1962, p. 855.
106. Russel, A.S. and Cochran, C.N., *Ind. Eng. Chem.*, 42, 1332, 1950.
107. Williams, A.M., Med. K., *Vetenskapsakad Nobilinst*, No. 2, p. 2, 1913.
108. Blackburn, A. and Kipling, J.J., *J. Chem. Soc.*, 1954, 3819.
109. Inner, W.B. and Rowley, H.H., *J. Phys. Chem.*, 51, 1154, 1947.
110. Inner, W.B. and Rowley, H.H., *J. Phys. Chem.*, 51, 1172, 1947.
111. Jones, D.C. and Outridge, L., *J. Chem. Soc.*, 1930, 1574.

112. Elton, G.A.H., *J. Chem. Soc.*, 1951, 2958.
113. Cronford, P.V. and Kipling, J.J., *Trans. Faraday Soc.*, 58, 24, 1962.
114. Kipling, J.J. and Wright, E.H.M., *J. Phys. Chem.*, 67, 1789, 1963.
115. Brooks, J.D. and Spotswood, T.M., *Proc. 5th Conf. on Carbon,* Pergamon Press, Oxford, 1962, p. 416.
116. Shooter, P.V., Quoted in *Adsorption from Solutions of Non-Electrolytes,* by J.J. Kipling, Acad. Press, London, 1965, p. 67, Ref. No. 53.
117. Puri, B.R. and Bansal, R.C., *Carbon,* 3, 523, 1967.
118. Bansal, R.C., Dhami, T.L., and Prakash, S., *Carbon,* 18, 395, 1980.
119. Traube, J., *Annalen,* 265, 27, 1891.
120. Freundlich, H., *Colloid and Capillary Chemistry,* Methuen, London, 1926.
121. Claesson, S., *Arkiv. Kemi, Min. Geol.*, 23, 1, 1946.
122. Baum, A. and Broda, E., *Trans. Faraday Soc.*, 34, 797, 1938.
123. Holmes, H.N. and McKelvey, J.B., *J. Phys. Chem.*, 32, 1522, 1928.
124. Jones, D.C. and Mill, G.S., *J. Chem. Soc.*, 1957, p. 213.

4 Carbon Molecular Sieves

Almost all porous carbons contain pores in a wide range of sizes and do not show any selectivity in the adsorption of molecules of varying dimensions. However, some special carbons have been prepared that are highly microporous and have a large surface area but still do not adsorb in appreciable amounts molecules larger than a particular size. Such carbons are called *molecular sieve carbons* (MSC). These carbons may be viewed as a form of activated carbons, distinct from peat-, coconut-based, and other conventional activated carbons. The microporous structure of such carbons is unique, because the slit-like apertures or *constrictions* of their micropores are of a size similar to the molecular dimensions of the adsorbing species. In the separation of gases, molecules that are smaller than the size of the micropore constriction rapidly diffuse through them into the associated micropore volume. On the other hand, a large molecule is denied access to the volume behind the constriction. A small change in the effective size of the constriction can affect the rate of diffusion of an adsorbing gas molecule to a considerable extent so that its nonactivated diffusion through the pore constriction now becomes activated. These carbons contain pores of molecular width, sometimes only a few Angstrom in size. Usually they form entrances to pores wider than themselves, but these still may be of molecular width. Such ultrafine pores are found in natural materials such as coals and dehydrated zeolites. They can also be formed during the carbonization of certain organic materials such as polymers, sugar, and wood under certain conditions. Naturally occurring coals have small surface areas and are associated with a number of metallic impurities that may start unnecessary side reactions. Therefore, polymer carbon molecular sieves, although expensive, are preferred because they have a high adsorption capacity.

Polymer carbons produced by the carbonization of different polymers and thermosetting materials have reproducible characteristics and uniform pore size and shape that they inherit from the regular structure of their polymer precursors.[1] Two widely investigated carbons of this class are those obtained by the carbonization of polyvinylidene chloride (PVDC) and Saran, a copolymer of vinylidene chloride and vinyl chloride. The porous structure and the adsorption properties of these carbons depend upon several factors that have been investigated by several researchers.[2-20] For example, Bailey and coworkers[11-13] observed that the microporous structure and the characteristics of a polymer carbon depend upon the morphology of the original polymer and the temperature at which the first 10% of the decomposition of the polymer has taken place. Adams et al.[14] prepared carbons from Saran powder and fibers, and found that the morphology of the polymer, the initial heat treatment temperature, and the environment under which the carbonization is carried out influence the properties of the resultant carbon. The charcoal obtained from Saran powder was highly porous and retained the morphology of the polymer. Barton et al.[2] and Boult et al.[15] studied the effect of heating rate on the development of microporosity. They observed that the charcoal prepared by slow pyrolysis of the polymer retained original morphology of the polymer, but no such retention was seen in the rapidly pyrolysed charcoal, which was attributed to the occurrence of fusion during the pyrolysis process in the latter case. This view was further supported by Bailey and Everett[12] but was doubted by Winslow and coworkers,[16] who believe that few valency angles and distances survive during the course of their preparation by pyrolysis. Howard and Szynaka,[3] however, are of the view that fusion during carbonization is not an essential step in the development of microporosity in polymer carbons.

Molecular sieve carbons contain ultrafine micropores that permit the penetration of smaller molecules but impede the entry of larger molecules. The diffusion of certain gases and vapors into these materials is an activated process and is strongly temperature dependent. Thus, Marsh and Wyne-Jones[17] observed that the uptake of nitrogen at 77 K for a polymer carbon was much smaller than its uptake of CO_2 at 195 K. Kipling et al.[18] observed that the surface area of a polymer charcoal prepared by carbonization at 700°C was very low when measured by adsorption of nitrogen at 77 K than when measured by the adsorption of CO_2 at 195 K.

Several workers also studied the adsorption of nitrogen and CO_2 at different temperatures on coals.[19–21] In each case the adsorption of CO_2 at higher temperature was found to be larger. In one case, Nandi and coworkers[20] have reported that the ratio of the specific surfaces determined by adsorption of CO_2 at 195 K and nitrogen at 77 K were 7:1 for one anthracite and 87:1 for another, although the CO_2 value for the first sample was only slightly larger than that of the second sample. This indicates that the second sample contained a greater proportion of the ultrafine pore entrances, which were sufficiently narrow to prevent the entry of nitrogen at 77 K.

4.1 PREPARATION OF CARBON MOLECULAR SIEVES (CMS OR MSC)

A molecular sieve consist of two parts: the aperture and the cavity. The aperture has to be small so that it can exercise a screening effect for larger molecules to prevent their entry into the pore cavity. The cavity, on the other hand, should be as large as possible because it will determine the adsorption capacity. Consequently, all the methods that are used to prepare CMS involve either creating constrictions in the pore aperture blocking of the pore entrances by some suitable element or compound so as to produce the screening effect. MSC are generally prepared from activated carbons by a post treatment that narrows the pore-size distribution to produce a material with a predominance of micropores in a limited range of size.

Two approaches that have been used to prepare MSC consist of controlling the dimensions of the pore entrances by either reducing the pore diameter by pore blocking or controlling the heat treatment temperature during carbonization. For example, Dacey and Thomas[22] carried out blocking of larger pores in an active carbon by the adsorption of monomeric vinyl chloride followed by thermal decomposition of the monomer. Moore and Trimm[23] and Cheredkova et al.[24] modified the gaseous adsorption properties of carbons by depositing carbon produced by the pyrolysis of benzene vapors. The resulting material was a carbon molecular sieve with pore diameter between 0.3 and 0.6 nm in which case nitrogen and oxygen could be adsorbed to different extents. Barton and Koresh[25] prepared molecular sieve carbons from macadamia nut shell carbons by thermal decomposition of benzene vapors on the carbon surface at 800°C for various time periods. It was found that the pore structure of the carbon could be modified to improve the separation of oxygen and nitrogen by the thermal degradation of benzene vapors on the carbon surface. The MSC so prepared were characterized by water-vapor adsorption isotherms and by immersion calorimetry. It was found that within the limits of precision, both the Gurvitch values for water vapors and heats of immersion values were constant and independent of benzene treatment time, indicating that pore closure was proceeding in such a way as not to diminish either the pore volume or the energy of interaction with the adsorbate. These workers concluded that the pyrolysis of benzene vapors, which confers molecular sieve properties on active carbons, appears to proceed mainly by the formation of constrictions at preferred sites in the pore structure.

Vnukov et al.[26] prepared molecular sieve carbons by depositing carbon on the active carbon surface by the pyrolysis of methane at temperatures between 600°C and 900°C. However, the deposition of the carbon resulted in a decrease in the pore volume. Sutt[27] prepared molecular sieves for separating gaseous or liquid mixtures by impregnating an active carbon with an organic polymer or by impregnating small amounts of an inorganic polymer. Moreno-Castilla[28] deposited sulfur in the pores of Saran charcoal by heat treatment of CS_2 or sulfur and obtained MSC with a mean pore size of 0.58 nm. Yoshio et al.[29] controlled the size of the micropores in a coconut-shell charcoal by heat treatment of the charcoal at 1020 K with a mixture of hydrocarbons, such as biphenyl and napthalene, or a mixture of coal tar and naphthalene. The resulting molecular sieve carbon was found to contain pores in the range of 0.28 to 0.24 nm.

Bansal and coworkers[30-32] and Sharma[33] prepared carbon molecular sieves from PVDC, Saran, and a commercially available active carbon by blocking their pores by the thermal decomposition of CS_2 at 600°C and depositing sulfur by the pyrolysis of benzene vapors and depositing carbon and by the impregnation of PVC, followed by its thermal decomposition. The treatment of Saran with CS_2 at 600°C results in the decomposition of CS_2 into carbon and sulfur that are deposited in the microporous structure of the carbon. This results in blocking some of the micropores and reducing dimensions of some others. The deposited sulfur could be gradually removed by heat treatment of the carbon with hydrogen gas at 500°C for different intervals of time. This treatment caused partial opening of the pores, depending upon the time of treatment with the hydrogen gas. The opening of the micropores could be controlled to get carbon molecular sieves with pore dimensions varying between 0.3 and 0.8 nm. The deposition of sulfur and its complete removal by hydrogen treatment was found to result in widening of the micropores.

The pyrolysis of benzene over the active carbon surface results in the deposition of the carbon on the surface of the substrate carbon as well as in the microporous system and at some preferred sites. The adsorption isotherms of organic molecules of varying sizes and shapes indicated that the carbon gets deposited preferentially in the pore entrances reducing entrance diameter resulting in the formation of carbon molecular sieves. Pore-size distribution curves indicated that the treatment with benzene between 3 and 6 hrs reduces the mean pore dimensions to 0.6 nm, and a larger time of treatment reduces pore entrances to less than 0.6 nm.

The blocking of pores by PVC impregnation was carried out by treatment of the active carbons with suspensions of PVC in alcohol under reflux followed by carbonization at 600°C. This resulted in the deposition of appreciable amounts of carbonaceous material into the microporous structure causing a reduction in pore dimensions and producing carbon molecular sieves. These MSC were found to permit the adsorption of smaller molecules such as benzene or cyclohexane but prevented the adsorption of larger molecules such as isooctane and α-pinene.

Lilan et al.[34] prepared MSC for air separation by the carbonization of two Chinese anthracite coals. The coals were ground, mixed with a binder, carbonized, and subsequently activated with CO_2. The activated material thus obtained was modified by the pyrolysis of toluene and tar fractions of oil at 1073 K to obtain molecular sieve materials. These MSC materials were found to have predominantly a micropore size of 0.33 to 0.40 nm and were used to separate air to a nitrogen purification of 98.6%.

Lizzio et al.[35] prepared MSC for gas separation processes using a Illinois high volatile content bituminous coal under several heat treatment and activation conditions. The pore structure of some selected chars was modified by carbon deposition by pyrolysis of methane for 0.5 hr at temperatures between 600°C and 1000°C. Micropore pore-size distribution as determined by nitrogen adsorption indicated the existence of pores with dimensions around 0.4 nm and could be used for separation of O_2-N_2, CO_2-CH_4 and CH_4-H_2 mixtures. The equilibrium and the kinetic data on the adsorption of different gas mixtures using different chars showed that the chars with most of the surface area in pores less than 0.4 nm were better in the kinetic separation of air into O_2 and N_2, because air separation depends on the fact that O_2 (mol. diam. 0.346 nm) diffuses into the carbon more rapidly than N_2 (mol. diam. 0.368 nm). All the five chars prepared in these investigations had some capability for air separation. On the other hand, the chars with a greater percentage of surface area in pores greater than 0.4 nm were more efficient in equilibrium-based separation of methane and hydrogen mixtures. This is due to the fact that the equilibrium adsorption as opposed to kinetics will allow larger CH_4 molecules, (mol. diam. 0.38 nm), given equal access to the pore structure, pores greater than 0.38 nm to be adsorbed to a greater extent than hydrogen (mol. diam. 0.29A°). Similarly several chars showed good potential for efficient separation of CO_2 (mol. diam. 0.33A°) and methane mixtures.

De Salazar et al.[36] prepared CMS by carbon deposition by the pyrolysis of benzene on an activated carbon. The carbon was prepared by the carbonization of peach stone followed by its activation with CO_2. The effect of preparation variables such as temperature, time, and benzene partial pressure on the characteristics of MSC was studied. The MSC obtained in this study were completely accessible to both dichloromethane and benzene as shown by similar values of immersion enthalpy for both liquids. However, the accessibility of a particular liquid into the MSC was found to depend upon the deposition time. A linear relationship was found between the decrease of accessibility of these two liquids to the microporosity of the MSC and the deposition time. The separation ability of different MSC for N_2/O_2 and CO_2/CH_4 mixtures was analyzed by measuring the adsorption kinetics of pure gases. It was observed that the carbon sample without any deposition was not selective for the separation of these mixtures, but the separation could be obtained after pyrolytic deposition of benzene. The adsorption of these pure gases on a CMS showed that oxygen adsorption was much faster then nitrogen adsorption and CO_2 was adsorbed in much larger amounts than methane. The separation ability of the different samples of CMS was related to the deposition time and temperature.

Daguerre et al.[37] studied the influence of precursor composition on the preparation of pitch-based carbons with molecular sieve properties and found that the removal of low molecular weight species by toluene fractionation gave activated carbons with highly microporous structures with gate effects. These molecular sieve properties appeared to depend on the composition of the precurser as well as on the degree of gasification. Pedrero and coworkers[38] prepared molecular sieve carbons from lignin impregnated with $ZnCl_2$. The concentration of $ZnCl_2$ and the weight of lignin were adjusted to obtain $ZnCl_2$-lignin weight ratio of 0.4 and 0.23. The impregnated samples were carbonized and activated in inert atmosphere at 623 K, 673 K, and 773 K for 2 hr. The resulting carbons were washed with HCl to remove excess zinc and then heat-treated at 1073 K in nitrogen.

These lignin based carbons were characterized by adsorption of N_2 at 77 K and CO_2 at 273 K. The kinetics of O_2/N_2, CO_2/CH_4, and C_6H_6/C_6H_{12} mixtures showed that the carbons showed molecular sieve properties and could be used to separate these mixtures, depending upon their temperature of activation. In later investigations these workers[39] modified the microporous structure of lignin-based activated carbons by the deposition of pyrolytic carbon by cracking benzene. The time and cracking temperature were found to be the principal parameters governing the deposition process. Benzene cracking temperature of 1073 K and a pyrolytic carbon deposition of 0.3% were enough to prepare a carbon with molecular sieve properties suitable for efficient separation of O_2-N_2 and CO_2-CH_4 mixtures.

Manso et al.[40] studied the formation of CMS by carbon vapor deposition (CVD) over activated carbons from four different rank coals. The deposition of carbon was carried out by pyrolyzing benzene vapors at 725°C. This produced gradual closing of the micropores, due to the formation of constrictions at their entrances. As a result the MSC with a narrow micropore-size distribution around 0.35 to 0.5 nm were obtained. Samples with diameters smaller than 0.33 nm obtained by a high degree of deposition were able to separate O_2/N_2 and CO_2/CH_4 mixtures.

Abdi et al.[41] prepared activated carbons from Persian nutshell by impregnating $ZnCl_2$ by heat treatment at 500°C. The microporosity of the activated carbon was modified by impregnating coal tar pitch dissolved in benzene and heat treatment at 700°. The pore-size distribution was determined using molecular probes CCl_4 (mol. diam. 0.6 nm), $CHCl_3$ (mol. diam. 0.46 nm), CH_2Cl_2 (mol. diam. 0.4 nm), and CS_2 (mol. diam. 0.37 nm). The results indicated that the pore diameter could be reduced to 0.4 nm (Table 4.1).

Ya and Zi[42] prepared carbon molecular sieves from cheap condensed petroleum cokes. The cokes were impregnated with potassium hydroxide and the resulting activated carbon micropore system was modified by cracking of methane or liquefied petroleum gas. The activation of the coke with potassium hydroxide produced high surface area activated carbons with pores of 0.85 nm average diameter. This pore diameter was reduced to between 0.58 and 0.33 nm on deposition of carbon, and the micropore volume remained almost unchanged.

TABLE 4.1
Adsorption of Molecules of Different Dimensions
by a Molecular Sieve Carbon

Compound Name	Molecule Diameter (Å)	Equilibrium Adsorption (cm3/g)
CCl$_4$	6.0	0.00
CHCl$_3$	4.6	0.00
CH$_2$Cl$_2$	4.0	0.12
CS$_2$	3.7	0.15

After Abdi, M.A., Mohadiarfar, M., Ahmadpour, A., and Mirhabibi, A.R., 25th Bienn. Conf. on Carbon 2001, Lexington, KY, July 14–19, 2001, Paper 22.1. With permission.

Alcaniz-Monge et al.[43] prepared MSC from a low rank coal and a low cost binder, coal tar pitch. The bituminous coal was steam activated and then impregnated with a suitable solution of a coal tar pitch (coal/pitch ratio 60:40 by weight), then molded in a monolith form and carbonized at temperatures between 700°C to 1200°C. The monolith obtained was characterized by TGA, XRD, and SEM, and the porous structure was determined by adsorption of nitrogen at 77 K and CO$_2$ at 273 K. The pitch acted as a binder and modified the porosity of the activated carbon, producing a material with molecular sieve properties. The pitch caused an efficient blockage of the pore entrances. The molecular properties were analyzed through the measurement of CH$_4$ and CO$_2$ adsorption kinetics (kinetic diameters 0.38 and 0.33 nm, respectively).

Valente et al.[44] prepared CMS from carbon Fiber felt obtained by the carbonization of oxidized PAN felt at 800°C in an inert atmosphere. The carbon felt was oxidized in an oxygen plasma excited by radio frequency (13.56 MHz) in a reactive ion etching reactor. Different oxygen flow rates of 30, 60, 90 ml/min. and different etching times up to 340 min. were used. The physical characteristics of the activated carbon samples indicated that the micropore production was dependent on the etching time; the gas-flow rate only slightly influencing the pore characteristics. The pore-size distribution for samples treated with 60 ml/min oxygen flow rate and for different etching time (Figure 4.1) show that all the pore sizes (minimum, average, and maximum) are displaced toward higher values as the treatment time is enhanced. It has been suggested on the basis of the SEM studies that the surface etching by plasma create pits on the carbon Fiber surface due to oxidation starting the micropore growth. The oxygen then attacks the pore walls enlarging the micropores. The micropores so created then coalesce and form the macropores. The method has been claimed to produce clean MSC by relatively easy and less expensive method.

Ahmadpour et al.[45] prepared MSC from a naturally occurring substrate (Iranian walnut shell) and a commercial activated carbon (Silcarbon), using four different methods based on the adjustment of pore openings in the activated carbon pore structure. In the first two methods, activated carbons were heated in a tubular furnace in

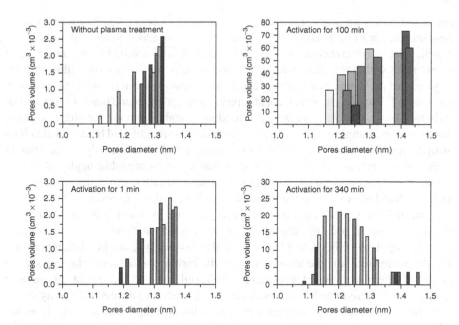

FIGURE 4.1 Pore-size distribution after different treatment times with the gas-flow rate of 60 l/min. (After Valente, C.O., Coutinho, A.R., Maciel, H.S., Petroconi, F.G., Otani, C., and Massi, M., Carbon '02 Intern. Conf. on Carbon, Beijung, Sept. 15–19, 2002. With permission.)

nitrogen in the temperature range of 700°C to 850°C. Acetylene gas or benzene vapors were then passed over the carbons with a flow rate of 100 ml/min. for different intervals of time. The carbons were then cooled under inert atmosphere. The carbon materials obtained contained different amounts of carbon deposited in the pores of the activated carbons, due to decomposition of acetylene or benzene. In the other two methods, the activated carbons were impregnated with different amounts of coal tar pitch, which had been extracted with benzene under reflux. The amount of impregnated pitch varied between 7% and 53% weight. The pitch-impregnated carbon samples were then heated to a maximum temperature of 850°C at the rate of 5°C/min and soaked at the highest temperature for different times. In another method, the carbon sample was prepared by impregnation of coal tar pitch by 21% weight was exposed to the benzene vapors of known concentration. All the samples so prepared were characterized by adsorption kinetics of O_2, N_2, CH_4, and C_2H_4. It was observed that carbon molecular sieves (CMS) with uniform pores smaller than 0.4 nm could be prepared by carbon vapor deposition (CVD) process in different stages. In the first step, pyrolyzing a heavy hydrocarbon resulted in the formation of a uniform pore structure in activated carbons. This pore size can be further reduced by pyrolyzing a light hydrocarbon such as benzene. It was found that one of the CMS sample prepared in a two-step process was quite efficient for the separation of O_2 and N_2.

Walker and coworkers[46–53] carried out several studies on the preparation of carbon molecular sieves using active carbons and cellular precursors. Schmitt and Walker[46] prepared CMS from a system consisting of furfuryl alcohol and chloroplatinic acid. This mixture, prior to its complete polymerization, was added to an activated carbon, which, upon polymerization and carbonization at 700°C, resulted in the formation of microcracks within the composite. These microcracks served as rapid diffusion pathways to the molecular-sized pores. The final sample was a molecular sieve carbon containing 1% by weight, which could separate straight-chain and branch-chain hydrocarbons. This molecular sieve was also used as a catalyst support for shape selective reactions. For example, selective hydrogeneration of a straight- and branch-chain olefin in hydrogenation was measured at 150°C using a bed of this catalyst. The straight chain olefin, 1-butene was effectively hydrogenated to n.butane while negligible hydrogenation or isomerization of 3-methyl-1-butene occurred because the branch-chain molecule had little excess to the Pt. catalyst present within the pores of the CMS.

The molecular sieve carbons were also prepared by altering the pore structure of microporous carbons by heat treatment or by mild gasification[48,49] or by chemisorbing propylene at 573 K to 873 K followed by its cracking at a higher temperature. The deposition of a suitable amount of pyrolytic carbon was achieved through several cycles of chemisorption of propylene[50] followed by its cracking at an elevated temperature. The number of such cycles required to achieve desired selectivity varied with the width of the pore structures present in the starting material. Propylene was selected as the preferred organic source of pyrolytic carbon because its ΔH for cracking is very low. The molecular sieve properties of various carbons were studied by measuring rates of oxygen, nitrogen and argon adsorption.

The chemisorption of propylene was carried out by exposing a degassed carbon sample kept at a desired temperature to propylene at 0.1 MPa pressure in a closed system. After exposure for a given period of time, the carbon was degassed at the adsorption temperature under vacuum for about 15 minutes to remove any physically adsorbed propylene. The carbon sample was heated at 1123 K, which resulted in cracking of the chemisorbed propylene to deposit pyrolytic carbon. These workers also prepared carbon molecular sieves from carbons impregnated with Ni.[52] As the pyrolysis of propylene on carbon surface is considerably enhanced in the presence of Ni, the deposition of pyrolytic carbon can be carried out at comparatively lower temperatures and, specifically, at the Ni sites on the carbon matrix. Thus, if Ni is impregnated on the outer surface of the active carbon, the pyrolytic carbon will be deposited closer to the outer surface of the host carbon, leaving behind sufficiently high adsorption capacity in the micropores.

The impregnation of Ni on the carbon surface was carried out by injecting a solution of $Ni(NO_3)_2.6H_2O$ into the carbon bed so that the carbon sample was just wet. This resulted in the adsorption of most of the solution. The loading of the carbon with Ni was varied by varying the concentration of the Ni solution. The impregnated carbon sample was dried at 383 K for 2 hr and then held at 773 K in N_2 for 4 hr in order to decompose the $Ni(NO_3)_2$ and then treated with H_2 at the same temperature for 4 hr to reduce nickel oxide into Ni. The sample was then degassed at 823 K or 923 K and exposed to a flowing stream of pure propylene in a bed and the propylene cracked to deposit pyrolytic carbon.

Several cellulosic precursors such as viscose rayon yarn, viscose tire cord, and coconut shells after modification have also been used for the preparation of carbon molecular sieves.[53] The cellulose precursors were modified with respect to their crystallanity by milling, using a rolling motion or ball milling with vibration and also by treatment with ethylene diamine. The modified cellulose fibers and powders were carbonized using a horizontal tube furnace fitted with a quartz tube. Purified nitrogen was passed through the quartz tube at the rate of 75 to 125 cc/min. The heating rate was maintained at 7°C to 8°C/min with a soaking time of 1.5 hr at the maximum temperature. The carbonization was carried out at different temperatures in the range of 400°C to 800°C. The amorphous and crystalline contents of the precursors as determined by water adsorption proved to be an important parameter in determining the molecular sieve character of the carbons. It was observed that cellulose precursors with amorphous and crystalline contents of about 50% yielded a 530°C carbon that showed negligible capacity for adsorption of isobutene and a substantial capacity for n.butane and CO_2.

Chihara and Suzuki[54] carried out adjustment of the micropore diameter in molecular sieve carbons to control the diffusivities of oxygen and nitrogen. The treatment consists of adsorption of ethylbenzene or styrene at low temperatures between 150°C and 400°C followed by decomposition of the hydrocarbon by heat treatment at 400°C. Diffusivities of oxygen and nitrogen, as measured chromatographically were found to decrease sharply with increase in the amount of carbon deposited. The ratio of the two diffusivities, however, remained unaffected. Darmstadt et al.[55,56] prepared microporous carbon molecular sieves by pyrolysis of sucrose adsorbed in SBA-15 and MCM-48 silica matrices. The pyrolysis was carried out at different temperatures between 700°C and 1100°C. The mesoporous carbon sieves were obtained by dissolving the silica matrix by hydrofluoric acid solution. Some of the samples were subsequently heat treated at temperatures ranging between 1100°C and 1600°C. The pore-size distribution in mesoporous sieves varied between 2 and 8 nm, depending upon the temperature of the heat treatment. However, the materials also contained narrow micropores with widths less than 0.9 nm. In the case of molecular sieves produced in SBA-15 matrix, the concentration of micropores decreased until at a temperature of 1600°C, a purely mesoporous material was obtained. The mesopore diameter also increased from 3.5 to 5.0 nm. The carbon sieves were found to have mesopore volume and BET surface area of up to 1.25 cm³/g and 1721 m²/g, respectively. The surface area of these materials was also calculated from the adsorption potential distributions and was found to be in the range 730 to 1400 m²/g, indicating that these materials have a high potential as adsorbents for large molecules and as catalyst support. It was also observed that for carbon sieves prepared in SBA-15 silica matrix, the concentration of the micropores could be regulated by the heat treatment temperature.

Vyas et al.[57] have reported suitable process conditions for the synthesis of CMS from bituminous coal. The coal was crushed, milled, and agglomerated with sulphate pulp-waste liquor or coal-tar pitch as the binders and then carbonized at 800°C for one hr in nitrogen atmosphere. Coke was deposited on the surface of the char by cracking acetylene or benzene at 800°C for 3 to 10 min and 10 to 30 min, respectively. Cracking of acetylene did not produce sieving effect, probably because of overcoking in deeper locations in the pores. However, the samples prepared by cracking of benzene were found to be highly suitable for CO_2-CH_4 separation. CMS produced using sulphate pulp-waste liquor as binder showed poor O_2/N_2 uptake ratios, and those prepared using coal-tar pitch as binder showed far better ratios. The O_2/N_2 separation ratios were about five times compared with the uncoked samples.

4.2 CHARACTERIZATION OF CARBON MOLECULAR SIEVE CARBONS

Molecular sieve carbons contain ultrafine micropores so that their size characterization, which is very important for the design and utilization of these porous carbon materials, cannot be carried out directly. In general, the characterization of porous carbon materials is carried out by the adsorption of N_2 at 77 K. However, the main disadvantages of N_2 adsorption at 77 K are the diffusional problems of the N_2 molecule into the narrow pores at this low temperature. Furthermore, there is additional difficulty in the adsorption because very low relative pressures are required to extend the range of studies to extremely narrow porosity. In order to overcome these problems, adsorption of CO_2 at 273 K, to cover the same range of relative pressures as in the case of nitrogen, has been suggested. Similarly high-pressure adsorption of methane has also been suggested. Emphasis, these days, however, has been directed in using certain molecular probe studies, using molecules of different shapes and sizes. These methods include adsorption of inorganic gases and organic vapors representing a wide range of sizes and shapes, immersional heats of wetting, or displacement densities in liquids with different molecular dimensions and activated diffusion.

Moore and Trimm[58] characterized carbon molecular sieves obtained by pore blocking by the pyrolysis of benzene by adsorption of CO_2 at 195 K and applying the Dubinin-Radushkevich equation. The pore size of the active carbon after pyrolysis of benzene decreased, although the pore volume remained the same. This indicates that the pyrolyzed carbon has been deposited largely in the mouth of the pore. These workers observed that although the adsorption of nitrogen and oxygen did not differ very much, there was a considerable difference in the adsorption of but-1-ene (critical diameter 0.51 nm) and 2.2 dimethylene propane (critical diameter 0.69 nm.) (Figure 4.2).

Bansal and Dhami[59] characterized molecular sieve properties of several polymer charcoals by studying the adsorption of nitrogen at 77 K and that of CO_2 at 273 K using volumetric methods (Figure 4.3 and Figure 4.4). The surface areas of different charcoals were calculated using Langmuir, BET, and Dubinin-Polanyi equations. The molecular areas of CO_2 and nitrogen were taken as 18.7Å^2 and 16.2Å^2 respectively.[60,61] The surface areas of polyvinylidene (PVDC), Saran (a copolymer of PVDC and PVC) and steam-activated Saran charcoals calculated from nitrogen adsorption at 77 K using any one of these three equations (cf Table 4.2), are comparable with the surface areas calculated from CO_2 adsorption at 273 K using the same equation. However, in the case of PF (polyfurfuryl alcohol) and UF (urea formaldehyde) charcoals, the N_2 surface areas are extremely small compared to their CO_2 surface areas, the two values differing by several orders of magnitude. This shows that UF and PF charcoals have ultrafine micropores.

The accessibility of nitrogen into these pores at the low temperature of adsorption (77 K) is hindered because of the activated diffusion effects. Thus, CO_2, which is adsorbed at a much higher temperature (273 K), can have access to a larger proportion of the pores in these charcoals. This was further supported by the results of surface

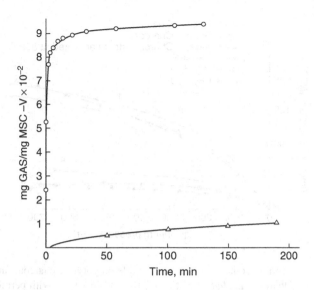

FIGURE 4.2 Adsorption of but-1-ene and 2,2 dimethyl propane on untreated MSC-V. T = 195 K. ○ = But-l-ene; △ = 2,2 dimethyl propane. (After Moore, S.V. and Trimm, D.L., *Carbon*, 15, 177, 1977. With permission.)

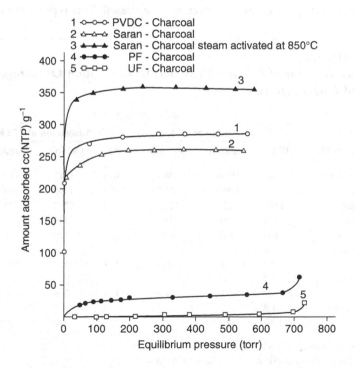

FIGURE 4.3 Adsorption isotherm of nitrogen on different polymer charcoals at 77 K. (After Bansal, R.C. and Dhami, T.L., *Indian J. Chem.*, 19A, 1146, 1980. With permission.)

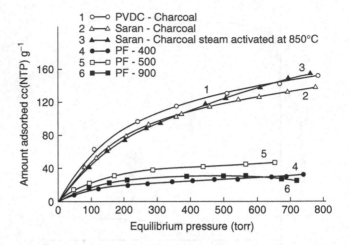

FIGURE 4.4 Adsorption isotherms of carbon dioxide on polymer charcoals at 273 K. (After Bansal, R.C. and Dhami, T.L., *Indian J. Chem.*, 19A, 1146, 1980. With permission.)

area of UF charcoals obtained by adsorption of N_2 at 195 K. The BET values shown in parenthesis (*cf* Table 4.2) increased considerably. At higher temperature the rate of diffusion of N_2 into the microcapillary pores increased so that the number of molecules entering the pores in a given time increased. This type of inaccessibility

TABLE 4.2
Surface Areas of Polymer Charcoals by Adsorption Of Nitrogen and Carbon Dioxide

	Surface Area (m^2/g)					
	N_2 Adsorption at 77 K			CO_2 Adsorption at 273 K		
Sample	BET	Dubinin	Langmuir	BET	Dubinin	Langmuir
PVDC-600	887	1076	1086	903	1017	1005
Saran-600	787	1091	1097	798	1065	1049
Saran-600[a]	1188	1298	1290	1138	1212	1250
PF-600	105	151	138	288	359	297
PF-900	106	133	141	163	229	296
UF-400	1.0					
	(23.2)[b]	1.6	1.6	103	150	132
UF-650	1.2					
	(27.2)[b]	2.9	3.2	198	246	232
UF-850	1.3					
	(21.5)[b]	2.3	2.8	225	344	325

[a] Steam activated at 850°C
[b] Values in parenthesis refer to N_2 adsorption at 195 K

After Bansal, R.C. and Dhami, T.L., *Indian J. Chem.*, 19A, 1146, 1980. With permission.

to nitrogen adsorption at 77 K resulting in low values of monolayer capacity has also been observed by several workers in coals[61,20] and several molecular sieve materials.[17,20] This indicates that these polymer charcoals can act as carbon molecular sieves.

Darmstadt et al.[55,56] characterized mesoporous carbon molecular sieves obtained by pyrolysis of sucrose in SBA-15 (CMK-3F series) and MCM-48 (CMK-1F series) silica matrices by adsorption of nitrogen at 77 K and observed that on head treatment up to 1600°C, the CMS obtained were associated with only mesoporous, almost all of the micropores were converted into mesopores (Figure 4.5a,b).

4.2.1 CHARACTERIZATION OF CARBONS BY ADSORPTION OF ORGANIC VAPORS

Bansal and coworkers[30–32] characterized molecular sieve carbons prepared after pore blocking by depositing sulfur by the decomposition of CS_2; by depositing carbon by the decomposition of benzene; by the decomposition of impregnated polyvinyl chloride (PVC); and by adsorption of molecules of varying sizes and shapes such as water, benzene, cyclohexane, heptane, isooctane, and α-pinene as molecular probes. Treatment of Saran charcoal with CS_2 at 600°C results in the decomposition of CS_2 into carbon and sulfur, which are deposited in the microporous structure of the substrate carbon. This results in blocking some of the micropores and reducing entrance diameters of some larger pores. The amount of deposited sulfur can be removed completely by treatment of the sulfurized charcoal with hydrogen gas at 600°C for 5 hr. When the time of treatment with hydrogen gas is less than 4 hrs, the adsorbed sulfur is removed only partially so that charcoal samples associated with varying amounts of adsorbed sulfur can be obtained by varying the time of treatment with hydrogen gas (Table 4.3).

The deposition of sulfur results in a slight increase in the adsorption of water vapors at lower relative vapor pressures (rvp less than 0.4), but there is a considerable decrease at higher relative pressures (Figure 4.6). Because the removal of associated oxygen by heat treatment at 600°C should result in a decrease in the adsorption of water vapor at lower relative pressures,[62–64] it appears that the deposition of sulfur causes a modification of the carbon micropores structure by blocking some of the larger micro capillary pores so that these pores are filled at lower relative pressures. The change in the shape of the isotherm from Type V to Type I on sulfurization also points to a reduction in the pore dimensions. Bansal et al.[64] have shown that the associated oxygen influences the adsorption of water vapor at relative pressures lower than 0.5m, whereas the porosity of the carbon is predominantly responsible for adsorption at higher relative pressures.

When the sulfurized charcoal is heat-treated in hydrogen for different intervals of time to eliminate the adsorbed sulfur partially, the adsorption of water vapor at higher relative pressures increases gradually with increase in the removal of sulfur (Figure 4.6). This indicates that more and more of the blocked pores are now available for water-vapor adsorption at higher relative pressures. The pore-size distribution curves for the different charcoal samples calculated from water adsorption isotherms using the Kelvin equation[65] (Figure 4.7) indicate that the average pore diameter is reduced from 1.2 nm to less than 0.4 nm on deposition of sulfur and increases to

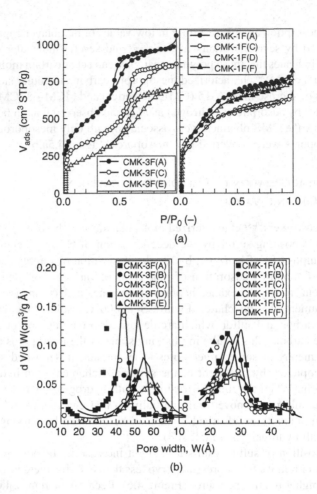

FIGURE 4.5 (a) Nitrogen adsorption isotherms at 77 K for heat-treated carbon sieves in SBA-15 (CMS-35 series) and MCM-48 (CMK-IF series) silica matrix. (After Darmstadt, H., Roy, C., Kaliaguine, S., Choi, S.J., and Ryoo, R., 25th Bienn. Conf. on Carbon, Lexington, K.Y., July 14–19, 2001, Paper 11.2 and Darmstadt, H., Roy, C., Kakaguine, S., Choi, S.J., and Ryoo, R., 25th Bienn. Conf. on Carbon, Lexington, K.Y., July 14–19, 2001, Paper 3.1. With permission.) (b) Pore-size distribution for heat-treated SBA-15 and MCM-48 series carbon sieves. (After Darmstadt, H., Roy, C., Kaliaguine, S., Choi, S.J., and Ryoo, R., 25th Bienn. Conf. on Carbon, Lexington, K.Y., July 14–19, 2001, Paper 11.2 and Darmstadt, H., Roy, C., Kakaguine, S., Choi, S.J., and Ryoo, R., 25th Bienn. Conf. on Carbon, Lexington, K.Y., July 14–19, 2001, Paper 3.1. With permission.)

between 0.6 nm and about 2.0 nm on removal of sulfur and by hydrogen treatment depending upon the time of treatment with hydrogen.

The adsorption of benzene decreased from 21% on the untreated charcoal to about 11% on treatment of the charcoal with CS_2. However, the amount adsorbed increased gradually as the adsorbed sulfur was gradually eliminated on treatment with hydrogen (Figure 4.8). In the case of the sulfurized sample heat-treated with

TABLE 4.3
Amount of Sulfur Adsorbed and Oxygen Associated with Different Carbon Samples

Particulars of Samples	Adsorbed Sulfur (g/100 g)	Oxygen Evolved on Outgassing at 1000°C (g/100 g)			
		CO_2	CO_2	H_2O	Total
Original Saran (Saran-600)	—	2.23	3.00	0.20	5.43
Saran-600 treated with CS_2 (Saran-CS_2)	4.53	1.23	2.40	—	3.63
CS_2-treated-Saran treated with hydrogen for 1 hr (Saran-CS_2-H_2 1 hr)	3.20	0.12	0.21	—	0.33
CS_2-treated-Saran treated with hydrogen for 2 hr (Saran-CS_2-H_2 2 hr)	1.60	—	—	—	—
CS-treated-Saran treated with hydrogen for 5 hr (Saran-CS_2-H_2 5 hr)	—	—	—	—	—

After Bansal, R.C., Bala, S., and Sharma, N., *Indian J. Technol.*, 27, 206, 1989. With permission.

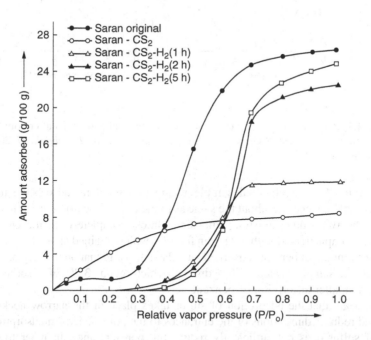

FIGURE 4.6 Adsorption of water vapor on Saran charcoal before and after pore blocking by deposition of sulphur. (After Bansal, R.C., Bala, S., and Sharma, N., *Indian J. Technol.*, 27, 206, 1989. With permission.)

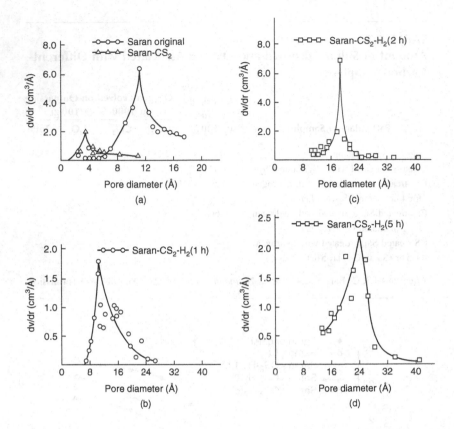

FIGURE 4.7 Pore-size distribution curves for Saran charcoal before and after pore blocking with sulphur. (After Bansal, R.C., Bala, S., and Sharma, N., *Indian J. Technol.*, 27, 206, 1989. With permission.)

hydrogen for 5 hr, a charcoal sample which contained no adsorbed sulfur, the adsorption of benzene was about the same as in the case of the original Saran charcoal sample, the two isotherms super imposing almost completely. In the case of the sulfurized sample treated with hydrogen for 2 hr, which retained about 1.6% adsorbed sulfur, the amount of benzene adsorbed was also similar to that on the original sample. However, the sample treated with hydrogen, which retains 3.2% of adsorbed sulfur, showed a smaller adsorption of benzene.

It appears that the adsorption of sulfur takes place in the narrow necks of the pores and reduces dimensions of the entrances to the pore cavity. An adsorption up to 1.6% of sulfur does not sufficiently reduce the pore entrance diameter to make it inaccessible to benzene molecule. However, when larger amounts of sulfur are retained, a fraction of the pores becomes inaccessible to even benzene molecules (molecular diam. 0.37 nm). This was further supported by the adsorption isotherms of organic molecules with larger molecular dimensions such as cyclohexane (mol. diam. 0.48 nm), n.heptane (mol. diam. 0.675 nm), isooctane (mol. diam. 0.68 nm), and α-pinene (mol. diam 0.80 nm). The adsorption of these molecules was found to decrease considerably

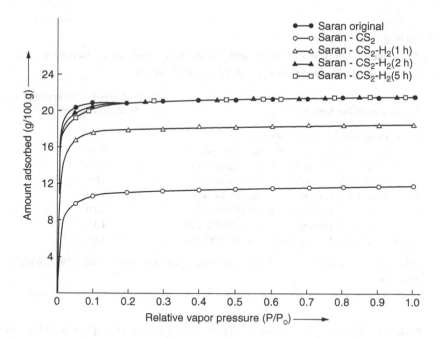

FIGURE 4.8 Adsorption isotherms of benzene vapors on Saran charcoal before and after pore blocking with sulphur. (After Bansal, R.C., Bala, S., and Sharma, N., *Indian J. Technol.*, 27, 206, 1989. With permission.)

at all relative pressures on pore blocking, although the quantum of decrease was different for different adsorbates. The adsorption decreased from 16.8% to 5% for cyclohexane, from 12.4% to 4.6% for n.heptane, from 12% to 2.4% for isooctane, and from 3.8% to almost no adsorption for α-pinene. When the adsorbed sulfur was gradually removed on treatment with hydrogen, there was a gradual increase in the adsorption of these vapors. An interesting feature of these isotherms was the case of the sulfurized charcoal retaining 1.6% sulfur. It was found that while this sample produced no sieving effect for benzene and cyclohexane molecules, it did produce considerable sieving effect for n.heptane, isooctane, and α-pinene molecules, indicating that the pore entrance diameter was reduced considerably.

The pyrolysis of benzene vapors over PVDC and Saran charcoals at 600°C was found to result in the deposition of carbon into the microporous structure as evidenced by an increase in weight of the carbon samples (Table 4.4). The increase in weight increased gradually as the time of benzene pyrolysis was enhanced from 40 min to 160 min, indicating that more and more of the pyrolytic carbon was being deposited. The adsorption of benzene vapors increased slightly in case of both PVDC and Saran charcoals given benzene pyrolysis treatment for 40 min (PVDC-40 and Saran-40 samples), remained more or less unchanged in the case of PVDC-80 and Saran-80 samples, and decreased slightly when the time of pyrolysis was enhanced (PVDC-120, PVDC-160, Saran-120, and Saran-160 Samples). The increase in adsorption of benzene in the case of PVDC-40 and Saran-40 samples has been

TABLE 4.4
Increase in Weight of Saran and PVDC Charcoals after Benzene Pyrolysis for Different Intervals of Time at 600°c

Particulars of the Sample	Weight Gained after Benzene Treatment (g/100 g)
Saran treated with benzene for 40 min (Saran-40)	2.76%
Saran treated with benzene for 80 min (Saran-80)	3.60%
Saran treated with benzene for 120 min (Saran-120)	4.40%
Saran treated with benzene for 160 min (Saran-160)	4.60%
PVDC treated with benzene for 40 min (PVDC-40)	3.40%
PVDC treated with benzene for 80 min (PVDC-80)	3.82%
PVDC treated with benzene for 120 min (PVDC-120)	4.33%
PVDC treated with benzene for 160 min (PVDC-160)	4.42%

After Sharma, N., Ph.D. thesis submitted to Panjab Univ., Chandigarh, India, 1991. With permission.

attributed largely to the elimination of acidic surface oxygen groups, which are polar and known to suppress the adsorption of benzene.[66–68] In the case of PVDC-80 and Saran-80 carbon samples, the increase in adsorption due to the elimination of acidic groups is compensated by a decrease in adsorption due to blocking of some of the micropores so that the net result was more or less unchanged adsorption. In case of the samples obtained by pyrolysis of benzene for larger intervals of time, the decrease in adsorption was due to the blocking of increasing number of micropores. The adsorption of cyclohexane was found to decrease continuously as more and more carbon was deposited, because this adsorption was not influenced by carbon-oxygen surface groups. The decrease in adsorption of cyclohexane is of the same order of magnitude as in the case of benzene, the maximum decrease being about 2% in both the charcoals.

Both n.heptane and isooctane were adsorbed almost in similar amounts (Figure 4.9 and Figure 4.10) on both PVDC and Saran charcoals, which could be attributed to similar molecular dimensions of the two adsorbates. However, the amounts adsorbed decreased on pore blocking, and the decrease at each degree of pore blocking was once again similar in magnitude (from about 13% to 2%). The adsorption of α-pinene also decreased on pore blocking, with little or no adsorption taking place in case of the PVDC-160 and Saran-160 samples (Figure 4.11 and Figure 4.12).

In order to examine that the adsorption of isooctane and α-pinene was a diffusion-controlled process, adsorption isotherms were determined at two different temperatures. It is interesting to note (Figure 4.12) that the adsorption of both isooctane and α-pinene increased with an increase in the temperature of adsorption, indicating that the adsorption was an activated process. This also indicated that the width of pore constrictions was only slightly greater than the diameter of these adsorbates so that the molecule experienced a very strong attractive force as it reached the constriction, and this delayed its entry into the pore cavity. The increase in the temperature of

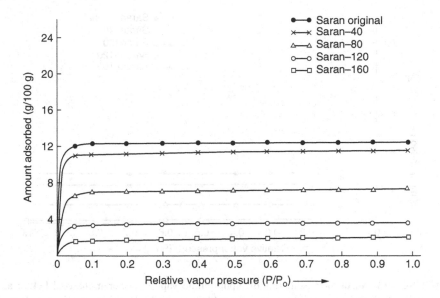

FIGURE 4.9 Adsorption isotherms of heptane vapors on Saran charcoal before and after pore blocking by benzene pyrolysis. (After Sharma, N., Ph.D. thesis submitted to Panjab Univ., Chandigarh, India, 1991.)

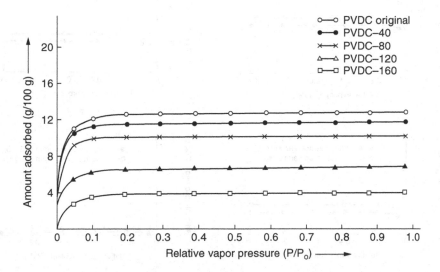

FIGURE 4.10 Adsorption isotherms of isooctane vapors on PVDC charcoal before and after pore blocking by benzene pyrolysis. (After Sharma, N., Ph.D. thesis submitted to Panjab Univ., Chandigarh, India, 1991. With permission.)

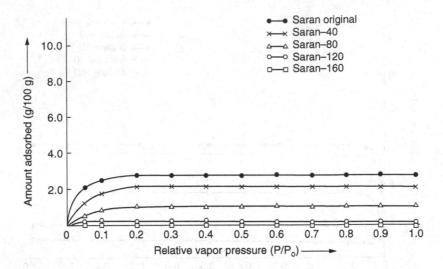

FIGURE 4.11 Adsorption isotherms of alpha pinine vapors on Saran charcoal before and after pore blocking by benzene pyrolysis. (After Sharma, N., Ph.D. thesis submitted to Panjab Univ., Chandigarh, India, 1991. With permission.)

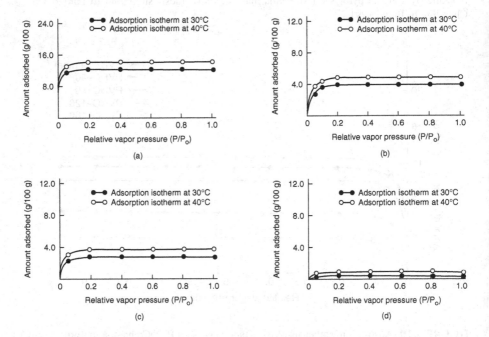

FIGURE 4.12 (a) Absorption isotherms of iso-octane vapours on Saran charcoal at different temperatures. (b) Absorption isotherms of iso-octane vapors on Saran-120 at different temperatures. (c) Absorption isotherms of a-pinene vapors on Saran charcoal at different temperatures. (d) Adsorption isotherms of a-pinene vapors on Saran-120 charcoal at different temperatures. (After Bansal, R.C. and Dhami, T.L., *Carbon,* 18, 207, 1980. With permission.)

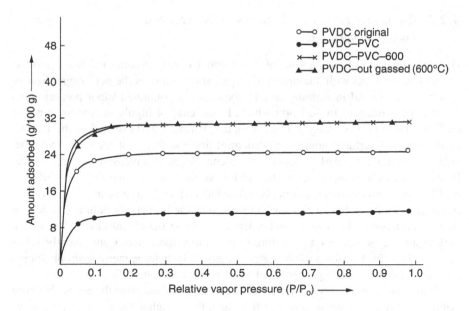

FIGURE 4.13 Adsorption isotherms of benzene vapors on PVDC charcoal before and after pore blocking by decomposition of PVC. (After Sharma, N., Ph.D. thesis submitted to Panjab Univ., Chandigarh, India, 1991. With permission.)

adsorption resulted in an increase in the rate of diffusion so that in a given adsorption time, the amount adsorbed increases with increase in the temperature of adsorption.

Thus, it is evident that the pyrolysis of benzene vapors at 600°C over carbons is capable of reducing their pore aperture dimensions or producing constrictions[58] so as to make them useful as carbon molecular sieves. The reduction in pore dimensions can be controlled by the duration of the pyrolytic treatment.

The pore blocking in PVDC was also carried out by impregnation of the charcoal sample with PVC followed by its pyrolysis in nitrogen at 600°C. The adsorption of benzene vapors decreased on treatment of PVDC charcoal with PVC but was increased when the pore blocked charcoal was heat-treated in nitrogen at 600°C (PVDC-PVC-600°C (Figure 4.13).The adsorption of benzene decreased from 24% to 11% on impregnation of the carbon with PVC and then increased to about 31% on pyrolysis of the PVC-impregnated sample at 600°C, and the isotherm superimposing on the PVDC sample degassed at 600°C (PVDC-600). In the case of larger molecules of n.heptane, isooctane, the adsorption was considerably smaller on the PVDC-PVC-600°C sample. The adsorption isotherm shifted bodily downwards at all relative pressures, with the maximum adsorption decreasing from 13% to 6% in the case of n.heptane and from 12.6% to 4.5% in the case of isooctane. There was little or no adsorption of α-pinene. This indicated that the impregnation of PVDC charcoal with PVC followed by pyrolysis at 600°C in nitrogen resulted in the deposition of carbonaceous material in the porous structure and reduced the pore diameters or created constrictions so that larger molecules could not enter the pore cavity, thus producing a molecular sieve effect.

4.2.2 CHARACTERIZATION OF CARBONS BY IMMERSIONAL HEATS OF WETTING

Heat of immersion is a measure of the thermal effects produced when a solid is brought into contact with a nonreacting liquid and comprises the net integral heat of adsorption produced in forming an adsorbed layer at saturated vapor pressure plus the heat of wetting of this adsorbed film. In the case of highly microporous active carbons, the heat of wetting also involves the thermal effects related to the wetting of the external surface (nonporous structure) and the filling of micropore volume. Heat of wetting in the earlier days has been considered as a measure of surface area. However, it has been observed that the heat of wetting for the same carbon is different in different liquids so that it can not be related to surface area. It is, in fact, a measure of the accessibility of the active carbon surface to the immersing liquid. The heat of immersion has also been found to be dependent upon the chemical structure of the carbon surface as well as on the nature of the immersion liquid. Consequently, it has been used for the determination of surface polarity, site heterogeneity, hydrophobicity, and several other surface properties of microporous carbons.[69–75]

Immersional heat of wetting has also been used to determine the accessible pore volume and size of pores in active carbons and, thus, to characterize molecular sieve carbons by molecular probe methods using molecules of different sizes and shapes as immersion liquids. Maggs and Robins[76] using a variety of commercial active carbons, and Atkins et al.[77] using a charcoal cloth having predominantly narrow pores and a sample of Amoco activated carbon with a wide distribution of micropores, determined heat of immersion in a number of organic liquids with varying molecular dimensions. The accessible area as a function of the molecular diameter of the immersion liquid decreased as the size of the molecule increased in all carbons and carbon cloth (Figure 4.14). Comparison of the accessible area with BET (N_2) areas showed that only a part of the area was available to even the smaller molecules such as benzene. Thus, the BET monolayer capacity, in fact, represents a pore volume capacity and that the heat of immersion values in different liquids can be used to measure the pore-size distribution and to characterize the carbons.

Zettlemoyer et al.[79] determined the heats of immersion of a polymer-based carbon carbosieve-S with narrow micropore-size distribution (about 0.5 nm) in water, methanol, and 2-propanol as a function of precoverage of the surface with the immersion liquid. The larger fall in the heat of immersion at low coverages was attributed to surface heterogeneity, whereas the larger difference between methanol and 2-propanol (Figure 4.15) in the initial stages of surface coverage was due to differences in their molecular diameters. Stoeckli and Kraehenbeuhl[79] established a link between the enthalpy of immersion and the parameters of Dubinin's theory of volume filling of micropores in the case of a large range of micropores (Figure 4.16). It is seen that the left-hand side distribution obtained from molecular sieve experiments agrees well with the curve obtained from Dubinin equation. These workers suggested that the enthalpy of immersion is a function of the characteristic energy.

Groszeck[80] studied the heats of immersion of several microporous carbons in n.heptane using flow calorimetry and found that the pattern of heat evolution indicated the extent of pore system available to the given compound. The heats of

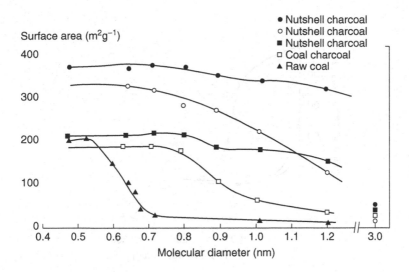

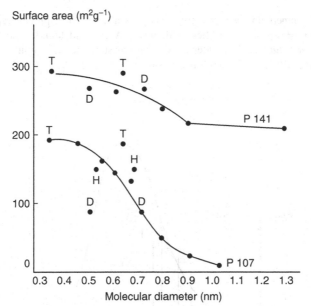

FIGURE 4.14 Accessible surface area as a function of the molecular dimensions of the immersion liquid for different charcoals. (After Maggs, F.A.F. and Robins, G.A., in *Characterisation of Porous Solids*, S.J. Gregg, K.S.W. Sing, and H.F. Stoeckli, eds., Soc. Chem. Ind., 1979, p. 59. With permission.)

preferential adsorption of the probe from carrier fluids, which can easily penetrate the carbon pores, indicated the properties of the pores capable of adsorbing molecules with various shapes and dimensions. Kraehenbeuhl et al.[81] and Daguerre et al.[82] also observed that immersional heats of wetting of carbons in liquids of increasing

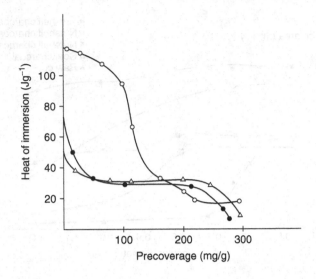

FIGURE 4.15 Immersion heats of Carbosieve-s as a function of precoverage. Δ = water;
● = methanol; O = 2-propanol. (After Zettlemoyer, A.C., Pendelton, P., and Micale, F.J., in
Adsorption from Solution, R.H. Ottewill, C.H. Rochestor, and A.L. Smith, eds., Academic
Press, New York, 1983, p. 113. With permission.)

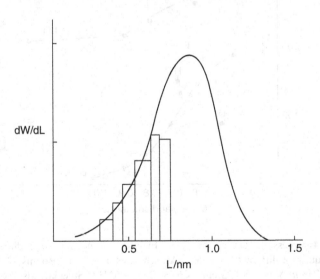

FIGURE 4.16 Comparison of micropore distribution of carbon CEP-59 calculated using
Dubinin theory of volume filling of micropores and the experimental data obtained from
molecular sieve experiments in the region of micropore width less than 0.75 nm. (After
Stoeckli, F.H. and Kraehenbeuhl, F., *Carbon*, 19, 353, 1982. With permission.)

molecular dimensions can lead to accurate micropore distributions. Gavrilov et al.[83] also observed that the specific heat of immersion of activated carbons in liquids of increasing molecular dimensions can lead to accurate micropore distributions. These workers also observed that the specific heat of immersion of activated carbons in benzene increased with an increase in the micropore dimensions. De Salazar et al.[36] observed that the pore blocking by deposition of carbon by CVD produces a decrease in enthalpies of immersion both in benzene and dichloromethane, and the decrease is higher in benzene. A linear relationship was found between the decrease of accessibility of these two liquids and the deposition time.

Bansal, and coworkers[31,32] determined heats of immersion in benzene, n.hexane, n.heptane, isooctane, and α-pinene for PVDC charcoals before and after pore blocking by pyrolysis of benzene vapors (Table 4.5), using a phase-change calorimeter containing diphenyl ether as the calorimetric fluid. The heat of immersion of original PVDC charcoal was maximum in benzene and minimum in α-prinere. This has been attributed to the smaller size of benzene molecule, which enables it to penetrate a larger proportion of the microporous structure. The α-pinene, on the other hand, has larger molecular dimensions and was, therefore, incapable of entering some of the finer micropores. It was interesting to note that n.hexane, n.heptane, and isooctane, which had similar molecular dimensions (0.67 to 0.68 nm), showed similar values of heats of immersion. This indicated that immersional heat of wetting was a function of the molecular dimensions of the immersing liquid and depended upon the pore-size distribution in carbons. When the pores were blocked by benzene

TABLE 4.5
Immersional Heats of Wetting (J/G) Charcoal before and after Pore Blocking by Pyrolysis of Benzene

	Benzene (.37 nm)	Hexane (.67 nm)	Heptane (.675 nm)	Iso-octane (.68 nm)	α-pinene (.80 nm)
PVDC	58.52	45.82	43.46	36.57	5.01
	(0.284)	(0.216)	(0.220)	(0.144)	(0.023)
PVDC-40	54.34	35.94	33.42	25.92	3.76
	(0.280)	(.208)	(.180)	(0.095)	(0.10)
PVDC-80	52.66	22.57	23.48	21.31	2.50
	(0.280)	(.132)	(0.148)	(0.095)	(0.10)
PVDC-120	52.25	10.36	11.24	11.45	0
	(0.272)	(.065)	(0.090)	(0.051)	(0.004)
PVDC-160	49.99	7.52	8.4	8.23	0
	(0.268)	(0.053)	(0.059)	(0.029)	0

Note: The values in parenthesis represent pore volume (ml/g)

After Sharma, N., Ph.D. thesis submitted to Panjab Univ., Chandigarh, India, 1991. With permission.

pyrolysis, the heats of immersion decreased and the extent of decrease depended upon the molecular dimensions of the immersion liquid. The decrease in the heat of immersion was only slight in the case of benzene; appreciable in the case of immersion in n.hexane, n.heptane, and isooctane; and very large in the case of immersion in α-pinene, which is comparatively a much larger molecule. The immersional heat of wetting in α-pinene was negligible in the case of PVDC-120 and PVDC-160 charcoal samples.

As the heat of immersion of a carbon in liquids of different molecular dimensions has been suggested[76,77] to be a measure of the accessibility of the pore volume of the carbon to the immersion liquid, apparent pore volumes for different PVDC charcoal for different liquid adsorbates were calculated from their vapor adsorption isotherms.[84] These values are included in Table 4.5 in parenthesis. A plot of the apparent micropore volume, calculated from adsorption at a relative pressure of 0.98, and expressed as volume of liquid, assuming the liquid to have normal liquid density, and the immersional heats of wetting in different adsorbates is linear (Figure 4.17), indicating that the heat of immersion in a given liquid is directly related to the pore volume accessible to that liquid adsorbate. A small scatter around the linear plot was attributed to very small thermal effect variations resulting from small interactions of the carbon surface with the immersing liquid. This clearly shows that the immersional heat of wetting in a given liquid can be used as a measure of its accessible pore volume. The accessible pore volume calculated from

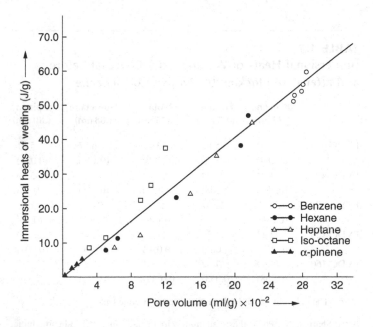

FIGURE 4.17 Immersional heats of wetting in relation to pore volume for PVDC charcoal before and after pore blocking by benzene pyrolysis. (After Sharma, N., Ph.D. thesis submitted to Panjab Univ., Chandigarh, India, 1991. With permission.)

the immersional heats of wetting using adsorbates of different molecular dimensions was further used to determine the pore-size distribution in a given carbon. These workers[31] found that in original PVDC charcoal about 80 to 90% of the pore volume consisted of pores with average dimensions between 0.6 and 0.8 nm, but less than 20% consisted of pores with dimensions larger than 0.8 nm. When the charcoal was given pore blocking treatment by pyrolysis of benzene, the deposition of carbon takes place preferentially in pores with diameters larger than 0.6 nm.

4.3 ADSORPTION BY CARBON MOLECULAR SIEVES

The development of ultrafine porous structure in active carbons (i.e., the preparation of molecular sieve carbons) has been the subject matter of a large number of investigations because these materials find applications in industrial separation processes. These materials have some distinct advantages over the zeolite sieves.[85]

- CMS can be used at temperatures higher than 700°C, and some are stable even up to 1400°C, so that separation processes can be carried out at higher temperatures.
- CMS are quite stable in strongly acidic solutions, whereas zeolite sieves become increasingly unstable as the pH of the medium progressively falls below 5.
- The hydrophobic character of CMS can be modified depending upon the starting material and the heat treatment conditions so that CMSs are more efficient than zeolites in removing substances from wet gas streams.

Furthermore, CMS obtained from certain carbonaceous materials such as coal can be obtained at very low cost and do not polymerize olefin material in petroleum industry, whereas zeolite sieves can cause such polymerization. CMS also have sufficient strength to be used in a fluidized process for which zeolites are unsuitable due to their inherent weakness.

Most of the work on the preparation of CMS has been carried out with a view to use these materials for the separation of nitrogen and oxygen although CMS can also be used for the separation of small amounts of impurities from industrial solvents. Chihara and Suzuki[54] used CMS prepared by pore blocking of activated carbons, by the decomposition of ethyl benzene, or styrene for the separation of O_2 and N_2. These workers measured the micropore diffusivities by chromatographic method and observed that absolute values of diffusivities for O_2 and N_2 decreased by more than one order of magnitude, although the ratios of the diffusivities for O_2 and N_2 remained more or less the same. Moore and Trimm[59] used CMS prepared by pore blocking by pyrolysis of benzene and having pore diameters reduced from 0.6 to 0.3 nm for the adsorption of O_2 and N_2 and observed that, while the adsorption of N_2 was reduced drastically when the deposition of carbon was about 16 mg/g, the adsorption of benzene remained more or less unchanged, indicating that the molecular sieve carbons could be used for the separation of air.

de Salazar et al.[36] studied the adsorption kinetics of N_2/O_2 and CO_2/CH_4 mixtures on CMS prepared by chemical vapor deposition by pyrolysis of benzene and

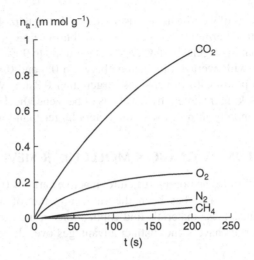

$n_a.(\text{m mol g}^{-1})$

FIGURE 4.18 Adsorption kinetics curves of N_2/O_2 and CO_2/CH_4 on CMS. (After de Salazar, C.G., Sepulveda-Estribano, A., and Rodriguez-Reinoso, F., 24th Bienn. Conf. on Carbon 1999, Ext. Abstr., p. 36. With permission.)

observed that the original sample was not selective for the separation of N_2/O_2 and CO_2/CH_4 mixtures, but the carbon molecular sieves prepared after deposition showed different adsorption kinetics (Figure 4.18). It can be seen that oxygen adsorption is much faster than nitrogen adsorption and that carbon dioxide is adsorbed to a much larger extent than methane. The separation ability of the different samples was found to depend upon the temperature and the time of deposition. The adsorption kinetics of CO_2 and CH_4 were also studied on molecular sieves prepared from coal.[43] The molecular sieve prepared by heat-treating a mixture of coal and pitch at 700°C showed adsorption of smaller molecule CO_2, while the adsorption of CH_4 was very small. The additive pitch acts as a binder and modifies the porosity.

Pedrero et al.[38,39] prepared CMS by chemical vapor deposition on a lignin-based microporous carbon. The textural characterization of the CMS was carried out by adsorption of N_2 at 77 K and CO_2 at 273 K. The sieving properties of the CMS were determined by the kinetics of adsorption of O_2/N_4 and CO_2/CH_4 mixtures. The adsorption capacities of the carbon for O_2 and CO_2 decreased slightly with deposition. When the pyrolytic carbon deposited was 0.3%, the adsorption of nitrogen was reduced drastically, and the adsorption of CH_4 was impeded. The oxygen selectivity (ratio of O_2 to N_2 adsorbed in 2 min.) was increased rapidly to a value of 6, while the decrease in the adsorption of N_2 was only 20%. Similar behavior was observed in the case of CO_2/CH_4 mixtures.

Ahmadpour and coworkers[45] used carbon molecular sieves prepared naturally occurring substrate (Iranian Walnut shell) and commercial activated carbon (Silcarbon) for air and hydrocarbon separation, using four different methods based on the adjustment of pore opening in the activated carbon structure. They studied the adsorption kinetics of O_2, N_2, CH_4, and C_2H_4 (Figure 4.19). Comparison of Figures 4.19 (a), (b), (c), and (d) indicates that none of these methods is able to separate O_2 from N_2 in air.

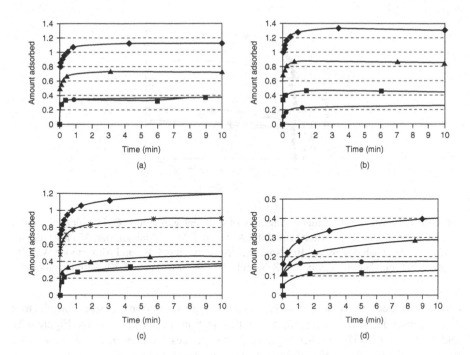

FIGURE 4.19 (a) Adsorption kinetics of gases for CMS sample prepared by acetylene decomposition at 298 K; Symbols: ● O_2, ■ N_2, p CH_4, ◆ C_2H_4. (b) Adsorption kinetics of gases for CMS sample prepared by benzene decomposition at 298 K; Symbols: ● O_2, ■ N_2, p CH_4, ◆ C_2H_4. (c) Adsorption kinetics of gases for CMS sample prepared by CTP impregnation (21% ratio) at 298 K; Symbols: ● O_2, ■ N_2, p CH_4, ◆ C_2H_4, * C_2H_6. (After Ahmadpour, A., Abedinzadegan, M., Mahadiarfar, M., Rashidi, A.M., Jalilian, A., and Mirhabibi, A.R., Carbon '02 Intern. Conf. on Carbon, Beijung, Sept 15–19, 2002.) (d) Adsorption kinetics of gases for CMS sample prepared by CTP impregnation (53% ratio) at 298 K; Symbols: ● O_2, ■ N_2, p CH_4, ◆ C_2H_4.

However, the CMS prepared by coal-tar pitch (CTP) impregnation by 21% has the ability to equilibrially separate CH_4 from C_2H_4 and CH_4 from C_2H_6 (Figure 4.19c). These workers on the basis of TEM images of the samples prepared using different methods suggested that it was not possible to narrow down the pore entrances of activated carbons to a large extent. They also observed that increasing the amount of CTP in the activated carbon reduces the adsorption capacity of the product. These workers concluded that to avoid reduction in adsorption capacity and to increase selectivity of the final product for different gaseous species, it is better to narrow down the pore opening in activated carbons to some extent by using heavy hydrocarbons in the first step, and to use smaller hydrocarbons such as benzene in the second step, to narrow down the pore openings further.

In another publication, Mirhabibi and coworkers[86] carried out separation of methane, ethane, and ethylene from nitrogen was carried out by adsorption on a carbon molecular sieve (Figure 4.20) with pore diameter less than 0.4 nm prepared from Iranian natural-nut shell. The kinetic adsorption curves shows that good selectivities

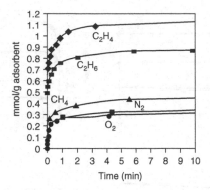

FIGURE 4.20 Adsorption kinetic diagrams of different gases on a CMS prepared from a Persian nutshell charcoal. (After Mirhabibi, A.R., Abdi, M.A., Mahadiarfar, M., Rashide, A., Jalilian, A. Aghahabazadeh, A., Brydson, R., Brown, A., Rand, B., and Ahmadpour, A., 25th Bienn. Conf. on Carbon, Lexington, K.Y., July 15–19, 2001, Paper 5.30. With permission.)

can be obtained. These workers also obtained high-resolution transmission electron microscopy (HRTEM) images from very thin sections of porous carbons (Figure 4.21 and Figure 4.22). The higher regions of the images have been considered to be the pores. These images were digitized using an image scanner, and a power spectrum was produced using a two-dimensional fast Fourier transform (FFT). The figures

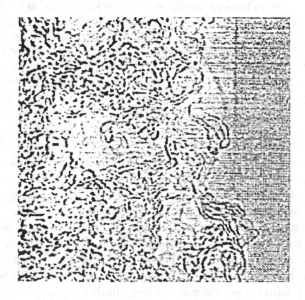

FIGURE 4.21 High resolution electron transmission electron microscopy (HRTEM) microstructure of original Persian nutshell charcoal. (After Ahmadpour, A., Abedinzadegan, M., Mahadiarfar, M., Rashidi, A.M., Jalilian, A., and Mirhabibi, A.R., Carbon '02 Intern. Conf. on Carbon, Beijung, Sept. 15–19, 2002. With permission.)

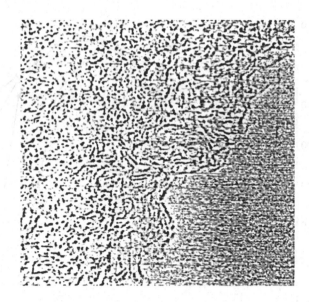

FIGURE 4.22 HRTEM microstructure of CMS prepared from Persian nutshell charcoal. (After Ahmadpour, A., Abedinzadegan, M., Mahadiarfar, M., Rashidi, A.M., Jalilian, A., and Mirhabibi, A.R., Carbon '02, Intern. Conf. on Carbon, Beijung, Sept. 15–19, 2002. With permission.)

show a wide pore-size distribution and the existence of a large number of mesopores in the microstructure of the original char (Figure 4.21), and a narrower pore-size distribution and the absence of micropores in CMS prepared from the original char (Figure 4.23).

Li et al.[87] studied the adsorption and diffusion properties of ethylene, carbon dioxide, and oxygen using a CMS with pore size ranging between 0.3 and 0.5 nm. It was found that ethylene had the largest equilibrium adsorption capacity and the slowest adsorption rate of the three gases so that ethylene could be separated from the gaseous mixtures via kinetic separation based on the difference of adsorption rates. The oxygen and carbon dioxide could be removed from the gaseous mixture effectively with pressure-swing adsorption. High purity ethylene with a concentration exceeding 98% was obtained using this carbon molecular sieve.

Pedrero et al.[39] prepared CMS by the chemical activation of lignin with zinc chloride at different degrees of impregnation followed by heat treatment at 1073 K. The carbons prepared with an impregnation ratio of 2:3 did not show any molecular sieve effect as determined by adsorption of N_2 at 77 K and carbon dioxide at 273 K. However, the carbon prepared with an impregnation ratio of 0.4 showed an appreciable increase in the adsorption of CO_2 and no increase in the adsorption of N_2, indicating the creation of narrow microporosity in which the diffusion of nitrogen gas was restricted at the low adsorption temperature of 77 K. These molecular sieve carbons were able to separate benzene from cyclohexane because they adsorbed large amounts of benzene and very little of cyclohexane. However, these carbons could not separate efficiently CO_2 from CH_4, although CO_2 has about the same mol. diam. 0.37 nm as benzene. This indicated that these carbons had slit-shaped narrow pores.

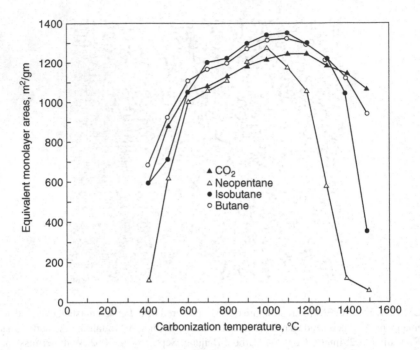

FIGURE 4.23 Surface areas of PVDC carbons heated to different temperatures. (After Walker, P.L., Jr., *Carbon*, 28, 261, 1990. With permission.)

Walker and coworkers[46-53,88,89] carried out a comprehensive research program on the preparation of CMS and their applications in the separation of gases and liquid hydrocarbon. According to them, the ability to separate different molecules depends upon the fact that small differences in pore size of the CMS or in molecular size produces large differences in the activation energy required for diffusion in the repulsive region of a potential energy diagram. They even calculated critical pore dimensions (Table 4.6) for the adsorption of a number of gases,[89] assuming the pore to be slit-shaped with its sides consisting of carbon basal planes. When the size of the slit widths is less than the sizes given in the table, the gas molecule shall undergo activated diffusion when the diffusion coefficient varies exponentially with temperature. However, when the slit widths are somewhat larger than the given sizes, the adsorbing molecule will undergo Knudsen diffusion in which case the diffusion coefficient varies with $T^{1/2}$.

Lemond et al.[88] while preparing CMS from PVDC, observed that maximum carbonization temperature determined the size of the aperture in CMS (Figure 4.23). The figure shows the effect of carbonization temperature on the surface areas of PVDC carbons obtained from the adsorption of four gaseous molecules of different minimum molecular dimensions adsorbed over a 1-hr period. It is seen that at lower temperatures the aperture opening is partially blocked because hydrochloric acid has not been completely released by the decomposition of PVDC. But when the carbonization temperature is between 600°C and 1000°C, the aperture size is maximum

TABLE 4.6
Critical Pore Dimensions
for Different Gases

Gas	Slit Width (nm)
CO	0.541
H_2	0.548
CO_2	0.542
O_2	0.544
N_2	0.572
Ar	0.575
He	0.508

After Walker, P.L., Jr., *Carbon*, 28, 261, 1990. Reproduced with permission from Elsevier.

so that there is no sieving effect shown between the smallest molecule CO_2 (cross-section ~0.33 nm). And neopentane (cross-section ~0.62 nm). At still higher carbonization temperatures, there is breakage of cross-links in the structure, its densification, and gradual reduction in aperture size, producing molecular sieve carbons.[89] At very high temperatures, there is annealing of the apertures so that these are inaccessible to all gases, although a large surface area is still present in the closed apertures. This closed surface area can be measure α by small angle x-ray spectroscopy (SAXS) studies. These workers separated branched-chain hydrocarbons from straight-chain hydrocarbons in a flow system using these PVDC carbon molecular sieves,[90] and compared the separation behavior of CMS with Linde 5, a zeolite sieve, and a BPL-activated carbon from Calgon corporation. The zeolite sieve, which is used commercially for the separation of hydrocarbons, showed the expected behavior on a breakthrough plot. The branched-chain hydrocarbon 2, 2, 4-trimethyl pentane was found to have negligible access to the pores in the zeolite and broke through the bed almost immediately, but the straight-chain hydrocarbon n.heptane diffuses through the pore of the zeolite rapidly and gets adsorbed in the cavities so that it showed a considerable breakthrough time. In the case of BPL activated carbon, the pores were sufficiently large and it showed almost the same breakthrough time for the two hydrocarbons. The CMS showed excellent sieving behavior for the two hydrocarbons, the breakthrough time for the branched hydrocarbon was short, and that for the straight-chain hydrocarbon was even longer than observed for a comparable weight of the zeolite.

More recently, there has been interest in the production and commercial utilization of CMS for the separation of nitrogen and oxygen from air. It is seen from the critical pore dimensions of pores given in (Table 4.6) that oxygen can enter slightly smaller slits (0.544 nm) than nitrogen (0.572 nm) without involving activated diffusion. The diffusion coefficients were obtained using unsteady-state diffusion measurements for some gases in a CMS prepared from a noncoking bituminous coal.

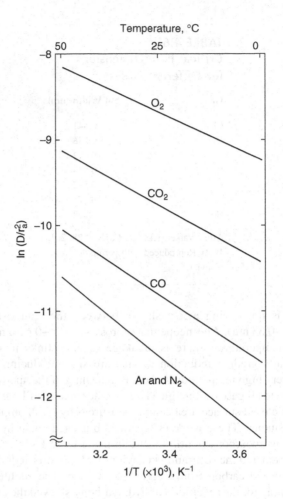

FIGURE 4.24 Arrhenius plots of diffusion parameters for diffusion of gases through a CMS. (After Walker, P.L., Jr., *Carbon*, 28, 261, 1990. With permission.)

The Arrhenius plots of diffusion parameter D/r_0^2 in units of 1/s were plotted instead of the diffusion coefficient (D), because the average diffusion length (r_0) within a particle of CMS is unknown (Figure 4.24). The diffusion parameters of O_2 and N_2 differ by orders of magnitude. This contrasted with the results obtained earlier by Chihara et al.[91] who found that the diffusion coefficients for O_2 and N_2 were essentially equal. This has been attributed to the fact that their CMS was exposed to H_2 gas at 150°C. This resulted in the chemisorption of hydrogen, which prevented the subsequent chemisorption of oxygen. This could effect the interaction energy between a diffusing species and a surface.

Verma[48] and Verma and Walker[49] investigated the molecular sieving properties of microporous carbons, such as a pelletized carbon molecular sieve (CMS-A), a granulated carbon (CMS-B), and two activated carbons, which had pore structure ranging between that of carbon molecular sieve (CMS) and the activated carbon (AC). The molecular sieve properties of these carbons were modified by thermal treatment, controlled gasification, or by deposition of carbon from pyrolysis of organic compounds. The sieving properties were investigated by adsorption and diffusion of Ar, N_2, and O_2. Sieving behavior was improved in the case of carbons containing a significant distribution of micropores with narrow size (< 0.5 nm). The diffusion rates of O_2 and Ar increased significantly on heating pelletized CMS up to ~973 K. This increase was thought to be the result of partial desorption of the carbon-oxygen surface groups and to some extent to the decomposition of the organic impregnant present on the as-received carbon surface. The increase in rate of Ar adsorption has been attributed to widening of some of the pores to a size > 0.38 nm in addition to the creation of some additional pore volume. However, the rates of diffusion of both O_2 and Ar decreased significantly at HTT up to 1273 K. This suggests that the increasing HTT above 973 k gradually reduces the size of the pore apertures to less than 0.38 nm and ultimately even to less than 0.28 nm. Although some carbon-oxygen surface groups may still be desorbing at temperature > 973 K, pore sintering becomes predominant.[92] The thermal treatment also affected the kinetic selectivity or selectivity ratio of O_2 to Ar adsorption (O_2/Ar) for specific adsorption times. The selectivity of O_2 and Ar was determined for 30 s and 5 min. adsorption times. It was observed that the selectivity ratio for 30 s decreased with HTT up to 873 K but increased sharply at HTT greater than 873 K. The selectivity ratio was maximum at ~1098 K and then decreased rapidly with HTT at higher temperatures. The heat treatment also had a marked effect on the O_2/Ar ratio for the other two molecular sieve carbons. The temperature at which maximum in the selectivity ratio occurred was found to depend upon a balance between the stability of the oxygen groups on the carbon and the resistance of the carbon to sintering. The thermal treatment, however, did not produce any molecular sieving properties in the case of activated carbon (AC).

Easler[93] and Hurt et al.[94] have shown that carbon gasification of disordered carbons (chars) can also lead to particle shrinkage as a result of gasification induced atomic rearrangement. This can result in the reduction of aperture size and loss of volume in open pores. In fact, the gasification of low-temperature disordered carbons can result in the partial removal of cross-linking atoms, allowing somewhat better alignment of small triagonally-bonded carbon crystallites. Gasification can also accelerate the normal sintering process that otherwise will take place at higher heat treatments. Thus, the overall effect of carbon gasification on pore size in molecular sieve carbons will depend upon a balance between the removal of carbon atoms around the apertures (aperture widening) and crystallite rearrangement enhancing sintering (aperture narrowing). Verma and Walker[48] tried gasification of molecular sieve carbons as a means to adjust the size of the apertures and observed that the net effect of carbon gasification was shrinkage of pores in a particular size range.

Modification of the pore structure of microporous carbons by depositing carbon from pyrolysis of propylene between 973 K and 1123 K was investigated to develop molecular sieve carbons[49,50] for the separation of O_2 and Ar. Propylene was preferred over other sources of pyrolytic carbon because its ΔH for cracking is very low and it is easy to handle under ambient conditions. Diffusion plots of uptake of O_2 and Ar on the modified CMS showed that uptake of O_2 is reduced only slightly on deposition of pyrolytic carbon, and that of Ar is reduced drastically. The uptake of O_2 at 8 min is reduced by about 29%, and that of Ar is reduced by more than 90%. The selectivity defined as the ratio of O_2 uptake to Ar uptake was increased from 1.2 to 11.2 at 30 sec and from 1.3 to 11.1 at 5-min adsorption periods. When the modified CMS was further heat treated at 1123 K for 1 hr under vacuum, the rate of diffusion for both the gases was slightly diminished, the effect being slightly greater for O_2. It has been suggested[49,50] that upon heat treatment at 1123 K, cracking of polymerized-chemisorbed propylene complex, remaining on the carbon surface following treatment at 973 k, takes place, which results in the deposition of more carbon and reducing further the sizes of some pores below 0.28 nm and thus diminishing O_2 uptake. The heat treatment at 1123 K may also cause some sintering of the pores, reducing their size below that of O_2. Some pore apertures that allowed rapid diffusion of Ar are also reduced to a size below 0.38 nm, reducing the rate of Ar adsorption.

The rate and amount of propylene cracking was found to depend upon the cracking conditions such as the temperature, time of cracking, and concentration of propylene as well as on the active surface area of the carbon. Thus, the conditions selected will be of paramount importance in determining the efficiency in converting microporous carbons into useful CMS. For example, when pyrolysis temperature is sufficiently low, the deposition rate of carbon is determined solely by the rate of cracking of propylene over the carbon active sites. The deposition is uniform throughout, and the desired sieving properties can be obtained with a maximum amount of carbon deposited. On increasing the temperature, there is a rapid increase in the rate of cracking of propylene, the increase in rate being more rapid then the initial rate of propylene diffusion. This results in a nonuniform deposition, the amount of carbon deposition being more on the exterior surface of the substrate carbon and less at its center. Thus, the deposition of carbon has to be controlled to get the desired sieving effect. At too high a temperature (Zone IV kinetics region), the deposition of carbon becomes too rapid, so it takes place mostly at the external surface, resulting in lack of molecular sieve properties in the carbon.

Verma and Walker[52] further continued their work on the preparation of carbon molecular sieves by propylene pyrolysis over nickel-impregnated activated carbons. Exploratory experiments showed that molecular sieves for the separation of O_2 and Ar could be prepared by depositing carbon from cracking of propylene over and in an activated carbon previously impregnated with Ni. The activated carbon was loaded with up to 5% by weight of Ni from a concentrated solution of $Ni(NO_3)_2 6H_2O$. The technique could result in an even distribution of Ni through the pore network of the carbon prior to carbon deposition, at least in the transitional pores, (2 to 20 nm) and macropores (> 20 nm). However, these workers found that extensive amounts of Ni were present at the surface of activated carbon particles. Although good separation of O_2 and Ar could be achieved, it was not clear whether this sieving was primarily due to the introduction of new apertures of size 0.54 to 0.57 nm within the pore system or by depositing on pore entrances at exterior surface of the particles. They, however, are of the opinion that the addition of a catalyst for the cracking of a hydrocarbon presents an additional useful variable for the production of CMS from activated carbons.

The effect of interactions of O_2 and H_2O with carbon molecular sieve surfaces on their performance was also studied by using O_2 and Ar as molecular probes by Verma and Walker.[51] These interactions seriously influence the performance of CMS, due to their strong interactions with the active centers on the CMS surface. Carbon molecular sieves were treated with Cl_2 and H_2 gas. The treatment with Cl_2 was carried out at 450°C for one hr to remove any physisorbed chlorine. The treatment with hydrogen was carried out in two ways. First a sample of CMS was treated by flowing H_2 (0.1 MPa) through the carbon bed maintained at 250°C for 40 hrs. In the second procedure a fresh sample of CMS kept at 150°C was exposed to 5.5 MPa pressure of H_2 for 72 hr. Treatment with H_2 at 5.5 MPa and 150°C was found to deactivate the carbon active sites, thus eliminating the strong interactions with O_2 and H_2O, and producing stable CMS surface so that they can be stored for extended periods of time.

Verma et al.[52] also studied the influence of temperature on the separation of O_2 and Ar by CMS. On decreasing the adsorption temperature from 298 K to 273 K, the rate of adsorption of O_2 remains more or less unchanged while the rate of uptake of Ar is considerably decreased (Figure 4.25). This results in enhancing the selectivity of the CMS. On decreasing the adsorption temperature to 189 K, the rate of Ar uptake is further reduced, while that of O_2 uptake, although reduced initially, is enhanced for longer time adsorption. This further increased the selectivity of the CMS. Thus, the separation of O_2 from Ar is enhanced by reducing the adsorption temperature. This is due to the fact that the activation energy of O_2 diffusion into CMS is lower than that of Ar.[94]

The molecular sieve behavior of a number of microporous activated carbons was studied using molecules of different sizes and shapes as molecular probes by

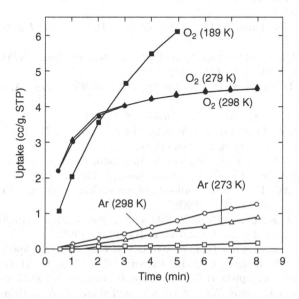

FIGURE 4.25 Adsorption of oxygen and Ar on CMS-A after heat treatment at different temperatures. (After Verma, S.K. and Walker, P.L., Jr., *Carbon*, 31, 1203, 1993. With permission.)

Carrott,[96] who suggested that there are at least two basic types of molecular sieve actions for the separation of small molecules. In the case of polymer carbons produced at low burn-off, a small micropore size is thought to be largely responsible for the molecular sieve action, but in certain others it is thought that constrictions at the micropore entrances are involved. It has also been found that there is an empirical relationship between log (η_o) and log (V_m), where η_o is the micropore capacity for a given adsorbate in mmol/g and V_m in cm^3 mol^{-1} is the molar volume of the adsorbate. Carrott has shown that the observed differences in the linearity and slope of the log (η_o) against log (V_m) plots can be rationalized on the basis of different types of molecular sieve behavior and that the analysis of the plots should, therefore, be able to provide useful information relating not only to the micropores themselves but also to the micropore entrances.

Thus, carbon molecular sieves are very good adsorbents for the separation of O_2, N_2, and Ar from air by pressure-swing adsorption (PSA). These carbon materials exhibit greater selectivity for O_2 because of their aperture-cavity type structure. The aperture determines the selectivity, which the cavity provides the high adsorption capacity. This high purity Ar is in great demand in several chemical, metallurgical, and electronic processes. High purity gases are also needed in numerous laboratory processes. Thus, by passing through a bed of molecular sieve carbons, the impurities such as CO_2 and O_2 are eliminated. Carbon molecular sieves can also be used to remove trace amounts of impurities from organic liquids and vapors.

REFERENCES

1. Winslo, F.H., Baker, W.O., and Yager, W.A., *Proc. 2nd Conf. on Carbon*, Buffalo, pp. 93, (1995).
2. Barton, S.S., Evans, M.J.B., and Harrison, B.H., *J. Colloid Interface Sci.*, 49, 462, (1974).
3. Howard, G.J. and Szynaka, S., *J. Appl. Poly. Sci.*, 19, 2633, (1975).
4. Fryer, J., *Carbon*, 19, 431, (1981).
5. Marsh, H., Crowford, D.O., O'Gradey, T.M., and Wennenberg, A., *Carbon*, 20, 419, (1982).
6. Bansal, R.C. and Dhami, T.L, *Carbon*, 15, 153, (1977).
7. Innes, R.W., Fryer, J.R., and Stoeckli, H.F., *Carbon*, 27, 71, (1989).
8. Stoeckli, H.F., *Carbon*, 28, 1–6, (1990).
9. Kartel, N., Puziy, A., and Strelkov, V., International Carbon Conference, Carbon '90, Paris, *Extended Abstracts and Programme*, pp. 110, (1990).
10. Bansal, R.C., International Carbon Conference, Carbon '90, Paris, *Extended Abstracts and Programme*, pp. 22, (1990).
11. Bailey, A. and Everett, D.H., *J. Polymer Sci.*, Part A, 27, 87, (1969).
12. Bailey, A. and Everett, D.H., *Nature*, 211, 1082, (1996).
13. Bailey, A. and Everett, D.H., Anniversary Meeting, (*Exceter*, April, 1967).
14. Adams, L.S., Boucher, E.A., and Everett, D.H., *Carbon*, 8, 761, (1970).
15. Boult, E.H., Campbell, H.G., and Marsh, H., *Carbon*, 7, 700, (1969).
16. Winslow, F.H., Baker, W.O., Pope, N.R., and Matreyek, W., *J. Polymer Sci.*, 16, 101, (1955).
17. Marsh, H. and Wyne-Jones, W.F.K., *Carbon*, 1, 269, (1964).

18. Kipling, J.J., Sherwood, J.N., Shooter, P.V., and Thompson, N.R., *Carbon,* 1, 321, (1964).
19. Youssef, A.M., *Carbon,* 13, 1, (1975).
20. Nandi, S.P. and Walker, P.L., Jr., *Fuel,* 43, 385 (1964).
21. Bansal, R.C., Dhami, T.L., and Prakash, S., *Carbon,* 18, 395, (1980).
22. Dacey, J.R. and Thomas, D.G., *Trans. Faraday Soc.,* 50, 740, (1954).
23. Moore, S.V. and Trimm, D.L., *Carbon,* 15, 177, (1977).
24. Cheredkova, K.I., Golovina, G.S., and Tolstykh, T. *Yu. Khim. Tverd. Topl.,* (Moscow), 1, 90–93, (1989).
25. Barton, S.S. and Koresh, J.E., *Extended Abstracts and Program,* 19th Bienn. Conf. on Carbon, pp. 6–7, (1989).
26. Vnukov, S.P., Polyakov, N.S., Dubinin, M.M., and Fedosaev, D.V., *Izv. Akad. Nauk SSSR, Ser. Khim.,* 2, 267–273, (1986).
27. Sutt., Robert, F., Jr., (Calgon Carbon Corp.), Eur. patent appl. EP 119, 924, (Cl. B01 J20/20), US appl. 474, 797, March 16, 1983.
28. Moreno-Castilla, C., Fernandez-Morales, I., Domingo-Garcia, M., and Copez-Garzon, F.J., *Chromatographia,* 20, (12), 709–712 (1985).
29. Yoshio, N., Yasunori, T., Hideki, S., and Takeshi, K., *Kagaku Kogaku Ronbunshu,* 15, (3), 489–496, (1989).
30. Bansal, R.C., Bala, S., and Sharma, N., *Indian J. Technol.,* 27, 206, (1989).
31. Bansal, R.C., Carbon, '90, Intern. Conference on Carbon, Paris, July 16–20, (1990), *Ext. Abstr.,* p. 22.
32. Bansal, R.C. and Dhami, T.L., *Proc. First 1st Indian Conf. on Carbon NPL,* New Delhi, Dec., 1982; Indian Carbon Soc. Publi., 1984, p. 129.
33. Sharma, N., Ph.D. thesis submitted to Panjab Univ., Chandigarh, India, 1991.
34. Lilan, L., Anjia, W., and Shucai, G., 19th Bienn. Conf., on Carbon, 1989, *Ext. Abstr.,* p. 24.
35. Lizzio, A.A. and Rostam-Abadi, M., 21st Bienn. Conf. on Carbon, 1993, *Ext. Abstr.,* p. 432.
36. de Salazar, C.G., Sepulveda-Estribano, A., and Rodriguez-Reinoso, F., 24th Bienn. Conf. on Carbon 1999, *Ext. Abstr.,* p. 36.
37. Daguerre, E., Guillot, A., and Stoeckli, H.F., 24th Bienn. Conf. on Carbon 1999., *Ext. Abstr.,* p. 434.
38. Pedrero, C., Cordero, T., Rodriguez-Mirosol, J., and Rodriguez, J.J., 24th Bienn. Conf. on Carbon 1999, *Ext. Abstr.,* p. 590.
39. Pedrero, C., Cordero, T., Rodriguez-Mirosol, J., and Rodriguez, J.J., 25th Bienn. Conf. on Carbon, Lexington, KY, July 14–19, 2001, Paper 38.3.
40. Manso, R., Pajares, J.A., Bronick. E., Jankowska, A., and Kaczmarezyk, J., 24th Bienn. Conf. on Carbon 1999, *Ext. Abstr.,* p. 686.
41. Abdi, M.A., Mohadiarfar, M., Ahmadpour, A., and Mirhabibi, A.R., 25th Bienn. Conf. on Carbon 2001, Lexington, K.Y., July 14–19, 2001, Paper 22.1.
42. Ya., L. and Zi,. Fang., 25th Bienn. Conf. on Carbon, Lexington, KY, July 14–19, 2001.
43. Alcaniz-Monge, J., Losano-Castillo, D., Cazoria-Amoros, D., and Linarer-Solam. A., 24th Bienn. Conf. on Carbon 1999, *Ext. Abstr.,* p. 682.
44. Valente, C.O., Coutinho, A.R., Maciel, H.S., Petroconi, F.G., Otani, C., and Massi, M., 'Carbon '02' Intern. Conf. on Carbon, Beijing, Sept. 15–19, 2002.
45. Ahmadpour, A., Abedinzadegan, M., Mahadiarfar, M., Rashidi, A.M., Jalilian, A., and Mirhabibi, A.R., 'Carbon '02' Intern. Conf. on Carbon, Beijung, Sept 15–19, 2002.
46. Schmitt, J.L. and Walker, P.L., Jr., *Carbon,* 9, 791, (1971).

47. Sehmitt. J.L. and Walker, P.L., Jr., *Carbon,* 10, 87, (1972).
48. Verma, S.K. and Walker, P.L., Jr., *Carbon,* 28, 175, (1990).
49. Verma, S.K. and Walker, P.L., Jr., *Carbon,* 29, 793, (1991).
50. Verma, S.K. and Walker, P.L., Jr., *Carbon,* 30, 829, (1992).
51. Verma, S.K. and Walker, P.L., Jr., *Carbon,* 30, 837, (1992).
52. Verma, S.K. and Walker, P.L., Jr., *Carbon,* 31, 1203, (1993).
53. Christner, L.G. and Walker, P.L., Jr., *Carbon,* 31, 1149, (1993).
54. Chihara, K. and Suzuki, M., *Carbon,* 17, 339, (1979).
55. Darmstadt. H., Roy, C., Kaliaguine, S., Choi, S.J., and Ryoo, R., 25th Bienn. Conf. on Carbon, Lexington, KY, July 14–19, 2001, Paper 11.2.
56. Darmstadt. H., Roy, C., Kakaguine, S., Choi, S.J., and Ryoo, R., 25th Bienn. Conf. on *Carbon,* Lexington, KY, July 14–19, 2001, Paper 3.1.
57. Vyas, S.N., Patwardhan, S.R., and Gangadhar, B., *Carbon,* 30, 605, (1992).
58. Moore, S.V. and Trimm, D.L., *Carbon,* 15, 177, (1977).
59. Bansal, R.C. and Dhami, T.L., *Indian J. Chem.,* 19A, 1146, (1980).
60. Toda, T., Halani, M., Toyada, S., Yoshida, Y., and Honda, H., Fuel, 50, 187, (1971).
61. Illey, M., Marsh, H., and Reinoso, F.R., *Carbon,* 11, 633, (1973).
62. Kiselev, A.V. and Kovaleva, N.V., *Zh. Fiz. Khim.,* 30, 2775, (1956).
63. Avgul, N.N., Dzhigit, D.M., Kiselev, A.V., and Shervakova, K.D., *Dokl. Akad. Nauk SSSR,* 91, 105, (1953).
64. Bansal, R.C., Dhami, T.L., and Prakash, S., *Carbon,* 16, 389, (1978).
65. Gregg, S.J. and Sing, K.S.W., in *Adsorption, Surface Area and Porosity,* (Acad. Press, London, (1982).
66. Bansal, R.C. and Dhami, T.L., *Carbon,* 18, 207, (1980).
67. Puri, B.R., Kaistha, B., Vardhan, Y., and Mahajan, O.P., *Carbon,* 11, 329, (1973).
68. Bansal, R.C., Dounet, J.B., and Stockli, F., in *Active Carbon,* Marcel Dekker, New York, 1988.
69. Chessick, J.J. and Zettlemoyer, A.C., *Advances in Catalysis,* 11, 263, (1959).
70. Wade, W.H. and Hackerman, N., *J., Phys. Chem.,* 64, 1196, (1960).
71. Wade, W.H., Teranishi, S., and Durham, J.L., *J. Colloid Interface Science.,* 20, 838, (1966).
72. Whalen, J.W., *J. Phys. Chem.,* 65, 1676, (1961).
73. Barton, S.S., Boulton, G.L., and Harrison, B.H., *Carbon,* 10, 391, (1972).
74. Barton, S.S. and Harrison, B.H., *Carbon,* 10, 745, (1972).
75. Barton, S.S. and Harrison, B.H., *Carbon,* 13, 47, (1975).
76. Maggs, F.A.F. and Robins, G.A., in *Characterisation of Porous Solids,* (S.J. Gregg, K.S. W. Sing, and H. F. Stoeckli, Eds.,) *Soc. Chem. Ind.,* 1979, p. 59.
77. Atkins, D., Mcleod, A.I., Sing, K.S.W., and Capon, A., *Carbon,* 20, 339, (1982).
78. Zettlemoyer, A.C., Pendelton, P., and Micale, F.J., in *Adsorption from Solution,* (R. H. Ottewill, C. H. Rochestor, and A. L. Smith, Eds.,) Academy Press, New York, 1983, p. 113.
79. Stoeckli, F.H. and Kraehenbeuhl, F., *Carbon,* 19, 353, (1982).
80. Groszeck, A.J., *Carbon,* 27, 33, (1989).
81. Kraehenbuehl. F., Stoeckli, H.F., Addoun, A., Ehrburger, P., and Donnet, J.B., *Carbon,* 24, 483, (1986).
82. Daguerre, E., Guillot A., and Stoeckli, F.H., 24th Bienn. Conf. on Carbon, 1999, *Ext. Abst.,* p. 438.
83. Gavrilov, D.N., Ivakhnyuk, G.K., Fedrov, N.F., and Kapitonenko, Z.V., *Zhur. Zh. Prikl. Khim,* (Leningrad), 58, (9), 1985.
84. Bansal, R.C. and Dhami, T.L., *Indian J. Technol.,* 23, 92, (1985).

85. Walker, P.L., Jr., Austin, L.G., and Nandi, S.P., in *Chemistry and Physics of Carbon,* (P.L. Walker, Jr., Ed.), Vol. 2, p. 257, Marcel Dekker, New York, 1966.
86. Mirhabibi, A.R., Abdi, M.A., Mahadiarfar, M., Rashide, A., Jalilian, A. Aghahaba-zadeh, A., Brydson, R., Brown, A., Rand, B., and Ahmadpour, A., 25th Bienn Conf. on Carbon, Lexington, KY, July 15–19, 2001, Paper P5–30.
87. Li, D., Wang, J., and Lu, J., 'Carbon '02,' Intern. Conf. on Carbon, Beijung, 2002.
88. Lemond, T.G., Metcalfe, J.E., III, and Walker, P.L., Jr., *Carbon,* 3, 59, (1965).
89. Walker, P.L., Jr., *Bull. College of Mineral Industries.* Penn. State Univ., University Park, PA, Jan. 1966 pp. 1–7.
90. Walker, P.L., Jr., *Mineral Industries*, Penn. State Univ., Jan. 1966, pp. 1–7.
91. Chihara, K., Suzuki, M., and Kawazoc, K., J. Colloid Interface Science., 64, 584, (1978).
92. Koresh. J. and Soffer, A., *J. Chem. Soc.,* Faraday I., 76, 2457, (1980).
93. Easler, T.E., Ph.D. thesis, Penn. State Univ., 1983.
94. Hurt, R.H., Dudek, D.R., Longwell, J.P., and Sarofin, A.F., *Carbon,* 26, 433, (1988).
95. Hoffman, W.P., Ph.D. thesis, Penn. State Univ., 1979.
96. Carrott, P.J.M., *Carbon,* 33, 1307, (1995).

5 Activated Carbon Adsorption Applications

Activated carbons are excellent and versatile adsorbents. Their important applications relate to their use in the adsorptive removal of color, odor, taste, and other undesirable organic and inorganic impurities from drinking waters; in the treatment of urban ground and industrial waste water; solvent recovery; air purification in inhabited spaces such as restaurants, food processing, and chemical industries; for the removal of color from various types of sugar syrups, oils, and fats; in the purification of many chemical, food, and pharmaceutical products; in respirators for work under hostile environments; and in a variety of other gas-phase applications. They are increasingly being used in the field of hydrometallurgy for the recovery of gold, silver, and other inorganics, and as catalytic and catalyst supports. Their use in medicine and health applications to combat certain types of bacterial ailment and for the removal of certain toxins is well known. These applications of activated carbon are of interest to most economic sectors and concern areas as diverse as the food, pharmaceutical, chemical, petroleum, mining, nuclear, automobile, and vacuum industries. Nearly 80% of the total activated carbon is consumed for liquid-phase applications, where both the granulated and powdered forms of active carbon are used. For gas-phase applications, granulated carbon is usually the choice. However, with the commercial production of fibrous activated carbons in the form of fibers and fabric, these materials may be in preference, especially for water treatment processes, because they produce low hydrodynamic resistance to flow and can easily be molded into any shape in the adsorption equipment.

The adsorbent properties of activated carbons are essentially due to their surface area, universal adsorption effect, highly microporous structure, and a high degree of surface reactivity. The availability of favorable pore size makes the internal surface accessible and enhances the adsorption rate. The most widely used activated carbons have a specific surface area of 800 to 1500 m^2/g. This surface area is contained predominantly within micropores that have effective diameters smaller than 2 nm. In fact, a particle of active carbon consists of a network of pores that have been classified into micropores (diameters < 2 nm), mesoporous (diameter between 2 and 50 nm) and macropores (diam. > 50 nm). The macropores do not contribute significantly toward surface areas but act as conduits for the passage of the adsorbate into the interior mesopore and the micropore surface where most of the adsorption takes place.

Although the adsorption capacity of active carbons is determined by their physical or porous structure, it is strongly influenced by the chemical structure of their surface. In graphites, for example, which have a highly ordered crystalline structure, the adsorption capacity is determined by the dispersion component of London forces. In the case of active carbons, however, the random ordering of the aromatic sheets causes a variation in the arrangement of electron clouds in the carbon skeleton, which results in the creation of unpaired electrons and incompletely saturated valencies that would undoubtedly influence the adsorption behavior. In addition, active carbons are generally associated with oxygen and hydrogen, which are present in the form of carbon-oxygen and carbon-hydrogen surface groups. These surface groups are bonded at the edges of the aromatic sheets. Because these edges constitute the main adsorption surface, these surface groups profoundly influence the adsorption behavior of active carbons. Besides, the active carbon surface has active sites in the form of edges, dislocations and discontinuities that determine the chemical reactions and the catalytic properties of active carbon.

5.1 LIQUID PHASE APPLICATIONS OF ACTIVATED CARBON ADSORPTION

5.1.1 FOOD PROCESSING

Activated carbon adsorption from liquid phase has found wide applications in several areas of food production and processing industries. The activated carbons remove undesirable odors, colors, and unwanted components of the solution, and improve the quality and consumability of the food material. The use of active carbons in food processing is continuously on the increase because the food processing and production industries in which the use of active carbon is well established are always expanding. Furthermore, the use of activated carbons is also being explored and expanding to areas of the food processes, where it was not previously used.

5.1.2 PREPARATION OF ALCOHOLIC BEVERAGES

Wood charcoal has since long been used in distilleries for refining neutral spirits, but it has now been almost completely replaced by activated carbons. Active carbon now finds increasing applications at different stages of the production of alcoholic beverages. In each case, the aim is to remove unwanted components to improve the taste, color, and other properties.

In the preparation of rectified spirit, three main fractions are collected during rectification of the crude spirit: the light fraction that contains low boiling aldehydes and other oxygen containing organic compounds; the medium fraction that is the spirit itself; and the heavy fraction that is mainly fusal oil. The middle fraction, rectified spirit, always contains traces of fusal oil, which produces an unpleasant taste and odor. By filtering through a bed of activated carbon, the last traces of fusal oil can be removed by adsorption, and a fine quality spirit, which can be used for preparing alcoholic beverages, is obtained.

Activated carbons used in the preparation of different wines must comform to special requirements. The carbon is generally added in the powdered form in a certain ratio and should have specific properties so that it can preferentially remove certain color but does not remove components that give desired characteristics to wine. It should be able to remove undesirable components originating from mildew, cork, yeast, and so on, which may give an unpleasant taste and odor. The activated carbon used should also be able to modify the color of the wine as required. Singleton and Draper[1] carried out detailed investigations into the use of activated carbons in wine making, using frontal chromatography and more than 40 different activated carbons to adsorb particular compounds present in wine. Thus, it is essential to carry out the analysis of wine and to select a proper activated carbon that can produce the best results.

In actual practice, powdered activated carbon is added to the wine or the must in amounts equal to about 0.05 to 1 g. of activated to each liter of wine, but to avoid loss of quality, the smallest effective dose should be used. The dose can be determined by laboratory tests on the wine. A suspension of carbon in a small amount of wine is first prepared and then added to the rest of the wine (must) contained in a vat. The adsorptive purification can be enhanced by stirring the contents of the vat or by bubbling air fed at the bottom of the vat. This process takes several hours, after which the spent-up activated carbon can be separated by sedimentation followed by filtration in a filter press or in a centrifuge.

Although treatment of wines with activated carbons results in a loss of quality due to the insufficient selectivity in the removal of certain wine components, this method of improving the taste and color of the wine constitutes an important step in the wine making process. It is expected that with the production of new grades of activated carbons with tailored properties, the treatment of wines with activated carbons will help in the production of better quality wines.

In the production of brandies, treatment with activated carbons helps in the removal of undesirable flavors arising from the presence of acids, furfural, and tannins that have been picked up during manufacture and storage. Activated carbon also reduces the amount of aldehydes, fusal oil, and other components in the raw distillate and also accelerates maturing. The addition of about 5 g of powdered carbon per liter of brandy is sufficient to give good quality brandy. But larger amounts of added activated carbon amounting to about 30 g per liter can remove traces of fusal oil and produce distinct improvements in taste.

Activated carbon is also used in the brewing of beer at a number of stages. For example, it is used for the purification of water, air, and carbon dioxide used in brewing. It is also used directly in the brewing process[2,3] to improve the quality of a defective beer by modifying the color of the beer and by removing the taste and odor of phenols and coloring matter and also those produced due to the autolysis of yeast and infections.[2,3] The treatment with carbon is carried out generally prior to bottling if the beer quality is not very bad, but in more serious cases about 2 to 2.5 g active carbon is added per liter of beer or, in some cases, even up to 10% of inaculum to the beer in the cask several days before bottling. It may, however, be mentioned that the amount of activated carbon used should be as small as possible because it can remove some useful components. This minimum dose of activated carbon can be determined by laboratory tests.

5.1.3 Decolorization of Oils and Fats

For decolorization of oils and fats, active carbon is not used alone but always in combination with certain bleaching clays. The composition of these mixture of adsorbents depends upon the oil or fat to be decolorized, and this is usually determined before the treatment. Although activated carbons are very effective materials for the treatment of oils and fats, and very small amounts can bring about good results, certain clays are added only due to economic considerations. The active carbon plays an important role in such a mixed adsorbent system because even its small amount contributes to a significant removal of the coloring matter present in the oil or the fat. Because of variations in oils and fats, and the need to determine for each lot the suitable dose of the mixture of adsorbents and processing conditions, the decolorization is usually carried out in a batch process. The deacidified oil is mixed with the optimum amount of the adsorbent mixture in a vacuum mixer to avoid air oxidation. The charge is stirred vigorously for 10 to 20 min at a temperature of 110 to 120°C. The oils are then separated from the adsorbent in a filter press. The oil retained in the filter cake (sludge) is extracted with petrol or any other solvent or an alkali solution. As the adsorption capacity of the adsorbent is not completely exhausted in one batch, it can be used for the preliminary purification of another batch of oil. The spent-up adsorbent is not regenerated but discarded. The decolorization of an oil cannot be carried out at temperatures above the room temperature when the vitamin A content of the oil is to be preserved. In such cases the oil can be thinned with a suitable solvent such as heptane or ethylene dichloride before the decolorization treatment at room temperature.

5.1.4 ACTIVATED CARBON ADSORPTION IN SUGAR INDUSTRY

The manufacturing of white sugar by decolorization of sugar solutions by adsorption on wood charcoal was first reported from a London refinery. Later, in 1812, bone charcoal was used in the preparation of refined sugar. However, with the discovery of decolorization of sugar juices using liming and carbonation, the use of bone charcoal was abandoned because of its cost. However, the search continued for a substitute of bone charcoal. The first industrial production of active carbon was started in 1890 to 1901 in order to replace bone char in the sugar refining industry. This active carbon was prepared by carbonizing mixtures of materials of vegetable origin in the presence of metal chlorides or by the action of CO_2 or steam on the charred materials. Activated carbons with better decolorizing power were prepared by the carbonization of wood and other high carbon content materials with zinc chloride. A carbon activated with steam, Eponite, was prepared and tested in Austria-Hungary in 1910 and 1911. Wyneberg and Sauer[4] obtained a patent for the use of Norit in sugar refining. During the next 15 years, there was a big upsurge in the manufacture of different active carbons using different raw materials and different manufacturing conditions.

Several adsorbents with ion exchange properties such as Cellactivit, prepared by the action of sulfuric acid on a cellulose material, were used.[5] This was nothing but a sulfonated charcoal, which showed both adsorption and ion-exchange properties. However, Cellactivit had poor storage properties and decreased the pH value of the decolorized solution that hampered the crystallization process. Similarly, several synthetic resins that were weakly basic or amphoteric ion exchangers were developed during 1939 to 1958. At present, the sugar industry is using powdered activated carbon, granular charcoal, and decolorizing ion exchangers for the decolorization of sugar syrup. Although in some sugar plants only one form of decolorizing adsorbent is used, in others better results are obtained with a combination of adsorbents at different stages of the production process.

Adsorption decolorization of sugar solutions is the final step in the purification process. The sugar juices are first cleared of the impurities by processes such as clarification (liming), saturation, affination, centrifugation, and filtration. These treatments help the activated carbon to retain its adsorption capacity. The activated carbon treatment is the last stage of the purification process before the sugar juices are boiled to produce white mother liquor from which white sugar can be obtained. In the manufacture of beet sugar, activated carbon is used for decolorizing thin juice, thick juice, and liquors.

Sugar solutions contain different types of coloring matter. They can be classified into (i) caramels (i.e., nitrogen-free coloring substances formed by the partial thermal decomposition of sugars containing phenolic and quinoid groups, (ii) melanoidines (i.e., nitrogen containing coloring substances formed by reactions of reducing sugars with amino compounds, and (iii) iron containing polyphenolic complexes. These coloring substances are present both as dissociated and nondissociated compounds, but frequently the anion types are prevalent. The substances with molecular masses between 8000 and 15000 occurring in colloidal form produce the most intense color in sugar solutions.

The decolorizing capacity of an active carbon depends upon the physical structure that involves pore size and the chemical structure that includes the acidity or alkalinity of the carbon surface as also on the nature of the coloring matter present in the sugar solution. Within the range of pH values met with in the sugar manufacturing process, lowering of the pH value usually improves decolorization. A pH of 4.5 is optimum for decolorization, but it cannot be used because inversion of sugar may take place. It is advisable to reduce the alkalinity of sugar juices by sulfitation before treating with active carbons. Sulfitation improves decolorization by active carbons. The decolorizing activity of an active carbon is measured in terms of the molasses number.

The treatment of sugar solutions with active carbons only slightly increases the purity of the solution, usually 0.1 percent at most, but gives the solution better optical appearance. It also markedly enhances the processing properties. For example, by removing surface active agents and colloidal substances, the surface tension of the solution is enhanced and its viscosity is decreased. These changes result in higher rates of sucrose crystallization and improve the separation of syrup from crystals during centrifugation.

5.1.4.1 Decolorization with Powdered Activated Carbons

Decolorization of sugar solutions using powdered active carbons can be carried out by two methods: contact batch method and the continuous layer filtration method. There is a third method, which is a combination of both the methods.

The contact batch method is a batch process in that a given amount (5 to 10 kg m^{-3} of syrup) of the activated carbon is added to the sugar syrup placed in a container.

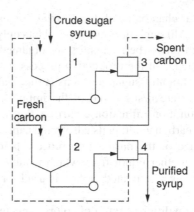

FIGURE 5.1 Schematic diagram of a two-step sugar syrup purification. (From Jankowska, H., Swiatkowski, A., and Choma, J., in *Active Carbon*, Ellis Howard, Ltd. England, 1991. With permission.)

The resulting suspension is kept at a temperature of 80 to 90°C for about 20 min, which is sufficient time to attain adsorption equilibrium. The suspension is then pumped into filter presses to remove the active carbon. During filtration the thickness of the filter cake (carbon bed) increases to 25 to 35 mm, and this results in an increase in the hydrodynamic resistance from 2×10^5 to 4×10^5 Pa.[6] Therefore, the pressure during filtration has to be increased from 2 atm to 4 atm. About 600 kg of carbon is collected in $1 m^3$ of filter volume. When the filter press is full, the active carbon is sweetened off by washing with 6 to 9 times its weight of water. The active carbon is then removed and the filter press assembled again.

In cane sugar processing plants working only with one liquor, the amount of active carbon used can be reduced to about half its amount by employing a plant such as that shown in Figure 5.1.[6] In such a system, the crude sugar liquor is brought into contact with active carbon from the second stage in which the pretreated liquor is further decolorized by fresh active carbon. The exhausted active carbon is discharged from the first stage and the decolorized liquor from the second stage.

Continuous layer filtration involves filtering the sugar syrup through a layer of activated carbon. Several types of filters are used, such as pressure leaf filters with metal frames on which a filter cloth that may be cotton, polyamide, or wire mesh is fixed; rotary leaf filters or bed filters in which the filtering medium is a ceramic or sintered plate, wire mesh, or finely perforated metal plate. The latter filters are usually coated with a layer of filter aid that may be a diatomaceous earth. A suspension of active carbon in water or liquor is passed through the filter until a uniform layer of active carbon bed 10 to 15 mm thick builds up.[7] The filter is then ready for filtration of the liquor that must flow to the filter at a uniform rate to avoid breaking the layer.

It is worth mentioning here that each of the two methods described above requires different activated carbons with different properties. For example, in the contact batch method, the active carbons used should have good filtering properties, because the flow rates here are about 10 times higher than in the continuous layer filtration method, although the decolorizing ability of the carbon may be lower. This method is very flexible, and dosage of the carbon can be easily varied depending upon the color of the syrup or juice. The continuous layer filtration, on the other hand, requires an active carbon with high decolorizing properties. In this case, the filtering properties of the active carbons are not important. Another important difference between the two methods is the consumption of the active carbon that amounts to 0.3 to 1.0% of refined sugar in the contact batch method and about 10 times smaller (i.e., 0.05 to 0.1%) in the continuous layer filteration method. However, the quality of the active carbon in the latter method has to be higher. The choice between the two methods depends upon the size of the plant and the economic considerations.

The combined method utilizes the advantages of the above two methods. The filter is precoated with a layer of fresh active carbon, and a dose of active carbon can be added to the sugar solution. The decolorized solution is refiltered through a bed of diatomaceous earth, which serves to remove active carbon that might have passed through a damaged filter, cloth, or channel in the filter cake. Without refilteration, the final product may have a gray color.

5.1.4.2 Decolorization with Granulated Activated Carbons

Granulated activated carbons are also frequently used for the decolorization of sugar solutions. The decolorization process can be carried out by several methods such as the fixed-bed method, the moving-bed method, and the countercurrent continuous methods.

In the fixed-bed method, the decolorization is carried out in a unit of cylindrical adsorption columns that enable simultaneous decolorization in one section of the plant, sweetening off of the adsorbent carbon in another, and its exchange for fresh or regenerated carbon in the third section of the plant. The liquor heated to about 80°C flows downward through the fixed bed of the activated carbon at a velocity between 1.2 and 2.5 m/hr for a period of 3 to 4 hr with a volumetric flow rate of 7 to 10 m^3 per hr. The exhausted carbon is sweetened off by washing with water. The washed carbon is then transported to the regeneration plant by a hydraulic system where the adsorbed salts are also washed out. The regeneration of the spent up granular carbon can be carried out by pyrolysis in an oxidizing atmosphere at about 1000 to 1100°C or by treatment with super heated steam at 350 to 380°C. The regenerated carbon generally has an adsorption capacity of 90 to 95% of the fresh carbon. The consumption of granular carbon is about 1% of the charge when used alone, and it decreases when it is used in combination with powdered carbon.[7]

In the moving-bed method, the adsorbent carbon is periodically removed in small proportions from the bottom of the filter during the decolorization operation, and a fresh portion of the adsorbent is simultaneously added to the top of the bed. The adsorbent bed is thus gradually displaced in the countercurrent direction with respect to the continuous flow of the sugar liquor.

In the countercurrent continuous method, which is very commonly used in many sugar processing plants, the sugar solution to be decolorized flows upward through the granular carbon bed that continuously moves downward in the adsorption column. The flow rate of the solution is so chosen that it expands the bed of the adsorbent but does not allow it to become fluidized. The main advantage of this expanded bed system is that the solution flows around the entire surface of the granules, but the granules are not free to move throughout the volume of the bed. Under the fluidized conditions, the granules of the adsorbent carbon exhausted to different degrees will continue to mix, thus reducing the adsorption capacity of the bed. In the unexpanded bed system, a portion of the granule surface shall remain unutilized because the particles are in contact with each other.[8,9] This method of sugar decolorization was first introduced in 1957 for decolorizing thick bed juices and has since been taken up by several plants.

The decolorization unit essentially consists of three decolorizing columns, one sweetening off column, a hydraulic transport system, a carbon dewatering vacuum filter, and adsorbent regeneration oven. The adsorbent carbon is continuously removed from the decolorization column, sweetened off with boiling water, and then hydraulically transported to a dewatering rotational filter. The spent-up carbon is fed into a rotary kiln by a vibrating transporter. The temperature in the rotary kiln is controlled at about 750°C. The regenerated carbon is returned by a hydraulic transport system to the decolorizing plant. The process is fully automatic.

The countercurrent continuous method is the most economical, because it utilizes fully the adsorption capacity of the carbon and requires much smaller space and building, and the amount of sweet water is reduced to a minimum. It also reduces sugar losses and the consumption of fuel.

5.1.5 APPLICATION IN CHEMICAL AND PHARMACEUTICAL INDUSTRIES

One of the oldest and the most widespread applications of activated carbon adsorption is the decolorization of organic components. During the processing of organic compounds, several types of polymeric impurities impart dark color and bring about serious technological problems, particularly during crystallization. Many times the solutions are colloidal, due to the presence of colloidal substances, and need clarification. When treated with activated carbons, the organic compounds are not only decolorized but also clarified, because activated carbons also have a tendency to flocculate. This decolorization and clarification is usually carried out in a batch process. A certain amount of powdered active carbon is added to the solution to be decolorized and after a period of time when the equilibrium is attained, the active carbon is separated from the solution by filtration or by centrifugation. The purification process is usually carried out at room temperature, because higher temperature decreases the decolorization efficiency of the carbon. High temperature is used only when the solutions are viscous so as to enhance the rate of diffusion. When possible, the highly viscous solutions can be diluted with solvents to make them less viscous so that decolorization or purification can be carried out at room temperature.

Several factors, such as the pH of the solution, the nature of the solvent used, and the nature of the carbon surface determine the effectiveness of the decolorization process. The ash content of the active carbon that determines the amount of minerals present in a carbon is also an important factor in the use of active carbons for the purification of organic solutions because these minerals can contaminate the product being purified. Cases have been found where the active carbons can initiate catalytic oxidation of certain organic compounds, resulting in their decomposition. In such cases, reducing agents such as sulfites or sulfur dioxide are added to reduce the oxidation processes. The doses of active carbons used are of considerable significance because the larger quantity of the carbon may adsorb some of the useful compounds from the solution. Thus, it is essential that a correct dose of an active carbon for a given solution is determined in laboratory tests.

This accurate dosing of activated carbon in the purifying solutions is particularly important in the pharmaceutical industry. The purification and decolorization of compounds such as glycerol, lactic acid and its salts, betaine, glutomic acid, and tatartic acid and their salts are generally carried out using activated carbons.

Under certain circumstances, the undesirable components from the products can be removed by adding active carbon to the reactants. An example is the production of estheric plasticizers. During their preparation, highly resinous substances are formed by side reactions, and these impart a dark color to the products, which is difficult to remove. By adding active carbon, the coloring matter is adsorbed immediately as it is formed and a pure product is obtained.

An important application of active carbon in the pharmaceutical industry is in the depyrogenation of solutions for hypodermic infections. Such solutions are many times contaminated by bacterial toxins which are not removed by filtration or destroyed by sterlization. These toxins have an acute reaction with organisms and enhance the body temperature. The active carbon, when used, has a depyrogenating effect and is, therefore, directly used for this purpose.[10–12] The active carbon depyrogenates the final solution and purifies the starting materials, such as distilled water. The carbon dose is usually 0.1 to 0.6%, which is removed by filtration after shaking the solution for 10 min. Only high purity carbons are used so as to not influence the reactions of the treated solution.

Recovery of dissolved substances from dilute solutions, such as of natural substances, however, cannot be carried out using active carbon adsorption on industrial scale, because these substances get bonded strongly on the carbon surface, and their elution is a difficult and costly affair and involves considerable losses. Chromatography has been used for separating alkaloids, vitamins, and plant pigments,[13,14] but extraction with a solvent is generally preferred. Similar is the case in the purification of antibiotics. Initially, penicillin was separated by adsorption on active carbon, followed by desorption by amyl acetate, and by an 80% solution of acetone in water and streptomycin by eluting with acidified methanol solution. However, a very accurate dose of the active carbons was of considerable importance because smaller amounts of carbon resulted in losses due to incomplete adsorption, and an overdose caused losses due to incomplete elution. Thus, in newer technologies the active carbons are rarely used in the production of antibiotics.

5.1.6 ACTIVATED CARBON FOR THE RECOVERY OF GOLD

The high selectivity of carbons for gold and silver in the presence of large concentrations of base metals such as copper, iron, nickel, cobalt, and antimony and their availability in abundance, in several different grades having different surface characteristics and at competitive prices, is going to play a significant future role in the recovery of gold and silver. The history of the recovery of gold by using activated carbons can be traced back to 1880, when a process for the recovery of gold from chlorination leach solutions by adsorption on wood charcoal was patented by Davis in the U.S.A. and Australia.[15] Soon after it was recognized that cyanide was a better solvent for gold, and chlorination was gradually replaced by cyanidization for the leaching process, Johnson[16] observed that the wood charcoal could also be used for the recovery of gold from cyanide leach liquors by adsorption. These workers continued their investigations into the possibility of using carbons for the recovery of gold. However, the charcoals available in those days did not possess the pore structure and the surface areas of the activated carbons that are available these days and thus showed very small adsorption capacity for gold. Furthermore, no suitable method was known at that time for the elution of gold from the carbon surface. The only method known was the melting of gold after burning the charcoal. This was not economically viable because of the production of a very large quantity of ash compared to the amount of gold recovered. Thus, carbon was not very seriously used for the recovery of gold until the early 1950's, when Zadra et al.,[17,18] working at the U.S. Bureau of Mines, developed a direct carbon-in-pulp (CIP) process in which the carbon granules were directly added to the cyanide pulp, moving in countercurrent principle. In this method, the gold-loaded carbon was recovered by screening.

CIP technology was the basis for the building up of the first commercial plant in the U.S.A. by the Homestake Mining Company[19] for the large-scale production of gold using activated carbons. With the success of this plant, and with the availability of large quantities of more suitable activated carbons with large surface area and more appropriate pore structure, a considerable interest was aroused in South Africa, and the South Africans built up several gold-recovery facilities based on the CIP process.

The activated carbon process has certain distinct advantages over the zinc concentration process that was being used. The first important advantage is that the adsorption of gold or silver is not significantly affected by the presence of other metallic impurities such as copper, nickel, antimony, and so on, in the leach liquor. Moreover, the carbon can be added directly to the cyanide pulp, thereby avoiding the filtration and clarification stages of the zinc concentration process and making the activated carbon process more economical. (Figure 5.2). Furthermore, several different elution procedures have been investigated to recover gold from the carbon and for regenerating the carbon and recycling it. Because activated carbons can adsorb gold, even when present in small concentrations (0.2 mg/L or even less). It is expected that activated carbons will have applications in the finishing process for the removal of dissolved gold cyanide from gold plant effluents.

5.1.6.1 Mechanism of Gold Recovery by Activated Carbon Adsorption

Although activated carbons are being increasingly used for the recovery of gold, the nature of the interaction between the activated carbon surface and the gold cyanide solution is not clearly understood. Different groups of investigators have advanced different mechanisms to explain their results, the more important being (1) the reduction

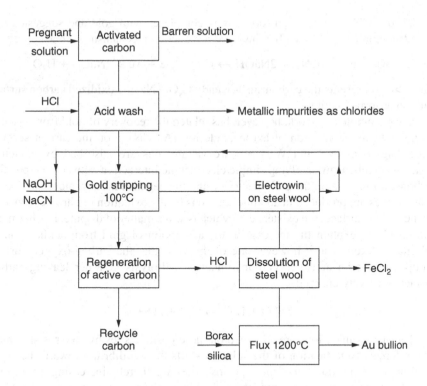

FIGURE 5.2 Removal of gold by activated carbon.

theory, (2) the ion-pair adsorption theory, (3) the aurocyanide anion adsorption theory, and (4) the cluster-compound adsorption theory. A detailed discussion of these theories is published elsewhere,[20] in this book and only a brief description of the processes involved in gold recovery shall be given in this section.

Brusov,[21] while working with the recovery of gold from chlorine bleach solutions, and Green[22] from cyanide solutions using activated carbons, observed that the color of the activated carbon changed to yellow, and gold could be seen on the surface of the carbon. Both these workers attributed this deposition of gold to the reduction of the chlorine or the cyanide on the carbon surface. Although Brusov attributed this reduction to the reduction potential of carbon being ~0.14 volts compared with a reduction potential of ~0.8 V for the chlorine bleach solution, Green was of the view that this reduction was due to the presence of CO and H_2 on the carbon surface. However, the reduction by CO and H_2 gases was rightly criticised by Allen,[23] because these gases are strongly bended on the carbon surface and are evolved only on heat treatment of the carbon at temperatures of 500°C and above. Puri and Bansal[24] and Bansal et al.[25] in their high-temperature evacuations of charcoals and carbon blacks have shown that CO and H_2 are evolved from the carbon surface on high-temperature evacuations at 500 to 950°C as a result of the decomposition of carbon-oxygen surface groups. Grabovskii and coworkers[26,27] observed that larger quantities of gold could be recovered from cyanide solutions in the presence

of CN⁻ or SCN⁻ ions that can form strong complexes with gold and suggested that
the adsorption of gold by carbon involved a reduction mechanism such as

$$>C_x + NaAu(CN)_2 + 2NaOH \rightarrow C_xO - Na + Au + 2NaCN + H_2O$$

where $>C_x$ represents the carbon surface and $>C_xO - Na$, the oxidized carbon surface
with an adsorbed cation.

A large amount of work, however, has related the recovery of gold from cyanide
solutions to the adsorption of aurocyanide ion $[Au(CN)_2]^-$ on the carbon surface.
According to Garten and Weiss,[28] aurocyanide ions are adsorbed by an anion
exchange mechanism involving simple electrostatic interaction between the positive
carbonium ion sites on the carbon surface and the negative aurocyanide ions. These
positive sites are produced in alkaline solutions by the oxidation of chromene groups
on the carbon surface, the existence of which is still a matter of dispute. Furthermore,
this could not explain the increase in the adsorption of gold from acidic cyanide
solutions. Dixon et al.[29] however, are of the view that the carbon-oxygen surface
groups present on activated carbons undergo hydrolysis in water leaving carbon
surface positively charged.

$$>C_xO + H_2O \rightarrow C_x^{2+} + 2OH^-$$

The aurocyanide ions are adsorbed on these positive sites by electrostatic interaction forces. Acidification of the solution shifts the equilibrium toward the right,
resulting in the production of more positive sites and thereby increasing the adsorption of gold cyanide. Clauss and Weiss,[30] while studying the adsorption of gold on
active carbons oxidized by nitric acid and hydrogen peroxide, and in the presence
of reducing agents such as hydrazine and hydroquinone in the cyanide solution,
observed that while the oxidation with nitric acid decreased the adsorption, oxidation
with hydrogen peroxide enhanced the adsorption. The presence of reducing agents
also decreased the adsorption. This indicated that all the oxygen groups present on
the carbon surface were not involved in the adsorption of gold cyanide. The gold
cyanide was adsorbed only on certain quinonic sites.

Puri and coworkers[31,32] and Goyal and coworkers[33,34] have shown that oxidation
of active carbons with nitric acid predominantely produces acidic surface groups such
as carboxylic and lactonic, and the oxidation with hydrogen peroxide produces more
quinonic groups than are evolved as CO on degassing. The decrease in adsorption in
the presence of hydrazine and hydroquinone in the cyanide solution is due to the
competition between the gold cyanide and these neutral molecules, because carbons
are known to adsorb such molecular species.[35] In fact, several organic molecules such
as ethanol, acetone, and methanol are used for the elution of gold from the active
carbon surface because these molecules are preferentially adsorbed. In later work,[36-38]
Dixon, Cho, and coworkers carried out a comprehensive study on the kinetics of
adsorption of both gold and silver cyanides on a coconut-shell charcoal as a function
of sodium calcium and free cyanide concentration, temperature, and charcoal content,
solution pH, and zeta potential. A comparative study of the adsorption of gold and
silver cyanides from their solutions containing similar concentrations of the two
cyanides[37] showed that gold cyanide was adsorbed three times more than silver cyanide.

This could not be explained on the basis of the Garten and Weiss mechanism,[28] because both $Au(CN)_2^-$ and $Ag(CN)_2^-$ are univalent negative ions. Similarly, the adsorption of silver cyanide $Ag(CN)_2^-$ at pH 11 to 11.5, where the zeta potential of the carbon is −50 to −60 mv, and where the carbon surface will have more negative than positive sites, could not be explained by the electrostatic forces.

Based on this work, Cho and Pitt[37] invoked the ion salvation theory of Bokros and coworkers,[39,40] and suggested that weakly hydrated ions such as $Au(CN)_2^-$ and $Ag(CN)_2^-$ were adsorbed specifically on the charcoal surface, although small anions such as CN^- stay in the outer part of the electrical double layer. This received support from their observation that the adsorption of $Ag(CN)_2^-$ anion was much larger than that of CN^- ions and that the hydrated Na^+ and Ca^{2+} cations were adsorbed only when they were present in the solution along with $Ag(CN)_2^-$ anions and not when present with smaller CN^- ions. Their adsorption model (Figure 5.3), which is primarily electrostatic and consistent with the generally accepted views on electrical double layer and the solvation theory, accounts for both the higher adsorption of the larger $Ag(CN)_2^-$ ions or by analogy $An(CN)_2^-$ ions and the beneficial effects of Na^+ and Ca^{2+} cations. The larger adsorption of $Au(CN)_2^-$ anions compared to that of $Ag(CN)_2^-$ ions from their solutions of similar concentrations was attributed to the larger size of the gold cyanide ions.

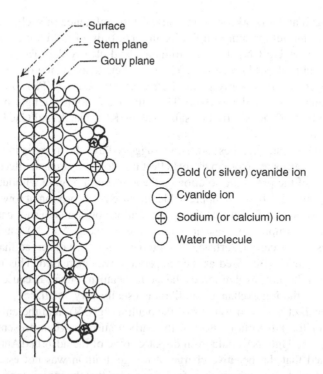

FIGURE 5.3 Schematic representation of the adsorption model for silver cyanide on activated carbons. (Adapted from Radovic, L.R., in *Chemistry and Physics of Carbon*, Vol. 27 L.R. Radovic, Ed., Marcel Dekker, New York, 2001. With permission.)

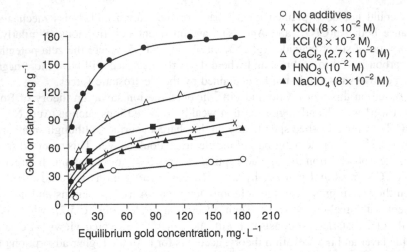

FIGURE 5.4 Adsorption isotherms of gold cyanide in the presence of equal strengths of different ions. (Adapted from McDougall, G.J., Hancock, R.D., Nicol, M.J., Wellington, O.L., and Copperthwaite, R.G., *J. South African Inst. Min. Metall.*, 80, 344, 1980. With permission.)

McDougall and coworkers[41] reinvestigated the influence of such factors as ionic strength, pH, and temperature on the loading not only of gold but also of a neutral molecular species $Hg(CN)_2$ in order to understand clearly the mechanism involved in the adsorption of gold on carbon. These workers also carried out analysis of the adsorbed layer and used x-ray photoelectron spectroscopy (XPS) to determine the oxidation state of the gold adsorbate. The influence of ionic strength on the adsorption was studied by the addition of salts such as KCl, NaCl, and $CaCl_2$ at the same ionic strengths.

Based on these studies, these workers[41] suggested that the increase in the adsorption of gold may be attributed more to the pH effect than to ionic strength effect (Figure 5.4). The influence of the pH of the solution on the adsorption of gold without the addition of a salt (Figure 5.5) shows an appreciable increase in adsorption at lower pH values. Thus, these workers inferred that the adsorption of gold on activated carbons cannot be attributed to simple electrostatic interactions between $Au(CN)_2^-$ anions and the positive sites on the carbon surface.[28,37,42] Furthermore, they observed that the adsorption of gold cyanide decreased as the temperature was increased. This indicated that the adsorption did not involve ion exchange mechanism either, because the effect of temperature on the ion exchange equilibria is usually very small.[29,43]

McDougall et al.[41] also found that the neutral $Hg(CN)_2$ and the anion $Au(CN)_2^-$ competed for the adsorption sites on the carbon surface when present together in the solution, and $Hg(CN)_2$ could even displace some of the adsorbed $Au(CN)_2^-$ anion. This indicated that the negative charge of the gold anion was not essential for the adsorption of gold. The authors, therefore, concluded that the initial stage of adsorption of aurocyanide on carbon involves adsorption of aurocyanide as the less soluble M^{n+} $[Au(CN)_2^-]_n$ complex, where M^{n+} could be an H^+ ion or a metal ion (Na^+, K^+, Ca^{2+}), which is followed by a reduction of the cyanide complex either into a substoichiometric

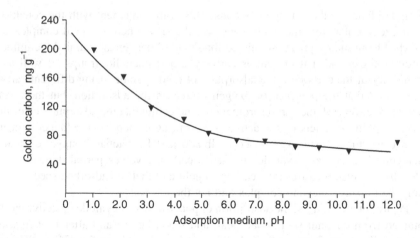

FIGURE 5.5 Adsorption of gold cyanide as a function of pH. (Adapted from McDougall, G.J., Hancock, R.D., Nicol, M.J., Wellington, O.L., and Copperthwaite, R.G., *J. South African Inst. Min. Metall.*, 80, 344, 1980.)

AuCN polymer with metallic gold in the matrix or into a cluster compound of gold, comprising both zero and one oxidation states. On the basis of the XPS spectra, this cluster compound was thought to be structurally similar to the well-known triphenyl phosphine compounds of gold [$Au_{11}(CN)_3$ [$P(C_6H_5)_3$]. In a later publication,[44] based on the comparison of the behavior of activated carbons with polymer adsorbents and ion exchange resins, these workers once again rejected the ion exchange and double-layer mechanisms, and suggested that the adsorption involved the formation of the ion-pair specie [M^{n+}]Au(CN)$_2^-$]$_n$. But they suggested that the condensed aromatic ring system may also contribute to the adsorption of aurocyanide by activated carbons by the donation of electrons to vacant orbitals in the gold atom.

The role of surface oxygen on the adsorption of gold cyanide by active carbons has also been investigated by several workers.[45–48] The uptake of gold cyanide and silver cyanide was less in the presence of oxygen, although the adsorption of neutral $Hg(CN)_2$ was unaffected. These workers invoked the chromene oxidation mechanism of Garten and Weiss,[28] and the Frumkin electrochemical model endorsed by Cho and Pitt,[37] both agreeing with the electrostatic and ion exchange mechanisms. Adams and coworkers,[49–53] based on the negligible effect of oxygen on the adsorption of gold cyanide, concluded that at high ionic strength, aurocyanide is extracted in the form of an ion pair M^{n+}[Au(CN)$_2^-$]$_n$, and at low ionic strengths a portion of the gold is adsorbed by electrostatic interaction with ion exchangeable sites formed by the oxidation of the carbon surface. Klauber,[54] based on his XPS analysis, however, suggested that the linear Au(CN)$_2^-$ anion gets adsorbed on and parallel to the graphitic planes of the carbon with the graphitic π electrons taking part in a donor bond to the central gold atom. The adsorbed specie is stabilized by a charge transfer to the terminal nitrogen atom. Similar studies using XPS[54,56–60] secondary ion mass spectrometry (SIMS),[55] Mossbauer spectroscopy[61,62] and FTIR, spectroscopy[45] have suggested that adsorption of gold cyanide does not take place on the surface oxygen group sites but on the graphic planes of the activated carbons. This is in agreement with the conclusions of Groszek et al.,[63] that the basal planes are the sites where the gold complexes are adsorbed from their aqueous solutions. Ibrado and Fuerstenau[64] studied the influence of total oxygen and that present as carboxylic and phenolic groups, as well as the aromaticity of the carbon on the adsorption of gold cyanide, using different carbons. They found that the presence of oxygen decreased the adsorption, but there was a good correlation with the carbon aromaticity, indicating that the adsorption takes place at the graphitic platelets of the active carbon in agreement with Klauber.[60] In their subsequent studies using spectroscopy, Ibrado and Fuerstenau[65] suggested that the interaction between aurocyanide and active carbon involves partial donation of the delocalized π electrons from the carbon to gold and that the adsorbed specie is most likely the unpaired dicyano complex and not the ion pair.

Papirer et al.[66] studied the adsorption of potassium aurocyanide on active carbons obtained from coconut shell, coal, peat, and wood before and after heat treatment at 800°C, heat treatment at 800°C followed by a washing with HCl (50%), heat treatment at 650°C followed by washing with 10% HCl, esterification with methanol, oxidation with nitric acid, and extraction with toluene. The amount of gold adsorbed was measured using UV spectrophotometry. The adsorption of potassium aurocyanide was determined by the number of basic sites on the carbon surface. In the case of low-temperature acidic carbons, the amount of basic sites increased on heat treatment in nitrogen at 800°C, as well as by oxidation with nitric acid followed by heat treatment at 800°C under nitrogen, thereby enhancing the adsorption of potassium aurocyanide. However, in the case of high-temperature carbons with a base-like character (H carbons), the same treatments decreased the adsorption. The number of basic groups produced was found to depend on the nature of the oxidizing conditions. However, these workers have suggested caution regarding the nature of the positive groups that determine the adsorption of potassium aurocyanide acting as a source of π electrons as suggested by electron spectroscopy.[67] They have also suggested the need to do more to confirm these hypotheses.

Kongolo et al.[68] emphasized the existence of a positive surface charge on the carbon surface caused by the protonation of carboxyl and phenolic groups. Although they did not elaborate their mechanism, it received some support from the work of Lagerge et al.,[69,70] who used several different types of carbons ranging from active carbons to graphite. They postulated the existence of ionized polar groups or structural deficiencies located at the edges of some graphitic ring systems that could provide positive surface charge for the irreversible adsorption of gold cyanide anions by electrostatic interactions,[71] although they did not identify such groups. Jia et al.[72] used more than a dozen active carbons, and one of them was oxidized with nitric acid and subsequently with ammonia at 800°C. The adsorption of aurocyanide decreased more on oxidation with nitric than that on the carbon oxidized and followed by ammonia treatment. However, the adsorption increased with an increasing degree of activation (in steam), and there was a good correlation with the total pore volume rather than with micropore volume, indicating that pore-size distribution is a more important factor for adsorption of gold cyanide. These workers observed that mesoporous with pore dimensions larger than 2 nm are responsible for the adsorption.

Because the cyanide species are environmentally less desired, Arriagada and Garcia[73] prepared thiourea complexes of gold and compared the adsorption capacities of aurocyanide complex (at pH = 10) and thiourea gold complex (at pH = 2 to 3) on activated carbons with well-defined surface pores and chemical surface structures. They observed that the presence of oxygen groups on the active carbon surface reduces the adsorption capacity of anionic gold cyanide complex, while they enhanced the capacity to adsorb cationic gold thiourea complex. Aggarwal et al.[34] and Goyal et al.[33,74] have shown that oxygenated activated carbons are associated with acidic carboxylic and lactonic surface oxygen groups that on ionization in aqueous solutions leave the carbon surface with negatively charged sites. These negatively charged sites will enhance the adsorption of cationic thiourea-gold complex and reduce the adsorption of anionic gold cyanide complex due to electrostatic interactions.

Thus, there is still need to produce activated carbons with adequate mesoporous structure and carbon surface chemistry for the recovery of gold.

5.1.6.2 Desorption of Gold from Active Carbon Surface

Several methods have been developed that can strip gold from the carbon surface almost completely. These methods are based on desorption of gold at high temperatures using dilute solutions of certain electrolytes or their mixtures, or they depend on the selective adsorption of certain organic solvents from the carbon-active sites at lower temperature. The more important of these processes are summarized in Table 5.1.

TABLE 5.1
Desorption of Gold from Carbon Surface

Process	Eluant	Gold on Carbon (ppm)	Temperature (°C)	Gold Desorbed (%)
Homestake (1970)	1% NaOH 0.2% NaCN	5,000–15,000	90	Very low
Davidson (1974)	Pretreated + water 5% K_2CO_3	30,000	90	91% in 19 bed volumes
	5% K_2CO_3 + 3% KOH	30,000	90	96% in 16 bed volumes
	10% K_2CO_3 + 5% KOH	30,000	90	94% in 7 bed volumes; 98% in 12 bed volumes
	5% K_2CO_3 + 10% KOH	30,000	90	94% in 7 bed volumes; 99% in 12 bed volumes; 100% in 22 bed volumes
Anglo-American (1976)	90% acetone ethanol methanol	2,900	70–90	30–70%
NIM (1978)	5% NaCN 1% NaOH Presoak and water	20,000	110	35%
Murdoch (1981)	20–40% organics	5,500	25	Complete

5.1.6.3   Desorption of Gold Using Inorganic Salts

Several inorganic salts of sodium and potassium and their mixtures were tested for the elution of gold from the active carbons surface. Gross and Scott[75] found that only potassium cyanide and sodium sulfide were effective, and about 50% of the adsorbed gold could be desorbed using sodium sulfide. The desorbing action of sodium sulfide was explained differently by different workers. Allen[23] suggested that the sodium sulfide changes the surface energy at the charcoal solution interface. Feldtman[76] ascribed the desorption to the interaction of sodium sulfide with the carboxyl addition compound of gold cyanide formed on the carbon surface. The Homestake process used a mixture of sodium cyanide and sodium hydroxide in dilute solutions, but the elute obtained was very dilute.

Davidson[43] observed that gold can be desorbed from the activated carbon surface by treatment of the gold-loaded carbon with alkali metal carbonates such as those of Na, Li, and K before elution with hot water. The elution of gold-loaded carbon with deionized water at 90°C could recover only 2% of the loaded gold. But when the same gold loaded carbon was first pretreated with a 10% solution of potassium carbonate and then eluded with dionized water at 90°C, the recovery of gold was enhanced to 84%. Davidson optimized conditions for the maximum recovery of gold from the active carbon surface and observed a pretreatment with a solution containing 10% potassium hydroxide and 5% potassium carbonate, followed by elution with hot water at 90°C. He was able to recover about 99% of loaded gold in 12 bed volumes and almost all of it in 22 bed volumes (Table 5.1). The desorption of gold could be further enhanced using lesser bed volumes but at higher temperature (Figure 5.6).

5.1.6.4   Desorption of Gold by Organic Solvents

Heinen et al.[77] and Martin et al.[78] observed that the desorption of gold was much more rapid and efficient at 60°C when carried out using 2% ethanol solutions than when carried out using aqueous cyanide solutions alone.[17,18] Tsuchida et al.[79] used a variety of organic solvents such as ketones, higher alcohols, dimethyl formamide, and methanol and found that aqueous acetonitrile was the best solvent for the recovery of gold. When a solution containing 40% v/v acetonitrile and 0.25 M sodium cyanide was contacted with gold-loaded carbon, more than 80% of the gold was recovered in less than 10 hr. The larger desorption capacity of the acetonitrile was attributed to the fact that it was adsorbed more strongly on the carbon surface from dilute solutions compared to other solvents. The elution of gold-loaded carbons with organic solvents in the absence of cyanide ions could not desorb any gold, suggesting that these solvents changed the activity of cyanide and aurocyanide[79] anions. Tsuchida[79] also determined the desorption of gold with a 40% solution of acetonitrile-water solution in the presence of different electrolytes and observed that both cyanide and lauryl sulfate anions desorbed gold readily and more efficiently and that the thiocyanate ion was more effective than the OH^-, NO_3^- and Cl^- anions (Table 5.2). These workers are of the view that the eluting anion displaces the aurocyanide anion from the carbon active sites, rather than forming a complex with gold cyanide.

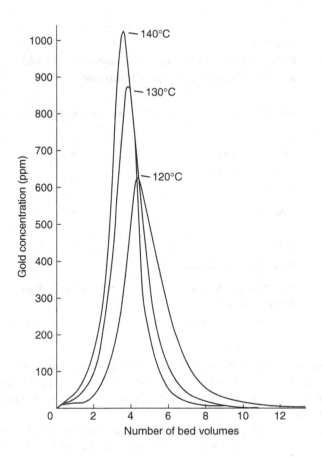

FIGURE 5.6 Effect of temperature of water on the elution of gold. (From Davidson, R.J., *J. South African Inst. Min. Metall.*, 75, 67, 1974. With permission.)

5.1.7 PURIFICATION OF ELECTROLYTIC BATHS

Electroplating is an important process that is used in electroplating, electrorefining, and for the recovery of metals from their ores. Electroplating is a useful surface coating technique that is used to protect metallic surfaces from corrosion and for decoration purposes. Certain plastic materials can also be electroplated to give them a metallic touch. In addition, the articles made from electroplated plastics are light in weight and present a more shining metallic look. An electroplating bath generally contains emulsified oils and fats, lacquers, wetting agents, and brighteners. After the electroplating process, the bath needs to be replaced for a new one. It contains organic impurities, such as emulsified oils and fats, residues of lacquers, decomposition products of wetting agents and brighteners, and many other impurities obtained from the surface that have been electroplated. These impurities accumulate in the bath and impair the quality of the electroplated surface. Suspended or insoluble impurities in the bath can be removed by filteration. However, for the soluble and emulsified substances, activated carbon treatment is most advantageous.

TABLE 5.2
Effect of Electrolytes on the Desorption of Gold
in 40% Acetonitrile-Water Solution[a]

Salt (0.25 M)	Recovery of Gold (%)
NaCN	77.0
n-$C_{11}H_{23}CH_2OSO_3Na$ (sodium lauryl sulfate)	66.8
NaSCN	17.1
NaOH	5
$NaNO_3$	1.7
NaCl	1.2
Blank	1.4

[a]Time: 24 hr; temperature: 28°C.

After Tsuchida, N., Ph.D. dissertation, Murdoch Univ., 1984.
With permission.

There are also cyanide baths that are used for the recovery of silver and gold. These baths contain cyanide anions and other salt anions, and several cations that are used for the efficient and more effective recovery of these precious metals and for electroplating purposes.

The first successful application of activated carbons was made for the purification of nickel plating baths. But it was then extended to the purification of acid copper plating, acid and alkaline tin plating, and to all cyanide baths. However, activated carbon is not suitable for chromium plating baths because activated carbon reduces hexavalent chromium to trivalent chromium.

The purification of electroplating baths can be carried out using a batch process or a continuous process. In the batch process, the bath solution is treated with activated carbon for 25 to 50 min and then filtered. It is ensured that it is clear and does not contain any activated carbon. The clear filtrate to be used again is the bath. The batch process can also be carried out by filtering the bath solution through a fabric coated with a suitable activated carbon. The frequency of the purification depends upon the operation of the electroplating bath, and the amount of carbon dose depends upon the degree of contamination of the bath, but usually it is 0.5 to 1.3 kg of activated carbon for each 100 liter of the bath solution.

In the continuous process of purification, the bath is continuously passed through a bed of activated carbon, which is changed from time to time. In this process the bath is almost free of undesirable impurities. However, it is not applicable to baths containing easily absorbable desirable components of the bath. This is especially true for silver plating solutions containing carbon disulfide.[7]

It may, however, be worth mentioning that activated carbons being universal adsorbents can partially remove useful components of the bath, such as brighteners and wetting agents, so that these ingredients need to be added to the bath. Furthermore, active carbons are good catalysts so that they can decompose or modify certain components of the bath such as thiourea and its derivatives, amino acids, and sodium bisulfite.

5.1.8 REFINING OF LIQUID FUELS

Gasoline contains several types of sulfur compounds, including mercaptans. These compounds can be removed from gasoline using activated carbons impregnated with sodium hydroxide. Usually active carbons containing 10 to 15% sodium hydroxide are used. The removal of mercaptans involves adsorption on the carbon surface and a chemical reaction with sodium hydroxide.

$$RSH + NaOH \rightarrow NaSH + N_2O$$

The gasoline is passed through a bed of carbon and sodium hydroxide. The carbon bed can be regenerated by passing steam through the bed and by recharging with sodium hydroxide.

5.2 GAS-PHASE APPLICATIONS

Activated carbons are good adsorbents for gases and vapors. Consequently, a significant fraction amounting to as much as 20% of the total production of activated carbons in industrialized countries is being used in gas-phase adsorption processes. The major areas of concern are the removal of hazardous substances from industrial exhaust gases, separation of gas mixtures, recovery of useful and valuable components from industrial exhaust gases, purification of process gases from undesirable impurities, and recovery of solvents. The activated carbon used for gas-phase applications is usually in the granulated form that can be used as such or after impregnation with one or more organic or inorganic compounds. When active carbon is used as such, the process involved is generally only physical adsorption, but with impregnated carbons both adsorption and a chemical reaction take place.

5.2.1 RECOVERY OF ORGANIC SOLVENTS

The recovery of organic solvents and the removal of certain volatile organic compounds (VOC) from industrial waste gases is one of the important and largest gas-phase application of active carbons and has significance both from the point of view of the recovery of valuable materials as well as from the point of view of the protection of the environment. VOC obtained in urban areas as a result of emissions from combustion sources such as internal combustion engines and power plants can combine with NO in the presence of sunlight to produce ozone. This ozone in the upper atmosphere, although is useful to prevent UV radiations from reaching the Earth's surface, it can cause acute respiratory problems for people, including decreased lung capacity and impairment of the immune defense system. Thus, the Clean Air Act of 1990 of the U.S. was amended, requiring the reduction of 149 VOCs that are detrimental to air quality.[90]

The recovery of useful components and the removal of harmful components from VOC present in industrial waste gases involves two different objectives. When the effluent stream contains a high concentration (1 to 2%) of the organic compound, then recovery and recycling is normally the principal objective. However, when the concentration of the pollutant is very low, of the order of 10 to 1000 ppm or even lower, then their capture and disposal is the preferred objective.

The principle of solvent recovery of activated carbon adsorption has been known for almost a century; a process was patented in 1905,[80] but in practice the method was used in 1916 when vapors of volatile substances were recovered by using on activated carbon prepared by using chemical activation with zinc chloride. The solvent recovery became very important during World War I, due to the shortage of solvents because the war industries needed them in large quantities.

Some of the more important industries that produce solvent-contained air streams are printing; dry cleaning; and the manufacture of paints, polymers, adhesives, celluloid, rubber (e.g., rubber-coated fabrics), rayon, and gunpowder; and extraction processes. The main solvents recovered by activated carbon adsorption are benzene, toluene, xylene, alcohols, acetone, petrol, ether, carbon disulfide, halocarbons (e.g., chloroform, carbon tetrachloride, trichloroethylene, methylene chloride, chloroben-zene, etc.). The major production facilities and the solvents recovered are listed in Table 5.3. In many cases the concentration of the organic solvent in waste gases is of the order of 1 to 2%.

When allowed to mix with the atmosphere, the highly volatile nature of some of the solvents can create several unacceptable problems, such as fire and explosion hazards, and pollution of the environment resulting in several health ailments. It is important, therefore, that the solvents be removed from the industrial exhaust streams

TABLE 5.3
Solvent Recovery Application of Active Carbon

Field of Application	Main Solvents
Production of smokeless gun-powder	alcohol, ether, acetone
Printing	xylene, toluene, benzene
Dry cleaning	petrol, benzene, trichloroethylene, trichloromethane
Rayon production	ether, alcohol, acetone, carbon disulphide
Rubber industry, rubber-coated fabrics, rubber-asbestos goods	petrol, benzene, toluene
Paint shops	thinners for nitrocellulose lacquers
Celluloid production	alcohol
Plastics and artificial leather production	alcohol, acetone, esters, ether
Extraction processes	petrol, benzene, trichloroethylene
Film and foil manufacture	ether, alcohol, acetone, methylene chloride
Adhesive plaster production	petrol

After Smisek, M. and Cerny, S., in *Active Carbon*, Elsevier Publishing, Amsterdam, 1970. With permission.

before putting them into the atmosphere. The solvent recovery by activated carbon adsorption is highly economical because the recovered solvents can be used again.

Several different technologies, such as thermal and catalytic incineration, biofiltration, condensation, and adsorption on activated carbons are available for the removal of solvents and other volatile organic compounds (VOC) from industrial exhaust, as well as those emitted from combustion sources such as internal combustion engineers and power plants. But the adsorption on activated carbons is more cost effective at low concentrations than the incineration technique and offers high efficiency for the removal of solvents and volatile organic compounds both at low and high concentrations. Biofiltration can be used with cost advantage when the concentration of the VOCs is low because at higher concentrations the efficiency of the process is low. Incinerations results in the decomposition of the VOCs producing relatively less harmful products such as CO_2 and H_2O, but the process is viable only when the VOC concentration is high and the removal recovery is not the requirement.

The solvent recovery by activated carbon adsorption is carried out by passing the solvent-laden air through a bed of granulated active carbon. The solvent is adsorbed on the carbon, and the clean air leaves the adsorber. The carbon bed becomes gradually saturated with the adsorbate, and when the concentration of the solvent in the purified gas stream exceeds the limit of acceptability, which normally is about 10% of the inlet concentration, the inlet gas stream is switched on to a second adsorber bed while the first carbon bed is regenerated. The adsorbed solvent is obtained from the saturated bed by passing super-heated steam, hot air, or nitrogen in a direction countercurrent to the air steam, although steam is the most commonly used stripping agent. The released solvent and the steam is condensed and separated. A typical steam demand is 0.3 kg steam per kg of activated carbon.[81] When the solvent is soluble in water, distillation is resorted to for separation. Multiple carbon beds can be used to make the process semicontinuous so that some of the beds are in use for adsorption while the others are being desorbed (Figure 5.7). Commercial activated carbon adsorbers can treat air volumes of 50 to 1500 m^3/ min.

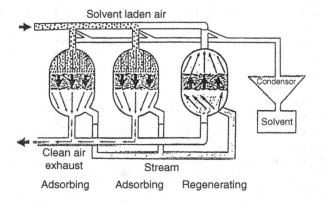

FIGURE 5.7 Solvent recovery process using activated carbon beds. (From Derbyshire, F., Jagtoyen, M., Andrews, R., Rao, A., Martin-Gullon, I., and Grulke, E.A., in *Chemistry and Physics of Carbon*, L.R. Radovic, Ed., Marcel Dekker, New York, 2001, Vol. 27. With permission.)

The recovery efficiency of the adsorption process varies between 92% and 98%. The process can be carried out at 100% efficiency, but then the cost becomes very heavy. The high efficiency of the activated carbon recovery process is also important from a safety point of view because by maintaining suitable ventilation, the composition of the air-vapor mixture can be kept well below the lower explosive limit. The explosion limits of some technically important substances[80] are given in Table 5.4. These data can help to maintain safe conditions when working with these substances. Although great efforts have been made to recover carbon disulfide in viscose production units, it could not be successfully done because active carbon oxidizes carbon disulfide into sulfur that gets deposited on the carbon surface as surface compounds, thereby decreasing active carbon efficiency. Similarly, the adsorption of trichloroethylene on the active carbons surface may form polymerization products.

TABLE 5.4
Detonation Limits of Technically Important Substances in Mixtures with Air

Substance	Lower Limit		Upper Limit	
	(vol. %)	(g/m³)	(vol. %)	(g/m³)
Acetaldehyde	4.1	75	55.0	1007
Benzene	1.4	45	6.7	217
1-Butane	1.8	48	8.4	103
n-Butyl alcohol	1.5	46	11.3	348
Carbon disulphide	1.3	41	44.5	1408
Carbon monoxide	12.5	145	74.2	863
Dichloroethylene	5.6	226	13.0	524
Dioxane	2.0	73	22.5	824
Ethyl acetate	2.5	91	15.5	567
Ethyl alcohol	4.3	82	18.9	362
Ethyl ether	1.9	58	48.5	1494
Ethyl chloride	3.8	102	15.4	413
Hexane	1.2	43	7.4	265
Hydrogen sulphide	4.3	61	45.5	615
Methane	5.2	15	14.0	98
Methyl acetate	1.1	95	15.6	480
Methyl alcohol	6.7	89	36.5	486
Methylethyl ether	2.0	50	10.1	252
Methylethyl ketone	1.8	54	9.5	285
n-Pentane	1.4	42	7.8	234
Propane	2.1	38	9.1	167
n-Propyl alcohol	2.1	52	13.5	337
Pyridine	1.8	49	12.4	407
Toluene	1.5	57		
o-Xylene	1.1	48	6.0	265

After Smisek, M. and Cerny, S., in *Active Carbon*, Elsevier Publishing, Amsterdam, 1970. With permission.

5.2.2 Removal of Sulfur Containing Toxic Components from Exhaust Gases and Recovery of Sulfur

Exhaust gases from several industrial and production units, where coal or sulfide ores are used, contain sulfur-containing toxic gases and vapors such as sulfur dioxide, hydrogen sulfide, carbon disulfide, and several organosulfur compounds. These industrial gases need to be desulfurized effectively before mixing with the environment. The removal of these sulfur-containing gases has two advantages: It protects the environment from pollution, and it provides commercial benefits in terms of the recovery of the valuable component, sulfur. When activated carbon is used to desulfurize the industrial gases, three different processes generally can take place:

- Physical adsorption of the sulfur-containing gases
- Chemisorption of the gases resulting in the formation of carbon-sulfur surface compounds
- The catalytic oxidation of the gases on the carbon surface resulting in the deposition of elemental sulfur.

The physically-adsorbed gases and vapors can be recovered from the carbon surface by evacuation or by passing an inert gas over the surface of the carbon; the elemental sulfur can be removed by extraction with various solvents. The chemisorbed sulfur that is present in the form of very stable carbon-sulfur surface compounds is difficult to recover. It cannot be removed by extraction or by heat treatment alone. It can, however, be conveniently removed by heat treatment in hydrogen gas at 500°C. The formation and properties of carbon-sulfur surface compounds formed on carbons as a result of treatment with sulfur dioxide, hydrogen sulfide, or carbon disulfide is beyond the scope of this book, but they have been dealt with in great detail in our earlier books, *Active Carbon*[82] and *Carbon Black*.[83]

5.2.2.1 Removal of Sulfur Dioxide from Waste Gases

The main sources of sulfur dioxide emission to the atmosphere are coal-fired production units and industrial facilities using sulfide ores such as power plants, ferrous and non-ferrous metallurgical industries, and several chemical and petrochemical industries.

Although the coal used in many of these industries is desulfurized, the process is not very effective. The sulfur contained in fuel and metal ores is usually given out as sulfur dioxide. The amount of sulfur dioxide emitted annually into the environment in highly developed countries amounts to as much as 30 to 40 million tons. Therefore, the recovery of sulfur from such huge amounts of sulfur dioxide does not only protect the environment but also is of considerable commercial importance. Although several different methods have been suggested for the recovery of SO_2 from exhaust gases, adsorption by activated carbon is considered to be more viable. When the flue gases do not contain any oxygen or water vapor, which is seldom the case, the removal of SO_2 takes place through physical adsorption on the activated carbon surface. The adsorbed sulfur can be removed from the carbon surface by either applying vacuum or desorbing it by passing another gas over it at the same temperature. Alternatively, the adsorbed SO_2 is oxidized to SO_3 in the presence of oxygen. Both SO_2 and SO_3 can be removed from the carbon surface, and the carbon is regenerated. Several factors influence the adsorption capacity of the activated carbon for oxides of sulfur, SO_x. Davini[84] observed that basic activated carbons preferred adsorption of sulfur oxides to the acidic carbons. Thus, heat-treated carbons that have lost their acidic surface functionality and predominantly contained basic surface groups were better adsorbents for SO_x oxides of sulfur.[85,86] Moreno-Costilla et al.[87] agreed with these views and further showed that activated carbons with narrow microporosity were better adsorbents for SO_x gases. However, the exhaust gases are usually associated with oxygen and water vapor in addition to other components. In such a case the physical adsorption of SO_2 is accompanied by its oxidation to SO_3 that reacts with water to produce sulfuric acid. The adsorbed sulfuric acid is recovered by washing the activated carbon with water. The adsorbent capacity is increased by a factor of 2 or 3 in the presence of oxygen and by a factor of 20 to 30 if water is also present (Table 5.5).

Another part of the sulfur formed by the reaction between the carbon surface and SO_2 gets bonded to the carbon surface in the form of stable carbon-sulfur surface compounds that can be neither desorbed by extraction with a solvent nor by heat treatment. These surface compounds can only be desorbed as H_2S on heat treatment in hydrogen gas, as mentioned earlier. The formation of these carbon-sulfur surface compounds reduces the availability of the carbon surface for any further adsorption, making the regeneration of the sulfurized activated carbon difficult. The proportion of these three forms in which SO_2 is adsorbed on the carbon surface depends upon the temperature and composition of the exhaust gases and the nature of the carbon surface. It, therefore, indicates that the experimental conditions for using activated carbon adsorption technique for the recovery of SO_2 should be such that there is

TABLE 5.5
Uptake of Sulfur Dioxide by Activated Carbon in the Presence of Oxygen and Water Vapor

Sample	Adsorbed Amount (mmol/g)		
	Only SO$_2$	5% O$_2$	5% O$_2$ 10% H$_2$O
A	0.10	0.34	2.3
FE-100–600			
B	0.18	0.56	3.8
FE-200–800			
C	0.13	0.26	2.7
FE-300–800			

From Derbyshire, F., Jagtoyen, M., Andrews, R., Rao, A., Martin-Gullon, I., and Grulke, E.A., in *Chemistry and Physics of Carbon*, L.R. Radovic, Ed., Marcel Dekker, New York, 2001, Vol. 27. Reproduced with permission from Marcel Dekker.

minimum formation of carbon-sulfur surface compounds. Generally, activated carbons with a moderately developed microporosity are preferred.

The adsorbed sulfuric acid tends to desorb slowly from the activated carbons surface so that the carbon surface becomes inactive when the acid occupies all the sites where oxidation of SO_2 can take place. Thus, the activated carbon needs to be regenerated, or the H_2SO_4 has to be removed at a rate fast enough to allow steady state removal of SO_2. The regeneration of the carbon can be carried out either by (a) long washing of the carbon with water continuously or cyclically when several carbon beds are employed for loading and regenerating in turn[88,89] or (b) by heat treatment of the loaded carbon in an inert gas when H_2SO_4 decomposes into SO_2 and H_2O and a concentrated stream of gas-phase SO_2 is obtained. In the latter process, however, the activity of the activated carbon surface to adsorb SO_2 may change with the number of regeneration cycles. The reduction or the enhancement in the carbon surface activity will depend upon the original pore structure of the carbon. In general, the activity of low surface area microporous carbons remains constant or increases with each regeneration cycle, due to its activation during thermal regeneration process but for carbons with well-developed porosity, the activity can decrease with each regeneration cycle.[90] The temperature at which regeneration is carried out also influences the activity of the activated carbon surface. Low temperature regeneration ($< 380°C$) does not regenerate the original sites and, therefore, decreases the activity of the carbon surface. Thus, the regeneration should be carried out at temperature above 400°C so that the oxygen groups are decomposed to produce the active sites.

In recent years, fibrous activated carbons (activated carbon fibers) have been prepared from several precursors such as polyacrylonitrile (PAN), coal tar pitch, and petroleum pitch, that have exhibited high activity for the conversion of SO_2 into H_2SO_4.[91] Furthermore, the catalytic activity of these carbon fibers has been found to increase with the heat treatment temperature, the extent of increase depending upon the type of the fiber and the heat treatment temperature and atmosphere.[92]

The breakthrough plots (Figure 5.8) for a PAN-based ACF sample using a simulated flue stream containing 1000 ppm of SO_2, 5% O_2, and 10% H_2O at different

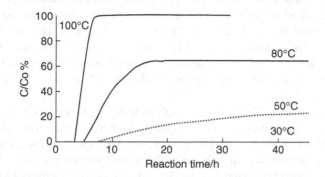

FIGURE 5.8 Breakthrough plots for PAN-based ACF at various temperatures (1000 ppm SO_2, 5% O_2 and 10% H_2O). (From Derbyshire, F., Jagtoyen, M., Andrews, R., Rao, A., Martin-Gullon, I., and Grulke, E.A., in *Chemistry and Physics of Carbon*, L.R. Radovic, ed., Marcel Dekker, New York, 2001, Vol. 27. With permission.)

temperatures shows that at 30°C almost 100% removal of SO_2 is achieved. The removal activity of the ACF surface decreases as the temperature is raised until at 100°C there is no conversion into H_2SO_4. This indicates that the presence of a thin layer of water on the carbon surface is essential for the formation of H_2SO_4. Furthermore, the increase in temperature decreases the solubility of SO_2 in water and results in lower uptake of SO_2 from the gas phase and a lower concentration of dissolved SO_2 in water on the surface of the carbon fiber.[93]

Several plants using activated carbons have been set up to recover SO_2 from flue gases.[94] These plants have many things in common, although they differ in details. For example, some plants use stationary while others use fluidized bed adsorbers. Similarly, there are differences in details of the regeneration of the activated carbon absorbers. The most commonly used plants, the techniques, and the conditions for the recovery of SO_2 from flue gases using activated carbons are summarized in Table 5.6.

In the Sumitomo process,[95] activated carbon is used at a temperature of 100 to 130°C for the adsorption of SO_2 in a moving bed, and the regeneration of the loaded carbon is carried out by using an inert gas at 350°C. In the BF method, special activated carbon prepared from hard coal is used for the adsorption of SO_2 at 100 to 160°C in a moving bed, and the regeneration of the loaded carbon is carried out at 650°C in a hot sand circuit, which produces a highly SO_2 enriched gas that is converted into elemental sulfur. The removal of SO_2 can also be carried out using a fluidized bed, where the loaded carbon is regenerated in several steps with hydrogen gas at 350°C. The flow sheet of BF process[96] for the recovery of SO_2 from flue gases is shown in Figure 5.9. The adsorption of SO_2 takes place in the annular moving bed reactor in which the activated carbon moves slowly from the top downward while the flue gases flow through the carbon from inside to outside. After the removal of dust, the sulfuric acid loaded carbon is led to a desorption unit using a hot sand circuit. During regeneration using an inert gas, gases containing about 30% of SO_2 are obtained. This SO_2 is reduced to elementary sulfur.

Joly and coworkers[97,98] studied the adsorptive removal of SO_2 from flue gases when present at low concentrations (< 30 ppm) using activated carbons and observed that SO_2 is adsorbed both reversibly (physisorbed) and irreversibly (chemisorbed). The presence of adsorbed water on the carbon surface helps the adsorption of SO_2. The chemisorbed SO_2 undergoes oxidation to SO_3 which, in the presence of water, forms H_2SO_4. This mechanism is similar to the mechanism of recovery of SO_2 from flue gases, when present in larger amount. The oxidative removal of SO_2 over pitch-based ACFs[99] also involves adsorption, oxidation, dehydration, and desorption of SO_2, SO_3, and H_2SO_4.

Liu and coworkers[100,101] prepared catalysts using waste semicokes such as anthracite semicoke, lignite semicoke, and bituminous semicoke for the removal of SO_2 from fuel gases. These semicoke catalysts were prepared by activating the granulated cokes with water in an autoclave and then modifying their surface properties by oxidation with nitric acid and hydrogen peroxide. The surface properties of the oxidized materials were further modified by heat treatment at 700°C and by loading of the surface with copper. These workers observed that the lignite semicoke after oxidation with 45% nitric acid and loaded with copper followed by heat treatment at 700°C produced the best catalyst for the removal of SO_2. The effects of water content in the flue gas,

TABLE 5.6
Processes for Adsorption of Sulfur Dioxide from Flue Gas

Process	Adsorptive	Adsorption Adsorbent	Temp (°C)	Technique	Regeneration	Stage of Development
Sulfreen	tail gas of a Claus plant SO_2 from	activated carbon + alkaline silicate	20–50	Fixed bed	Inert gas at 450°C	Industrial operation
Sulfacid	SO_2 from chemical industries	activated carbon + metal catalyst	80	Fixed bed	Water shrinking	Industrial operation
Hitachi	SO_2 from flue gases	activated carbon	greater than 100	Fixed bed	Water scrubbing	Prototype plant 400,000 m³/h S.T.P
Sumitomo	SO_2 from flue gases	activated carbon	100–130	Moving bed	Hot-Gas 350°C	Prototype plant 170,000 m³/h S.T.P
BF	SO_2 from flue gases	"Aktivkoks" (activated carbon from hardcoal)	100–160	Moving bed	Hot-Sand 650°C	Prototype plant 30,000 m³/h S.T.P
Westvaco	SO_2 from flue gases	activated carbon	greater than 100	Fluidized bed	H_2 350°C	Prototype plant 30,000 m³/h S.T.P

After Juntgen, H., *Carbon*, 15, 273, 1977. Reproduced with permission from Elsevier.

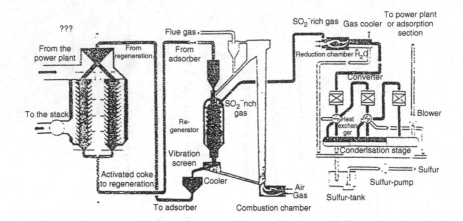

FIGURE 5.9 Flow sheet of BF process for desulfurization of flue gas. (After Juntgen, H., *Carbon*, 15, 273, 1977.)

reaction temperature, and space velocity on the desulfurizing property were investigated to determine the optimum operating conditions. It was observed that a temperature of 90°C, water content of 7% and a space velocity of 830 per hr showed good performance of the catalyst for SO_2 removal. Under these optimum conditions, the carbon catalyst retained about 9.6 g of SO_2 per 100 g of carbon.

The results on the oxidized samples showed that the removal of SO_2 was enhanced by the presence of basic groups, while the acidic groups showed no such effect, indicating that adsorption of SO_2 takes place on the basic sites on the carbon surface. In case of the carbon samples loaded with copper, the retention of SO_2 was considerably enhanced that has been attributed to the oxidation of SO_2 to form $CuSO_4$. Copper-loaded carbons showed no relationship between the amount of sulfur retained and surface basicity or surface acidity of the carbon. However, in the case of carbons without any copper loading, the retention of sulfur was related linearly to the surface area (BET) and the pore volume. These workers concluded that the porous structure of the carbon determines the removal efficiency of a carbon when it has a surface area less than 700 m^2/g; the surface chemical structure determines the removal of SO_2 when the surface area of the carbon is large.

5.2.2.2 Removal of Hydrogen Sulfide and Carbon Disulfide

Hydrogen sulfide and carbon disulfide are two other obnoxious sulfur compounds that are present in the exhaust gases emitted to the environment by the rayon industry. The amount of these two gases together could be more than 10,000 mg/m^3. The removal of H_2S has been carried out by adsorption in different solutions, but the removal has not been complete. Therefore, attempts have been on to develop new technologies for this removal. In contrast, CS_2 can be effectively removed by adsorption on activated carbon. During the last three decades, technologies have been developed through which both CS_2 and H_2S can be removed simultaneously by using activated carbons.

In one of these methods, the adsorptive removal of both H_2S and CS_2 is carried out by passing the flue gases through two different static beds of two different activated carbons. One of the bed contains an active carbon with larger pores in the mesopore range of 3.0 to 8.0 nm and usually impregnated with iodine. H_2S is adsorbed and oxidized on the surface of the carbon producing elemental sulfur that is deposited in the mesopores. The second bed contains a microporous active carbon where adsorption of CS_2 can take place. When the carbon beds are saturated, the regeneration of the active carbon is carried out. The CS_2 is desorbed by passing super-heated steam through the bed. The vapors are condensed and separated as described in the solvent recovery section. The elemental sulfur obtained as a result of oxidation of H_2S is recovered from the carbon surface by extraction with liquid CS_2. Some partial oxidation of H_2S may result in the formation of H_2SO_4 on the carbon surface, which can be converted into ammonium sulfate by reacting with gaseous ammonia passed through the carbon bed. The ammonium salts can be easily washed from the carbon surface with water in the regeneration cycle. In another method, a single static bed of mesorporous activated carbon impregnated with a small amount of iron is used. Both oxidation of H_2S and adsorption of CS_2 take place in the bed. The oxidation of H_2S occurs near the entry of the waste gases through the bed, and the physical adsorption of CS_2 takes place afterward. The elemental sulfur and the adsorbed CS_2 are recovered from the carbon surface in a multistep regeneration cycle.[94,102,103]

The catalytic oxidation of H_2S by oxygen on the surface of carbon has been studied by a number of workers[104–106] but the exact reaction mechanism is not well understood. The major reaction, however, is the formation of water and elemental sulfur.

$$H_2S + \frac{1}{2}O_2 \rightarrow H_2O + \frac{1}{2}S_2$$

However, depending upon the reaction conditions and the nature of the activated carbon surface, several side reactions can occur, which give rise to certain unwanted products such as SO_2 and H_2SO_4. Therefore, the reaction conditions or the active carbons are chosen to eliminate, or at least reduce, the formation of these unwanted products to a negligible amount. Many times certain types of reaction promoters such as ammonia have been used. The reaction is highly exothermic and, therefore, tends to increase the temperature of the carbon bed. Thus, it is recommended that the catalytic oxidation by active carbon be carried out when the concentration of the H_2S in the waste gas does not exceed 3 to 5 gm^3.

It is generally agreed that the catalytic oxidation of H_2S in the presence of oxygen by an active carbon takes place on the carbon active sites and that the rate of the reaction depends on the concentration of these sites as well as the concentration of H_2S in the gaseous mixture. There are, however, different views regarding the mechanism of the process. According to Puri and coworkers[104] and Cariaso and Walker,[105] the reaction takes place through the chemisorption of oxygen on the active sites, which then reacts with H_2S in the reaction mixture. Steijns et al.[107] postulate that the oxidation involves reaction between the chemisorbed oxygen and the dissociatively adsorbed H_2S, producing sulfur and water. The molecular oxygen in the gas phase then reacts with adsorbed sulfur, producing SO_2, which then oxidizes H_2S. However, this view was contested by Puri et al.,[104] who found only small amounts of SO_2 that could not contribute to the oxidation of H_2S.

Hedden et al.[108] proposed that the oxidation of H_2S takes place through the formation of a water film on the carbon surface in which both H_2S and O_2 dissolve. The oxygen gets chemisorbed on the active carbon surface and decomposes into active species, and reacts with the hydrosulfide species formed by the dissociation of H_2S to produce hydroxyl ion and sulfur, which is deposited on the carbon surface.

Kim et al.[109] studied the adsorption of H_2S on activated carbons prepared by activation and carbonization of rice hulls with KOH and impregnated with Na_2CO_3 and KIO_3. The impregnated activated carbons showed excellent adsorption capacity for H_2S. The amounts of H_2S adsorbed as obtained from breakthrough curves for Na_2CO_3 and KIO_3 impregnated carbons increased with increase in the amount of the impregnant (Table 5.7). The adsorption capacity for the KIO_3^- impregnated carbon was higher than for the same amount of impregnant in the case of $Na_2CO_3^-$ impregnated activated carbon. The optimum contents for $Na_2CO_3^-$ and KIO_3 impregnant were 5 and 2.4% weight, respectively. The improved adsorption capacity for the impregnated carbons has been attributed to the oxidation of H_2S and the chemical reaction with the impregnant.

Svetlana et al.[110] studied the adsorption of sulfur dioxide hydrogen sulfide and carbon disulfide from single components as well as from complex gas-air mixtures using static and dynamic conditions on several commercial activated carbons, an activated carbon fabric with a developed porous structure, and thermally treated semicokes that had a small pore volume but considerable concentration of basic centers. These carbon sample were also oxidized and impregnated with copper, iron, cobalt, and nickel ions by ion exchange techniques. To increase the concentration of basic groups, some of the carbons were treated with ammonia. The adsorption of SO_2 increased with the porosity of the carbon material but was strongly influenced by the chemical nature of the carbon surface. The adsorption of SO_2 also increased with increase in the concentration of basic sites (Figure 5.10) as well as by the impregnation of metals on the carbon surface. The difference in the adsorption of SO_2 by different impregnants was attributed to the difference in the strength of the metal-SO_2 coordination bonds in the surface complex.[111] Similar results were obtained for the adsorption of H_2S from gas air mixtures. The adsorption of H_2S decreased when the carbon fabric was oxidized and increased on impregnation of the oxidized carbon fabric. The adsorption is more when the impregnant is iron than when Cu is the impregnant (Figure 5.11). The adsorption of carbon disulfide vapors decreased with oxidation of the carbon surface but was increased substantially on impregnation of the oxidized samples with cations.

In addition to adsorption, SO_2, H_2S, and CS_2 also undergo catalytic conversions on the impregnated metal sites. The desorption experiments carried out by these workers showed[110] the role of chemical interactions and surface complexing in the adsorption of these sulfur compound on the cation impregnated carbons. It was found that only a small fraction (10 to 15%) of the adsorbed compound was desorbed from the cation modified carbon surface. The major fraction remained strongly bonded to the carbon surface or was catalytically converted, as evidenced by the presence

TABLE 5.7

Influence of the Amount and the Nature of the Impregnant on the Surface Properties of Activated Carbon

Sample Application	As-received Activated Carbon	Na$_2$CO$_3$ Impregnated AC				KIO$_3$ Impregnated AC			
		2.4 wt%	5.0 wt%	7.3 wt%	10.0 wt%	2.4 wt%	5.0 wt%	7.3 wt%	10.0 wt%
Surface area (m^2/g)	2,684	2,413	2,334	2,258	2,122	2,513	2,451	2,346	2,240
H$_2$S adsorption capacity (mg/g)	36	46	75	56	65	97	82	90	89
Pore diameter (Å)	16.85	16.84	16.87	17.10	16.86	16.94	16.88	16.82	17.09
Pore volume ~20 (Å) 20–2000 (Å)	1.131	1.016	0.973	0.965	0.894	1.065	1.035	1.010	0.957
(cc/g)	0.198	0.202	0.201	0.198	0.199	0.200	0.198	0.198	0.191

After Kim, J.S., Jung, S.W., Kim, M.C., and Kim, M.S., Carbon 02 Intern. Conf. on Carbon, Beijung, 2002, Paper P795D 111. With permission.

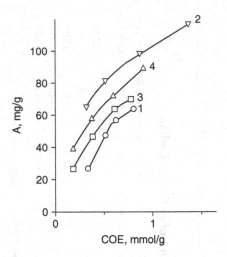

FIGURE 5.10 Amount of sulfur dioxide adsorbed from its mixtures with air on carbons associated with different surface groups. (1) activated carbon AR-3, (2) activated carbon treated with ammonia, (3) activated carbon fiber ACF, (4) thermally treated semicoke TSC. (After Svetlana, S., Kartel, N.T., and Tsyba, N.N., Carbon '02, Intern. Conf. on Carbon, Beijung, 2002, Paper PI 68 D. 063. With permission.)

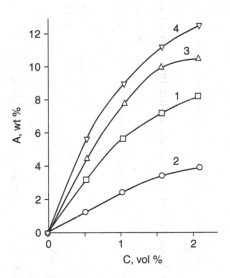

FIGURE 5.11 Amount of hydrogen sulphide adsorbed from its mixture with air on different carbons. (1) activated carbon fiber, ACF (2) oxidized activated carbon fiber OCF, (3) oxidized activated carbon fiber impregnated with copper, (4) oxidized carbon fiber impregnated with iron. (After Svetlana, S., Kartel, N.T., and Tsyba, N.N., Carbon '02, Intern. Conf. on Carbon, Beijung, 2002, Paper PI 68 D. 063. With permission.)

of oxidized products in the evolved gaseous products or on the carbon surface. The desorption experiments also showed that the adsorption of CS_2 on the unmodified carbons was purely physical as it could be completely recovered on treatment with nitrogen at 100°C, indicating the absence of a chemical conversion. However, at temperatures higher than 120°C, catalytic oxidation of CS_2 takes place evolving sulfur oxides and carbonic acids into the gaseous products. This catalytic oxidation of CS_2 takes place much more slowly on unmodified carbons than on cation-modified carbons. It has been suggested that CS_2 is adsorbed on the cation-modified carbons in the same manner as the other sulfur compounds (SO_2, H_2S) via ligand adsorption with the formation of the –C-O-M ... (CS_2)n metal complex (where M represents the metal cation). This complex plays an important role in adsorption and catalytic oxidation.

5.3 ACTIVATED CARBON ADSORPTION IN NUCLEAR TECHNOLOGY

Nuclear technology involves several operations, the more important being the production of nuclear fuels, processing of spent fuels, and operation of nuclear reactors. All these operations require a thorough purification of the exhaust and circulating gas. Because many of these processes involve inert gases, their removal can best be carried out by activated carbon adsorption. For example, helium, an inert gas, is circulated as a protective gas in nuclear reactors cooled and moderated by heavy water[112] and carries with it deuterium and oxygen formed by radiolysis, and thereby prevents the formation of an explosive mixture. During exchange of fuel elements, argon gets mixed up with the air entering the system and accumulates and affects the operation of the reactor so that the circulating helium needs to be decontaminated continuously or periodically. The concentration of argon in the protective gas should not exceed 10 ppm.

Contaminated helium is first led to a reactivator in which deuterium and oxygen combine over a catalyst to form heavy water and then pass through an activated carbon bed cooled in liquid nitrogen. Active carbon adsorbs argon and nitrogen as well as radioactive xenon and krypton that may have leaked into the protective gas from faulty fuel elements.[113] The loaded activated carbon is regenerated by raising its temperature and evacuation. The temperature during regeneration is raised slowly so that the desorbed gases can be separately collected. The protective gas helium is collected first and returned to the cycle while the products of radioactive decay are compressed and collected in pressure vessels for disposal. The adsorption on activated carbon in the field of nuclear technology is carried under very specific conditions, and the process is expensive because it is usually carried out at low temperatures and it must take into account the heating of the adsorbent carbon due to the heat of the radioactive decay of the adsorbed species.

The nuclear fission reaction in nuclear plants also produces radioactive iodine isotopes I^{131} and I^{133}, which are present in the coolant release and in the ventilation system both in the elemental form and in the form of its compounds e.g., methyl iodide. Activated carbons impregnated with potassium iodide and similar compounds[114] and with amines including several[115,116] pyridines have been widely used for the retention of radioactive iodine compounds. These carbons remove very low levels of iodine and its compounds from gas streams with high efficiency, even in the presence of high humidity. The performance of the impregnated carbon varies with the history of its formation and by ageing. Billinge et al.[114] compared the efficiency of potassium iodide impregnated coconut and coal-based charcoals for capturing radioactive methyl iodide from gas-cooled nuclear reactors. The coal-based impregnated charcoal was more efficient than the impregnated coconut charcoal. The adsorption of methyl iodide was related to the amount of the associated oxygen and the nature of the surface oxygen groups. The surface oxygen groups that were acidic and evolved as carbon dioxide on degassing increased the retention efficiency more than the surface oxygen groups that were evolved as carbon monoxide. The impregnated potassium iodide reacted with the surface oxygen groups on carbons and modified their desorption behavior, thereby improving the efficiency of the carbon to retain radioactive methyl iodide.

Baker and Poziomek[115] studied the reactivity of charcoals impregnated with five different pyridines toward methyl iodide. Different pyridines reacted differently and showed varying efficiencies for the removal of methyl iodide, the most efficient being the pyridine (Table 5.8). When the same charcoal sample was impregnated with different amounts of 4-aminopyridine and 4-vinylpyridine, the retention of

TABLE 5.8
Influence of Impregnation of Charcoal on the Retention of Methyl Iodide

Impregnant	Concentration of Impregnant (mmol/g)	Methyl Iodide Retained (%)
4-Vinylpyridine	0.048	76.5
	0.096	73.3
	0.190	71.3
	0.286	77.0
	0.475	77.7
4-Aminopyridine	0.053	53.5
	0.106	57.5
	0.212	61.7
	0.318	69.2
	0.530	75.5
Pyridine	0.253[a]	80.2
4-Cynopyridine	0.192[a]	54.8
4-n-Propylpyridine	0.165[a]	48.7
4-Vinylpyridine	0.190[a]	71.3
4-Aminopyridine	0.212[a]	61.7

[a]Corresponds to a 2% by weight loading on the charcoal.

After Baker, J.A. and Poziomek, E.J., *Carbon*, 12, 45, 1974. Reproduced with permission from Elsevier.

methyl iodide increase from 53 to 75% with increase in the concentration of 4-aminopyridine on the surface of the carbon but remained more or less unchanged (~71%) with increase in the concentration of 4-vinylpyridine on the carbon surface (Table 5.12). Generally, carbons impregnated with methyl iodine are used in the filtering devices for the purification of air in nuclear reactors.

5.4 ACTIVATED CARBON ADSORPTION IN VACUUM TECHNOLOGY

Oil- or mercury-diffusion pumps are generally used for the creation of vacuum in the laboratory. However, some mercury or oil vapors are always introduced into the vacuum system that may interfere with the processes to be carried out in the evacuated vessel. It is a common observation that in a vacuum system using a mercury-diffusion pump, the glass tubes connecting the diffusion pump with the evacuated vessel turn blackish due to the deposition of mercury vapors in the cooler portion. Therefore, in order to get a clean vacuum system, adsorption of gases and vapors on activated carbons cooled to a low temperature is the answer. The use of activated carbon for vacuum generation does not need a complicated apparatus. It can just be a glass vessel filled with granulated activated carbon cooled with liquid air in a Dewar flask. The activated carbon can be first evacuated, preferably at 300°C, to remove physisorbed gases and vapors, cooled, and then connected to the system to be evacuated. It has been claimed[117] that pressures as low as 10^{-6} mm of Hg can be obtained with 30L/s as the rate of evacuation. As activated carbons are high-capacity reliable adsorbents, they are very suitable as complements to diffusion pumps, although they can also be used alone. The saturated activated carbon can be easily regenerated by bringing it to room temperature or high temperature at which the gases and vapors are desorbed.

Liquid air-cooled activated carbons can also be used for purification of helium enriched with the isotopic H^3e for use in a cyclotron[118] from impurities due to air leaking into the system.

5.5 MEDICINAL APPLICATIONS OF ACTIVATED CARBON ADSORPTION

Carbon in its various forms such as pyrolytic carbon, glassy carbon, carbon fiber, carbon fiber reinforced composites, and activated carbon has been a potential and important material in medical science and in medicinal applications. This is due to its properties such as good biocompatibility, nontoxicity, no immune reaction with the body, low density, chemical inertness, low coefficient of friction, elastic modulus similar to that of bone, and high adsorption capacity. Some of these applications of carbon materials have been discussed elsewhere[118] and are beyond the scope of this book. In this section we shall look into some of the medicinal applications based on adsorption by active carbons. Adsorbent carbons in the form of charcoal and activated carbons have been in use for medicinal and health applications for centuries and have been listed in the pharmacopoeia as antidotes and intestinal adsorbents.

The students of Hippocrates recommended the dusting of wounds with charcoal to remove their unpleasant smell. Old Egyptians also used charcoal about 1550 BC for putrifying wounds to remove odors, and for problems in the gastic intestinal tract as an effective treatment against bacteria and toxic materials. A necessary condition for the healing effect of active carbon is for it to pass through the digestive system sufficiently rapidly so that the desorption of harmful species from it in the distal regions by changed pH and the resorption effect of the intestinal mucosa is prevented. Active carbon is, therefore, generally used in combination with a laxative, preferably salinic, which is not adsorbed well and, therefore, does not decrease the effectiveness of the carbon. Hippocrates and Pliny in 460 BC recorded the use of charcoal for treating epilepsy, chlorosis, and anthrax.[90] However, systematic research in the antitoxic properties of active carbons began at the beginning of the twentieth century. The adsorptive removal of toxic substances such as phenols, alkaloids, barbiturates, insecticides, and heavy metal salts from industrial and domestic waste water has been widely studied and is discussed later in this book.

The antitoxic properties of activated carbon adsorption received a big boost in the early 1950s, when Arwell[119] proposed that the blood of poisoned patients could be purified by passing it through a column filled with active carbon. This was more applicable for those cases in which the oral application of active carbon was not effective. Greek scientist Yatzidis and coworkers developed this concept and saved the lives of two patients who were poisoned with barbiturates. Yatzidis[120–122] designed equipment that he named the *Carbon Kidney*. It consisted of a 600-cm³ cylindrical column filled with 200 gm of active carbon of 0.4 to 1.4 mm particle size and equipped with 2 filters. The column was coupled to a pump, and its inlet and outlet were connected to the patient's artery and vein, respectively. Yatzidis was of the view that the active carbon could be used not only for the removal of externally introduced poisons, but also for toxins that appear in the blood due to the malfunctioning of the kidneys. Several other workers[123–126] examined the prospects of the application of this Carbon Kidney in clinical practice, but they observed that together with the noxious substances, the Carbon Kidney could adsorb certain vital components of the blood. The active carbon pellets trapped some blood clots, and thus part of the patient's blood was lost.

Hagstam et al.,[124] while purifying the blood of rabbits poisoned with barbiturates using Carbon Kidney, observed that although the poison was removed from the blood, the active carbon powder was observed in the liver, lungs, spleen, brain, and kidneys of the rabbits. This was attributed to the formation of a certain amount of powdered carbon in every type of granulated carbon. This powdered carbon is formed due to the attrition of the carbon pellets during the passage of blood through the carbon bed, during washing, and during transport. This formation of powdered carbon in every active carbon is the main disadvantage in the widespread application of the Carbon Kidney for the purification of the blood of living organisms. Thus, investigations were initiated to develop active carbon treatments to prevent the formation of powders from carbon pellets and in turn to eliminate the process of blood clotting.

Lavine and La Course[126] and Sparks[127] proposed placing a semipermeable membrane on the active carbon pellet surface. The thickness of the semipermeable membrane was about 50 nm, which ensured a good rate of diffusion of adsorbed metabolites. In addition to the semipermeable membrane, the carbon pellets were covered with polymer membranes, which markedly improved the mixing of blood with the carbon. Chang[128] coated active carbon granules with natural polyampholyte-albumins and designed a blood-purifying apparatus using these coated carbon granules. The apparatus was successfully used in the therapy of acute poisoning and chromic kidney failure. Chang observed that a column containing 30 g of active carbon pellets coated with albumin was more efficient than the currently used artificial kidney. However, the coating of pellets with albumin is expensive, and the resulting adsorbent carbon is not very stable.

Nikolaev et al.[129] using different activated carbons with varying surface characteristic, studied their interactions with blood and their other biological characteristics in order to understand the suitability of an adsorbent carbon for removing toxic substances, both exogenous and endogenous, from blood. The more important exogeneous and endogenous toxic substances that need to be removed simultaneously from blood are, according to Bruskina,[130] given in Table 5.9. Walker[131] determined the adsorption capacity of an active carbon for certain low molecular-weight toxic substances in terms of their blood concentration (Table 5.10). Strelko[132] and Nikolaev and Strelko[133] prepared an active carbon from styrene-divinyl benzene copolymer, which is in the form of exceptionally smooth and hard spherical granules and which was not coated with any surface film. This active carbon proved to be very useful for blood purification. This active carbon is cheaper than any other carbon used for blood purification and is free of any surface treatment, so many of the side effects are avoided.

Kartel and coworkers[134–137] prepared synthetic activated carbons SCN and SCS that were in spherical form and used them for hemoperfusion (i.e., for the purification of human blood by adsorption of toxic and harmful substances in the extracorporal circle).[138,139] The preparation of these synthetic carbons are parts of Ukrainian and Russian patents.[140,141] According to Kartel and coworkers, activated carbons for blood purification should have a high adsorption capacity, be sterile, and be free of toxic, pyrogenic products, and dust particles and be hemocompatible (i.e., they should not destroy or adsorb platelets, especially thrombocytes and leucocytes) and should not initiate the mechanism of curtailing blood and shift acid-alkaline balance of blood. The SCN and SCS activated carbons were found to have homogeneous and ideal spheres that provided these materials uniform packing density in the purifying column as well as a minimal hydrodynamic resistance to the flow of blood and a lower factor of friction between the granules.

Sem micrographs of SCN and SCS synthetic activated carbons are compared with conventional activated carbons[134] obtained from peat coal or wood in Figure 5.12. The figure clearly shows that the synthetic carbons have a smooth external surface, ensuring favorable conditions for hemocytes membranes that are very sensitive to heterogeneity of the contact surface and that their surface does not need any additional covering by biocompatible polymer films. The bottommost micrographs show that the internal structure of synthetic activated carbon spheres has well-developed channels and pores with sizes considerably smaller than the size of the platelets. The surface characteristics of some hemosorbents are compared in Table 5.11.

TABLE 5.9
Toxic Species Removal from Blood or Lymph Following Disease or Poisoning and Their Effects on Different Parts of the Body

Substances Removed	Liver Failure	Kidney Failure	Exogenous Poisoning	Radiation Burns in Oncology	General Disturbances in Acid-base Equilibrium of Salt Exchange	Stored Blood
Ions K^+, Cl^-, $H_2PO_4^- HCO_3^-$				+	+	+
Nitrogen-containing compounds:						
ammonia		+		+		+
creatine	+	+	+	+		
urea	+	+	+	+		
uric acid	+	+	+	+		+
Organic acids:						
lactic				+	+	+
pyruvic				+	+	+
acetoacetic					+	
–hydroxybutyric					+	
aminoacids	+			+		
Proteinaceous substances:						
oligomers of nucleic acids				+		
fragments of cell membranes	+			+		+
$$$				+		
$$$ haemoglobin				+		+
bilirubin	+		+			
Exogenous phenols	+		+	+		
Cytostats	+			+		

After Bruskina, T.N., *Adsorbtsiya Adsorbenty*, 9, 77, 1981. With permission.

Because blood is a multicomponent system containing a large number of ionized and molecular species differing in size, mass, cell structure, and colloidal particles, several processes such as polymerization of blood proteins, reaction with biosurfactants, and other accompanying processes besides adsorption can take place on the surface of adsorbent carbon. These processes can influence the adsorptive removal of toxics and other harmful components from blood during hemoperfusion. Furthermore, experiments concerning destruction of thrombocytes and deposition of platelets and leucocytes on the adsorbent carbon surface can only be carried out on animal blood.

TABLE 5.10
Adsorption Capacity of Active Carbon
for Several Different Body Species

Compound	Blood Concentration mg dm⁻³	Adsorption mg (200 g)⁻¹
Creatine	110–250	600–1500
Uric acid	50–150	400–1200
Indican	4–5	25–30
Phenols	30–40	150–175
Guanidine	30–40	85–110
Organic acids	15–20	30–35
Urea	7–4	2–3
Barbiturates	75–300	500–2000
Salicylates	250–750	2000–3000

After Walker, J.M., Benti, E., Van Wagenon, R.D., *Kidney Inter.*, Suppl. 7, 320, 1976. With permission.

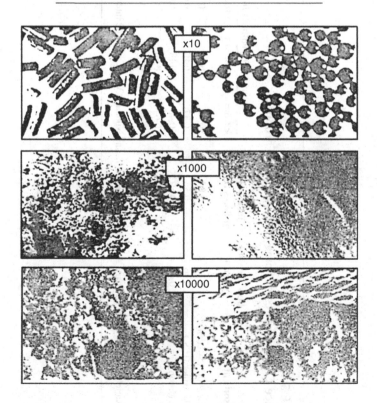

FIGURE 5.12 SEM microphotographs of technical (left) and synthetic (right) active carbons. (After Kartel. N.T., Strelko, V.V., Puziy, A.M., Mikhalovsky, S.V., and Nikalaev, V.G., Carbon '02, Intern. Conf. on Carbon, Beijung, 2002, Paper 21.4. With permission.)

TABLE 5.11
Surface Characteristics of Some Homosorbent Carbons

Hemosorbent	Bulk Density (g/cm³)	Total Volume of Pores (cm³/g)	Volume Pore with Radii			Sorption Volume of Pores on Benzene (cm³/g)	Specific Surface Area on Argon (m²/g)
			<3 nm	3–100 nm	>100 nm		
Adsorba 300C (Sweden)	0.43	0.85	0.43	0.06	0.36	0.49	920
Hemodetoxifier (U.S.)	0.46	0.56	0.37	0.10	0.09	0.39	690
DHP-1 (Japan)	0.45	0.69	0.48	0.18	0.03	0.52	810
SCN	0.34	1.59	0.64	0.86	0.09	0.97	1320
SCS	0.36	1.15	0.44	0.59	0.12	0.93	1380

After Kartel. N.T, Strelko, V.V., Puziy, A.M., Mikhalovsky, S.V., and Nikalaev, V.G., Carbon 02 Intern. Conf. on Carbon, Beijung, 2002, Paper 21.4. With permission.

Kartel et al.[134] carried out the purification of dog blood with respect to the removal of creatinine, medinal, and substances of middle molecular mass (MMS) (peptides), using synthetic activated carbons SCN and SCL 120 g in a 350 cm^3 column. Medinal and creatinine were introduced in the dog blood to an initial concentration of 100 mg/L. The adsorption column was connected with an AV shunt and a dose peristatic pump. The perfusion test was carried out for 60 min with a volumetric rate of 160 ml/min.

The concentration of creatinine, medinal, and MMS was determined at the inlet and the exit of the perfusion column after 15, 30, 45, and 60 min, using standard methods and techniques. Simultaneously, the loss of platelets and leucocytes, as well as of free hemoglobin, from blood plasma was carried out.

The results obtained by using 120 g of synthetic activated carbons SCN and SCS in 350 cm^3 column after averaging data obtained at different time intervals (Table 5.12) when compared with the data obtained from literature using similar commercial columns for hemoperfusion clearly show the superiority of the synthetic carbons, both for detoxification and hemocompatibility. Based on their investigations and experience of use of hemosorbents SCN and SCS in medical practice, a number

TABLE 5.12
Purification of Blood by SCN and SCS Synthetic Carbons and Some Commercial Hemosorbents with Respect to Different Body Species

Parameter	SCN	SCS
During 1 hr perfusion, % from volumetric rate:		
- on Medicinal ($C_m = 100$ mg/l)	>70	>75
- on Creatinine ($C_m = 100$ mg/l)	>80	>90
- on MMS (500–1500 Da)	>60	>50
Influence on Hemocytes (under conditions of 10% Hemodilution & regional decalcination):		
- loss of leukocytes, %	<10	<10
- loss of platelets, %	<15	<15
- arising free haemoglobin. mg/l	<100	<100

	"Haemocol 100"	"Adsorba 300C"	"Hemodetoxifier"
Clearance. ml/min (as perfusion rate of 100 ml/min):			
- on creatinine	50	55–75	40–55
- on pentobarbital	40	—	—
- on phenobarbital	60	55	72.5
- on "middle" molecules	—	60	—
Total loss of hemocytes (%)	up to 30	up to 30	26–53

After Kartel. N.T., Strelko, V.V., Puziy, A.M., Mikhalovsky, S.V., and Nikalaev, V.G., Carbon '02, Intern. Conf. on Carbon, Beijung, 2002, Paper 21.4. With permission.

TABLE 5.13
List of Diseases and Other Body Disorders in Which SCN and SCS Synthetic Carbons Have Been Effective after Hemoperfusion

Liver and Kidney Diseases	Endogenic Toxicoses	
-hepatic coma	-peritonitis	-pancreatitis
-cirrhosis	-sepsis	-wound infection
-hepatitis A, B	-burns	-gestosis
-biliary syndrome	-oncology toxicosis	-ischemia
-acute and chronic renal insufficiency	-atherosclerosis	-hypercholesterolemia
-glomerulonephritis		

Acute Poisonings with Autoimmune and Psychoneurological Diseases

-barbiturates	-allergosis	-bronchial asthma
-phosphorous-containing organic	-psoriasis	-systemic lupus
compounds	-arthritis	-erythema
-heavy metal salts	-delirium tremens	-abstinent syndrome
-mushroom toxins	-rhesus conflict	-schizophrenia
-alkaloids	-narcomania	

After Kartel. N.T., Strelko, V.V., Puziy, A.M., Mikhalovsky, S.V., and Nikalaev, V.G., Carbon 02 Intern. Conf. on Carbon, Beijung, 2002, Paper 21.4. With permission.

of medical institutions in Ukraine and other CIS countries have recommended the use of these carbons in the treatment of a number of diseases (Table 5.13).

Activated carbons are also used for deodorizing gaskets for colostomy patients.[142] In hematology, activated carbons can be used to adsorb Ig G antibodies[143] and toxic agents from plasma[43] Activated carbons filters are used as a part of the water purification system in hemodialysis[145] and as microfilters for blood transfusion.[146] Some activated carbon capsules are also used as implants in the body,[147,148] and in some cases activated carbons with adsorbed drugs on their surface are used as drug delivery systems for release of the drug in the body.[149] Active carbon in large doses has also been used in cases of acute gastritis and entertis.[150,151]

Chen et al.[152,153] prepared activated carbon fibers (ACF) from sisal fiber as substrate, impregnated them with different concentrations of zinc chloride, and then heat treated them in nitrogen at 800°C. Some of these zinc chloride-impregnated carbon fibers were treated with $Ag(NH_3)_2^+$ to deposit silver. These ACFs were tested for their antibacterial activity against *E. coli* and *S. aureus* bacteria. The ACF sample impregnated with zinc showed strong activity against *E. coli* and could kill almost all bacteria with the initial concentration of about 2×10^6 cfu/ml (*cfu* is a clone-forming unit) (*cf*, Table 5.14). However, the antibacterial activity was independent of the concentration of ZnO present on the ACF surface. The antibacterial activity was suggested to be due to the formation of H_2O_2 by reaction of ZnO with water and air. When a sample of ACF impregnated with 10% $ZnCl_2$ was heat-treated at 800°C, the sample showed stronger activity against *S. aureus* bacteria and could kill

TABLE 5.14
Antibacterial Activity of Zinc Oxide Impregnated Activated Sisal Carbon Fiber (SACF) against *E. Coli*

Samples	Amount of ZnO on ACF (wt%)	Residual Bacteria (cfu/mL)
ZnO-SACF-5%	6.63	230
ZnO-SACF-40%	3.50	0
ZnO-SACF-50%	4.95	360

After Kartel. N.T., Strelko, V.V., Puziy, A.M., Mikhalovsky, S.V., and Nikalaev, V.G., Carbon 02 Intern. Conf. on Carbon, Beijung, 2002, Paper 21.4. With permission.

all the bacteria compared with the sample heat-treated at lower temperatures that were not as effective. The ACF samples containing silver deposited on their surface along with ZnO were found to improve the antibacterial activity against *S. aureus* considerably (Table 5.15). Yamamoto[154] also studied the antibacterial activity of ZnO-supported activated carbons obtained by the carbonization at 500 to 900°C in nitrogen of ion-exchange resin having carboxyl group as the exchangeable functional group against *S. aureus* (*Staphylococcus aureus*), and *E. coli* (*Escherichia coli*) bacteria. These bacteria were cultured in brain-heat infusion broth (BHI) at 37°C for 24 hr. The bacterial culture was suspended in a sterile physiological saline at a final concentration of 10^2 cfu/cm^3. The antibacterial activity of the carbon sample was evaluated by measuring the changes in electrical conductivity with bacterial growth. The active oxygen in the form of H_2O_2 in the physiological saline was measured by oxygen electrode. The antibacterial activity for *S. aureus* was stronger than for the *E. coli* bacteria (Figure 5.13). Furthermore, the activity was stronger in

TABLE 5.15
Antibacterial Activity of Sisal-Activated Carbon Fiber against *S. Aureus* Bacteria Before and After Impregnation with Zinc Oxide and Silver

Samples	ZnO (wt%)	Ag (wt%)	Residual Bacteria (cfu/mL)
ZnO-SACF-Ag	3.99	0.05	120
ZnO-SACF	5.84	0	
SACF	0	0	
Blank ref.	0	0	

After Kartel. N.T., Strelko, V.V., Puziy, A.M., Mikhalovsky, S.V., and Nikalaev, V.G., Carbon 02 Intern. Conf. on Carbon, Beijung, 2002, Paper 21.4. With permission.

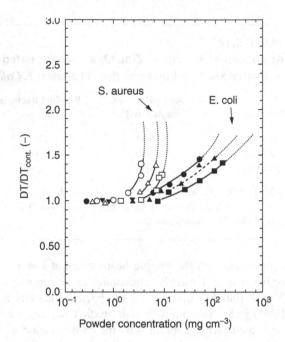

FIGURE 5.13 Comparison of antibacterial activity for *S. aureus* and *E. coli* for different activated carbons. ○,● = 500°C sample; △,▲ = 700°C sample; □,■ = 900°C sample. (After Yamamoto, O., Carbon '02, Intern. Conf. on Carbon, Beijung, 2002, Paper 21.3. With permission.)

the case of the samples carbonized at 500°C and weaker in case of the 900°C carbonized sample. The antibacterial activity could be attributed to two factors: the pH value in the medium and the active oxygen generated from ZnO. Because the pH values measured in this study were almost the same, 5.6 to 5.7 in all three carbons, the antibacterial activity was due to the formation of H_2O_2 from ZnO in carbon samples. These workers measured the amount of H_2O_2 generated in each carbon sample and found that the amount of H_2O_2 generated for the same amount of each carbon was maximum in the case of the carbon sample carbonized at 500°C (Figure 5.14). The reason that the carbon samples showed stronger activity for *S. aureus* than for *E. coli* is assumed to be the greater sensitivity of *S. aureus* for H_2O_2. The activity of carbon fibers immersed in 1.0% solution of $AgNO_3$ also showed antibacterial activity against both *S. aureus* and *E. coli* strains.[155]

In order to control drug overdosage, liver disease, and kidney disease, the adsorption of several bioactive drugs from aqueous solutions such as analgesics,[156,157] steroids,[158] aminoacids,[159] hypnotics,[156,159,160] and tranquilizers[139] has been studied. The adsorption is completely reversible and follows Type I adsorption isotherm. Enzymes also adsorb reversibly, but they exhibit decreased activity in the adsorbed state.[161] In some cases, physical adsorption is followed by chemisorption as in the case of aminopyrine,[156] or by carbon catalyzed hydrolysis as in the case of diuretic benzothiodizines,[163] or with nitrogen heterocyclic anti-inflammatory drugs.[164] The adsorption of barbiturates on microporous carbons is considerably influenced by the

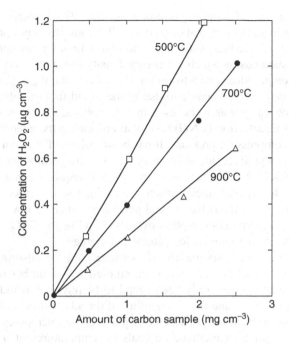

FIGURE 5.14 Amount of hydrogen peroxide produced with the amount of carbon added. (After Yamamoto, O., Carbon '02, Intern. Conf. on Carbon, Beijung, 2002, Paper 21.3. With permission.)

solvation effects. There is almost a fivefold increase in the apparent monolayer capacity by changing the solvent from aqueous buffer to 80% ethanol.[160,165]

5.6 ACTIVATED CARBON ADSORPTION FOR GAS STORAGE

Petroleum is an important and widely used source of energy in the world because it is available in liquid form and can, therefore, be conveniently stored. However it is environmentally unfriendly because it produces oxides of sulfur, oxides of nitrogen, and carbon monoxide that are hazardous to humans, animals, and vegetable matter. Furthermore, the resources of petroleum are depleting as the need for it is ever increasing, due to the ever-increasing number of automobiles on the roads. Consequently, a considerable amount of research is being focused on developing alternative source of energy. Natural gas is one such alternative transportation fuel that is environment friendly and can be used for automobiles.

Natural gas is composed mostly of methane (85 to 90%) with only small amounts of ethane, propane, and butane. It is almost free of harmful contaminants and can burn efficiently, and may release only small amounts of harmful gases and vapors. However, the low energy density of natural gas retards its wide application as automobile fuel. One liter of petroleum on combustion produces about 3.5×10^4 KJ energy, but one liter of natural gas produces only 40 KJ energy on combustion, which is approximately 0.1% that of petroleum.[166] Thus, the major problem for using natural gas as a vehicle fuel is its storage to obtain high-energy density. At present there are three methods by which the energy density of natural gas can be enhanced: liquefaction, compression, and adsorption. Liquefied natural gas (LNG), although it has an energy density comparable to that of the liquid fuel (gasoline), it is a costly affair and is too dangerous to be used in automobiles. Compression to a smaller degree without liquefaction (CNG) results in enhancing the energy density in proportion to the compression pressure. It has been estimated[167] that a compression of about 25 mPa can produce adequate energy density for natural gas to be used as an automobile vehicle, but a pressure of this degree requires container vessels that would be very heavy and thick, which need added safety risks. In contrast, the adsorbed natural gas (ANG) has several advantages such as low self-weight, high security, and low energy consumption over the LNG or the CNG and is, therefore, considered a promising method for natural gas storage.

The natural gas consists mainly of methane, and the adsorption of methane on coals and activated carbons has been studied by a number of workers. The adsorption on coal has generally been found to be physical as indicated by small heats of adsorption[168,169] and the reversibility of the adsorption isotherms.[170,171] The adsorption at the ambient temperature takes place in the micropores. Ruppel et al.[172] studied the adsorption of methane on coals at temperatures of 0 to 50°C and at pressures from 1 to 15 mPa, and applied Langmuir and Polanyi adsorption models. Ruppel found that the adsorption could be well explained by the Langmuir equation.

Clarkson et al.[173] studied the adsorption of methane on wet coals at elevated temperature and pressure above the critical point and applied the BET, DR (Dubinin-Radushkevich), and Dubinin-Astakhov (D.A.) equations to the adsorption data and compared the results with the Langmuir equation. The three-parameter DA equation gave the best curve fit to the high-pressure methane adsorption data. The two-parameter DR and BET equations were also found to be better than the Langmuir equation. The nonvalidity of the Langmuir equation to the adsorption data was attributed to the failure of the assumption of an energetically homogenous surface that was not strictly valid in these cases.

Xing and Yan[174] determined the adsorption of methane at ambient temperature on activated carbons prepared from petroleum coke. A certain volume of the dried activated carbon sample was placed in a storage vessel that was evacuated to 6×10^{-2} Pa. Thereafter, the methane at a specific pressure was introduced into the storage vessel for 10 minutes. The amount of methane adsorbed was determined by the increase in weight of the vessel. The activated carbon in the vessel was then exposed to atmosphere to find out the amount of methane released. The difference in weight of the vessel after adsorption and after exposure to atmosphere gave the storage capacity of the carbon for methane. The relationship between methane effective adsorption capacity and storage pressure (Figure 5.15) showed that the adsorption capacity increased with increase in the storage pressure. The results also showed that the activated carbon prepared at the lowest temperature possessed the largest adsorption (storage) capacity. This was attributed to the high density and the small micropore

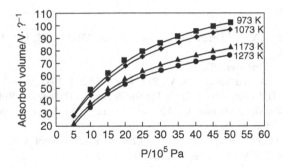

FIGURE 5.15 Methane effective adsorption capacity vs. storage pressure. (After Xing, W. and Yan, Z.F., Carbon '02, Intern Carbon Conf., Beijing 2002, Paper PI 94D 106. With permission.)

size in the low-temperature activated carbons. The results could fit very well in the Freundlich and Langmuir adsorption equations but showed deviations with the BET equation of multiplayer adsorption.

Quinn and MacDonald,[175] while studying the storage of methane using activated carbons at ambient temperature, observed that there was negligible adsorption of methane in macropores and that the macropore volume could be considered as void space in which the methane can be present at the density of the compressed gas. This indicated that for efficient and optimum storage capacity of an activated carbon for methane, the macropores and interstitial voids should be minimum, to allow sufficient passage for the entry of methane into the adsorbent. Furthermore, the micropores should be of dimensions that permit high storage density and should not be so narrow as to limit the release of methane during adsorption.

Another important parameter that limits the storage of methane by activated carbons is the exothermic nature of the adsorption process. This heat released during the adsorption process can raise the temperature to about 140°C.[176] This rise in temperature can, however, be controlled by the rate of filling of the storage vessel. Conversely, the heat would be required to release the adsorbed methane to maintain a proper rate of release of the gas. Thus, these factors have to be kept in mind while designing the storage vessel and while choosing the type of activated carbon to be used. Derbyshire et al.[167] has suggested that monolithic carbons possess superior heat-transfer characteristics.

Hydrogen is another gas which, when properly stored, can be used as a fuel, and sufficient interest has been indicated in the adsorbent properties of activated carbons toward hydrogen gas. For example, Amankwah et al.[177] studied the adsorption of hydrogen at temperatures of 150 to 160 K using an activated carbon and found that about 17 kg H_2/m^3 could be stored at 54 atm. These workers found that hydrogen storage by adsorption was much less expensive than the pressurized hydrogen storage or liquefied-hydrogen storage methods. Dillon et al.[178] while working with carbon nanotubes, observed that carbon nanotubes with 2.0 nm diameters were better materials for the storage of hydrogen and could store larger amounts of hydrogen gas than the activated carbons.

REFERENCES

1. Singleton, V.A. and Draper, D.E., *Am. J. End.*, Viticult, 13, 114, 1962.
2. Nawak, G., *Brauwelt,* 95, 760, 1955.
3. Kaiser, K., *Brauwelt,* 97, 1203, 1957.
4. Wyneberg, A. and Sauer, J.N.A., Brit. Patent 21204, 1911.
5. Smit, P., Tijdscher, *Algem. Tech. Vereen Bectwortelskikerfabr.,* Reff., 31, 91, 1936.
6. Jankowska, H., Swiatkowski, A., and Choma, J., in *Active Carbon*, Ellis Howard, LTD, England, 1991.
7. Smisek, M. and Cerny, S., in *Active Carbon,* Elsevier Publ. Co., Amsterdam, 1970.
8. Groswenor, W.M., U.S. Patent 2954305, 1960.
9. Groswenor, W.M., *Int. Sugar J.,* 340, 368, 1961.
10. Rajkowski, S., Nowak, K., and Pichocki, T., *Pharma. Zentralholle*, 1958, p. 97.
11. Howard, F. and Spooner, E.C.R., *Chem. and Ind.,* 1946, p. 186.
12. Charonnet, R. and Lechat, P., *Pharma. France,* 8, 171, 1950.
13. Willia, R. and Hightowar, J.V., *Chem. Engg.,* 55, 133, 1948.
14. Shearon, W.H. and Gee, O.F., *Ind. Engg. Chem.*, 41, 218, 1949.
15. Davis, W.N., U.S. Patent 227963, 1880.
16. Johnson, W.D., U.S. Patent 522260, 1894.
17. Zadra, J.B., U.S. Bureau of Mines, Washington, D.C., R.I. No. 11672, 1950.
18. Zadra, J.B., Angel, A.K., and Heinen, H.J., U.S. Bureau of Mines, Washington, D.C., R.I. No. 4843, 1952.
19. Hall, K.B., *World Min.,* 27, 44, 1974.
20. Bansal, R.C., Dennet, J.B., and Stoeckli, F., in *Active Carbon,* Marcel Dekker, New York, 1988.
21. Brusov, G. Z., *Chem. Ind. Kolloids,* 5, 137, 1909.
22. Green, M., *Trans. Inst. Min. Metall.,* 23, 65, 1913–1914.
23. Allen, A.W., *Metall. Chem. Engg.,* 18, 642, 1918.
24. Puri, B.R. and Bansal, R.C., 1, 451, 1964.
25. Bansal, R.C., Dhami, T.L., and Prakash, S., *Carbon,* 15, 157, 1977.
26. Grabovskii, A.I., Ivanova, L.S., Korostyshevskii, N.B., Shirshov, V.M., Storozhuk, R.K., Matskevich, E.S., and Arkadakskaya, N.A., *Zh. Prikl. Khim.* 49, 1379, 1976.
27. Grabovskii, A.I., Grabshek, S.L., Ivanova, L.S., Storozhuk, R.K., and Shirshov, U.M., *Zh. Prikl. Khim.,* 50, 522, 1977.
28. Garten, V. A. and Weiss, D. E., *Rev. Pure Appl. Chem.,* 7, 69, 1957.
29. Dixon, H., Cho, E., and Pitt, C.H., AICRE Meeting, Chicago, Nov. 26–Dec. 2, 1976.
30. Clauss, C.R.A. and Wiess, K., Pretoria CSIR Report No. CENG 206, Sept. 1977.
31. Puri, B.R. and Mahajan, O.P., *J. Indian Chem. Soc.,* 39, 292, 1962.
32. Puri, B.R. and Bansal, R.C., *Carbon,* 1, 457, 1964.
33. Goyal, M., Rattan, V.K., Aggarwal, D., and Bansal, R.C., Colloids and Surfaces, A. *Physicochemical and Engineering Aspects*, 190, 229, 2001.
34. Aggarwal, D., Goyal, M., and Bansal, R.C., *Carbon,* 37, 1989, 1999.
35. Puri, B.R., in *Chemistry and Physics of Carbon*, Vol. 6, P.L. Walker Jr., ed. Marcel Dekker, New York, 1970.
36. Cho, E., Dixon, S.N., and Pitt, C.H., *Metall. Trans.,* 10B, 185, 1979.
37. Cho, E. and Pitt, C.H., *Metall., Trans.,* 10B, 159, 1979.
38. Cho, E. and Pitt, C.H., *Metall., Trans.,* 10B, 165, 1979.
39. Bokros, J.O.M., Devanathan, M.A.V., and Miller, K., *Proc. Roy. Soc.,* London, A. 274, 55, 1963.
40. Anderson, T.N. and Bokros, J.OM., *Electrochem. Acta*, 9, 347, 1964.

41. McDougall, G.J., Hancock, R.D., Nicol, M.J., Wellington, O.L., and Copperthwaite, R.G., *J. South African Inst. Min. Metall.,* 80, 344, 1980.
42. Kuzminykh, V.M. and Tyurin, N.G., *Izv. Vyssh Uchebn. Zaved. Tsved. Metall.,* 11, 65, 1968.
43. Davidson, R.J., *J. South African Inst. Min. Metall.,* 75, 67, 1974.
44. Adams, M.D., McDougall, D.J., and Hancock, R.D., *Hydrometallurgy,* 19, 95, 1987.
45. Fleming, C.A. and Nicol, M.J.J., *South African Inst. Min. Metall.,* April 1984.
46. Van der Merwe, P.F. and Van Deventer J.S., *J. Chem. Eng. Commun.,* 65, 121, 1988.
47. Tsuchida, N. and Muir, D.M., *Metall. Trans.,* 17B, 529, 1986.
48. Peterson, F.W. and Van Deventer, J.S., *J. Chem. Eng. Sci.,* 46, 3053, 1991.
49. Adams, M.D., *Hydrometallurgy,* 25, 171, 1990.
50. Adams, M.D. and Fleming, C.A., *Metall. Trans.,* B20, 315, 1989.
51. Adams, M.D., Friedl, J., and Wagner, F., *Hydrometallurgy,* 31, 265, 1992.
52. Adams, M.D., *Hydrometallurgy,* 31, 121, 1992.
53. Adams, M.D., Friedl, J., and Wagner, F., *Hydrometallurgy,* 37, 33, 1995.
54. Klauber, C., *Surface Science,* 203, 118, 1988.
55. Groenewold, G.S., Ingram, J.C., Appelhans, A.D., Delmore, J.E., and Pesic, B., *Anal. Chem.,* 67, 1987, 1995.
56. Mc Dougall, G.J., Hancock, R.D., Nicol, M.J., Wellington, D.L., and Copperthwaite, R.G., *J. South African Inst. Min. Metall.* Sept. 1980, p. 344.
57. Cook, R., Crathorne, E.A., Monhemius, A.J., and Perry, D.L., *Hydrometallurgy,* 22, 171, 1989.
58. Cook, R., Crathorne, E.A., Monhemius, A.J., and Perry D.L., *Hydrometallurgy,* 25, 394, 1990.
59. Klauber, C. and Verson, C.F., *Hydrometallurgy,* 25, 387, 1990.
60. Klauber, C., *Langmuir,* 7, 2153, 1991.
61. Cashion, J.D., Cookson, D.J., Brown, L.J., and Howard, D.G., in *Industrial Applications of the Mossbauer Effect,* G.J. Long and J.G. Stevens, Eds., Plenum Press, New York, 1986, p. 595.
62. Cashion, J.D., McGrath, A.C., Volz., P., and Hall, J.S., *Trans. Inst. Min. Metall.,* 97C, 129, 1988.
63. Groszek, A.J., Partyka, S., and Cot, D., *Carbon,* 29, 821, 1991.
64. Ibrado, A.S. and Fuerstenau, D.W., *Hydrometallurgy,* 30, 243, 1992.
65. Ibrado, A.S. and Fuerstenau, D.W., *Miner. Eng.,* 8, 441, 1995.
66. Papirer, E., Polania-Leon, A., Donnet, J.B., and Montaganon, P., *Carbon,* 33, 1331, 1995.
67. Jones, W.G., Klauber, C., and Linge, H.G., 19th Bienn. Conf. on Carbon, 1989, *Ext. Abst.,* p. 38.
68. Kongolo, K., Kinabo, C., and Bahr, A., *Hydrometallurgy,* 44, 191, 1997.
69. Lagerge, S., Zajac, J., Partyka, S., and Groszek, A.J., *Langmuir,* 15, 4803, 1999.
70. Lagerge, S., Zajac, J., Partyka, S., Groszek, A.J., and Chesneau, M., *Langmuir,* 13, 4683, 1997.
71. Radovic, L.R., in *Chemistry and Physics of Carbon,* Vol. 27 L.R. Radovic, Ed., Marcel Dekker, New York, 2001.
72. Jia, Y.F., Steele, C., Hayward, J., and Thomas, K., *Carbon,* 36, 1299, 1998.
73. Arriagada, R. and Garcia, R., *Hydrometallurgy,* 46,171, 1997
74. Goyal, M., Rattan, V.K., and Bansal, R.C., *Indian J. Chem. Technol.,* 6, 305, 1999.
75. Gross, J. and Scott, J.W., U.S. Bureau of Mines, Washington, D.C. Tech., Paper No. 378, 1972.
76. Feldtmann, W.R., *Trans. Inst. Min. Metall.,* 24, 329, 1914–1915.

77. Heinen, J.J., Petterson, D.G., and Lindstrom, R.E., in *World Mining and Metallurgy Technology,* Vol. 1 Weiss, Ed. AIME, New York, 1976, p. 551.
78. Martin, J.P., Davidson, R.J., Duncanson, E., and Nkosi, N., *Anglo-American Res. Lab. Report No. 6,* Johannesberg, South Africa, 1970.
79. Tsuchida, N., Ph.D. dissertation, Murdoch Univ., 1984.
80. Smisek, M. and Cerny, S., in *Active Carbon,* Elsevier Publ. Co., Amsterdam, 1970, p. 162.
81. Pamale, C.S., Connell, W.L.O., and Basdekis, H.S., *Chem. Engg.,* 86, 58, 1979.
82. Bansal, R.C., Donnet, J.B., and Stockli, F., *Active Carbon,* Marcel Dekker, New York, 1988.
83. Donnet, J.B., Bansal, R.C., and Wong, M.J., *Carbon Black,* J.B. Donnet, R.C. Bansal, and Wang, M.J., eds. Marcel Dekker, New York, 1993.
84. Davini, P., *Fuel,* 68, 145, 1989.
85. Menendez, J.A., Phillips, J., Xia, B., and Radovic, L.R., *Langmuir,* 12, 4404, 1996.
86. Hutchins, R.A., *Chem. Engg.,* 87, 101, 1980.
87. Moreno-Castilla, C.F., Carrasco-Mastin, E., Utrera-Hidalgo, E., and Rivera-Ultrilla, J., *Langmuir,* 9, 1378, 1993.
88. Gangwal, S.K., Howe, G.B., Spivey, J.J., Silverston, P.L., Nudgins, R.R., and Metzinger, J.G., *Environmental Progress,* 12, 128, 1993.
89. Haure, P.M., Hudgins, R.R., and Silverston, P.L., *AIChE J.,* 35, 1437,1989.
90. Derbyshire, F., Jagtoyen, M., Andrews, R., Rao, A., Martin-Gullon, I., and Grulke, E.A., in *Chemistry and Physics of Carbon,* L.R. Radovic, Ed., Marcel Dekker, New York, 2001, Vol. 27.
91. Fei, Y., Sun, Y.N., Givens, E., and Derbyshire, F., ACS FCD, 40, 1051, 1995.
92. Andrews, R., Raymundo-Pincro, E., and Derbyshire, F., *Proc. Carbon 98,* Strasbourg, France, 1998, p. 275–76.
93. Hartman, M. and Coughlin, R.W., *Chem. Engg. Sci.,* 27, 867, 1972.
94. Juntgen, H., *Carbon,* 15, 273, 1977.
95. Ando, J., *GasJapan,* 1973, p. 119.
96. Knaublasch, K., Schwarte, J., Grochowski, H., and Juntgen H., *VBB-Tagyung Kraftwerk and Umwelt,* 1975.
97. Joly, J.P., Perrad, A., and Martin C., Carbon '02, Intern. Conf. on Carbon, Beijing, 2002, Paper 37.2.
98. Martin C., Joly, J.P., and Perrad, A., Carbon 01 Intern. Conf. on Carbon, Lexington, KY, 2001, *Ext. Abst.* P1-09.
99. Tada, K., Enjoji, T., Yoon, S., Mochida, I., Yasutake, A., and Yoshikawa, M., Carbon '02, Intern. Conf. on Carbon Beijing, 2002, Paper 4.3.
100. Liu, Q., Li, C., and Li, F., *Carbon,* 41, 2217, 2003.
101. Liu, Q., Guan, J.S., Li, J., and Li, C., *Carbon,* 41, 2225, 2003.
102. Kienle Von, H. and Bader, E., Actikohle and ihr Industrielle Anwendung, *Ferdinand Enke Vorlag,* Stuttgart, 1980.
103. Keltsev, N.V., *Podstawi Techniki Adsorpcyjnog (Principles of Adsorption Technology),* WNT Warsaw, 1980.
104. Puri, B.R., Kumar, B., and Kalra, C., *Indian J. Chem.,* 9, 970, 1971.
105. Cariaso, O.C. and Walker, P.L., Jr., *Carbon,* 13, 233, 1975.
106. Klein, J. and Henning K.D., *Fuel,* 63, 1064, 1984.
107. Steijns, M., Derks, F., Verlonp, A., and Mars, P.J., *Catalysis,* 42, 87, 1976.
108. Hedden, K., Henber, L., and Rao, B.R., VDI-Bericht No. 253, S. 37/42, Dusseldorf, VDI-Verlag, 1976.
109. Kim, J.S., Jung, S.W., Kim, M.C., and Kim, M.S., Carbon '02, Intern. Conf. on Carbon, Beijing, 2002, Paper P795D 111.

110. Svetlana, S., Kartel, N.T., and Tsyba, N.N., Carbon '02, Intern. Conf. on Carbon, Beijung, 2002, Paper PI 68 D. 063.
111. Tarkovskaya, I.A., Satvitskaya, S.S., Lukyanchuk, V.M., and Tarkovskaya, G.V., *Khim Tekhnol. Vody* N15, 578, 1993.
112. Lehmer, O., *Atomwirtsch*, 5, 356, 1960.
113. Kohrt, U., *Chem. Ing. Techn.*, 33, 135, 1961.
114. Billinge, B.H.M., Docherty, J.B., and Bervan, M.J., *Carbon*, 22, 83, 1984.
115. Baker, J.A. and Poziomek, E.J., *Carbon*, 12, 45, 1974.
116. Collins, D.A., Taylor, L.R., and Taylor, R., In *Proc. Ninth AEC Air Cleaning Conf.*, CONF. 660904, Vol. 1, 1957, p. 59.
117. Lazarev, B.G. and Fedorova, M.F., *Zh. Techn. Phys.*, 29, 862, 1950.
118. Donnet, J.B and Bansal, R.C., in *Carbon Fibers*, Marcel Dekker, New York, 19.
119. Arwell, N., *Acta Med. Scand.*, 180, 593, 1952.
120. Yatzidis, H., *Lancet*, 2, 216, 1965.
121. Yatzidis H., *Nephron*, 1, 310, 1964.
122. Yatzidis, H., *Proc. Europ. Dial. Trans. Ass.*, 1 83, 1964.
123. Dunca, G. and Kolff, W.J., *Trans. Amer. Soc. Artif. Intern. Organs.*, 11, 179, 1965.
124. Hagstam, K.E., Larsson, L.E., and Thysell, H., *Acta Med. Scand.*, 180, 593, 1966.
125. Andrade, J.D., Kunitoma, K., and Van Wagenen, R., *Trans. Amer. Soc. Artif. Intern. Organs.*, 17, 222, 1971.
126. Lavine, S.N. and La Course, W.C., *J. Biomed. Mater. Res.*, 1, 275, 1967.
127. Sparks, R.E. et al., *Trans. Amer. Soc. Artif. Intern. Organs.*, 15, 333, 1969.
128. Chang, T.M.S., *Artificial Cells*, Thomas, Springfield, IL, 1972.
129. Nikolaev, V.G., Strazhesko, D.N., and Strelko, V.V., *Adsorbtsiya Adsorbenty*, 4, 24, 1976.
130. Bruskina, T.N., *Adsorbtsiya Adsorbenty*, 9, 77, 1981.
131. Walker, J.M., Benti, E., Van Wagenon, R.D., *Kidney Inter.*, Suppl. 7, 320, 1976.
132. Strelko, V.V., Galinskaya, V.I., and Davydov, V.I., *Adsorbtsiya Adsorbenty*, 4, 29, 1976.
133. Nikolaev, V.G. and Strelko, V.V., *Gemosorbtsiya na Aktivnykh uglakh (Chemisorption on Active Carbon)*, Naukova Dumka, Kiev, 1979.
134. Kartel. N.T., Strelko, V.V., Puziy, A.M., Mikhalovsky, S.V., and Nikalaev, V.G., Carbon '02, Intern. Conf. on Carbon, Beijung, 2002, Paper 21.4.
135. Kartel, N.T., Puziy, A.M., Strelko, V.V., in *Studies in Surface Science and Calolysis*, 62, Characterisation of Porous Solids II, Elsevier Science. Pub., Amsterdam, 1991, p. 439.
136. Kartel, N.T., *J. Thermal Analysis and Calorimetry*, 62, 499, 2000.
137. Kartel. N.T., in *Eurocarbon 2000*, 1st World Conf. on Carbon, Berlin, Vol. II, p. 603, 2000.
138. Strelko, V.V. and Kartel, N.T., in *Fundamentals of Curative Means Development*, Kharkiv, Osnova, 1998 p. 490.
139. Strelko, V.V. and Kartel. N.T., Carbon 01 Intern. Conf. on Carbon, Lexington, KY, 2001, CD ROM, ISBN No. 9674971-2.4.
140. Strelko, V.V. and Kartel, N.T., Ukraine Patent No. 3396, 1994.
141. Strelko, V.V. and Kartel, N.T., Russian Patent No. 1836138, 1993.
142. Savrikoc, E.V. Frumin, L.E., Odaryuk, T.S., Tsarkov, P.V., Sadovnichii, V.A., and Eropkin, P.V., *Biomed Eng.*, 21, 140, 1987.
143. Kliukmaun, H. and Behm, E., Biomater Artif. Cells Artif. Organs, VII Intern. Symp. on Hemoperfusion, Kiev, USSR, 15, 41, 1987.
144. McPhillips, D.M., Aemer, T.A., and Owen, D. R., *J. Biomed. Mater. Res.*, 17, 983, 1983.

145. Leuhmaun, D.A., *Med. Instrum.*, 20, 74, 1986.
146. Belyakov, N.A., Belkin, A.L., Vladyka, A.S., Gurevich, V.Y., Nikolaev, V.G., Oxmak, A.R., Petash, V.V., Serzhanter, I.P., and Shchegrinov, N.N. *Eng.*, 18, 130, 1984.
147. Mori, Y., Nagaoka, S., Tanzowa, H., Kikuchi, Y., Yamada, Y., Hagiwara, M., and Idezuki, Y., *J. Biomed. Mater. Res.*, 16, 17, 1982.
148. Halloway, C.J., Hartsick, K., Brunner, G., and Gaeger, I., Biomater. *Med. Dev. Artif. Organ.*, 9, 167, 1981.
149. Jacob, E.J., U.S. Patent 4606, 354, Aug. 19, 1986.
150. Krautz, J.C. and Carr, C.J., in *Pharmacology Principles of Medical Practice,* Williams and Wilkins, Baltimore, 1958.
151. Kuschinasky, G., *Taschenbuch der Modernan Zrzneibehandlune,* Thieme Verlag, 1956.
152. Chen, S., Liu, J., and Zeng, H., Carbon '02, Intern. Conf. on Carbon, Beijung, 2002, Paper 21.3.
153. Chen, S., Liu, J., and Zeng, H., *New Carbon Materials,* 17, 26, 2002.
154. Yamamoto, O., Carbon '02, Intern. Conf. on Carbon, Beijung, 2002, Paper 21.3.
155. Park, S.J., Jang, J.S., and Jung, C.H., Carbon '02, Intern. Conf. on Carbon, Beijung, 2002, Paper PI88 DO94.
156. Ohkura, T., Veda, H., Namfer, N., and Nagai, T., *Chem. Pharm. Bull.*, 28, 612, 1980.
157. Nakano, N.I., Shimamori, Y., Umeshashi, M., and Nakano, M., *Chem. Pharm. Bull.*, 32, 699, 1984.
158. Lathe, D.H., Mackay, D., and Marsh, H., *Carbon*, 23, 343, 1985.
159. Nagami, H., Nagai, T., Fukuoka, E., and Uchida, H., *Chem. Pharm. Bull* ., 16, 2248, 1968.
160. Phillipson, C. and Schmiedel, R., *Pharmazia*, 35, 224, 1980.
161. Nagami, H., Nagai, T., and Uchida, H., *Chem. Pharm. Bull.*, 17, 176, 1969.
162. Miyawaki, O. and Wingard, L.B., Jr., *Biotech. Bioeng.*, 26, 1364, 1984.
163. Veda, H., Nambu, N., and Nagai, T., *Chem. Pharm. Bull.*, 28., 3426, 1980.
164. Noda, J., Sakamoto, H., and Nagai, T., *Chem. Pharm. Bull.*, 23, 4451, 1975.
165. Phillipson, C. and Schmeidel, R., *Pharmazie*, 38, 181, 1983.
166. Cracknell, R.F., Gordon, P., and Gubbins, K.E., *J. Phys. Chem.*, 97, 494, 1992.
167. Derbyshire, F., Jagtoyen, M., Andrews, R., Rao, A., Martin-Gullon, I., and Grulke, E.A., in *Chemistry and Physics of Carbon,* L.R. Radovic, Ed., Marcel Dekker, New York, 2001, Vol. 27.
168. Gregg, S. J. and Sing, K.S.W., in *Adsorption, Surface Area, and Porosity,* 2nd edit., Academic Press, New York, 1982.
169. Yang, R.T. and Saunders, J.T., *Fuel,* 64, 616, 1985.
170. Mavor, M.J., Owen, L.B., and Pratt, T.J., SPE 20728, Paper presented at the SPE 65th Ann. Techn. Conf. and Exhibition, New Orleans LA, Sept. 23–26, 1990, p. 157.
171. Harpalani, S. and Pariti, U.M., Proc. 1993 Intern. Coal Bed Methane Symposium, Univ. of Albana–Tuscaloosa, May 17–21, 1993, p. 151.
172. Ruppel, T.C., Grein, C.T., and Beinstock, B., *Fuel*, 53, 152, 1974.
173. Clarkson, C.R., Bustin, R.M., and Levy, J.H., *Carbon*, 35, 1689, 1997.
174. Xing, W. and Yan, Z.F., Carbon '02, Intern Carbon Conf., Beijung 2002, Paper PI 94D 106.
175. Quinn, D.F. and Mac Donald, J.A., *Carbon*, 30, 1097, 1992.
176. Parkyns, N.D. and Quinn, D.F., in *Porosity in Carbons,* J.W. Patrick Edit Adward Arnold, London, 1995, p. 291
177. Amamkwah, K.A.G., Noh, J.S., and Schwarz, J.A., *Hydrogen Energy*, 14, 437, 1989.
178. Dillon, A.C., Jones, K.M., Bekkedahl, T.A., Kiang, C.H., Bethane, D.S., and Heben, M. J., *Nature*, 386, 377, 1997.

6 Activated Carbon Adsorption and Environment: Removal of Inorganics from Water

Increasing industrialization, uncontrolled use, and exploitation of natural resources during the last several decades have caused major devastation and degradation of the Earth. Human activities have put a considerable pressure on the availability of basic human necessities such as clean water and clean air. Waste water from various industries, municipal corporations, urban and rural runoff, coupled with the increasing use of various chemicals, surfactants, fertilizers, pesticides, and herbicides in agriculture, and the decomposition of vegetable and animal matter discharge varying amounts of these and other chemicals into ground and surface water, making it unfit for human and animal consumption. With the rapid development of chemical, polymer, processing, and petroleum processing industries, there is a rapid increase in the amount and the variety of chemicals that are thrown into waters. This increase is attaining hazardous conditions, especially in big cities where the population is large, the demand for water is very high, and chemical industries and other industries are developing at a faster rate. In addition, the chlorination of waste water also introduces several harmful chemicals in drinking water. Some of these organic and inorganic compounds, when present in water, are toxic, carcinogenic, and mutagenic, and cause several ailments in humans.[1-4]

More than 800 specific organic and inorganic chemicals have been identified in various drinking waters. The U.S. Congress in 1974 passed the Safe Drinking Water Act to ensure that the public water supply system meets the minimum protection standards for public health. The major pollutants identified in water were the trihalomethanes (THM) and synthetic organic chemicals, which are carcinogenic. In 1976 the National Cancer Institute of America published a report that linked the presence of chloroform in water to cancer in rats and mice. Although the effect of chloroform on human beings was not well documented, it was suspected that the presence of chloroform might cause cancer.

Several biological and chemical methods such as filtration, coagulation, oxidation, and ion-exchange have been used for the treatment of waste water, but the continuing increase in the variety and amount of hazardous chemicals present in our lakes rivers and sometimes ground water reserves make these conventional methods inefficient and sometimes even ineffective. Consequently, the development of new and more effective technologies becomes essential. Many laboratory tests and field operations carried out in several countries have shown that the use of activated carbon adsorption is perhaps the best broad-spectrum technology at the present moment for the removal of both organics and inorganics from water.[5,6]

Activated carbon can be used in its various forms: the powdered form, the granulated form, and now the fibrous form. Powdered activated carbons generally have a finer particle size of about 44 μm, which allows faster adsorption,[7] but they are difficult to handle when used in fixed beds. Furthermore, they cause a high pressure drop, due to blockage in the fixed beds. They also have a problem in the regeneration of the bed. The granulated carbons generally have granules of 0.6 to 4.0 mm in size and are hard, abrasion resistant, and relatively dense to withstand operating conditions. Although granulated carbons are expensive compared to powdered carbons, they do not cause a high hydrodynamic problem and can be more easily regenerated and used again.[8] The granulated activated carbons can be formed into a module that serves as the adsorber bed, which can be removed after saturation and regenerated by heat treatment in steam at temperatures of 200 to 800°C to desorb organics and other contaminants.[9] The fibrous activated carbons are expensive materials to be used for waste water treatment, but they have the advantage of capability to be molded easily into the shape of the adsorption system and producing low hydrodynamic resistance.

The adsorptive removal of organics and inorganics from waste water by activated carbons depends upon the surface area, the pore volume, and the pore-size distribution in carbons. Although adsorption capacity of an activated carbon depends on these parameters, it is strongly influenced by the surface chemistry of the activated carbons. Most of the as-received activated carbons are hydrophobic and are associated with small amounts of neutral carbon-oxygen surface groups. Such activated carbons are more suitable for the adsorption of neutral or nonpolar organic compounds and they show little affinity for polar and ionic pollutants. An example is the adsorption of phenol for which the surface chemistry[10-14] of a carbon is more important than the surface area.

Generally speaking, all the activated carbons are associated with a certain amount of chemisorbed oxygen. This amount can be enhanced by oxidation of the carbon surface with oxidizing agents such as nitric acid, hydrogen peroxide, ammonium persulfate, oxygen, or ozone. The chemisorbed oxygen on the carbon surface is present in the form of carbon-oxygen surface groups. Two types of carbon-oxygen surface groups have been identified: one desorbed as CO_2 on heat treatment in vacuum or in inert gas, and the other which is desorbed as CO. The former groups are thermally less stable and are desorbed in the temperature range 350 to 750°C. These are acidic in character and make the surface hydrophilic and polar and have been postulated as carboxylic, lactonic, and phenolic groups. The other groups that are desorbed as CO are neutral and make the surface hydrophobic and nonpolar. These groups have been postulated as quinone groups. The polar acidic surface groups enhance the adsorption of polar chemical compounds or ionic species.

Activated carbon treatment of water can be divided into three categories of water. The first one is the water for drinking and other human consumptions purposes. Under this category falls the treatment of municipal water supply to an acceptable quality. This involves large-scale treatment of water from rivers, lakes, and reservoirs for distribution to the community. The function of the activated carbon in this treatment generally involves removal of species that adversely affect the taste and odor and contaminants that are hazardous to health. The second category of water that needs to be given activated carbon treatment is the water required for different industrial processes such as in heat exchangers, steam generators and cooling towers. The treatment of such water involves the removal of compounds which cause corrosion. The third and the most important area in the treatment of waste water is the effluent from different process industries, because this water is either recycled in the plant or discharged into the ground or watershed. Before its discharge, this water should be free of toxic chemical compounds, which can pollute ground water. Thus, activated carbon plays an important role in the removal of organic and inorganic compounds from waste water.

6.1 ACTIVATED CARBON ADSORPTION OF INORGANICS FROM AQUEOUS PHASE (GENERAL)

The application of activated carbons in the adsorptive removal of inorganics from water has been the subject matter of numerous investigations. The adsorption of inorganic compounds has two aspects: the removal of heavy metal pollutants from waste water and the recovery of certain metals such as gold and silver. Furthermore, the loading of activated carbons with certain metals produces metal-impregnated activated carbons, which are very useful as catalysts for the decomposition of gases and vapors in industrial effluent streams as well as for protection from hazardous war gases such as HCN, $CNCl_2$, $COCl_2$, or nerve gases.

The ability of active carbons to adsorb inorganic species, specifically the metal cations, may be attributed to their ion exchange properties due to the presence of certain heteroatoms as oxygen, nitrogen, halogens, and sulfur on the activated carbon surface. These heteroatoms are bonded to the carbon surface in the form of carbon-oxygen, carbon-nitrogen, carbon-halogen, and carbon-sulfur surface functional groups. The formation and properties of these surface groups have been discussed exhaustively in several books and review articles.[15-22] Although all of these surface groups influence the adsorption of inorganics from aqueous phase, the carbon-oxygen surface groups are the most influencing and important. The amount of these carbon-oxygen surface groups can be enhanced significantly by surface oxidation of carbons, the more important oxidizing agents being, nitric acid, ammonium persulfate, hydrogen peroxide, and sodium hypochlorite in aqueous solutions, and oxygen or air at temperatures of 300 to 400°C in the gaseous phase. These oxidative treatments result in the formation of two types of carbon-oxygen surface groups, one of which is acidic and the other is neutral in character. The acidic surface groups are polar and enhance the ion-exchange properties of the carbons, thereby increasing the adsorption of cations. The oxidized activated carbons are gaining importance for the removal of trace amounts of heavy metals from waste water, for treatment of drinking water, recovery of valuable metals such as gold and silver, and for purification of inorganic salt solutions. For example, Linstedt et al.,[23] Cheremisinoff and Habib,[24] Argaman and Weddle,[25] and Netzer et al.[26] found that activated carbons were potential adsorbents for the removal of heavy metals from water. Faust and

Aly[27] and Huang[28] discussed the removal of several cations such as barium, iron, vanadium, and selenium, and several chlorine, fluorine, and cyanide compounds. More recently, activated carbons have also been used for the adsorptive removal of sulfides,[29] nitrates,[30] chlorides, chlorites, and chlorates,[31-33] cyanides,[34,35] boric acid, borax,[36] and borates,[37] as well as certain metals such as lithium,[38,39] cerium,[40] iron,[41,42] strontium,[43] and dysposium[44] from the aqueous phase.

Saito[45] used activated carbon and sulfonated coal to remove copper, cadmium, and iron from waste water. Kaya and Akyol[46] studied the removal of Co, Cu, and Cd from aqueous solutions as a function of concentration, temperature, adsorbent adsorbate ratio, and contact time using bentonite activated with H_2SO_4. Naganuma et al.[47] examined the adsorption of Cu and Zn on topical peat soils as a function of pH and Cu and Zn concentration. Rivera-Utrilla and Ferro-Garcia[48] studied the adsorption of Na^+, Cs^+, Ag^+, Sr^{2+}, and Co^{2+} on activated carbons associated with carbon-oxygen and carbon-nitrogen surface groups. Na^+ and Cs^+ were adsorbed strongly by carbons containing acidic surface groups, while Ag^+, Sr^{2+} and Co^{2+} were adsorbed by carbons associated with acidic as well as basic surface groups. Corapcioglu and Huang[49] found that the carbon type, the solution pH, and the surface loading of the carbon influenced the amount of the metal removed. These workers showed that the adsorption involved hydrogen bonding.

The adsorption of the Cu^{2+}, Zn^{2+}, and Pb^{2+} ions from aqueous solutions on oxidized anthracite at 20°C was studied by Petrov et al.[50] The metal ion uptake increased with increasing pH of the solution. The adsorption of individual metal ions decreased when the other metal ions were present in equal concentration. But the total metal uptake was considerably higher than that of the single ion species. The adsorption has been explained on the basis of the chemical and the porous structure of the carbon surface. Usmani et al.[51] observed that the adsorption of metal ions by activated carbons prepared from Eucalyptus comaldulensis was considerably influenced by the pH of the aqueous solution. Although the adsorption of Cu and Zn cations was insignificant below pH = 4.0, that of Cr was maximum at pH = 2.0. Andreev et al.[52] studied the adsorption of Cu, Zn, Pb, Fe, Cd, Ni, Co, Mg, Ag, Cr, and Mn on carbon fibers and on several other fibrous ion exchangers, while Mathur et al.[53] investigated the removal of many of these metals from sewer water by adsorption on flyash and blast furnace slag. About 99% removal of metal ions from industrial waste water was achieved by South African bituminous coal treated with bases.[54] Abdul-Shafy et al.[55] investigated the adsorption of Cd and Pb from aqueous solutions on granulated and powdered activated carbon, in batch as well as continuous systems, and observed that the adsorption followed Freundlich isotherm equation. Tewari et al.[56] studied the removal of Cr(VI), Cu, and Ni ions from dilute solutions in the concentration range 5 to 50 mg/L by adsorption on activated carbon in the pH range 5.5 to 8.0. The removal of Cr(VI) exhibited a peak at pH 5.5 and Cu at pH = 8.0, the adsorption of nickel ions was appreciable both in acidic and basic pH ranges, although the rate of adsorption of Ni was faster in the alkaline medium. Jankowska et al.[57] observed that oxidation of the activated carbons with air at 420°C enhanced its adsorption capacity for chromium and copper.

The adsorption of Cr(VI) and Cr(III) from aqueous solutions on different types of activated carbons,[58-66] on bituminous coal,[67] and on flyash was also studied.[68]

These workers observed that each carbon adsorbed appreciable amounts of both Cr(VI) and Cr(III). But the adsorption was a function of pH of the solution. Both adsorption and reduction of Cr(VI) occurred when the pH was in the acidic range. The adsorption capacity of different carbons varied between 30 and 99%.

Huang and Ostovic[69] investigated the adsorption of Cd(II) ions on different activated carbons as a function of pH of the solution and nature of the carbon surface. The amount of Cd adsorbed increased with an increase in pH up to 8.0 and with the addition of chelating agents. Basic carbons adsorbed more cadmium than the acidic carbons. Dobrowolski et al.[70] while studying the adsorption of Cd(II) ions on oxidized and subsequently degassed carbon samples from aqueous solutions, observed that the extent of adsorption depended upon the presence of carbon-oxygen surface groups on the carbon surface. The adsorption of Cd from aqueous solutions on weathered coal was examined by Liu et al.[71] They derived mathematical equations and related the adsorption rate with the adsorption temperature.

The removal of Ni(II) ions from industrial waste water has been studied, using different adsorbents such as activated carbons,[56,57] flyash,[53,72] and blast furnace slag.[53,73] In the case of removal by flyash, the influence of initial Ni concentration, residence time, temperature, and pH was studied.[72] Increase in temperature and pH, and a decrease in initial nickel concentration, favored the removal of Ni(II) ions. The maximum adsorption of nickel was observed at pH = 7. Blast furnace slag was found to remove about 30% of Ni(II) ions from aqueous solutions in the concentration range 10^{-4} to 10^{-3} mol/L.[53,73] Shakir and coworkers[74,75] investigated the removal of cobalt from radioactive process waste water using activated charcoal as the adsorbent and gelatin as a collector. These workers observed that the optimum pH range was 7.5 to 10.0 for the removal, and about 90% removal could be achieved under optimum conditions. The presence of strongly ionized surfactants decreased the removal while the presence of weakly ionized surfactants increased the removal of cobalt.

Moore[76] studied the removal of Cu(II) ions from synthetic sea water by adsorption on different activated carbons and observed that the adsorption was strongly influenced by pH and the nature of the carbon. The adsorption of Cu(II) ions increased with increase in the ionic concentration of the solution on addition of NaCl.[77] This increase in adsorption was attributed to the formation and adsorption of negatively charged Cu(II) chloride complexes. The effect of parameters such as contact time, pH and the concentration of the solution in the range 10 to 1000 ppm on the adsorption of Cu(II) ions has also been reported.[78] Low et al.[79] found that the dye modification of coconut husk enhanced its ability to absorb Cu.

The adsorption of Zn^{2+} ions and its complexes was reported by Mu and Young.[80] The extent of adsorption mainly depended on the solution pH, added inorganic ions, and on the presence of oxygen-containing groups on the carbon surface. The adsorption increased with increase in the surface acidity of the carbon surface. Bencheikh[81] examined the removal of Zn(II) ions from waste water by adsorption on a peat sample and obtained a removal efficiency of 93 to 96%. The adsorption obeyed Langmuir kinetics.

The adsorption of vanadium (IV) and vanadium (VI) was studied by Kunz et al.[82] and by Koustyuchenko et al.[83] using activated carbon adsorbents. Kunz used active carbon Filtrasorb-400 for the removal of vanadium from sodium metavanadate solution and found that more than 90% of the vanadium could be removed from a solution containing 50 mg/L of the metal. The efficiency of the carbon was enhanced when larger quantities of the carbon were used. More thorough investigation on vanadium removal by activated carbons was carried out by Koustyuchenko et al.[83] The adsorption was strongly influenced by the pH of the solution. The adsorption pattern was very similar for different carbons from dilute solutions and from brine, and depended little on whether vanadium was in the tetravalent or pentavalent state. For all carbons the maximum adsorption of vanadium was observed at pH 2.5 to 3.0. However, when the carbon was oxidized with air, H_2O_2, or HNO_3, it showed a much larger adsorption capacity. The removal of Sr(II) ions from electrolytic aqueous solutions by charcoal in the presence of several cations as a function of temperature, contact time, and the concentration of the adsorbate was examined by Riaz et al.[84,85] The adsorption data obeyed Langmuir and Frendlich isotherms. The Sr(II) was preferentially removed from the solution containing cations such as rhodium, rubidium and cesium. The thermodynamic parameters ΔH_o and ΔS_o were calculated. The isosteric heats of adsorption calculated using Clasius Clapeyron equation showed that the carbon surface was heterogeneous.

The adsorption of tungsten W(VI) on activated carbons from its solutions in NaCl and HCl as a function of pH was studied by Gurwagen and Pienarr.[86] The adsorption was quite large at pH, less than 4.5 when a loading of 2.8 millimoles of W(VI) per gram of activated carbon was obtained. The uptake of tungsten involved chemisorption on sites where surface oxygen was present on the activated carbon. Palladium (Pd) is adsorbed as palladium complex $[PdCl_3(OH)]^{2-}$ and $[Pd (NH_3)n]^{2+}$ on activated carbons,[87] when adsorption is carried out from aqueous solutions of palladium in HCl, H_2SO_4, NH_4OH, and NaOH. When oxidized carbons are used as adsorbents, there is no reduction of the $[Pd(NH_3)n]^{2+}$ complex. Qadeer and Hanif[88,89] investigated the adsorption of uranium on activated charcoal in the presence of different cations such as Cd^{2+}, Zn^{2+}, Cr^{3+}, y^{3+}, and Se^{3+}. The cations reduced the adsorption of uranium, depending upon their ionic potentials. The cations with larger values of ionic potential reduced the adsorption more than the cations with smaller ionic potential (Z/r value). The studies on the influence of temperature and concentration on adsorption showed it to be a two-step process. The rate of the first step was faster than that of the second step. The adsorption of La(III) ions on activated carbon made from bamboo showed that the adsorption was influenced by pH of the solution and was related to the amount of carbon-oxygen groups present on the carbon surface.[90]

The reduction-adsorption capacity for Pt (IV) on activated carbon fibers obtained from viscose rayon, sisal, and pitch was studied by Fu et al.[91] by chemical analysis of the reaction solutions and XPS studies of the deposited platinum. The reduction-adsorption capacity increased with the time of activation of the carbon fiber and was the highest in the case of viscose-based carbon fibers. Viscose-based carbon fibers were capable of reducing appreciable amounts of Pt (IV), even without activation. The state of the platinum product depended on the reaction conditions. Under acidic reaction conditions, the deposited platinum was mainly in the elemental state. However, under the alkaline conditions, the liquid-phase hydrolysis of Pt (IV) becomes significant, and most of the platinum deposited on the activated carbon fiber surface was present as PtO, which was not further reduced to the metallic state at room temperature. But heat treatment of a PtO loaded-carbon fiber sample at 400°C or above in vacuum changed the oxide into metallic platinum.

Activated carbons have also been used for the adsorptive removal of mercury (Hg) from aqueous phase.[92-98] The adsorption was found to be a function of pH,[92-95] the adsorption increasing with a decrease in pH of the solution. The amount of Hg removal increased by a factor of two, when the pH was decreased from basic to acidic range. The removal efficiency of a carbon for Hg(II) ions also depended on the nature of the carbon surface as well as on the nature of the activation treatment given to the carbon.[95] Activated carbons prepared from wood, coconut shell, and coal by steam activation showed a high adsorption capacity for Hg(II) ions below pH 5 in HCl medium, but the removal efficiency was low at pH > 5. However, activated carbons prepared from wood by chemical activation showed a high adsorptive ability, even at pH > 5. The removal of Hg(II) by activated carbons is also enhanced by the addition of certain chelating agents and by sulfurization of the carbon surface.[97] The adsorption of methyl mercuric iodide from aqueous solutions on Filtrasorb 200 in the concentration range 10 to 320 ppm at 25°C was studied by Ammons et al.[98] The adsorption isotherm was linear up to a concentration of 20 ppm and nonlinear, and reversible above 20 ppm. Neither the Langmuir nor the Freundlich isotherm equations fitted the adsorption data. The adsorption of Hg(II) from aqueous solutions on a sample of activated carbon fiber in a flowing system showed that the dispersed ferric oxide loaded on the carbon fiber surface enhanced the adsorption of Hg(II).[99]

The removal of uranium from aqueous phase using activated carbons has been studied by several workers.[100-103] Saleem et al.[100] while using a commercial activated carbon observed that the removal was maximum at pH between 4 and 5 because of the availability of free uranium ions. The adsorptive removal decreased on the addition of anions such as nitrate or oxalate ions. Abbasi and Streat[101] found that the oxidation of a carbon with nitric acid enhanced the adsorption capacity of uranium in the acidic solutions. However, none of these workers invoked the influence of carbon-oxygen surface groups on the adsorption of uranium. Dun et al.[104] and El-Bayoumy and El-Kolaly[105] studied the removal of molybdenum from aqueous solutions using activated carbons and observed that the adsorption depended largely on the pH of the solution. The uptake of molybdenum decreased by several orders of magnitude when the pH of the solution was increased from acidic to alkaline range. This change in adsorption capacity has been attributed to the carbon surface charge, which may produce electrostatic attractive interactions.

Thus, it is apparent that activated carbons are unique adsorbents that have great potential for the removal of metal ions from waste water and from potable drinking waters. This is due to the fact that activated carbons have a high surface area, a highly microporous character, and a high degree of chemical reactivity of their surface. Furthermore, the surface of an activated carbon can be modified by the formation of different types of carbon-oxygen, carbon-sulfur, carbon-nitrogen surface groups, by impregnation or by degassing to make them suitable for the removal of a particular group of pollutants. The surface of activated carbons can also be made acidic or basic in character by suitable treatments. The importance of activated carbons for the treatment of waste water can be judged from the fact that several reviews have appeared on the subject. Several symposia have been held on the application of activated carbon for the removal of organics and inorganics from water, and special sessions at International Carbon Conferences are devoted to this aspect of activated carbon applications. It is interesting to note that the adsorbent properties of activated carbons have become so important that about 30% of the total papers presented at the recent International Conference on Carbon at Beijing, China were devoted to adsorption and surfaces.

We shall now discuss in detail the use of activate carbons in the removal of certain metal ions which are encountered in most waters. The parameters involved in their removal, the surface properties of activated carbons which influence their removal and the mechanisms involved will also be examined.

6.2 ACTIVATED CARBON ADSORPTION OF COPPER

Copper is present in the waste water of several industries.[106-111] It is also a micronutrient in agriculture and can, therefore, accumulate in surface waters. Although copper is an essential mineral for human beings, its excessive intake results in its accumulation in the liver and produces gastrointestinal problems. Consequently, it is essential that potable waters be given some treatment to remove copper and other heavy metals before domestic supply. Furthermore, there is a need to develop methods to recover copper metal from acid mine wastes[112] and acidic corrosion of pipes.[113] Several types of adsorbents, such as activated carbon, synthetic goethite,[114] coconut husk and its dye-coated reactive form,[79] and inorganic chemically-active adsorbent beds[112] have been used for the removal of copper from aqueous solutions. But activated carbons, especially oxidized activated carbon, possess some unique properties due mainly to the presence of oxygen containing surface groups on their surface. These surface groups make the carbon surface polar and enhance its ion-exchange capacity and make activate carbons potential adsorbents for the removal of metal ions from industrial and domestic waste waters.

Petrov et al.[50,115] studied the adsorptive removal of several metal ions such as Zn, Cd, Pb, and Cu from aqueous solutions on anthracite before and after thermal oxidation in air. The metal uptake increased with increasing pH of the solution. The uptake was only slight at a solution pH of 1, but it increased considerably in the pH range of 3 to 4. The oxidation of the carbon enhanced the adsorption of metal ions because of the creation of acidic surface groups on the oxidized anthracite surface. In a later study using different carbons associated with varying amounts of carbon-oxygen surface groups, these workers emphasized the importance of acidic surface groups on the carbon surface in determining the adsorption of Cu(II) and other metal ions. These workers also studied the influence of electrolytes in the solution on the adsorption and observed that the presence of electrolytes decreased the adsorption of the metal ions. Mangalardi et al.[116] and Andreeva et al.,[52] while studying the removal of copper ions from water using a powdered activated carbon at different pH values of the solution, observed that the removal of Cu(II) ions was considerable at pH values lower than 7 and that the effluent was left with only 0.1 mg/L of copper. They compared the adsorption capacity of activated carbon with Fuller's earth and bentonite and found that the latter adsorbents were effective only at pH greater than 8. Mathur and coworkers[53] observed that flyash from power plants was also quite effective in the removal of heavy metals, including copper from sewage waste water.

Tarkovskoya et al.[117] examined the sorptive and cation-exchange properties of several carbonaceous materials such as activated oxidized anthracite, semicoke, brown coal, and several modified coals, and observed that these materials were effective to varying degrees for the removal of metal cations. The adsorptive removal of copper and several other metals from aqueous solutions on peat, lignite, and activated carbons was investigated by Allen and coworkers.[118,119] These workers found a wide variation in adsorption on different adsorbents and attributed it to the difference in the heterogeneity of the surface in these materials. However, they did not suggest any role of carbon-oxygen surface groups, although surface heterogeneity of the surface in carbons results from these surface groups. Johns et al.[120] while studying the adsorption of Cu(II) ions on activated carbons prepared from a wide range of agricultural byproducts before and after oxidation with oxygen gas, observed that the carbon-oxygen surface groups played an important role in the adsorption of copper ions from aqueous solutions. However, the role of carbon surface chemistry on the adsorption of copper was brought out more clearly by Biniak et al.[121] They carried out the adsorption of Cu(II) ions from aqueous solutions on activated carbon after modifying its surface by degassing in vacuum at 1000 K, by oxidation with concentrated nitric acid and by treatment with ammonia at 1170 K.

The oxidative treatment with nitric acid created acidic groups, and the treatment with ammonia created basic surface groups. The degassing in vacuum desorbed the carbon-oxygen surface groups. The acidic and basic surface groups were determined by selective neutralization of bases proposed by Boehm.[122] These workers found no relationship between uptake of Cu(II) ions and the surface area of the carbons but the adsorption was strongly influenced by the presence of acidic and basic groups present on the carbon surface and the pH of the solution (Table 6.1). The adsorption

TABLE 6.1
Effect of pH and Surface Oxygen Groups on the Adsorption of Cu(II) Ions

| Sample[b] | pH[c] | Functional Group Concentration (meq/g)[a] | | | | | Total O/N (wt%) | pH[d] | Uptake (mmol/g) |
		–COOH	–COO–	>C–OH	>C=O	Basic			
D–H	10.1	0.00	0.01	0.12	0.09	0.42	0.6/0.2	4.97	0.43
D–O	3.08	0.72	0.38	0.56	0.39	0.13	10.8/0.6	2.45	0.25
D–N	10.4	0.00	0.04	0.10	0.21	0.62	0.4/1.9	5.05	0.32

[a] Determined using Boehm's duration method; see Seron, A., Benaddi, H., Beguin, F., Frackoviak, E., Bretelle, J.L., Thiry, M.C., Bendosz, T.J., Jagiello, J., and Schwaz, J.A., *Carbon,* 34, 481, 1996.
[b] Commercial Carbo-Tech activated carbon (AC) treated in vacuum at 1000 K (D–H) in conc. HNO_3 (D–O) or ammonia at 1170K (D–N).
[c] 1 g of AC in 100 cm^3 of 0.1 M Na_2-SO_4 (pH = 6.42).
[d] Initial pH of 0.05 $CuSO_4$.

Source: Adapted from Biniak, S., Pakula, M., Syzmanski, G., and Swiatkowski, A., *Langmuir,* 15, 6117, 1999. With permission.

of copper ions increases linearly with pH in the pH range 1 to 6 on the degassed sample (D-H sample) increases sharply between pH 1 and 3 in the case of the oxidized sample (D-O sample), and it increases gradually with increasing pH in the case of the ammonia treated sample (D-N sample). In acidic solutions where the pH was adjusted by adding sulfuric acid, the uptake of Cu(II) ions increased in carbon samples modified by oxidation (D-O) and ammonia treatment (D-N sample), probably as a result of the interaction of Cu(II) ions with the surface functional groups. In strongly acidic solutions, (pH = 1), copper ion adsorption was more for the ammonia-modified sample, whereas in a less acidic solution (pH = 3), it is the oxidized sample that shows higher adsorption. In the case of almost neutral solutions (pH adjusted by the addition of NaOH solution), on the other hand, the uptake of Cu(II) ions was the highest in the case of degassed activated carbon sample (Figure 6.1).

The oxidized activated carbon sample showed a jump in adsorption in the pH range 1 to 3, which indicated that the point of zero charge of the oxidized carbon sample lies between pH 1 and 3, which was in agreement with the studies of Radovic and coworkers.[123–125] It appears that at low pH values there is electrostatic repulsion between Cu(II) ions and the positively charged surface of oxidized carbon. As the pH increases, the hydrogen ions from the carbon surface are replaced by cations and the Cu(II) ion adsorption increases rapidly.[126] The adsorption of Cu^{2+} ions is, therefore, influenced principally by the presence of surface functional groups. In order to determine the state of the adsorbed copper species, XPS studies of some carbon species were carried out before and after adsorption of copper ions. On the

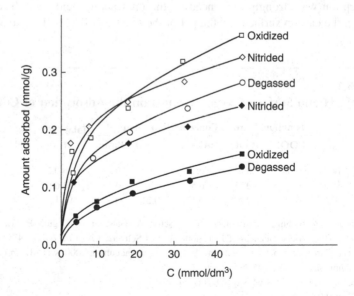

FIGURE 6.1 Adsorption isotherms of Cu(II) ions on modified carbons. Solid line pH = 3, dashed line pH = 1. (Source: Biniak, S., Pakula, M., Syzmanski, G., and Swiatkowski, A., *Langmuir*, 15, 6117, 1999. With permission.)

basis of these spectroscopic studies, Biniak et al.[121] are of the view that in case of the adsorption of copper on heat treated carbons (D-H sample), the dominant role is played by dipole-dipole (π-d) interactions between graphene layers and the metal ionic species, and the spontaneous electrochemical reduction of copper ions.

$$>C\!\!-\!\!H + Cu^{2+} \rightarrow C\text{-}H \cdot Cu^{2+} \text{ (dipole-dipole, } \pi\text{-d interactions)}$$
$$>C^{\bullet} + Cu^{2+} \rightarrow C \cdot Cu^{+} \rightarrow \oplus >C \cdot Cu^{\circ}$$
$$>C\!\!-\!\!O^{\bullet} + Cu^{2+} + H_2O \rightarrow C\!\!-\!\!OH\!\!-\!\!CuOH + H^{+}$$
$$>C = O + Cu^{2+} + H_2O \rightarrow >C \cdot Cu(OH)_2$$

where symbol ($\bullet$) denotes the ion radical.

In the case of oxidized activated carbon samples (D-O sample), the carbon-oxygen surface function groups play a dominant role. These functional groups undergo surface ionization and ion-exchange reactions between the H^+ on the carbon surface and Cu^{2+} ions in the solution.

$$>C\!\!-\!\!COOH + Cu^{2+} \longrightarrow >C\!\!-\!\!COOCu^{+} + H^{+}$$
$$(>C\!\!-\!\!COOH)_2 + Cu^{2+} \longrightarrow (>C\!\!-\!\!COO)_2\,Cu + 2H^{+}$$
$$>C\!\!-\!\!OH + Cu^{2+} \longrightarrow >C\!\!=\!\!O \cdot Cu + H^{+}$$

Heat treatment with ammonia results in the incorporation of nitrogen atoms into the graphene layers, which act as ligands and play a significant role in the adsorption of Cu(II) ions (D-N samples).

$$>C\!\!-\!\!NH_2 + Cu^{2+} + H_2O \rightarrow >C\!\!-\!\!NH_2 \cdot Cu(OH)^{+} + H^{+}$$
$$>N: + Cu^{2+} + H_2O \rightarrow N\!\!-\!\!Cu(OH)^{+} + H^{+}$$
$$>N: + Cu^{2+} + H_2O \rightarrow \oplus >NOH \cdot Cu^{\circ} + H^{+}$$

where symbol $\oplus$ denotes a positive hole in the carbon structure.

Kahn and Khattak[78,127] studied the adsorption, of Cu(II) ions from copper sulfate solutions on a carbon black Spheron-9 under varying conditions of time of contact, carbon dosage, pH, and solution concentration. The adsorption equilibrium was established within one hr in the concentration range 10 to 100 ppm. The adsorption isotherms obey Freundlich equation within the concentration range studied. The adsorption of Cu(II) ions was low in the pH range 2.0 to 2.2 but increased with pH, attaining a maximum value in the pH range 4.2 to 5.1. The low adsorption in the pH range 2.0 to 2.2 was attributed to high solubility and ionization of copper salt in the acidic medium. In this pH range, copper would be present predominantly as Cu(II) ions, which are likely to be adsorbed to a lesser extent on the carbon black surface, most of which is made up of nonpolar basal planes of the microcrystallites. The maximum adsorption in the pH range 4.2 to 5.1 might be due to partial hydrolysis of Cu(II) ions, resulting in the formation of $Cu(OH)^+$ ions and $Cu(OH)_2$, which would be adsorbed to a greater extent on the nonpolar carbon surface compared to Cu^{2+} ions. Furthermore, the low solubilities of hydrolyzed copper species may be another reason for maximum adsorption in the pH range 4.2 to 5.1. The low value of copper adsorption at pH greater than 6 is due to the precipitation of copper as

$Cu(OH)_2$, which lowers the concentration of the solution. It also lowers the availability of the carbon surface, because a large portion of the carbon surface is occupied or blocked by the precipitated $Cu(OH)_2$.

Goyal et al.[128,129] studied the adsorptive removal of copper from aqueous solutions in the concentration range 40 to 100 mg/L using two samples of granulated and two samples of fibrous activated carbon (activated carbons fibers) associated with varying amounts of different types of carbon-oxygen surface groups. The influence of pH of the solution on the adsorption of Cu(II) ions on the four samples is shown in Figure 6.2. The pH of the solution was controlled by the addition of HCl or NaOH. The uptake adsorption of Cu(II) ions at pH below 2 is small in both the granulated (GAC-S and GAC-E) as well as fibrous (ACF-307 and ACF-310) activated carbons but more so in the case of granulated carbons. However, on increasing the pH of the solution, a maximum in adsorption is obtained at pH 3 in the case of fibrous carbons and at pH 4 for the granulated carbons. At pH higher than 6, the adsorption studies could not be carried out because of the precipitation of copper as $Cu(OH)_2$.

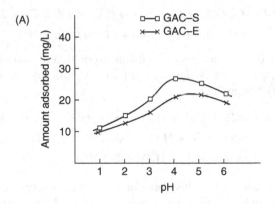

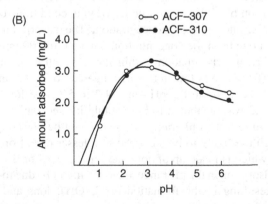

FIGURE 6.2 (**A**), (**B**) Adsorption isotherms of Cu(II) ions on activated carbons as a function of pH of the solution. (Source: Goyal, M., Rattan, V.K., Aggarwal, D., and Bansal, R.C., *Colloids and Surfaces, A. Physico-Chem. and Engg. Aspects,* 190, 229, 2001. With permission.)

These workers are of the view that a change in pH of the solution results in a change in the carbon surface charge due to the formation of different ionic species. A sudden increase in the adsorption of Cu(II) ions in the pH range 3 to 4 indicates that the zero point of charge (ZPC) of the carbon surface lies between these values. The slightly low range in pH for maximum adsorption of copper ions in the case of activated carbon fibers may be attributed to the fact that the fibrous carbon suspension in water show lower pH values than the granulated activated carbon suspensions. At pH values lower than ZPC (pH 3), there is excessive protonation of the carbon surface, which gives it a positive charge. This enhances the electrostatic repulsive interactions between the carbon surface and the positively-charged copper cations. As the pH of the solution is increased beyond ZPC, the preponderance of OH⁻ ions in the solution phase generates a competition between the carbon surface and the solution OH⁻ ions for the positively-charged copper cations (Cu^{2+}) resulting in a decrease in the adsorption of Cu(II) ions on the carbon surface.

The adsorption isotherms of Cu(II) ions from aqueous solutions in the concentration range 40 to 100 mg/L (Figure 6.3) are Type I of the BET classification, showing initially a rapid adsorption tending to be asymptotic at higher concentrations. The pH of each solution was maintained at pH = 5 by the addition of HCl. The adsorption of Cu(II) ions is smaller in the case of fibrous carbons compared with granulated carbons, the difference being a factor of 3. This could not be explained on the basis of surface area alone because ACF-310, which has about the same BET surface area as GAC-E (Table 6.2), adsorbs smaller amounts of Cu(II) ions than GAC-E. This could also not be explained on the basis of the total amount of oxygen associated with the carbons because fibrous activated carbons, which showed lower adsorption, were associated with larger amounts of associated oxygen (Table 6.2).

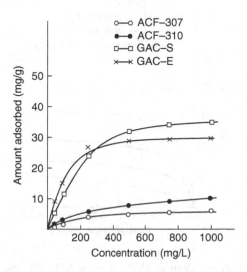

FIGURE 6.3 Adsorption isotherms of Cu(II) ions on different as-received activated carbons. (Source: Goyal, M., Rattan, V.K., Aggarwal, D., and Bansal, R.C., *Colloids and Surfaces, A. Physico-Chem. and Engg. Aspects*, 190, 229, 2001.)

TABLE 6.2
Surface Area and Carbon-Oxygen Surface Groups Evolved As CO and CO on Degassing As-Received Activated Carbons at 950 °C

Sample Identification	BET (N_2) Surface Area (m^2g^{-1})	Oxygen Evolved As (g 100 g^{-1})			
		CO_2	CO	H_2O	Total
ACF-307	910	1.00	5.30	1.30	7.60
ACF-310	1184	1.90	4.20	1.40	7.50
GAC-S	1256	2.10	1.05	1.24	4.39
GAC-E	1190	2.13	1.66	1.33	5.12

Source: Goyal, M., Rattan, V.K., Aggarwal, D., and Bansal, R.C., Colloids and Surfaces, A. *Physico-Chem. and Engg. Aspects*, 190, 229, 2001. Reproduced with permission from Elsevier.

The adsorption of Cu(II) ions was found to increase on oxidation of the carbons with nitric acid, ammonium persulfate, and oxygen (Figure 6.4 and Figure 6.5) the magnitude of increase being different for the three oxidative treatments. The uptake of Cu(II) ions was considerably larger when the oxidation was carried out with nitric acid. Incidentally, the treatment with nitric acid was a stronger oxidative treatment and resulted in the fixation of considerably larger amounts of oxygen on the carbon surface (Table 6.3). The amount of associated oxygen increased from 5 to 7% to

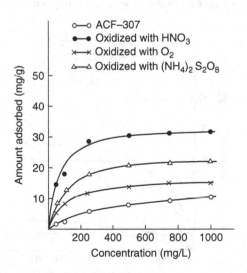

FIGURE 6.4 Adsorption isotherms of Cu(II) ions on ACF-307 before and after oxidation with different oxidizing agents. (Source: Goyal, M., Rattan, V.K., Aggarwal, D., and Bansal, R.C., Colloids and Surfaces, A. *Physico-Chem. and Engg. Aspects* , 190, 229, 2001. With permission.)

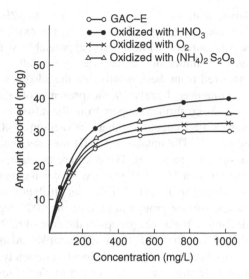

FIGURE 6.5 Adsorption isotherms of Cu(II) ions on GAC-E before and after oxidation with different oxidising agents. (Source: Goyal, M., Rattan, V.K., Aggarwal, D., and Bansal, R.C., Colloids and Surfaces, A. *Physico-Chem. and Engg. Aspects*, 190, 229, 2001. With permission.)

TABLE 6.3
Carbon-Oxygen Group Evolved As CO and CO on Degassing Different Oxidized Activated Carbons at 950°C

Sample	Oxygen Evolved (g/100 g) as			
	CO_2	CO	H_2O	Total
ACF-307, oxidized with HNO_3	12.90	7.47	2.40	22.77
$(NH_4)_2S_2O_8$	5.40	7.51	4.91	17.82
H_2O_2	2.55	7.42	2.10	12.07
O_2	3.11	8.71	1.20	13.02
ACF-310, oxidized with HNO_3	11.96	7.20	2.20	21.36
$(NH_4)_2S_2O_8$	4.70	6.30	2.10	13.10
H_2O_2	1.70	6.20	2.10	10.00
GAC-S oxidized with HNO_3	12.20	7.60	1.50	21.30
$(NH_4)_2S_2O_8$	6.34	6.53	1.25	14.12
GAC-E oxidized with HNO_3	12.40	11.20	1.92	25.52
$(NH_4)_2S_2O_8$	4.63	5.61	1.73	11.97
O_2	2.17	5.97	1.13	9.27

Source: Goyal, M., Rattan, V.K., Aggarwal, D., and Bansal, R.C., Colloids and Surfaces, A. *Physico-Chem. and Engg. Aspects*, 190, 229, 2001. Reproduced with permission from Elsevier.

20 to 22% on oxidation with nitric acid and only to 10 to 12% on oxidation with ammonium persulfate and oxygen gas. Furthermore, the oxidation with nitric acid enhanced the amount of surface acidic groups appreciably while the increase on oxidation with oxygen or ammonium sulfate was only small.

These workers inferred from these results that the adsorption of Cu(II) ions by activated carbons is influenced largely by the presence of acidic carbon-oxygen surface groups. This received further support from the adsorption isotherms on the oxidized carbon samples after outgassing them in vacuum at 400, 650, and 950°C (Figure 6.6 and Figure 6.7). The uptake of Cu(II) ions decreased gradually as the temperature of degassing was increased. The decrease in adsorption was only slight between 0.2 and 0.5% for both ACF-307 and GAC-E when outgassed at 400°C, was appreciably large on degassing at 650°C. This was attributed to a smaller (~15%) elimination of the acidic surface groups in case of the 400°-degassed samples and larger (~85%) elimination of the acidic groups in the case of 650°-degassed samples (Table 6.4 and Table 6.5). The 950°-degassed samples adsorbed even smaller amounts of copper ions because they were almost completely free of any acidic surface groups that provide sites for the adsorption of Cu(II) ions.

A plot between the surface acidity, as determined by titration with 0.1 N solution of sodium hydroxide, and the maximum adsorption of Cu(II) ions on GAC-E and ACF-307 at 1000 mg/L concentration of the solution before and after oxidation and degassing treatments (Figure 6.8), showed that the points for the two carbons could not be collected along a single straight line. However, the plots for a given carbon showed a linear variation of adsorption with surface acidity, thus giving rise to two linear plots, one for each carbon. The difference in the behavior of the two carbons

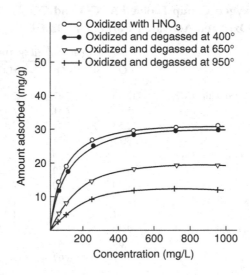

FIGURE 6.6 Adsorption isotherms of Cu(II) ions on nitric acid oxidised ACF-307 before and after degassing at different temperatures. (Source: Goyal, M., Rattan, V.K., Aggarwal, D., and Bansal, R.C., Colloids and Surfaces, *A. Physico-Chem. and Engg. Aspects*, 190, 229, 2001. With permission.)

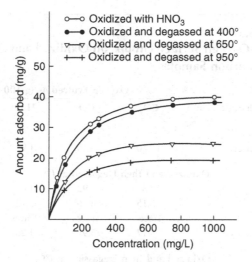

FIGURE 6.7 Adsorption isotherms of Cu(II) ions on nitric acid oxidised GAC-E before and after degassing at different temperatures. (Source: Goyal, M., Rattan, V.K., Aggarwal, D., and Bansal, R.C., Colloids and Surfaces, *A. Physico-Chem. and Engg. Aspects*, 190, 229, 2001. With permission.)

TABLE 6.4
Carbon-Oxygen Surface Groups on Oxidised and Degassed ACF-307 Carbon Samples

Sample Identification	Oxygen Evolved As (g 100 g⁻¹)			
	CO_2	CO	H_2O	Total
ACF-307				
HNO_3 oxidized	12.90	7.47	2.40	22.77
Oxidized and Then Degassed at (°C)				
400	10.85	7.35	0.85	19.05
650	2.15	6.86	0.12	9.13
950	—	Traces	—	—
Oxygen oxidized	3.11	8.21	1.20	12.02
Oxidized and Then Degassed at (°C)				
400	2.95	8.15	0.82	11.92
650	0.42	7.81	0.21	8.44
950	—	Traces	—	—

Source: Goyal, M., Rattan, V.K., Aggarwal, D., and Bansal, R.C., Colloids and Surfaces, *A. Physico-Chem. and Engg. Aspects*, 190, 229, 2001. Reproduced with permission from Elsevier.

314

Activated Carbon Adsorption

TABLE 6.5
Carbon-Oxygen Surface Groups on Oxidized and Degassed
GAC-E Carbon Samples

Sample Identification	Oxygen Evolved As (g 100 g^{-1})			
	CO_2	CO	H_2O	Total
GAC-E				
HNO$_3$ oxidized	12.40	6.20	1.92	20.52
Oxidized and Then Degassed at (°C)				
400	10.85	5.92	1.02	17.14
650	2.15	6.86	0.12	9.13
950	Traces	Traces	Traces	Traces
Oxygen oxidized	3.17	5.97	1.26	10.40
Oxidized and Then Degassing at (°C)				
400	2.86	5.56	1.00	9.42
650	0.58	4.78	0.62	5.98
950	—	Traces	—	—

Source: Goyal, M., Rattan, V.K., Aggarwal, D., and Bansal, R.C., Colloids and Surfaces, A. *Physico-Chem. and Engg. Aspects*, 190, 229, 2001. Reproduced with permission from Elsevier.

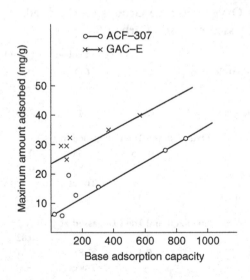

FIGURE 6.8 Relationship between the maximum amount of Cu(II) ions adsorbed and base adsorption capacity. (Source: Goyal, M., Rattan, V.K., Aggarwal, D., and Bansal, R.C., Colloids and Surfaces, A. *Physico-Chem. and Engg. Aspects*, 190, 229, 2001. With permission.)

in giving two different straight lines was attributed to the difference in the micro-porous character of the two carbons.

The activated carbon adsorption of EDTA-chelated copper complexes has also been investigated by several workers as a function of pH.[49,108,126,130,131,132] It was observed that the adsorption of copper was pH dependent and that electrostatic interactions played an important role. Ferro-Garcia et al.,[126] while working with olive stone-derived microporous activated carbons suggested that the adsorption of Cu-EDTA complexes could be rationalized by postulating that the Cu-EDTA complexes being larger in size were accessible to only some of the micropores. Chang and Ku,[130,131] while studying the adsorption and desorption of Cu-EDTA chelated com-plexes by activated carbon columns, suggested that the chelated metal ions are adsorbed by bonding of either the metal or the ligand directly to the carbon surface. These studies indicated that the adsorption behavior of chelated copper on activated carbon over the entire pH range depended upon the surface charge characteristics of the activated carbon and the distribution of metal and ligand species. The adsorbed amount of copper and EDTA varied significantly over the entire solution pH range and that of Cu-EDTA chelate was higher than that of free ions in the solution. This was in agreement with the results obtained by Bhattacharya and coworkers.[132]

6.2.1 MECHANISM OF COPPER ADSORPTION

The activated carbon surface has unsaturated $>C= C$ bonds, which on oxidation can add oxygen, resulting in the formation of carbon-oxygen surface groups. Two types of surface oxygen groups are formed: acidic and nonacidic. The acidic surface oxygen groups are polar in character and have been postulated as *carboxyls* and *lactones*; the nonacidic groups are identified as *quionones*. When an oxidized acti-vated carbon is placed in water, the acidic surface groups undergo ionization, pro-ducing H^+ ions.

$$>C\text{-}COOH \rightarrow >C\text{-}COO^- + H^+$$

The H^+ ions are directed toward the aqueous phase, leaving the carbon surface with negatively-charged $>C\text{-}COO^-$ sites. These negatively-charged sites generate a com-petition between the H^+ ions and the solution Cu^{2+} ions for the carbon surface. As the availability and concentration of these negatively-charged sites is increased on oxidation of the activated carbons, there is an increase in the adsorption of Cu(II) ions on the oxidized carbons. On degassing at gradually increasing temperatures, these acidic surface oxygen groups are gradually eliminated from the carbon surface. This results in a decrease in the concentration of surface $>COO^-$ sites, thereby decreasing the adsorption of Cu(II) ions. When the oxygen groups are removed almost completely on degassing at 950°C, the carbon surface loses its negative character and may even become positively charged so that the repulsive interactions between the carbon surface and the Cu(II) ions become more predominant, resulting in a decrease in the adsorption of Cu(II) ions. The pH of the 950°-degassed carbons is also more than 7 so that the possibility of the precipitation of copper as $Cu(OH)_2$ is enhanced. This can also result in a decrease in the adsorption of Cu(II) ions on the carbon surface.

6.3 ACTIVATED CARBON ADSORPTION
OF CHROMIUM

Chromium is present in effluent waters of several different industries such as leather tanning, metal finishing, wood preservation, pigments, chemical manufacturing process plants, and electroplating units. Although essential for several human metabolic processes, it is hazardous when present in larger concentrations because it affects human physiology, accumulates in the food chain, and causes several ailments. Cr(VI) is more hazardous to health because of its high toxicity. The stricter environmental regulations to the discharge of heavy metal cations make it necessary to develop processes for its removal from waste water. The various treatment techniques available for the removal of chromium from water are reduction, precipitation ion-exchange, and adsorption. Activated carbons have been considered as potential adsorbents for the removal of chromium metal from water and, consequently, a large number of investigations have been carried out. Furthermore, activated carbon adsorption of chromium is an important step in the production of chromium catalysts supported on activated carbons.

Narayana and Krishnaiah[133,134] and Kannan and Vanangamudi[135] used activated carbon, lignite coal, and bituminous coal for the adsorptive removal of Cr(VI) ions from aqueous solutions at different pH values of the solution and observed that both adsorption and reduction of Cr(VI) takes place. The adsorption of Cr(VI) was maximum at pH = 3; the reduction was maximum at pH = 1. Ouki and Newfeld[136] used a column adsorption technique for the removal of Cr(VI) from aqueous solutions in the concentration range 250 to 300 mg/L and observed that the adsorption of Cr(VI) increased significantly in the range of acidic pH values. These workers also observed that at acidic pH values, a radox reaction occurred on the carbon surface active sites when Cr(VI) was reduced to Cr(III). These active sites were in turn oxidized, resulting in an increased adsorption capacity.

Cici and Kales,[137] while studying the adsorption of Cr(VI) on rice-husk charcoal activated with zinc chloride at two different temperatures, found that 30 to 99% of Cr(VI) could be removed from solutions containing 20 to 200 mg/L of Cr(VI) ions. Andreev et al.[52] used several different activated carbon fibers for the removal of heavy metals under static conditions at different solution pH values and observed that the fibrous activated carbons were very effective for the adsorptive removal of heavy metals from waste water. The adsorption of chromium followed the Freundlich isotherm equation below 20°C. Dikshit et al.[138] on the other hand, used highly microporous bituminous coal for the removal of low concentrations of Cr(VI) from highly acidic waste water by the batch technique. They studied the kinetics of the adsorption process and found that the adsorption followed first-order kinetics. The adsorption was suggested to be a diffusion process involving micropores. Golub and Oren[139] used electrochemical methods to reduce Cr(VI) to Cr(III) on a graphite felt electrode and observed that although low pH of the solution favors the reduction, high pH favors the precipitation of copper hydroxide. Similar experiments using a high surface area carbogel[140] indicated that the process involves chemisorption of Cr(VI) on the carbogel anode and not the double-layer formulation.

Lalvani et al.[141] compared the adsorption capacity of Cr(VI) and Cr(III) from solutions for a commercial activated carbon with a carbon adsorbent produced in the form of soot in a graphite electrode arc. The activated carbon showed high adsorption capacity for Cr(III) ions and no adsorption for the hexavalent chromium anion Cr(VI). The arc soot carbon, on the other hand, selectively adsorbed Cr(VI) anions from solution, whereas only a negligible amount of the Cr(III) cation was adsorbed, the amount adsorbed depending on the pH of the solution. These workers postulated that the adsorption involved ion-exchange mechanism and attributed the high selectivity to positive charges on the carbon surface. Tereshkova,[142] on the basis of her studies on the adsorption of Cr(III) and Cr(VI) ions on fibrous activated carbons, did not discuss the influence of pH on the adsorption. Huang and Wu[66,143] used a calcined charcoal and Filtrasorb-400 activated carbon for the removal of Cr(III) and Cr(VI) ions from aqueous solutions. The removal of Cr(VI) ions involved two processes, namely reduction of Cr(VI) to Cr(III) and the adsorption of Cr (VI) and its reduced form Cr(III). The removal was dependent on the pH as well as the chromium content of the solution. The adsorption of Cr(VI) was much larger than the adsorption of Cr(III). The adsorption increased with increase in pH attained a maximum value of about pH = 6 and declined at higher pH values. This was attributed to the reduction of Cr(VI) into Cr(III) on the surface of the carbon in acidic solutions. The main Cr(VI) species involved in adsorptive removal were $HCrO_4^-$ and $Cr_2O_7^{2-}$. However, at total Cr(VI) concentration, and pH values generally found in chromium plating industrial waste water, $HCRO_4^-$ ions played a major role in the adsorptive removal of chromium. Adsorption data could be described by the Langmuir isotherm equation.

Perez-Candela et al.[144] investigated the adsorption of Cr(VI) ions on a wide variety of activated carbons from aqueous phase and found that there was a large increase in adsorption as pH decreased. This was attributed to the reduction of Cr(VI) into Cr(III). Kim[145] and Nagasaki and Terada[146] observed that the reducing action of the activated carbon was decreased when the proton concentration of the solution was about the same as the Cr(VI) ion concentration in the solution. However, Huang and Bowers[147] failed to confirm this observation. These workers observed[147] that the adsorption and reduction were occurring simultaneously and were responsible for the removal of Cr(VI). These workers also found that when the carbon was oxidized with nitric acid, and the solution containing Cr(VI) also contained chlorine water, the removal of Cr(VI) was decreased considerably. This was attributed to the fact that these oxidations resulted in the chemisorption of appreciable amounts of oxygen, which enhanced the acidity of the carbon surface. The carbon surface, therefore, had greater affinity for the reduced form of the metal Cr(III). Yoshida et al.[58] observed that Cr(VI) was readily adsorbed by activated carbons as anionic species such as $HCrO_4^-$ or CrO_4^{2-}; Cr(III) ion was scarcely adsorbed. The adsorbed Cr(VI) anions could be recovered by washing with sodium hydroxide or hydrochloric acid solutions. They also observed that in acidic solutions Cr(VI) was easily reduced into Cr(III) in the presence of activated carbon. Such a reduction of Cr(VI) into Cr(III) was also reported by Roerma et al.[148] Miyagaiwa et al.[149] and Nagasaki[146] also found that the removal of chromate ions by activated carbon from aqueous solutions was a function of pH of the solution, the best pH range being acidic range pH 3 to 7. The role of pH in the adsorption of both Cr(III) and Cr(VI) ions by activated carbons was also studied by Leyva-Remos.[150,151] They found that the adsorption of Cr(VI) increased significantly when the pH of the solution decreased from 10 to 6, but that of Cr(III) increased with increase in pH from 2 to 5 and decreased drastically at pH = 6. They attributed these changes in adsorption of Cr(III) and Cr(VI) to the formation of different chromium complexes in the solution.

The influence of carbon-oxygen surface functional groups on the adsorption of Cr(III) and Cr(VI) ions has also been studied by a number of workers.[61,152–156] Puri and Satija[156] studied the interaction of charcoal coated with varying amounts of carbon-oxygen surface groups with acidified potassium dichromate solution and observed that the charcoals which were associated with acidic oxygen surface groups were more effective in reducing Cr(VI) into Cr(III) than those charcoals that were free of acidic oxygen groups. The oxygen rendered available during the reaction was partly evolved as CO_2 and partly chemisorbed by the carbon. The reaction involved two stages:

$$K_2Cr_2O_7 + H_2SO_4 \rightarrow K_2SO_4 + Cr_2O_3 + H_2O + 3(O)$$

and

$$Cr_2O_3 + 3H_2SO_4 \rightarrow Cr_2(SO_4)_3 + 3H_2O$$

At lower concentrations of potassium dichromate, the reaction occurred through the first stage only because the amount of Cr_2O_3 formed was found to be almost exactly equivalent to that of potassium dichromate in the solution (Table 6.6). However, as

TABLE 6.6
Reduction of Acidified Potassium Dichromate by the Original Sugar Charcoal

Concentration of $K_2Cr_2O_7$	Amount of $K_2Cr_2O_7$ Present in 100 ml of Solution Mixed with 1 g Charcoal (mEq/g)	$K_2Cr_2O_7$ Reduced (mEq/g)	Cr_2O_3 Obtained (mEq/g)
0.02 N	2	2	2.0
0.03 N	3	3	2.9
0.05 N	5	4.8	4.3
0.10 N	10	8.4	4.0
0.20 N	20	12.0	2.1
0.30 N	30	17.4	1.25
0.40 N	40	22.2	1.00
0.50 N	50	23.0	0.36
0.60 N	60	25.0	0.34
0.70 N	70	25.8	0.20
0.80 N	80	26.50	0.00
0.90 N	90	28.50	0.00
1.00 N	100	29.70	0.00

Source: Puri, B.R. and Satija, B.R., *J. Indian Chem. Soc.*, 45, 298, 1968. With permission.

the concentration of the dichromate in the solution increased, the second reaction, involving the formation of $Cr_2(SO_4)_3$, was observed to take place to increasing extents, until at concentrations higher than 0.7 N the formation of Cr_2O_3 ceases completely (Table 6.6).

Jayson et al.[61] studied the adsorption of chromium ions from aqueous solutions on a sample of activated carbon cloth at different temperatures and pH values, using a radioactive tracer technique. The adsorption of Cr(VI) from chromate solution was always greater by a factor of ~10 than that of Cr(III) from chromic solutions at all temperatures, solution concentrations, and pH values studied. This has been attributed to the smaller radii of chromate Cr(VI) ions (0.5 nm) compared to that of chromic Cr(III) ions (0.9nm) resulting from the strongly bound water molecules with the cations. Thus, the chromic Cr(III) cations show a cross-sectional area of approximately three times the area of a chromate ion. In addition, the activated carbon cloth has pores some of which have radii less than 1 nm, which will limit its access to the hydrated chromic ions into the micropores, particularly when the pores are slit shaped. The increase in the adsorption of Cr(III) ions with increase in the concentration of the solution indicated that the adsorbed hydrated cations lose their hydration sphere by dehydration and are reduced in size so that some of micropores become accessible to them. Besides the size of the ions and the micropores, the electrostatic attraction or repulsion between the carbon surface and the ions in solution also plays an important role in the adsorption of chromium from aqueous solution. These workers are of the view that at very low pH values (pH less than 2.7) the activated carbon cloth has a positive charge that reaches zero at pH = 2.7 (ZPC)[157] so that at pH greater than 2.7, the carbon surface in aqueous solutions will acquire a negative charge. Thus, at low pH the repulsive interactions between the positively-charged surface and the cations are followed by attractive interactions between the negatively charged carbon surface and the solution cations as the pH is increased above 2.7 to a maximum of pH 4.4 and subsequent repulsive interactions at higher pH values (Figure 6.9). In the case of adsorption of chromate and dichromate [Cr(VI)] ions, which carry two

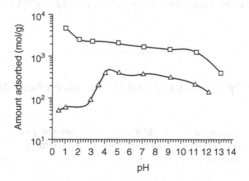

FIGURE 6.9 The effect of pH on the adsorption of chromium from 0.2 mol dm^{-3} chromic(III) (Δ) chloride and 0.15 mol dm^{-3} sodium chromate(VI) ($\square$) solutions onto activated charcoal cloth. (Source: Jayson, G.G., Sangster, J.A., Thompson, G., and Wilkinson, M.C., *Carbon*, 31, 487, 1993. With permission.)

negative charges across the whole range of pH values ($Cr_2O_7^{2-}$ in acid and CrO_4^{4-} in alkaline solutions), the attractive electrostatic interactions at low pH values are followed by repulsive interactions at higher pH values of the solution. However, the larger adsorption of chromate (VI) ions than the chromic [Cr(III)] ions as mentioned earlier is due to the smaller size of the chromate (VI) ions.

Moreno-Castilla and coworkers[154,155] studied the influence of carbon-oxygen surface groups on the adsorption of Cr(III) and Cr(VI) from aqueous solutions containing a certain amount of NaCl on activated carbon before and after oxidation with nitric acid, followed by its degassing in nitrogen at 873 K. The adsorption capacity of the oxidized carbon sample was much higher than that of the unoxidized sample although the surface area, and the porosity of the oxidized carbon was lower than that of the unoxidized sample (Table 6.7). The increase in uptake of Cr(III), in the case of the oxidized sample, has been attributed to the presence of acidic carbon-oxygen surface groups produced during the oxidation with nitric acid. These acidic surface groups dissociate in aqueous solution, producing specific sites for the adsorption of Cr(III) ions. When these acidic surface groups were removed by the treatment with nitrogen gas at 873 K, the adsorption capacity for Cr(III) was drastically reduced, although the surface area and the porosity of the sample were enhanced. The increase in adsorption capacity for Cr(VI) ions from 6.9 to 15.5 mg/g (cf, Table 6.7) on oxidation of the carbon with nitric acid has been attributed to the reduction of Cr(VI) into Cr(III) in the presence of the oxidized carbon sample. This reduction of Cr(VI) into Cr(III) in the presence of the oxidized carbon sample was supported by FTIR studies of chromium-supported carbon samples. The pH of the solution was quite conducive to the reduction of Cr(VI) into Cr(III) and, therefore, the chromium was adsorbed as a cation Cr(III) on the carbon surface. The presence of NaCl in the solution lowered the adsorption of Cr(VI) because of the competition of the chromate Cr(VI) and chloride anions for the carbon surface. Park and coworkers[153,158] studied the rates of adsorption of Cr(VI) on activated carbons and carbon fibers from aqueous solutions and observed that the adsorption rate increased with increase in the amount of acidic surface oxides, which enhance the electrostatic attractive interactions between the carbon surface and the Cr(VI) ions in the solution.

TABLE 6.7
Adsorption Capacity of Oxidized and Nonoxidized Activated Carbon Samples

Carbon Sample	Amount of Oxygen Present (g/100g)	BET (N_2) Surface Area (m^2/g)	Pore Volume V_{H_2O} (cm^3/g)	Maximum Adsorption Capacity (mg/g)	
				Cr(III)	Cr(VI)
Unoxidized	16.0	1089	0.654	2.7	6.9
Oxidized	38.9	164	0.188	25.3	15.5
Oxidized and then Degassed	19.4	555	0.215	0.2	1.8

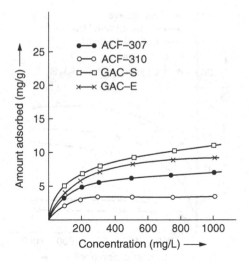

FIGURE 6.10 Adsorption isotherms of Cr(III) on different as-received activated carbons. (Source: Aggarwal, D., Goyal, M., and Bansal, R.C., *Carbon*, 37, 1989, 1999. With permission.)

However, these workers did not mention the pH of the solution that determine the carbon surface charge.

Aggarwal et al.[159,160] carried out a systematic study of the adsorption of Cr(III) and Cr(VI) ions from aqueous solutions on two samples of granulated activated carbon and two samples of activated carbon fibers associated with varying amounts of carbon-oxygen surface groups and having different surface areas. The amount of carbon-oxygen surface groups was enhanced by oxidation with HNO_3, H_2O_2 and ammonium persulfate in solution phase and oxygen gas at 350°C and decreased by degassing the oxidized carbons at gradually increasing temperatures in vacuum. The adsorption isotherms without adding any buffering regents to avoid the influence of external electrolyte on the as-received samples are Type I of the BET classification and obeyed the Langmuir adsorption equation (Figure 6.10). The granulated activated carbons (GAC-E and GAC-S) adsorbed several times more Cr(III) ions than the activated carbon fibers (ACF-307 and ACF-310), indicating that ACFs were highly microporous and contained a larger proportion of very fine micropores that were not accessible to highly hydrated $[Cr(H_2O)_6]^{3+}$ ions with a molecular diameter of 0.922 nm.[161]

The uptake of Cr(III) increased on oxidation of the activated carbons, the extent of increase depending upon the nature of the oxidative treatment (Figure 6.11). The increase in adsorption of Cr(III) is at a maximum in the case of the sample oxidized with HNO_3 and at a minimum in the case of the oxidation with H_2O_2. This has been attributed to the fact that the treatment with nitric acid is a stronger oxidative treatment and results in the formation of maximum amounts of (Table 6.3) acidic carbon-oxygen surface groups[128,161,162,163] compared with the other oxidation treatments. When the amount of these acidic surface groups was gradually decreased on degassing at gradually increasing temperatures (400°, 650°, and 950°C) the adsorption of

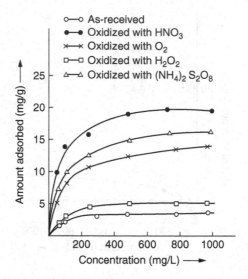

FIGURE 6.11 Adsorption isotherms of Cr(III) on ACF-307 before and after oxidation. (Source: Aggarwal, D., Goyal, M., and Bansal, R.C., *Carbon*, 37, 1989, 1999. With permission.)

Cr(III) gradually decreases with an increase in the temperature of degassing (Figure 6.12). The decrease in Cr(III) uptake was small on the 400°-degassed samples because only about 15% of the acidic surface oxygen groups were eliminated at this temperature. However, on degassing at 650°C, the uptake of Cr(III) ions decreased by several orders of magnitude because almost 85 to 90% of the acidic oxygen groups

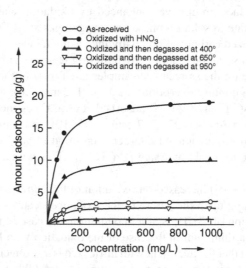

FIGURE 6.12 Adsorption isotherms of Cr(III) on ACF-307 before and after oxidation and degassing. (Source: Aggarwal, D., Goyal, M., and Bansal, R.C., *Carbon*, 37, 1989, 1999. With permission.)

TABLE 6.8
Langmuir Adsorption Constants and Surface Area Occupied by Cr(III) Ions on Different Activated Carbons

Sample	X_m (mg/g)	K (l/g)	\$\$\$ (m²/g)
GAC-S	13.31	0.07	103.1
GAC-E	10.52	0.08	80.5
ACF-307	7.08	0.10	58.8
ACF-310	3.52	0.07	27.3

Key: \$\$\$ is the maximum amount of Cr(III) ions adsorbed by activated carbon. K is Langmuir's constant and \$\$\$ is surface area of activated carbon covered by Cr(III) ions.

Source: Aggarwal, D., Goyal, M., and Bansal, R.C., *Carbon*, 37, 1989, 1999. Reproduced with permission from Elsevier.

were removed from the carbon samples. Thus, these workers concluded that the adsorption of Cr(III) is determined largely by the presence of acidic surface groups.

These workers[159] calculated the surface area covered by Cr(III) ions from the Langmuir linear plots, using 0.922 nm as the molecular diameter of these ions. It was observed that only a small fraction of the BET surface area in all the carbons is occupied by chromium ions (Table 6.8). The area covered by Cr(III) ions increased with oxidation of the carbon surface, indicating that the adsorption of Cr(III) ions takes place on certain sites. These sites appear to be acidic oxygen surface groups present on the carbon surface.

The adsorption of Cr(VI) ions was also studied[159] from aqueous solutions of potassium dichromate on all four activated carbon samples. The adsorption of Cr(VI) ions is comparatively much larger than that of Cr(III) ions under similar conditions (Figure 6.13). This was attributed partly to the smaller size of the Cr(VI) ions in aqueous solutions so that it can enter a larger proportion of the micropores and partly to the fact that the Cr(VI) ions is adsorbed as an anion. Because the pH of the carbon suspensions of as-received carbons is in the range 7.2 to 10.0, there is a possibility of the existence of some positively-charged sites where negatively-charged Cr(VI) anions can be adsorbed. In contrast with the adsorption of Cr(III) ions, the adsorption of Cr(VI) ions decreases on oxidation of the activated carbons (Figure 6.14).

Furthermore, the decrease in adsorption is at a maximum in the case of the samples oxidized with nitric acid. This decrease in uptake of Cr(VI) has been attributed to the formation of acidic carbon-oxygen surface groups, which impart negative character to the carbon surface. In addition, the oxidation of the activated carbon also produces quinonic groups on the carbon surface, which can cause reduction of Cr(VI) into Cr(III) ions. These workers feel that although two different processes—one resulting into a decrease and the other causing an increase in the removal of Cr(VI) ions— can take place simultaneously on the oxidation of the carbons, they find that the net result of these two processes is a decreased adsorption. This indicates that the increase in the removal of Cr(VI) ions from the solution by reduction is only small.

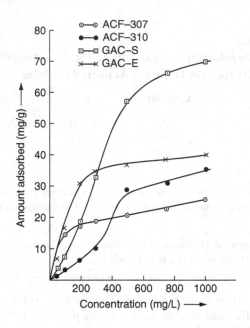

FIGURE 6.13 Adsorption isotherms of Cr(VI) on different as-received activated carbons. (Source: Aggarwal, D., Goyal, M., and Bansal, R.C., *Carbon*, 37, 1989, 1999. With permission.)

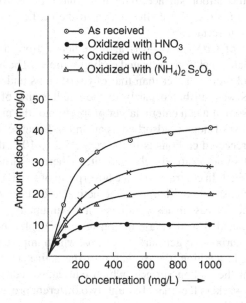

FIGURE 6.14 Adsorption isotherms of Cr(VI) on GAC-E before and after oxidation. (Source: Aggarwal, D., Goyal, M., and Bansal, R.C., *Carbon*, 37, 1989, 1999. With permission.)

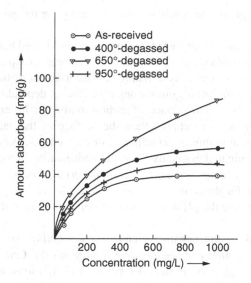

FIGURE 6.15 Adsorption isotherms of Cr(VI) on GAC-E before and after degassing. (Source: Aggarwal, D., Goyal, M., and Bansal, R.C., *Carbon*, 37, 1989, 1999. With permission.)

This they attributed to the fact that the optimum pH for the reduction of Cr(VI) into Cr(III) is about 5, and the solutions of oxidized carbons have a pH between 3 and 4.

When the oxidized activated carbon samples were degassed at 400°, 650°C, the removal of Cr(VI) increased appreciably and was maximum in case of the 650°-degassed samples (Figure 6.15). This has been attributed partly to the elimination of acidic carbon-oxygen surface groups and partly to the pH values of the degassed carbon samples (about 5.5). This pH value is very favorable for the reduction of Cr(VI) into Cr(III), because the standard reduction potential of the system Cr(VI)–Cr(III) in acid medium is +1.195 V against a quinhydrone electrode potential of +0.699 V. This reduction process of Cr(VI) into Cr(III) has been supported by FTIR studies carried out by Bautista-Toledo.[154] In the case of 950°-degassed carbon samples, the removal of Cr(VI) decreases because there are no interactive acidic oxygen groups or quinone groups to cause reduction of Cr(VI) into Cr(III) on the surface of these carbons. Whatever adsorption takes place on the 950°-degassed carbons may be attributed to the porosity of these carbons.

6.3.1 MECHANISM OF ADSORPTION OF CR(III) IONS

The chromic ions in aqueous solution exist as hydrated $[Cr(H_2O)_6]^{3+}$ ions. The water molecules in the hydrated ion can be exchanged with the hydroxyl ions; with a change in the pH of the solution, the number exchanged depends upon the pH of the solution as

$$[Cr(H_2O)_6]^{3+} \xrightarrow{pH=3-4} [Cr(H_2O)_5]^{2+} \xrightarrow{pH=6-7} [Cr(H_2O)_4]^{1+} \xrightarrow{pH=8} [Cr(H_2O)_2(OH)_4]^{1-}$$

Thus, a change in pH of the solution causes a change in the positive charge on the Cr(III) cation.

The acidic carbon oxygen surface groups on the oxidized active carbon surface, which have been postulated as carboxylic and lactonic groups, ionize in aqueous solution, producing H$^+$ ions. As a result of this ionization, the activated carbon surface in aqueous solutions acquires a negative charge, depending upon the amount of acidic surface groups. The surface of carbon oxidized with nitric acid, therefore, shall have a larger negative charge than the surface of the carbon oxidized with hydrogen peroxide, ammonium persulphate, or oxygen gas. When these acidic surface groups are eliminated gradually from the active carbon surface on degassing at gradually increasing temperatures, the carbon surface becomes lesser and lesser negatively charged. Furthermore, the presence of acidic surface groups on the carbon surface tends to change the pH of the solution depending upon the number of acidic groups.

These changes in the negative charge on the carbon surface as a result of oxidation and the changes in the positive charge on the Cr(III) ions in aqueous solution result in an increase in the adsorption of Cr(III) ions, because the electrostatic attractive interactions between the carbon surface and the Cr(III) ions present in the solution are enhanced. When the oxidized carbons are degassed, these electrostatic attractive interactions between the carbon surface and the chromic ions in the solution gradually decrease and result in a decrease in the adsorption of chromic ions. When all the carbon-oxygen surface groups (acidic as well as nonacidic) are removed on degassing at 950°C and the carbon surface is almost completely free of any associated oxygen, there is little or no adsorption of Cr(III).

6.4   ACTIVATED CARBON ADSORPTION OF MERCURY

Mercury has unique properties and finds widespread applications, ranging from dental filling to thermometers. Chloralkali manufacturers are the major users of mercury. It is estimated that about 1 kg of mercury is lost from the process for each 1000 kg of chlorine produced. The other major industries that discharge elemental mercury and its salts such as mercurous salts, methyl mercury, and certain phenyl derivative are pulp and paper industry, chemical processors, agricultural industries, and specialized battery producers. Mercury and its compounds are cumulative toxins that are harmful to most forms of life even when taken in very small amounts. In human beings, mercury compounds can bind and deactivate recycling processes of amino acids, proteins, and certain enzyme groups, resulting in several types of ailments that in many cases may be fatal. It is also well known that mercury is present in trace quantities in coals,[165,166] which on burning of coal in power plants and heat generators is discharged into the atmosphere as vapors. These mercury vapors dissolve in oxygenated rain water and are discharged into streams. Methyl mercury, which is the most toxic of all the mercury compounds, is also formed under microbial degradation of elemental mercury dissolved in oxygenated waters and settled at the bottom of the lakes and streams and accumulates in fish. Thus, it is essential that potable waters be free of any dissolved mercury salts. As activated carbons have been found to be potential adsorbents for the removal of trace metal cations from municipal water supply, a considerable amount of work has been carried out on the activated carbon adsorption of mercury and its compounds.

The adsorption of Hg(II) from aqueous solutions using activated carbons was found to be influenced by the pH of the solutions.[92–94] The adsorption increased as the pH of the solution was decreased, the acidic pH range being most favorable. When the pH of the solution was brought from 9 to the acidic range, the adsorption of Hg(II) was almost doubled. Yoshida et al.[58,95] studied the adsorptive removal of Hg(II) using several activated charcoals and observed that the removal efficiency of a carbon depended on the nature of the carbon and the nature of the activation treatment, which it received during its preparation. The activated carbon obtained from wood, coconut shell, and coal and activated in steam were found to have a higher adsorption capacity for Hg(II) from solutions with pH below 5. The ability decreased with increase in the pH of the solution above 5. The wood charcoal prepared by chemical activation with zinc chloride was found to have a higher adsorption capacity of Hg(II) ions, even at pH higher than 5. These workers, though, did not appreciate the differences in the chemical structures of the carbon surface prepared from different-source raw materials and by different activation treatments, they did suggest that the mechanism of adsorption of Hg(II) ions was different in acidic (HCl) and alkaline (NaOH) medium. The Hg(II) was adsorbed as $HgCl_4^{2-}$ complex in the HCl medium, and the adsorption was reversible both in the case of steam-activated and chemically activated carbons, but in the alkaline medium, pH > 7, the steam-activated carbons showed irreversible adsorption of Hg(II) as Hg(OH) accompanied by its reduction to Hg on the surface of the carbon.

Homenick and Schnoor[94] evaluated the influence of process variables such as pH, chelate type, and activated carbon dose on the adsorption of Hg(II) from mercuric chloride solutions by activated carbon (Filtrasorb-300). The presence of chelating compounds such as Ammonium 1-pyrolidine dithio carbonate (APDC) and triethylene tetramine significantly enhanced the uptake of Hg(II) ions. The increase in adsorption of mercury was 40% for a 50 g/L of triethylene tetramine chelate, but the same amount of APDC removed virtually all the Hg(II) present in the solution. The adsorption of Hg(II) was also considerably enhanced when the activated carbon Filtrasorb was used after presoaking in CS_2 and drying. It was postulated that during the CS_2 treatment, sulfide and other sulfur compounds are formed on the carbon surface, which then reacted with the Hg(II) ion. These workers suggested the following reaction as the probable mechanism for the adsorption (chemisorption) of mercury on the CS_2-treated activated carbon.

$$>C-\overset{\overset{\displaystyle S}{\|}}{C}+HgCl_2\ (aq) \xrightarrow{\ \ pH=10\ \ } >C-C\overset{\displaystyle S}{\underset{\displaystyle S^-}{\diagup}}HgCl + Cl^-$$

At lower pH (pH = 4), the removal of mercury involved physical adsorption or reduction of Hg(II) in addition to the above chemisorption process. The pH of the aqueous solution was also found to have a profound influence on the amount of Hg(II) adsorption by an activated carbon. The adsorption was several times more at a lower pH value of the solution (pH = 4) than at a higher pH value (pH = 10). It has been suggested that at the lower pH value the carbon-surface charge due to the presence of surface oxides is neutralized by hydrated protons, and this facilitates the pore diffusion of mercury chloride complexes, while at the higher pH value the basic surface charge of the carbon may have repelled the mercury complexes.

Sinha and Walker[97] found that sulfurized carbons could be used for the removal of mercury vapors from contaminated streams at 150°C. They sulfurized a Saran carbon to sulfur loading between 1.0 and 11.8% by weight of sulfur by the oxidation of H_2S on the carbon surface at 140°C in a fluidized bed. A mercury-contaminated airstream was passed through 1 g. of the sulfur-loaded carbon with a contact time of 0.5 sec. These workers observed that while the mercury from the unsulfurized carbon bed broke through immediately and built up to 95% of the inlet concentration in less than 30 sec, for the sulfurized carbons the breakthrough occurred in the effluent stream between 7 and 15 min, and the rate of mercury built up in the effluent was very low (Figure 6.16). In the case of the carbon containing 1% sulfur, the breakthrough time was 7 min, and the rate of build-up was such that only 0.8% of the inlet mercury was breaking through the bed after 80 min. In the case of carbons loaded with larger amounts of sulfur, the breakthrough occurred later, but the rates of build-up were also somewhat higher. The much larger capacity of the sulfurized carbons at 150°C for mercury was attributed to a reaction between the mercury and the sulfur on the carbon surface. These workers also studied the effect of moisture in the mercury-contaminated stream at 150°C for the carbon bed containing 1% sulfur. The breakthrough time for mercury was 2 min, and the initial rate of build-up in the effluent was five times more rapid than for the dry stream. However, after 8 min, the mercury build-up rate in the effluent decreased and approached that of the dry stream. They did not find any success in the removal of Hg(II) ions from aqueous solutions by Saran carbon sulfurized by H_2S oxidation.

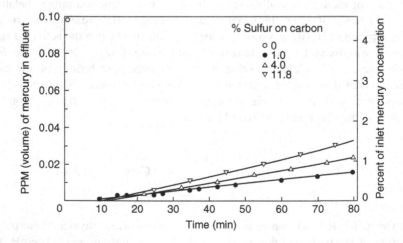

FIGURE 6.16 Effect of sulfur content in carbon on the removal of mercury vapors from air. (Source: Sinha, R.K. and Walker, P.L., Jr., *Carbon*, 10, 754, 1972. With permission.)

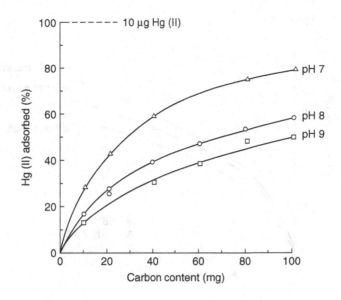

FIGURE 6.17 Removal of mercury as a function of carbon content at different pH values. (Source: Thiem, L., Badorek, D., and O'Conner, J.T., *J. Am. Water Works Assoc.*, 68, 447, 1997. With permission.)

Thiem et al.[93] found that powdered activated carbons can be used for the adsorptive removal of Hg(II), the amount removed depending upon the pH of the solution and the amount of the carbon (Figure 6.17). About twice as much mercury was removed at pH 7 as at pH 9. Increasing the hydroxyl ion concentration evidently decreased the adsorption of mercury by the carbon. The addition of chelating agents such as tannic acid, EDTA, or citric acid enhanced the adsorption. The addition of as small as 0.02 mg/L of the chelating agent enhanced the adsorption of Hg(II) from 10 to 30%, depending upon the pH of the solution and the amount of the carbon used. Of the three chelating agents that were studied, tannic acid was the most effective in enhancing the removal of mercury. At pH 7, the addition of only 1 mg/L of tannic acid improved the removal of Hg(II) by the activated carbon. More than 70% removal was attained by a carbon dose of 20 mg/L, and 40 mg/L of activated carbon removed 85% of mercury.

Although the higher concentrations of tannic acid showed increased removals over those produced by activated carbon alone, the increase was not as high as in the case of 1 mg/L concentration of tannic acid. The presence of calcium ions in aqueous solutions also enhanced the adsorption of Hg(II) ions (Figure 6.18). The mercury adsorption increased by between 10 and 20% as the concentration of calcium ions was increased from 50 mg/L to 200 mg/L. This was attributed to a reaction between the calcium ion in solution with the surface groups on the carbon surface in such a way that new adsorption sites are created in the process. When both calcium ions and tannic acid were present, the removal of Hg(II) ions was almost doubled, even with smaller amounts of the activated carbon.

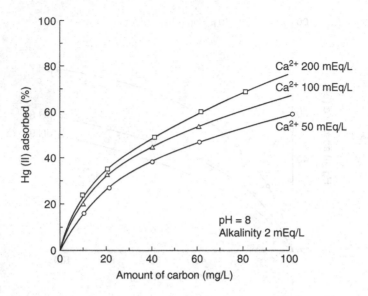

FIGURE 6.18 Effect of presence of calcium ions on the adsorption of mercury from aqueous solutions by carbon. (Source: Thiem, L., Badorek, D., and O'Conner, J.T., *J. Am. Water Works Assoc.*, 68, 447, 1997. With permission.)

Lopez-Gonzalez et al.[167] studied the adsorption of $HgCl_2$ from aqueous solutions on active carbons associated with carbon-oxygen and carbon-sulfur surface complexes. The adsorption of Hg(II) decreased on oxidation of the as-received carbon with hydrogen peroxide. However, when the oxidized carbon was degassed on heat treatment in helium at 873 K, the adsorption capacity of the original carbon was restored. The adsorption of the as-received or oxidized carbons was also enhanced on sulfurization of the carbon by saturation with CS_2 at 288 K followed by heat treatment at 773 K (Figure 6.19). These workers are of the view that oxidation of the activated carbon with H_2O_2 results in the loss of phenolic and hydroquinonic groups by conversion into carboxylic groups. They proposed that the adsorption of Hg(II) on activated carbons takes place as molecular $HgCl_2$ and by the reduction Hg(II) into Hg(I) on phenolic and hydroquinonic sites. The adsorption of Hg(II) decreases on oxidation as these sites are lost, and the adsorption is restored when these sites become available after heat treatment in helium.

$$>C\text{---}OH + HgCl_2 \rightarrow >C = O + Hg_2Cl_2 + HCl$$
$$(Aqueous)$$

In the case of the sulfurized carbon samples also, the reduction of Hg(II) into Hg(I) enhances the adsorption of Hg(II) ions. The observations and the postulations made by Lopez-Gonzalez et al.[167] in respect of the decrease in adsorption of Hg(II) ions on oxidation with H_2O_2 and the increase in adsorption of Hg(II) ions at higher pH values, appear to be contrary to the general behavior of oxidized carbons that

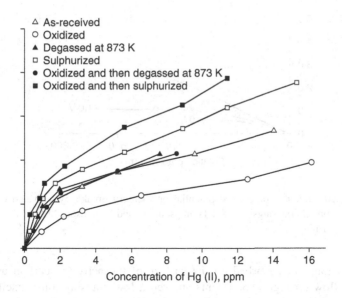

As-received
Oxidized
Degassed at 873 K
Sulphurized
Oxidized and then degassed at 873 K
Oxidized and then sulphurized

Concentration of Hg (II), ppm

FIGURE 6.19 Effects of oxidation and sulfurization on the uptake of mercury by activated carbon. (Source: Lopez-Gonzalez, J.D., Moreno-Castilla, C., Guerrero-Ruiz, A., and Rodriguez-Reinoso, F., *J. Chem. Technol. Biotechnol.*, 32, 575, 1982. With permission.)

show enhanced adsorption of metal cations due to the formation of negatively-charged sites on the active carbon surface. Adams,[168] while examining a carbon surface contacted with $HgCl_2$ solution by an electron microscope, also observed the presence of mercury, confirming the reduction of $HgCl_2$. Similar results were obtained by Huang and Blankenship[164] while studying the adsorption of Hg(II) ions on the two L-type and one H-type activated carbons. These workers observed that the removal of Hg(II) ions involved both adsorption and a reduction process. The acidic carbons showed a better removal capacity for Hg(II) ions over a pH range between 3 and 11, due to its larger reduction capacity.

Jayson et al.[169] studied the adsorption of Hg(II) ions from aqueous solutions of mercury acetate on an activated carbon cloth using shaking and flow-through techniques in the pH range 2.6 to 5.5. The adsorption of Hg(II) could not be studied at pH higher than 5.5 because of the precipitation of mercury as Hg(II) oxide. The adsorption capacity of the activated carbon cloth increased as the pH of the solution increased from 3 to 5.5. The isoelectric point for the carbon cloth was at pH 2.7[157] so that any increase in pH beyond 2.7 increased the negative character of the carbon surface. This resulted in an increase in the attractive electrostatic interactions between the positively-charged Hg(II) ions and the negatively-charged activated carbon surface. These workers calculated the area covered by the Hg(II) ions using a radius of 0.111 nm[170] for the Hg(II) ion and a radius of 0.56 nm for the hydrated Hg(II) ion, assuming the ion to be surrounded by a single sphere of hydration. The surface occupied using the former radius was only a small fraction of the BET (N2) area, while using the latter value of the radius for the hydrated Hg(II) ion gave a better value: 1186 m^2/g against BET surface area of 1300 m^2/g. This indicated that

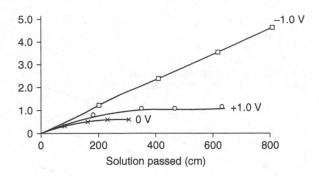

FIGURE 6.20 Influence of applied potential on mercury uptake by activated carbon cloth. (Source: Jayson, G.G., Sangster, J.A., Thompson, G., and Wilkinson, M.C., *Carbon*, 25, 523, 1987. With permission.)

the Hg(II) ions are adsorbed on the surface of the activated carbon as hydrated ions. The flow-through experiments showed a lower adsorption capacity because of a decreased concentration gradient and limitations due to film diffusion. However, the latter could be overcome by increasing the flow rate. These workers also studied the influence of applying electrical potential to the activated carbon cloth during the flow-through experiments and observed that the uptake of Hg(II) increased, and about 90% removal of Hg(II) was achieved with a negative potential of −1.0 V with reference to a calomel electrode (Figure 6.20). Under these conditions the carbon cloth continued to remove Hg(II) ions from the solution, because the diffusion constant had attained a constant value. This made the process a potential technique for cleaning up the industrial effluents contaminated with mercury.

Gomez-Serrano et al.[171] compared the adsorptive removal of Hg(II) ions by a sulfurized activated carbon treated with H_2S at 900°C and by the activated carbon sample degassed in nitrogen at 900°C, and observed that the adsorption increased with increase in pH between 2 and 5. Both sulfurization and degassing enhanced the adsorption of mercury. They explained their results on the basis of an acid-base mechanism involving adsorbate-adsorbent interactions and ignored the influence of carbon-oxygen surface groups on the adsorption of Hg(II) ions. Ma et al.,[172] however, found that the adsorption of Hg(II) from aqueous solutions on activated carbons was maximum at intermediate pH ranges. Namasivayam and Periasami[173,174] examined the influence of several experimental conditions such as agitation time, metal ion concentration, adsorbent dose, and pH of the solution on the adsorption of Hg(II) on activated carbons obtained from coir pith and bicarbonate-treated peanut hulls and a commercial activated carbon. In their earlier work on commercial and peanut-hull carbons, they found that the adsorption decreased at pH < 4 in both the carbons but showed a different trend at pH > 4 (Figure 6.21). The decrease in adsorption at pH < 4 was attributed to the formation of $HgCl_2$, because the trend of change above pH 4 was not clearly elucidated. In a later publication using coir pith carbon, they observed that the adsorption of Hg(II) increased with pH from 2 to 5 and remained almost constant at higher pH values up to pH = 11.

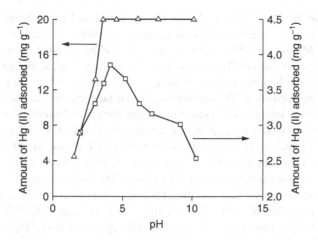

FIGURE 6.21 Effect of pH on the adsorption of mercury by activated carbons. (Source: Namasivayam, C. and Periasami, K., *Water Res.*, 27, 1663, 1993. With permission.)

Sasaki and Kimura[175] studied the adsorption of Hg(II) ions from aqueous solutions on palm-shell activated charcoal impregnated with immobilized polythiourea. The charcoal showed a higher capacity for the removal of Hg(II) ions but the increase in adsorption depended upon the degree of immobilization of the loaded thiourea. An aqueous solution containing 10 ppm of Hg(II) when passed through the polythiourea loaded carbon was able to reduce the mercury content in the elute to below 5 ppb, the amount permitted by the law. The adsorption of Hg(II) ions from solution has also been found to be enhanced by loading the activated carbons with zirconium[176] and dispersed iron,[177] the presence of the metal considerably enhancing the adsorption affinity. Shirakashi et al.,[178] in their studies on the adsorptive removal of Hg(II) ions on activated carbons from coal and coconut shell from aqueous solutions in the presence of bromine ions, found that the Hg^{2+} ions underwent reduction at pH = 7, although the reduction at pH = 1.4 was not clarified. The main adsorbed species were $[HgBr_2]$ and $[HgBr_3]^-$ at pH 7 and pH 1.4. The amount of $[HgBr_3]^-$ adsorbed increased with an increase in Br^- ion concentration in the solution.

Ammons et al.[98] studied the adsorption of methyl mercuric chloride from aqueous solutions on Filtrasorb-200 activated carbon at room temperature and found that the adsorption isotherms were nonlinear and did not fit either the Freundlich or the Langmuir isotherm equation. The adsorption was, however, reversible. Carrot et al.[179] carried out adsorptive removal of mercury(II) ions on five different commercial activated carbons associated with varying amounts of acidic and basic surface complexes. These workers observed that significant amounts of $HgCl_3^-$ and $HgCl_4^{2-}$ were adsorbed, but there was little adsorption of neutral $HgCl_2$ or Hg^{2+} ions. In the case of the basic carbons, the adsorption of the anions decreased with increase in the pH of the solution, but the uptake of cations was very small at pH = 0.2. They are of the view that the adsorption of Hg(II) ions is controlled by the charge of the carbon surface, giving rise to electrostatic attractive or repulsive interactions.

Rangel-Mendez and Streat[180] have recently studied the adsorption of Hg(II) ions from aqueous solutions of $HgCl_2$ on a commercial granulated activated carbon before and after electrochemical oxidation under a current of 3 mA m^2 in the presence of 0.5 M HCl as electrolyte. The carbon-oxygen surface groups present on the carbons were determined using Boehm selective neutralization technique. The adsorption of Hg(II) on the oxidized sample was about 10 times that on the as-received sample at pH = 5. The adsorption capacity increased with increase in pH of the solution because of the availability of higher amounts of dissociated functional groups for adsorption and ion-exchange as pH is increased. The dissociation of the acid or basic groups on the carbon surface and the pH of the solution changed the surface charge density. The isoelectric point (IEP) or the zero point charge (ZPC), as calculated from the zeta potential measurements at different pH values (Figure 6.22), were found to be at pH 2.9 and 0.96 for the oxidized and as-received activated carbons. This indicated that the carbon surface was positively charged below ZPC because of the protonation of the acidic groups, and it became negatively charged above IEP (ZPC) due to the ionization of acidic oxygen surface groups. This promoted electrostatic attractive interactions between the negatively-charged carbon surface and the positively-charged Hg(II) ions. Thus, these workers are of the view that at pH values between 2.5 and 5, the mercury can be removed from the aqueous solutions partly by adsorption of cation species such as Hg^{2+}, $HgCl^+$, and $Hg(OH)^+$ due to electrostatic attractive interactions, and partly by reduction of $HgCl_2(aq)$ to $Hg_2Cl_2(s)$ by the phenolic and hydroquinonic groups. The speciation diagrams of aqueous solution containing $10^{-4}M$ $HgCl_2$ (Figure 6.23) up to pH 5 contained about 99% $HgCl_2$ (aq), so that the reduction of Hg(II) into Hg(I) contributes to the removal of mercury from the solution.

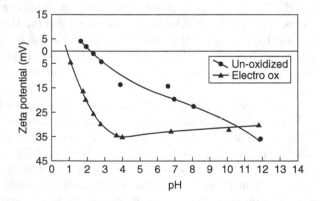

FIGURE 6.22 Electrophoretic mobility measurements of as-received and oxidized carbons. (Source: Rangel-Mendez, J.R. and Streat, M., *Carbon '01 Intern. Conf. on Carbon* , Lexington, KY, 2001. With permission.)

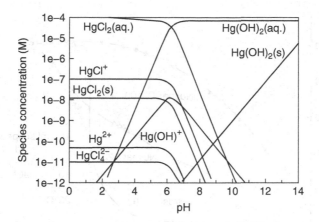

FIGURE 6.23 Speciation diagram for 0.4M mercuric chloride aqueous solution. (Source: Rangel-Mendez, J.R. and Streat, M., *Carbon '01 Intern. Conf. on Carbon*, Lexington, KY, 2001. With permission.)

6.5 ADSORPTIVE REMOVAL OF CADMIUM FROM AQUEOUS SOLUTIONS

The first systematic studies on the adsorptive removal of Cd(II) ions from aqueous solutions by activated carbons were reported by Huang and Ostovic.[69] These workers studied the influence of parameters such as pH of the solution, the nature of the carbon surface, the carbon dose, and the influence of addition of chelating agents on the adsorption of Cd(II) using several different activated carbons. The amount adsorbed increased with increase in the pH of the solution. The amount of Cd(II) adsorbed was very small at pH values less than 3 but increased significantly at higher pH values and became more or less constant at pH greater than 8. But the range of pH value at which maximum adsorption occurred was dependent on the nature of the activated carbon surface. The basic carbons adsorbed much larger amounts of Cd(II), even at lower pH values than the acidic carbons (Figure 6.24). Thus, Nuchar–C 190N activated carbon that was basic in character adsorbed about 3 times more Cd(II) compared to Nuchar-722 and a still larger amount compared to Filtrasorb-400 at pH = 5 from the same concentration of cadmium ions in the solution. The amount of Cd(II) ions adsorbed was also a function of the amount of Cd(II) in the solution and the carbon to cadmium ratio. By increasing the carbon dose, the removal of Cd(II) could be increased by three times. The adsorption efficiency of Filtrasorb-400 activated carbon increased appreciably by the addition of chelating agents such as nitrilotriacetate (NTA) and ethylene diamine tetracetate (EDTA). The addition of 1% chelating agent per unit weight of Cd(II) in the solution increased the removal efficiency by 20 to 40% at a pH of about 7. This enhanced adsorption was attributed to the formation of univalent cadmium complex anions, which were adsorbed on the positive sites on the carbon surface due to attractive electrostatic interactions. This was followed by the association of the Cd^{2+} ions with the adsorbed anions. In a subsequent publication, Huang and Smith,[181] using

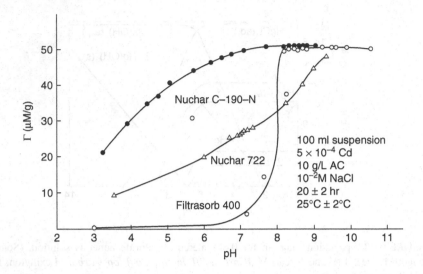

FIGURE 6.24 Influence of pH on the adsorption of Cd(II) by different activated carbons. (Courtesy C.P. Huang. With permission.)

four commercial activated carbons with isoelectric points (IEP or ZPC) varying between 3.8 and 7.1 pH, the activated carbons with lower IEP values were found to be more effective for the removal of Cd(II) from the solutions. These studies showed that powdered activated carbons that had low IEP values showed better adsorption capacity for Cd(II) than the granulated activated carbons that had higher IEP values. They suggested that certain functional groups on the powdered carbon surface show greater chemical affinity for Cd(II) species in the solution.

Dobrowolski et al.[70], studied the removal of Cd(II) ions from aqueous solutions on a Merck de-ashed activated charcoal and after modification by oxidation with hydrogen peroxide and by degassing in argon at 1400 K. The adsorption isotherms of Cd(II) obeyed the Langmuir equation. The relative adsorption isotherms showed that at low concentrations of Cd(II) ions in the solution, the adsorption was maximum in the case of the oxidized carbon sample and minimum in the case of the degassed sample. This was attributed to a larger pH change caused by the exchange of Cd(II) ions in the solution with the H^+ ions of the acidic surface groups in the oxidized carbon sample. The charge density versus pH curves for the three carbon samples showed that the surface of the degassed carbon sample had a positive charge over most of the pH range. This positive charge was stabilized by the presence of excess of OH^- ions in the diffuse external part of the double layer. This caused the solubility products of $Cd(OH)_2$ to be exceeded resulting in the precipitation of cadmium. De-ashed and oxidized carbon samples showed a positive surface charge only over very low pH values, and the surface charge changed from positive to negative as the pH increased. These workers suggested that this change of charge from positive to negative was due to the participation of different surface groups in the adsorption process at different pH values. However, they did not define the range of pH and

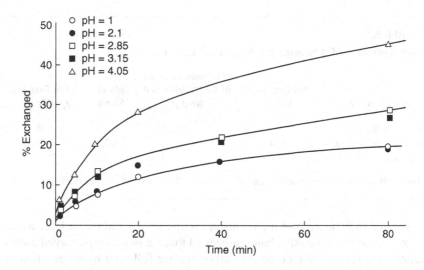

FIGURE 6.25 Isotopic exchange kinetics as a function of pH of $CdCl_2$ solution. (Source: Dobrowolski, R., Jaroniec, M., and Kosmulski, M., *Carbon*, 24, 15, 1986. With permission.)

the nature of the surface groups that contribute to this phenomenon. These workers also carried out the isotopic ion exchange reactions with the cadmium-loaded carbon samples.

The kinetics of isotopic exchange reactions showed that the exchange was very fast in the initial stages at high pH values (Figure 6.25). This was attributed to the exchange of Cd(II) adsorbed at easily accessible surface sites and micropores. The slow exchange reaction at low pH values was due to exchange with the Cd(II) physisorbed deep into the pores. Thus, these workers attributed adsorption of Cd(II) on activated carbon surface to physisorption in the micropores and to exchange mechanism. Similar Cd(II) adsorption studies were carried out by Polovina et al.[182] using an activated carbon cloth, whose surface was modified by oxidation with air, H_2O_2, and HNO_3 and by degassing in argon at 1000°C. The acidic carbon-oxygen surface groups were determined using Boehm's selective neutralization technique. These workers suggested that the adsorption of Cd(II) ions involved exchange mechanism with the acidic surface groups, although only a small fraction of the acidic functional groups had been exchanged. But their results on the 1000°-degassed carbon sample that had no acidic surface groups but showed adsorption of Cd(II) ions could not be explained. This is probably due to the fact that these authors did not take into account the charge on the carbon surface as a result of pH change of the solution.

Bhattacharya and coworkers,[132,183] Reed and Nonavinakere,[184] and Rubin and Mercer[185] studied the influence of chelating agent EDTA on the adsorptive removal of Cd(II) from aqueous solutions by activated carbons. Although Rubin and Mercer found little effect of the chelating agent on the adsorption, Bhattacharya and coworkers did find an increase in adsorption in the presence of EDTA, at least in the case of basic carbons. These coworkers even proposed the formation of an electron donor-acceptor

TABLE 6.9
Adsorption of Cadmium on Activated Carbons

Carbon Sample	Surface Area (m²/g)	Maximum Amount of Cd(II) Adsorbed (mg/g)	pH of Slurry	Ion Exchange Capacity (meq/g)
Peanut hull carbon	208	~20	6.7	0.49
Commercial Activated Carbon	354	~1.8	8.2	~0.00

Source: Periasamy, K. and Nanasivayam, C., *Ind. Eng. Chem. Res.,* 33, 317, 1994. With permission.

complex of the chelate and the carbonyl groups on the carbon surface. Reed and Nonavinakere,[184] on the other hand, suggested that the negatively-charged cadmium-chelate complex is adsorbed on the carbon surface followed by the reaction of the positively charged Cd^{2+} ions with the carbon-ligand complex. Periasami and Namasivayam[186] found that the adsorption capacity of Cd(II) on a peanut-hull char-coal was much higher than that of a commercial activated carbon, although the surface area of the commercial activated carbon was larger (Table 6.9). These workers have suggested that electrostatic interaction forces and specific chemical interactions play an important role in the adsorption of Cd(II).

Ferro-Garcia et al.[187] carried out adsorption of Cd^{2+} ions as well as Cu^{2+} ions and Zn^{2+} ions on activated carbons obtained from almond shells, olive stones and peach stones with ash contents less than 0.25%. The pH of the solution was found to have a great effect on the adsorption of Cd^{2+} ions. There was little or no adsorption at low pH values, but the adsorption increased sharply in the pH range 3 to 5, attaining almost a constant value at higher pH values (Figure 6.26). These results

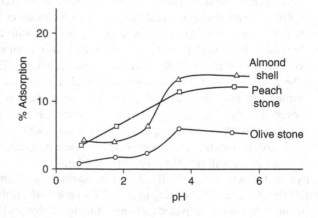

FIGURE 6.26 Adsorption of Cd(II) ions by activated carbons as a function of solution pH. (Source: Ferro-Garcia, M.A., Rivera-Utrilla, J., Rodriguez-Gordillo, J., and Bautista-Toledo, I., *Carbon*, 26, 363, 1988. With permission.)

TABLE 6.10

Langmuir Equation Constants for Adsorption of Cd(II) on Different Activated Carbons at Two Different Temperatures

Activated Carbon From	Temperature (K)	x_m(vmol/g)	K (L/g)
Almond shell	293	22.2	1.3
	313	17.9	1.2
Olive stone	293	52.6	0.6
	313	51.4	0.4
Peach stone	293	29.1	1.0
	313	18.9	0.9

Source: Adapted from Ferro-Garcia, M.A., Rivera-Utrilla, J., Rodriguez-Gordillo, J., and Bautista-Toledo, I., *Carbon*, 26, 363, 1988. Reproduced with permission from Elsevier.

have been explained on the basis of the changes in the carbon surface charge with changes in pH of the solution. At low pH the carbon surface has a positive charge, and the electrostatic repulsive interactions between the cations and the positively charged surface will take place. As the pH of the solution was enhanced, the Cd^{2+} ions replaced H^+ ions from the carbon surface and increased the adsorption. The sharp increase in the adsorption at pH 3 to 5 was attributed to the zero point of charge (ZPC) of the carbons lying between these pH values. The adsorptive isotherms of Cd^{2+} ions were Type I of the BET equation and fitted the Langmuir equation.

The data obtained from the Langmuir plots at two temperatures for the three carbons (Table 6.10) showed that the maximum adsorption capacity varied between 52.6 μmol/g for almond shell carbon to 17.9 μmol/g for olive-stone carbon. The Xm and K values indicated that the adsorption was exothermic and, therefore, was increased on decreasing the temperature. The adsorption of Cd(II) ions also increased on the addition of ions such as Cl^-, CN^-, and SCN^- but decreased in the presence of EDTA. These were attributed to the formation of some complex anions,[188] which would be adsorbed on the activated carbon surface. The lower adsorption in the presence of EDTA could be due to its larger size so that its complex with cadmium was inaccessible to some of the micropores. Although these workers suggested that the adsorption behavior of these activated carbons toward adsorption of Cd(II) ions depended upon the chemical nature of these surfaces, they did not correlate it with the nature and the concentration of the carbon-oxygen surface groups present on these carbons. Marcias-Garcia et al.[189] used a variety of chemically-modified activated carbons for the removal of Cd(II) from solutions.

Brennsteiner et al.[190] used vapor-grown carbon fibers, a coal-derived foam, and carbon nanofibers for electrochemical removal of heavy metals, including Cd^{2+} ions. These workers suggested that the adsorptive removal was due to the development of porosity[189] and high surface area[190] of the carbon materials. They did not consider the role of surface chemistry of the activated carbon or the pH of the solution in influencing the adsorption of Cd(II) by carbons. Jia and Thomas[191] studied the

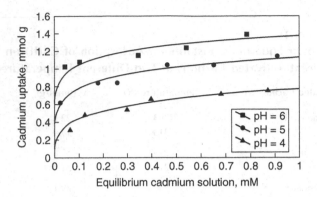

FIGURE 6.27 Adsorption isotherms of Cd on modified WHK carbon at different solution pH. (Source: Ferro-Garcia, M.A., Rivera-Utrilla, J., Rodriguez-Gordillo, J., and Bautista-Toledo, I., *Carbon*, 26, 363, 1988. With permission.)

adsorption of Cd^{2+} ions from aqueous solutions on a coconut-shell charcoal, after introducing different types of carbon-oxygen surface groups, and observed that the adsorption was considerably enhanced. This increase in adsorption was attributed partly to the increase in the ion-exchange capacity of the carbon and partly to the physisorption of hydrated cadmium ions.

Rangel-Mendez and Streat[180] examined the adsorption of Cd^{2+} ions from $CdCl_2$ aqueous solution using on activated carbon before and after its electrochemical oxidation in the presence of 0.5 M KCl solution as electrolyte. The adsorption capacity of the carbon increased with oxidation. The adsorption capacity also increased with the pH of the solution (Figure 6.27). This was attributed to the greater availability of ionized functional groups for ion exchange with the metal ions as pH increases. At low pH values, the carbon surface is positively charged due to excessive protonation. Therefore, the repulsive electrostatic interactions between the carbon surface and the Cd^{2+} ions lower the adsorption. As the pH increases beyond Isoelectric point (IEP,) which is at pH 2.19 and 0.96 for the as-received and oxidized carbons, the carbon surface becomes negatively charged, which enhances the attractive electrostatic interactions between the negatively-charged surface and the positively-charged Cd^{2+} cations, and increases the adsorption. The distribution and concentration of different cadmium species in the solution at different pH values (Figure 6.28) shows that at low pH values the cadmium species present in the solution up to pH = 7.5 exist as Cd^{2+} (~60%) and Cd Cl+ (~40%) so that the adsorption is likely to involve ion exchange due to electrostatic interactions. At higher pH values the adsorption may partly be due to precipitation of $Cd(OH)_2$).

6.6 ACTIVATED CARBON ADSORPTION OF COBALT FROM AQUEOUS SOLUTIONS

Cobalt is one of the heavy metal toxic pollutants that is present in trace amounts in natural waters, the amount usually being less than 1 μg/L.[192] Consequently, its removal by adsorption using suitable adsorbents has been studied by a number of

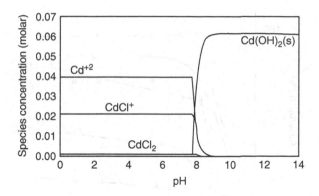

FIGURE 6.28 Speciation diagram of 0.1M cadmium chloride aqueous solution. (Source: Ferro-Garcia, M.A., Rivera-Utrilla, J., Rodriguez-Gordillo, J., and Bautista-Toledo, I., *Carbon*, 26, 363, 1988. With permission.)

investigators.[193–195] Kaya and Akyol[46] studied the adsorption of Cd(II) ions from aqueous solutions by activated bontonite as a function of temperature, contact time, and concentration of the solution, and observed that appreciable amounts of the metal could be removed. Shakir et al.[74] examined the removal of Co(II) ions from radioactive process waste water using activated charcoal as the adsorbent and gelatin as a collector. The effect of the pH of the solution and the carbon and collector dose on the adsorption was determined. Removals better than 97% could be achieved under optimum conditions in the pH range 7.5 to 10.0. The results were discussed in terms of the hydrolysis of the metal ions. In a later publication these workers[75] used Kaolinite for the removal of Co(II) ions from aqueous solutions in the presence of cationic and anionic surfactants. The removal of Co(II) increased with increase in the pH of the solution. The presence of strongly ionized cationic as well as anionic surfactants decreased the removal, and the presence of weakly ionized surfactants enhanced the removal of Co(II) ions.

Netzer and Hughes[196] examined the influence of pH on the adsorption of Co(II) ions by a number of activated carbons and found that the maximum adsorption occurred at a pH of 4. These workers invoked the effect of neither the carbon-oxygen surface groups nor of the electrostatic interactions and suggested that the adsorption was dependent only on the pH of the solution. Huang et al.[197] on the other hand, while using an *H* and an *L* carbon for the removal of Co(II) ions at different pH values, suggested that the removal involved the formation of Co(II) complexes with the phenolic groups especially on the L carbon surface.

Rivera-Utrilla and coworkers[48,198,199] carried out adsorption studies of Co(II) ions from aqueous solutions using activated carbons prepared from almond shell activated to different degrees by activation in CO_2 at 1123 K, and after the formation of carbon-oxygen acidic surface groups on treatment with nitric acid, hydrogen peroxide, and in air, and carbon-nitrogen basic surface groups on treatment with ammonia. The surface acidity and the surface basicity of each carbon was determined by titration with sodium hydroxide and HCl solutions, respectively. The adsorption isotherms of Co(II) ions on three typical samples of carbons, namely commercial

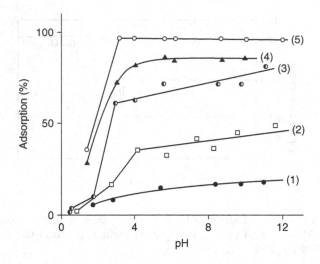

FIGURE 6.29 Adsorption of Co(II) ions on different activated carbons as a function of solution pH. (1) A-5 (2) A-8 (3) A-24 carbons activated in CO_2 at 1123 K, (4) as-received activated carbon, (5) A-5 oxidized with HNO_3. (Source: Rivera-Utrilla, J. and Ferro-Garcia, M.A., *Carbon*, 25, 645, 1987. With permission.)

Merck activated carbon and two almond shell carbons (one activated in CO_2 for 14 hr, and the other activated in CO_2 and then oxidized with HNO_3) showed that the adsorption was considerably larger on the nitric acid oxidized sample. These workers also studied the influence of solution pH on the adsorption of Co(II) ions (Figure 6.29) in the pH range 1 to 12 using some selected activated carbons. The adsorption of Co(II) ions increased sharply in the pH range 1 and 4 on all the activated carbons. At higher pH values while the adsorption attained a constant value in the case of nitric acid oxidized (~97%) and commercial carbon (~85%), there is a continuous increase in the case of carbons activated in CO_2. At low pH values (pH < 3), the activated carbon surface has a positive charge and this resulted in the repulsive electrostatic interactions between the positively-charged surface and the positively-charged Co(II) cations. As the pH increases the carbon surface attains a negative charge due to the ionization of acidic carbon-oxygen surface groups that resulted in the attractive electrostatic interactions, thereby increasing the adsorption of Co^{2+} ions. At higher pH values hydrolysis of Co(II) may become important resulting into the formation of $Co(OH)^+$ ions around pH 10[188,200] and into its adsorption on the carbon surface.

These workers also found that the adsorption of Co(II) was higher on the ammonia-treated samples compared with the CO_2-activated carbon samples. It has been suggested that in the former case, the adsorption process involves the formation of some cobalt-amino complexes such as $[Co(NH_3)_n]^{3+}$ in which the cobalt is present as Co^{3+} ions so that the adsorptive removal of Co(II) is higher. The adsorption of Co(II) on activated carbons increased considerably in the presence of anions such as Cl^-, Br^-, CH_3COO^-, NO_3^-, and $S_2O_3^{2-}$, and decreased in the presence of tartrate and citrate ions. It was suggested that the former anions were adsorbed on the carbon surface, enhancing the negative charge density, which increased the capacity of the surface to adsorb Co(II) cations. However, the presence of the latter ions (i.e., tartrate and citrate ions) resulted in the formation of stable complexes with cobalt cations, which may not be accessible to many of the smaller micropores in some activated carbons due to their larger size.

Bansal and coworkers[201] and Aggarwal[202] studied the adsorption of Co(II) ions from a solution of cobalt acetate on four granulated and fibrous activated carbon samples. All four samples adsorbed appreciable amounts of Co(II) ions, although the amount adsorbed was different for the different samples. The amount adsorbed, however, could not be explained on the basis of surface area alone. The adsorption fitted well with the Freundlich isotherm equation and gave linear plots on log c versus log amount adsorbed (Figure 6.30). The values of the constant n of the Freundlich equation were calculated and found to vary with the heterogeneity of the carbon surface, due to the presence of carbon-oxygen surface groups. The adsorption of Co(II) increased on oxidation of the carbons with nitric acid and oxygen gas. The increase in adsorption was much larger on oxidation with nitric acid than their oxidation with oxygen gas (Figure 6.31). This was attributed to the fact that nitric acid was a stronger oxidative treatment and resulted in the formation of larger amounts of acidic carbon-oxygen surface groups, chemical groups that provide sites for the adsorption of Co^{2+} ions. When these carbon-oxygen chemical groups were removed gradually by degassing in vacuum at increasing temperatures, the adsorption of Co(II) showed a gradual decrease (Figure 6.32).

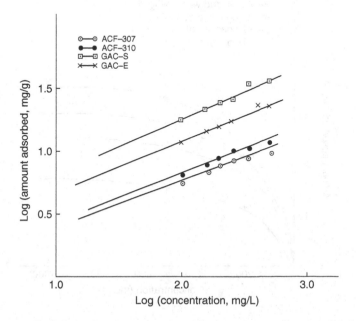

FIGURE 6.30 Freundlich adsorption isotherms of Co(II) ions on different activated carbons. (Source: Aggarwal, D., Ph.D. thesis, Panjab Univ., Chandigarh, India, 1997. With permission.)

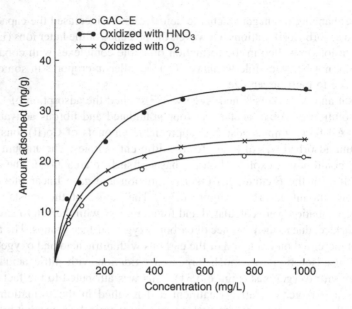

FIGURE 6.31 Adsorption isotherms of Co(II) ions on GAC-E before and after oxidation. (Source: Aggarwal, D., Ph.D. thesis, Panjab Univ., Chandigarh, India, 1997. With permission.)

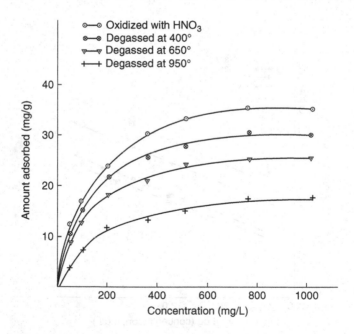

FIGURE 6.32 Adsorption isotherms of Co(II) ions on oxidized GAC-E before and after degassing. (Source: Aggarwal, D., Ph.D. thesis, Panjab Univ., Chandigarh, India, 1997. With permission.)

The oxidation of activated carbons and the formation of acidic carbon-oxygen surface groups that have been postulated as carboxyls and lactones changed the pH of the carbon surface and hence of the solution. The pH of as-received activated carbon fiber sample (ACF-307) was ~7, and that of the granulated activated carbon sample (GAC-E) was ~10. On oxidation with nitric acid, the pH of the two samples was reduced to between 4 and 5. As the ZPC of these activated carbons was between 2 and 3, the surface of these carbons will show a positive charge below ZPC, resulting in repulsive electrostatic interactions between the positively-charged carbon surface and the Co^{2+} ions. At pH greater than ZPC, the carbon surface attained a negative charge due to ionization of acidic surface groups. This resulted in the dominance of attractive electrostatic interactions between the carbon surface and the positively-charged Co^{2+} and $(CoOH)^+$ ions in the solution. When the acidic carbon-oxygen surface groups were removed almost completely as in the case of 950°-degassed samples, and the pH of the solution was higher, the cobalt acetate hydrolyzed to produce $Co(OH)_3^-$ or $Co(OH)_4^{2-}$ ions. The electrostatic interactions of these ions with the carbon surface decreased the adsorption of Cobalt (II).

Takiyama and Huang[203] studied the influence of pH of the solution on the adsorption of several metal ions including Co(II) from chlorate solutions on an acidic and a basic activated carbon and observed a gradual increase in the adsorption in the pH range 2 to 11. However, for the acidic carbon the adsorption increased sharply in the pH range 2 to 6 while it increased sharply in the pH range 6 to 9 in the case of the basic activated carbon. These workers inferred that the surface acidity of an activated carbon plays an important role in the removal of cations from aqueous phase.

Kim et al.[204] has recently studied the electrosorption of Co(II) ions from aqueous solutions on a cellulose-based activated carbon fiber KF-1500 after modifying the ACF surface by dipping in 1M HNO_3 and 1M NaOH solutions at room temperature. The zeta potential changes of different ACF samples with pH of the solution (Figure 6.33) showed that the zero point charge (pH$_{ZPC}$) of the untreated sample was

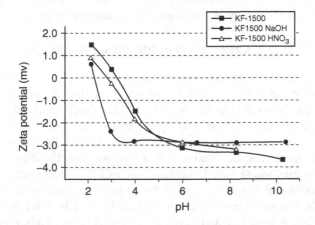

FIGURE 6.33 Zeta potential of different activated carbon fibers before and after different treatments. (Source: Kim, H.S., Ryu, S.K., Park, K.K., Lee, J.B., and Jung, C.H., *Carbon '02' Intern. Conf. on Carbon,* Beijung, 2002, Paper PI 91 DO 98. With permission.)

TABLE 6.11
Surface Acidic Groups on Activated Carbon Fiber KF-1500

ACF	Functional Group (meq/g)			Total Acidity (meq/g)
	Carboxyl	Lactone	Phenol	
KF-1500	0.044	0.487	0.314	0.850
KF-1500-HNO$_3$	0.785	0.816	0.633	2.230
KF-1500-NaOH	0.012	0.542	0.125	0.680

Source: Kim, H.S., Ryu, S.K., Park, K.K., Lee, J.B., and Jung, C.H., *Carbon 02 Intern. Conf. on Carbon,* Beijung, 2002, Paper PI 91 DO 98. With permission.

at pH = 3 and this was shifted to pH 2.5 on treatment of the ACF with nitric acid and to pH 2.1 on treatment with NaOH. The total surface acidity of the ACF increased 2.6 times by HNO$_3$ treatment and decreased by treatment with NaOH (Table 6.11). The electrosorption capacity of the ACF for Co(II) ion increased with treatment of the ACF with nitric acid and sodium hydroxide. This was attributed to the increase in the amount of carboxyl and lactonic groups on these treatments. The pH range of aqueous solution for the removal of Co(II) ions was also broadened by these treatment due to a change in ZPC. Paajanon et al.[205] has reported the removal of traces of radioactive cobalt (Cobalt-60) from nuclear waste solutions at pH 5.0, 6.7, and 9.1 using activated carbons prepared from peat, coconut shell, or coal but Tekker et al.[206] used activated carbons obtained from rice hulls to study the effect of pH, cobalt-concentration, contact time, and temperature on the removal of Co(II) from waste water. The results obtained were similar to those obtained earlier, showing an increase in adsorption of Co(II) with increasing pH. The optimum conditions for the removal time of Co(II) ions were pH 6 to 7, contact time 40 min, and carbon dosage 1.5g per 50 ml cobalt solution. These workers[205,206] suggested that the adsorption of Co(II) ions probably involves an ion-exchange mechanism.

6.7 ACTIVATED CARBON ADSORPTION OF NICKEL

Electroplating is one important process involved in surface finishing and metal deposition for better life of the article and for decoration. Although several metals can be used for electroplating, the most commonly used metals are copper, nickel, and chromium, the choice depending upon the specific requirements of the article. During rinsing of the electroplated articles and washing of the electroplating tanks, considerable amounts of the metal ions find their way into the effluent. Thus, waste water of industries where nickel plating is one of the process contains Ni(II) ions. It has been estimated that the amount of Ni(II) in waste water varies between 6 and 12 mg/L, which is much above the safe limit and causes harmful effects on human beings, animals, aquatic life, and vegetation. Several types of adsorbents are chemically active inorganic beds,[112] ungranulated blast furnace slag,[73] flyash,[53,72] and activated carbons.[49,56,67,132,207]

Tewari et al.[56] examined the removal of Ni(II), Cu(II), and Cr(VI) ions from dilute aqueous solutions in the concentration range 5 to 50 mg/L by adsorption on an activated carbon in the pH range 5.5 to 8.0. The adsorption was maximum at pH 5.5 for copper and 8.0 for Cr(VI), but the adsorption was appreciable both in the acidic and basic pH ranges for Ni(II) ions. However, the rate of adsorption for NI(II) ions was faster in the alkaline solutions. Singh and Rawat[67] used low-grade bituminous coal before and after oxidation with H_2O_2 and MnO_2 for the removal of Ni (also Cu, Zn, and Cr) from aqueous solutions. The rate of adsorption and the adsorption capacity of the coal enhanced significantly upon oxidation. The adsorption data followed the Langmuir equation. The adsorptive removal of Ni(II) ions from aqueous solutions was studied by Corapcioglu and Huang[49] using several different activated carbons at different pH values of the solutions and attributed Ni(II) ion adsorption to the surface chemical functional groups that ionize and exchange H^+ ions with the Ni^{2+} cations. Seco et al.[207] while studying the adsorption of Ni(II) ions (also of Cu, Zn, Cd) as well as some binary systems on a commercial activated carbon at different pH values of the solution, observed that the adsorption increased monotonically with pH of the solution. They explained their results on the basis of a decrease in the positive charge on the carbon surface and to a decrease in competition between the H^+ and the metal species for the surface sites. In the case of the binary Cu-Ni system, it was found that Ni^{2+} was not as strongly attracted by the carbon surface as the Cu^{2+} ions.

Jevtitch and Bhattacharya,[183] while examining the adsorptive removal of Ni(II) and Cd(II) cations on an activated carbon from aqueous solution at pH 7.5 to 8.0, observed that the adsorption decreased in the presence of ethylene diamine (EDTA), although in a subsequent study Bhattacharya and Cheng[132] working at pH 5 of the solution observed that the adsorption of both Ni(II) and Cd(II) ions increased in the presence of EDTA and attained maximum adsorption capacity of the activated carbon. They suggested that at this pH (pH = 5), both Ni(II) and Cd(II) are present in solution as negatively-charged complexes. As the activated carbon surface at this pH values has a positive charge (ZPC pH = 9.0), the electrostatic attractive interactions enhance the adsorption of both Ni(II) and Cd(II). Reed and Nonavinakere[184] also studied the influence of pH on the adsorption of Ni(II) ions in the presence and absence of EDTA on activated carbons at different pH values between 2 and 11. According to them the activated carbon surface (ZPC pH = 10.4) has a positive charge below pH 10.4 and a negative charge above this pH value. The EDTA complex of nickel in the solution, on the other hand, carries a negative charge so that there are electrostatic attractive interactions at lower pH values resulting in increased adsorption. As the pH value was increased, the postive carbon surface charge decreased and ultimately became negative, resulting in a decrease in the attractive interactions and becoming repulsive at higher pH values. The authors, however, cautioned that in addition to these electrostatic interactions, some other factos also play an important role in the adsorption of metal ions. Brennsteiner et al.[190] also found that the electrochemical removal of nickel by activated carbons was dependent on the pH of the solution, the maximum removal being at a pH = 7, but they did consider the part played by carbon-oxygen surface groups.

Goyal et al.[208,129] determined the adsorption isotherms of Ni(II) ions from aqueous solutions on two samples of granulated (GAC) and two samples of activated carbon fibers (ACF), associated with varying amounts of carbon-oxygen surface groups. The adsorption isotherms were Langmuir in shape, showing an initial rapid adsorption tending to be constant at higher concentrations. The adsorption of Ni(II) ions increased on oxidation of the carbon with nitric acid and oxygen gas at 350°C, the magnitude of increase being different for the two oxidation treatments. The adsorption increased from 2.7% to 4.4 and 3.1% in the case of GAC-E, and from 0.9 to 2.9 and 1.65 for ACF-307, respectively, on oxidation with nitric acid and oxygen gas. (Figure 6.34). This increase in adsorption of Ni(II) ions has been attributed to an increase in the amount of acidic carbon oxygen surface chemical groups on the activated carbon surface. This received further support from the adsorption of Ni(II) ions on the oxidized samples after degassing at gradually increasing temperatures of 400°, 650°, and 950°C (Figure 6.35). This treatment eliminates varying amounts of the carbon-oxygen surface groups more so acidic carbon-oxygen surface groups. The decrease in adsorption was only small (between 0.2 and 0.3%) for both the carbons on degassing at 400°C because this temperature eliminated only a small proportion of the acidic surface groups (~15%). However, the decrease in adsorption of Ni(II) was considerably large on the 650°-degassed carbon samples because at this temperature a major portion (~85%) of the acidic groups had desorbed, although these samples still retained a large portion of the nonacidic oxygen groups.

The presence of acidic groups on the carbon surface (postulated as carboxyls and lactones) and their ionization in water changes the pH of the carbon suspension. The pH of as-received ACF-307 and GAC-E samples were 7 and 10 respectively. The pH of these carbons was reduced to between 3 and 4 on oxidation with nitric

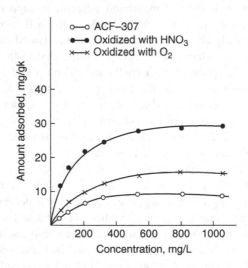

FIGURE 6.34 Adsorption isotherms of Ni(II) ions on ACF-307 before and after oxidation. (Source: Goyal, M., Rattan, V.K., and Bansal, R.C., *Indian J. Chem. Technol.*, 6, 305, 1999. With permission.)

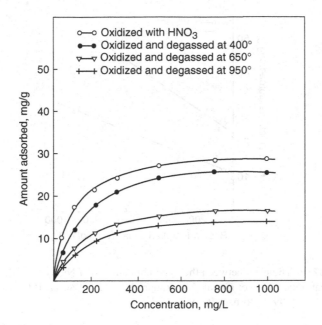

FIGURE 6.35 Adsorption isotherms of Ni(II) ions on HNO$_3$-oxidised ACF-307 before and after degassing. (Source: Goyal, M., Rattan, V.K., and Bansal, R.C., *Indian J. Chem. Technol.*, 6, 305, 1999. With permission.)

acid and to between 4 and 5 on oxidation with oxygen gas. The effect of the pH of the solution on the adsorption of Ni(II) ions in the case of ACF-307 and GAC-E activated carbons (Figure 6.36) showed that the maximum adsorption occurs in the pH range 3 to 5.

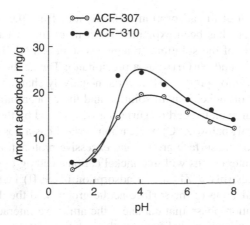

FIGURE 6.36 Adsorption isotherms of Ni(II) ions on ACF-310 as a function of pH of the solution. (Source: Goyal, M., Rattan, V.K., and Bansal, R.C., *Indian J. Chem. Technol.*, 6, 305, 1999. With permission.)

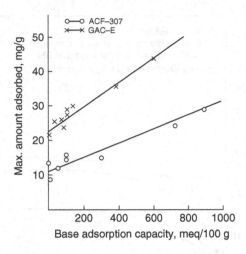

FIGURE 6.37 Relationship between the maximum amount of Ni(II) ions adsorbed and base adsorption capacity. (Source: Goyal, M., Rattan, V.K., and Bansal, R.C., *Indian J. Chem. Technol.*, 6, 305, 1999. With permission.)

A relationship between the maximum amount adsorbed and total surface acidity (as determined by titration with NaOH solution) of different as-received, oxidized and degassed ACF-307 and GAC-E activated carbons (Figure 6.37) shows that the points for the two carbons could not be collected along a single straight line. However, the plots for a given carbon show almost a linear variation of adsorption with total surface acidity. Thus, giving rise to two linear plots, one for each carbon. The difference in behavior of the two carbons in giving two different linear plots was attributed to the difference in the microporous structure of the two activated carbons.

The mechanism of the adsorption of Ni(II) ions from the aqueous solution by the activated carbons has been explained in terms of the surface chemistry of the carbons and the pH of the solution. In aqueous solutions NI(II) may exist as Ni^{2+}, $Ni(OH)^+$, $Ni(OH)^{2-}_4$ and $Ni(OH)_2$ (as a precipitate): The acidic carbon-oxygen surface group on ionization produce H^+ and the negatively charged surface sites. However, the presence of nickel ions in solution and the concentration of negative sites or the negative charge on the carbon surface is determined by the pH of the solution. At low pH value (pH below ZPC), the carbon surface has a positive charge because of low ionization of the surface groups and excessive protonation of the surface and creates repulsive interactions with the nickel cations leading to low adsorption. But at higher pH values (pH > ZPC), the adsorption of Ni(II) increases because of a high degree of ionization of the surface acidic groups and the dominance of nickel cations in the solution phase that enhances the attractive interactions. At very high pH values, the dominant nickel ions in the solution are anionic complexes that produce repulsive interactions and decrease the adsorption. Similarly, the elimination of acidic carbon-oxygen surface groups on degassing decreased the negatively charged sites on the carbon surface and resulted in a decrease in the adsorption of

Ni(II) on the carbon surface. Thus, the adsorption of Ni(II) on the activated carbon surface is determined by the concentration of acidic surface groups that also influence the pH of the solution.

6.8 REMOVAL OF LEAD FROM WATER

Lead is a pollutant that is present both in drinking water and in air. In air, it is derived from lead emissions from automobiles because it is used as an antiknocking agent in the form of lead tetraethyl in gasoline. The problem attains alarming proportions in industrialized countries and in big urban conglomerates, where a large number of vehicles are on the roads. In water, lead is released in effluents from lead treatment and recovery industries, especially from lead battery manufacturing units. In several regions lead is present in water in amounts much larger than the toleration limit of 50 μg/L.[157,192,209] In the human body, lead tends to form complexes that interfere with several synthetic and metabolic processes. Consequently, the removal of lead from water using activated carbons has been the subject matter of several investigations.

Netzer and Hughes[196] studied the adsorption of Pb(II) ions from aqueous solutions with different pH values on activated carbons and found that the adsorption involved both the hydrolysis and the precipitation of the metal, depending on the pH of the solution. They did not discuss the importance of carbon-oxygen surface groups on the adsorption of Pb(II). Corpacioglu and Huang[49] observed that the adsorption of Pb(II) was larger at higher pH values of the solution and involved complex ion formation and precipitation of the lead ions as Pb(OH)$_2$. Cheng et al.,[210] while studying the adsorption of Pb(II) ions on activated carbons associated with different carbon-oxygen surface groups at different pH value of the solution, however, suggested that the adsorption was a function of the pH of the solution and the nature of the carbon-oxygen surface groups. The surface groups showed varying degrees of protonation at different pH values of the solution. Taylor and Kuennen[211] also observed that the carbon-oxygen surface groups, which they called *surface oxides*, and the pH of the solution played an important role in the adsorption of Pb(II) ions by activated carbons from aqueous solutions. In their dynamic adsorption studies, these workers found that the breakthrough times both for H- and L-type active carbons at pH 6.5 were considerably reduced when the carbons were deoxygenated in nitrogen at 600°C. However, they did not dwell upon the nature of these surface oxides, although they did state that the change in pH of the solution affects the carbon surface charge, which is responsible for the electrostatic interactions between the carbon surface and the lead ions in the solution.

Ferro-Garcia et al.[209] studied the removal of lead from aqueous solutions by activated carbons prepared from almond shells, (Sample A), olive stones (Sample H) and peach stones (Sample P), and by conventional activated carbon (Sample M). The adsorption isotherms were Type II of the BET classification (Figure 6.38). The adsorption capacity of Pb^{2+} ions varied between 17.5 mg/g for carbon P to 22.7 mg/g for carbon A. The adsorption increased with increase in the surface area of the activated carbon. The surface areas occupied by the Pb^{2+} ions were calculated using 0.802 nm as the diameter of the ion, assuming it to be hydrated with hydration

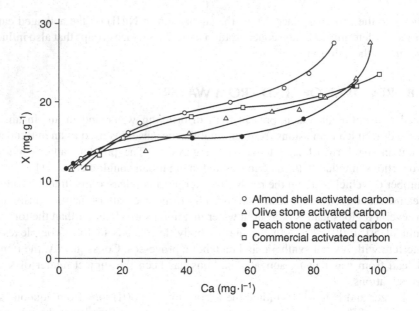

FIGURE 6.38 Adsorption isotherms of Pb(II) ions on different activated carbons. (Source: Ferro-Garcia, M.A., Rivera-Ultrilla, J., and Bautista-Toledo, I., *Carbon*, 28, 545, 1990. With permission.)

number between 4 and 7.5.[212] The lead ions covered only a small fraction of the BET surface area, which was attributed to the fact that many of the micropores were not accessible to larger hydrated Pb(II) ions.

These workers also carried dynamic adsorption studies using different columns of activated carbons to find out the operational parameters for the optimum removal of Pb(II) ions and determined the effect of the presence of different anions on the bed characteristics. The addition of Cl^-, NO_3^-, and ClO_4^- as sodium salts enhanced the amount of Pb(II) ions removed at specific breakpoints, the increase being larger in the case of ClO_4^- ions. These results have been explained on the basis of electrostatic and chemical interactions. According to these workers, the surface of these carbons has a positive charge and, therefore, adsorbs these anions. These adsorbed anions enhance the density of negative charge, thereby enhancing the electrostatic attractive interactions, which increases the adsorption of Pb(II) ions. In the case of Cl^- ions, there is formation of lead complex anions,[213,214] which may also favor the adsorption process. The presence of carbon surface charge could also explain the increase in adsorption of Pb^{2+} ions, because the carbons surface charge is a function of the solution pH.

Sohail and Qadeer[215] studied the adsorption of Pb(II) ions from aqueous solutions on commercial activated carbons and determined optimum conditions for maximum adsorption in terms of the time of contact, solution concentration, and temperature. The adsorption was found to obey first-order kinetics. The magnitude of adsorption energy calculated from D-R equation was found to be 7.61 KJ/mol, which showed that the adsorption involved weak bonding between the Pb^{2+} ions and the activated carbon surface, Yang et al.[216] used two activated carbons obtained from coal for dynamic adsorption of Pb(II) ions. The adsorption data conformed to Freundlich isotherm equation. Reed et al.[217–221] carried out batch and column studies to determine the influence of pH and hydraulic loading rate on the removal of Pb(II) from aqueous solution, using a granulated activated carbon. The removal of lead increased with increasing pH of the solution. However, they attributed the removal to adsorption and precipitation on the surface and within the pores. Although they found a large difference in the adsorption on pretreated and as received activated carbon samples, they did not accept the role played by electrostatic interactions between the carbon surface and the solution Pb(II) ions.

6.9 ADSORPTIVE REMOVAL OF ZINC

Zinc is among the most important heavy-metal pollutants in source and treated water.[222,223] It is leached in water mainly from corrosion of galvanized metals[113] and when used as a micronutrient in agriculture. Its removal from water using different types of activated carbons has therefore been the subject matter of several investigations.[50–53,115,179,187,224,225]

Mu and Yang[80] studied adsorption of Zn^{2+} and its complexes on an activated charcoal as a function of solution pH, added inorganic ions and carbon-oxygen surface groups. The adsorption was found to depend on the pH of the solution and increased with increase in the surface acidity of the carbon surface. Bencheikh[81] observed that the removal of zinc from waste water by adsorption on Sphegnum peat could reach a value above 90%. The adsorption data fitted the Langmuir isotherm equation. Petrov et al.[50] and Budinova et al.[115] while studying the adsorption of Cu^{2+}, Pb^{2+}, Cd^{2+}, and Zn^{2+} from aqueous solutions on oxidized anthracite, observed that the metal uptake increased with increasing pH of the solution but the adsorption of individual metal ions decreased when the other three cations were present in equal concentration.

Corapcioglu and Huang[49] studied the adsorption of Zn^{2+} on a number of activated carbons and found that the adsorption depended on the pH of the solution. However, their results on different carbons did not show any consistency. The maximum adsorption capacity was different at different pH values on different carbons. Mishra and Chaudhary[224] found the adsorption to be higher in the pH range 3 to 5. Carrot et al.[179] studied the adsorptive removal of Zn^{2+} ions on five different activated carbons associated with varying amounts of acidic and basic carbon-oxygen surface groups. These workers suggested that the adsorption of Zn^{2+} ions takes place on acidic groups sites. The acidic groups undergo ionization producing negatively-charged sites where adsorption of zinc cations can take place as a result of electrostatic attractive interactions. In the case of active carbons containing basic groups, the adsorption occurs through the formation of negatively-charged hydroxy complexes of zinc, which are picked up by the protonated positively-charged basic sites. Usmani et al.[51] reported that the adsorption of zinc on an activated carbon prepared from eucalyptus was insignificant below a pH of 4 but increased at higher pH values. Andreeva et al.[52] used activated carbon fibers for the removal of several metals ions including zinc from aqueous solution, and Mathur et al.[53] studied the removal of Zn^{2+} from sewer water using flyash and furnace slag.

Tai and Streat[226] studied the adsorption of zinc on an activated carbon before and after oxidation with nitric acid. The adsorption capacity of the carbon for Zn^{2+} ions increased by 10 times on oxidation, although the BET surface area and the micropore volume of the carbon were decreased. Karounou and Saha[227] used activated PAN carbon fiber (ACF) for the removal of zinc from aqueous solution. The ACF was oxidized with nitric acid and ozone to create acidic carbon-oxygen functional groups on the ACF surface. The ACF surface morphology was determined by scanning electron microscopy and x-ray photoelectron spectroscopy. The adsorption of Zn^{2+} ions at pH 6 showed that the oxidation of the ACF sample with nitric acid or ozone enhances the adsorption capacity considerably. The adsorption capacity for Zn^{2+} was increased by eight times on oxidation of the ACF sample with nitric acid for two hr. The increase was about four times on oxidation with ozone. The increase in adsorption of Zn^{2+} ions has been attributed to the formation of acidic carbon-oxygen surface groups, as shown by the neutralization of sodium hydroxide. The ACF oxidized with nitric acid showed the highest base neutralization capacity.

Ferro-Garcia et al.[187] studied the adsorption of Zn(II) ions on three activated carbons obtained from almond shells, olive stones, and peach stones, and activated by CO_2 at 1123 K. The effect of the solution pH on the removal of zinc ions (Figure 6.39) showed that the adsorption was almost negligible at pH values lower than 2. The adsorption increased sharply in the pH range 3 to 5 and remained more or less constant at higher pH values. These results have been attributed to the carbon surface charge, which was different at different pH values. The results indicated that at low pH values (pH < 2) the carbon surface has a positive charge and resulted in repulsive electrostatic interactions between the positively charged surface and the Zn^{2+} cations. When the pH was enhanced, there was an exchange of H^+ from the surface with the Zn^{2+} ions from the solution, resulting in an increase in adsorption. These workers were of the view that the sharp increase in adsorption in the pH range of 3 to 5 may be due to the fact that the zero point charge (ZPC) for these carbons

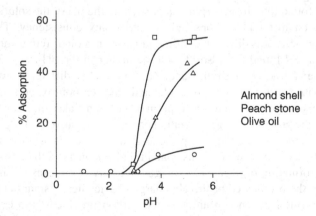

FIGURE 6.39 Adsorption of Zn(II) ions on activated carbons as a function of solution pH. (Source: Ferro-Garcia, M.A., Rivera-Utrilla, J., Rodriguez-Gordillo, J., and Bautista-Toledo, I., *Carbon*, 26, 363, 1988. With permission.)

lies between these pH values. Further, the adsorption of Zn(II) ions was found to increase on the addition of anions such as Cl⁻, CN⁻, and SCN⁻, and the addition of EDTA decreased the adsorption. The increase in adsorption on the addition of anions was attributed to the formation of complex anions of zinc, which could be more suitable for the adsorption on the positively charged carbon surface. The decrease in adsorption on the addition of EDTA was due to the large size of the EDTA-Zn complex, which was not accessible to many of the micropores.

6.10   ACTIVATED CARBON ADSORPTION OF ARSENIC

Arsenic is a toxic element found in natural water and in several solid and liquid industrial wastes[228] and gets leached into ground water, causing significant contamination. Several methods such as complexation,[229] coagulation and precipitation,[230] and ion-exchange[231] have been used for the removal of arsenic ions from waste water. But more recently several types of adsorbents such as activated carbon, flyash,[232] and impregnated silica gel[233] have been used for the removal of arsenic from water. Gupta and Chen[234] studied the effect of pH on the adsorption of As(III) and As(V) from aqueous solutions using activated alumina, bauxite, and activated carbon as adsorbents. The adsorption of As(V) was amximum at pH between 3 and 5 for the activated carbon and at pH < 7 for bauxite and activated alumina. They also observed that the adsorption of As(V) was less on the activated carbon than on bauxite or alumina. Although these workers explained the adsorption on the inorganic adsorbents on the basis of the surface charge, they did not suggest any such explanation for adsorption on the activated carbon. They suggested that the lower adsorption of As(V) on activated carbon was simply due to the lower affinity between the carbon surface and the nonionic, divalent, or trivalent forms of As(V). Kamegawa et al.,[235] while studying the adsorption of arsenic on five different activated carbons, observed that the adsorption was maximum in the acid pH range (pH = 5 to 6) and that the adsorption capacity of most carbons was larger for As(V) than for As(III). This was attributed to the larger capacity of these carbon to oxidized As(III). These workers also suggested that the arsenic was adsorbed in the form of complex ions.

Huang and Fu,[236] while examining the influence of solution pH on the adsorption of arsenic using fifteen different carbons, observed that the arsenic removal was a function of the pH. The optimum pH for the removal of arsenic was 4.0, where the adsorption was one order of magnitude higher than at other pH values. This was attributed to the fact that the concentration distribution of As(V) and the development of carbon surface charge were related to the pH of the solution. Rajakovic[237] also observed that the adsorption of both As(III) and As(V) was pH dependent because the pH of the solution determined the change in the ionic species. The adsorption was maximum in the pH range 4 to 9. It was also observed that the impregnation of the activated carbons with copper enhanced the adsorption of As(III) due to chemisorption or precipitation of $CuHAsO_3$. A similar increase in the adsorption of As(V) was observed by Huang and Van[238] by the impregnation of carbon with Fe^{2+} ions. It was suggested that the carbon surface develops a positive charge by the impregnation and results in the formation of ferrous arsenate surface complexes.

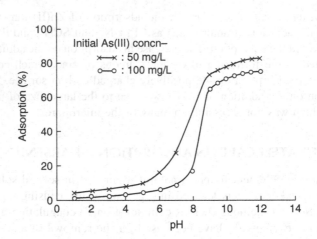

FIGURE 6.40 Effect of pH on activated carbon adsorption of As(III) ions. (Source: Raji, C. and Anirudhan, T.S., *J. Sci. Ind. Res.*, India, 57, 10, 1998. With permission.)

Raji and Anirudhan[239] used copper-impregnated sawdust carbon for the adsorptive removal of As(III) from aqueous solutions and studied the influence of such factors as solution pH and concentration, reaction time, and temperature on the extent of removal. The extent of adsorption of As(III) increased from 3.2% to 84.0% by increasing the pH from 1 to 12 (Figure 6.40). It has been suggested that at pH = 12, the predominant As(III) species are the anions H_2AsO_{3-} and $HASO_3^{2-}$, which undergo chemical reaction with the copper present on the carbon surface and form a low soluble precipitate $CuHASO_3$. The mechanism for the removal of As(III) by chemsorption has been suggested to be

$$>C-Cu(OH)^+ + H_2AsO_3^- \xrightarrow{\text{pH}=12} >C-Cu\ HAsO_3 + H_2O$$

copper-impregnated carbon surface

and

$$>C-Cu(OH)_2 + HAsO_3^{2-} \longrightarrow >C-Cu\ HAsO_3 + 2HO^-$$

At neutral and acidic pH values, these workers suggested that As(III) is predominantly present as H_3AsO_3, and this neutral molecule was not removed by adsorption or surface precipitation as the carbon surface was positively charged. The removal of As(III) was time dependent and followed the first-order kinetics with respect to initial concentration as well as temperature (Figure 6.41). The first-order rate constant was calculated using the equation

$$\log (qe - q) = \log qe - \frac{K}{2.303t}$$

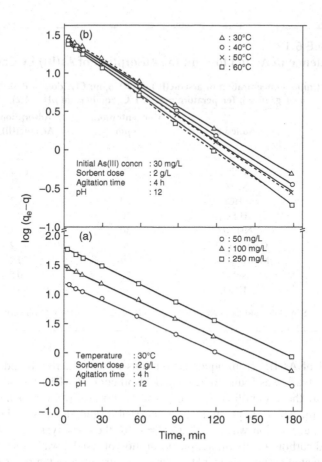

FIGURE 6.41 Kinetics of adsorption of As(III) ions by activated carbon at different (a) concentration and (b) temperatures. (Source: Raji, C. and Anirudhan, T.S., *J. Sci. Ind. Res.*, India, 57, 10, 1998. With permission.)

where q_e and q are the amounts of As(III) adsorbed at equilibrium and time t. The K values found from the slopes agreed very well for different initial concentrations of 50, 100, and 250 mg/L and at temperatures of 30°, 40°, 50°, and 60°C. The presence of diverse ions such as Cl^-, NO_{2-}, CH_3COO^-, ClO_4^-, and CO_3^{2-} in the solution did not affect the adsorption of As(III) and the spent-up carbon could be regenerated by treating the carbon with a solution of 15% H_2O_2 in 0.2 M HNO_3 and could be reused.

Lorenzen et al.[240] studied the adsorptive removal of arsenic using three activated carbons prepared from different source raw materials and observed that all three activated carbons impregnated with copper were capable of removing arsenic effectively from aqueous solutions at a slightly acidic pH of 6. They also observed that the adsorption was not related to the surface area of the carbon but was enhanced by the presence of ash in the activated carbon. Sen and De[241] determined the effect of several parameters such as, the time of equilibration, pH, and adsorbent dose on

TABLE 6.12
Influence of Added Ions on the Adsorption of As(III) by Carbon

[Initial concentration of arsenic(III) = 35.1 ppm; CFA dose = 0.98 ±
0.01 g/50 ml; Temperature = 30 ± 1°C; Equilibrium pH = 4.2]

Ions	Added as	Concentration, ppm	Adsorption of Arsenic(III), %
No ion	—	—	72.2
NO_3^-	HNO_3	20	67.7
NO_2^-	$NaNO_2$	40	69.9
Cl^-	HCl	20	69.2
SO_4^{2-}	H_2SO_4	20	69.6
NH_4^+	NH_4NO_3	30	70.1
Ca^{2+}	$Ca(NO_3)_2 \cdot 4H_2O$	20	72.2
Mg^{2+}	$Mg(NO_3)_2 \cdot 6H_2O$	20	72.2
Na^+	$NaNO_3$	20	70.5
K^+	KNO_3	20	69.8

Source: Sen, A.K. and De, A.K., *Indian J. Technol.*, 25, 259, 1987. With permission.

the removal of As(III) from aqueous solutions using coal flyash and an activated carbon. The flyash was found to be a good adsorbent of As(III). The adsorption data conformed to the Freundlich equation. The adsorption of As(III) increased with increase in the solution pH upto a certain value and decreased thereafter. The maximum in adsorption was found at pH = 4 for the coal flyash and at pH = 3 for the activated carbon. Furthermore, the adsorption of As(III) was larger on the flyash than the activated carbon above pH 4. This was attributed to the polar nature of the silica-oxygen bond in SiO_2 that was present in flyash. The polar silica-oxygen bond tended to accumulate H^+ ions at low pH values (in the acid medium), and the positively-charged surface of the coal flyash adsorbed arsenate ion. When the pH was increased, there were fewer H^+ ions on the surface and the adsorption decreased. The addition of diverse ions in the solution was found to have little or no effect on the adsorption of As(III) ions from the solution (Table 6.12).

6.11 ADSORPTIVE SEPARATION OF CATIONS IN TRACE AMOUNTS FROM AQUEOUS SOLUTIONS

It has been shown in the previous sections of this chapter that activated carbons are good adsorbents for the removal of metal ions from aqueous solutions. However, the efficiency of different activated carbons varies considerably for the different metal ions. It has been observed that the pore-size distribution in carbons and the chemical nature of the carbon surface play an important role in determining the difference in the behavior of different carbons. In order to lay down some parameters that may be helpful in separating different cations present in trace amounts in water, Rivera-Ultrilla and coworkers[187,198] carried out adsorptive removal of a large number of cations using the same activated carbons before and after modification of their surface by oxidation and by ammonia treatment. The activated carbons used were obtained from almond shell, peach stones, and olive stones, and activated in CO_2 at 1123 K. The surface modification of these activated carbons was carried out by oxidation with HNO_3 and H_2O_2 in the liquid phase and by treatment with air and ammonia gas.

The rate of adsorption of CS^+, Tl^+, Sr^{2+} and Co^{2+198} from aqueous solutions on the almond shell activated carbons activated for 8 and 14 hr in CO_2 and on a commercial activated carbon increased sharply in the beginning and ultimately attained a constant value (Figure 6.42). The amount adsorbed was at a maximum in the case of Tl^+ and at a minimum in the case Cs^+ on all three samples. These workers are of the view that the small adsorption in the case of Cs^+ may be due to the small polarizing power of Cs^+ in aqueous solution as deduced from its charge-radius ratio, which resulted in very weak carbon-cesium interaction forces so that the Cs^+ ions were adsorbed only on centers with high negative charge density. However, these workers could not explain the big difference in the adsorption capacity of carbons for Cs^+ and Tl^+ ions (~100%) because both the ions have similar ionic radi (Tl^+ 0.144 nm, Cs^+ = 0.169 nm), hydrated radii (Tl_{aq}. = 0.330 nm, Cs^+_{aq} = 0.329 nm), and similar charge. Furthermore, the adsorption of Tl^+ from the solution was higher than the adsorption of Sr^{2+} and Co^{2+} from the solution. This was attributed to the precipitation of thallium under the experimental conditions. This was confirmed by the adsorption values obtained using different activated carbons.

In a subsequent publication Rivera-Ultrilla et al.[48] carried out adsorption studies of Na^+, Cs^+, Ag^+, Sr^{2+}, and Co^{2+} using almond shell-activated carbons treated with air, H_2O_2, HNO_3, and ammonia. The adsorption of the carbon for Cs^+ increased considerably from 7% for the original sample to 80% on oxidation of the carbon

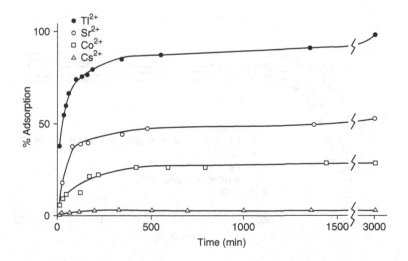

FIGURE 6.42 Adsorption of different cations on an activated carbon as a function of equilibrium time. (Source: Rivera-Utrilla, J., Ferro-Garcia, M.A., Mata-Arjona, A., Gonzalez-Gomez, C.J., *Chem. Technol. Biotechnol.*, 34a, 243, 1984. With permission.)

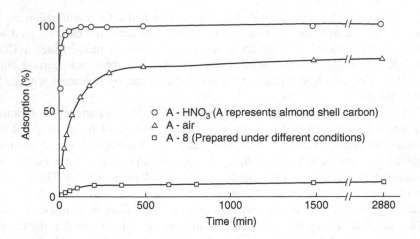

FIGURE 6.43 Adsorption of Cesium ions on modified activated carbons as a function of equilibrium time. (Source: Rivera-Utrilla, J. and Ferro-Garcia, M.A., *Adsorp. Sci. Technol.*, 3, 293, 1986. With permission.)

with air and to almost 100% on oxidation with HNO_3 (Figure 6.43). In the case of Sr^{2+}, the adsorptive removal varied between 43% for the H_2O_2 treated and NH_3-treated samples to 98% for the sample oxidized with nitric acid. When the cation was Ag^+, all the samples showed a high adsorption capacity varying between 80 and 95% (Figure 6.44). The removal of Na^+ ions could only be carried out by the oxidized carbons. Whereas the sample oxidized with nitric acid could remove more than 90% the Na^+ ion, the sample oxidized in air removed only about 50% of the metal cation.

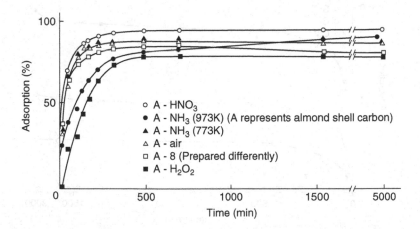

FIGURE 6.44 Adsorption of Silver ions (Ag(I)) on modified activated carbons as a function of equilibrium time. (Source: Rivera-Utrilla, J. and Ferro-Garcia, M.A., *Adsorp. Sci. Technol.*, 3, 293, 1986. With permission.)

The high adsorption capacity of Ag^+ ions by all the activated carbons was attributed to the reduction of Ag^+ ions to metallic silver by the hydroquinone groups present on the carbon surface, which in turn are oxidized to quinone groups. This redox process is supported by the standard reduction potentials of Ag^+ ($Ag^+ + e \rightarrow$ Ag, $E_o = 0.7996$ V) and quinhydrone electrode, $E_o = 0.6995$ V. The increase in adsorption of Ag^+ ions by the ammonia-treated sample was attributed to the formation of silver amino complexes which are quite stable under the conditions used in these studies.

A comparison of the amounts of Cd^{2+}, Zn^{2+} and Cu^{2+} ions adsorbed on three activated carbons from aqueous solutions[187] showed that at a given equilibrium concentration, the adsorption increased in the order $Cu^{2+} > Zn^{2+} > Cd^{2+}$. This was also explained on the basis of their ionic radii. Because in the aqueous solution these cations are hydrated having a hydration number of 6,[188, 200] the radii of these hydrated cations should be $Cu^{2+}_{aq} = 0.419$ nm, $Zn^{2+}_{aq} = 0.430$ nm, and $Cd^{2+}_{aq} = 0.426$ nm.[242] Thus, the Cu^{2+} ions will be more accessible to certain pores than the other cations i.e., Zn^{2+}, and Cd^{2+}, resulting in the larger adsorption of Cu^{2+} ions. The lower adsorption of Cd^{2+} ions compared to Zn^{2+} ions may be due to its smaller polarizing power so that carbon-cadmium interactions were weaker than the carbon-Zn(II) interactions. The Cd(II) will be adsorbed only on those sites where the negative charge density was higher than that necessary to retain Zn(II).

These results are of practical significance, because adsorptive separation of different cations when present in trace amounts in aqueous solutions can be carried out using different modified activated carbons. For example, Tl^+ can be separated from Cs^+ in an aqueous solution by using an activated carbon when about 98% of Tl^+ can be removed, while most of Cs^+ remains in the solution. Similarly, Ag^+ ions can be removed from the solution by more than 98%, using nitric acid oxidized carbons. These workers also determined the extent of meso-and micropores in different carbons and observed that the micropore surface was not accessible to these cations, not only because of their size but also because of electrostatic interactions.

6.12 MECHANISM OF METAL ION ADSORPTION BY ACTIVATED CARBONS

An activated carbon in contact with a metal salt solution is a two-phase system consisting of a solid phase, which is the activated carbon surface, and a liquid phase which is the salt solution. The solution contains varying amounts of different metal ion species and their complexes so that the interface between the two phases will behave as an electrical double layer and determine the adsorption processes taking place in the system. The adsorptive removal capacity of an activated carbon for metal cations from the aqueous solutions generally depends on the physicochemical characteristics of the carbon surface, which include the surface area, pore-size distribution, electrokinetic properties, and the chemical structure of the carbon surface, as well as on the nature of the metal ions in the solution.

Activated carbon surfaces are almost invariably associated with a certain amount of chemisorbed oxygen and hydrogen. These heteroatoms are derived from the source raw material and become a part of the chemical structure as a result of imperfect carbonization, or they become chemically bonded to the carbon surface during activation or subsequent treatments. In addition, the activated carbons may also become associated with atoms of nitrogen, halogens, and sulfur, but only on treatment with suitable reagents. These heteroatoms are present on the carbon surface in the form of carbon-oxygen, carbon-hydrogen. Carbon-nitrogen and carbon-sulfur surface compounds, also called *surface complexes* or *surface groups*. However, carbon-oxygen surface groups are by far the most important chemical structures that are present on all activated carbons and influence the adsorption characteristics under all practical situations.

Using several physical, chemical, and physiochemical techniques (discussed in Chapter I of this book), two types of carbon-oxygen surface functional groups have been identified. One type is the acidic surface groups which are evolved as CO_2 on evacuation or degassing in an inert atmosphere in the temperature range 300 to 750°C. The other type of of carbon-oxygen groups are basic surface groups that are evolved as CO in the temperature range 500 to 950°C. The acidic surface groups are polar in character and render the carbon surface hydrophilic. These groups have been identified as carboxyls and lactones. The basic surface groups have been postulated as pyrone and chromem structures. In addition, there are nonacidic surface groups that have been postulated as quinones.

In the solution phase, the metal salt undergoes ionization and hydrolysis, which results in the formation of free and complex cationic and anionic species. However, the preponderance of any one or several cations or anions depends upon the concentration and the pH of the solution. For example, for a divalent metal salt MCl_2, the ionization of the salt produces M^{2+} cations, while its hydrolysis can produce $(MOH)^+$, $M(OH)_2$ aq. $M(OH)_{3-}$ and $M(OH)_4^{2-}$ ions. In addition, small amounts of polynuclear species such as $M_2(OH)_3^+$ and $M_4(OH)_4^{4+}$ may also be formed at higher concentrations of the salt in the solution and at suitable pH values.

The perusal of the literature on the adsorptive removal of metal cations in general and review of the detailed studies of the adsorption of some metals[243,244] from aqueous solutions on activated carbons obtained from a variety of source materials and with different history of preparation clearly indicates that the adsorption has been explained differently by different workers. However, the overwhelming number of investigators are of the view that the more important parameters that influence and determine the adsorption of metals from aqueous solutions are the carbon-oxygen surface functional group present on the activated carbon surface and the pH of the solution. These two parameters determine the nature and the concentration of the charge on the activated carbon surface and the amount and the concentration of the ionic species in the solution. Electrokinetic studies have shown that the nature and the concentration of the carbon surface charge can be modified by changing the pH of the solution. For each carbon there is a pH value where the carbon surface corresponds to zero point charge (pH_{ZPC}). The carbon surface has a positive charge below ZPC and a negative charge above ZPC up to a certain range of pH of the solution.

The origin of the positive charge has been attributed to the presence of the basic surface groups, the excessive protonation of the surface at low pH values of the solution, and to graphene layers that act as Lewis bases, resulting in the formation of acceptor-donor complexes with water molecules. However, the information regarding the relative contribution of basic oxides and graphene layers has not been very well elucidated in the literature. Radovic[244] has rightly pointed out that the repulsive electrostatic interactions between the adsorbate metal ions and the graphenes layers are more "detrimental for adsorbent effectiveness" than the repulsion from the basic surface oxygen groups. The negative charge on the carbon surface at higher pH values (>pH = ZPC) is due to the ionization of the acidic surface oxygen groups.

The above description of the activated carbon solution system makes it easier to visualize the mechanisms involved in the adsorptive removal of metals from aqueous solutions by activated carbons. When in solution the acidic surface oxygen groups present on the activated carbon surface undergo ionization producing H^+ ions, the degree of ionization depending upon the pH of the solution. The degree of ionization is very low at pH values lower than ZPC and is very high when the pH of the solution is above ZPC. This ionization can be shown as

$$>C-COOH \longrightarrow >C-COO^- + H^+$$

$$>C \underset{OH}{\overset{COOH}{<}} \longrightarrow >C \underset{COO^-}{\overset{COO^-}{<}} + 2H^+$$

$$>C \underset{OH}{\overset{COOH}{<}} \longrightarrow >C \underset{O^-}{\overset{COO^-}{<}} + H^+$$

The H^+ ions are directed toward the liquid phase, leaving the carbon surface with negatively-charged sites. This results in attractive electrostatic interactions between the negatively-charged surface sites, and the positively-charged metal cations in the solution. When the activated carbon surface is given an oxidative treatment, the number and the concentration of the acidic surface groups and hence those of the negatively charged sites increase, resulting in higher attractive interactions and an increase in the adsorption of metal cations. Similarly, when these acidic surface groups are removed from the carbon surface by degassing, the attractive electrostatic interactions as well as the adsorption decrease, the decrease in adsorption depending upon the decrease in the number of acidic groups eliminated from the carbon surface. Bansal and coworkers[128,129,159,160,163] carried out systematic studies on the adsorptive removal of several metal cations on different granulated and fibrous activated carbons by enhancing and removing gradually the amounts of acidic carbon-oxygen surface groups. These workers found even a linear relationship between the amount of metal ion adsorbed and the surface acidity of the given carbon. In the case of heat-treated carbons, Biniak et al.[121,243] while studying the adsorption of copper, have suggested that the mechanism could involve dipole-dipole (π-d) interactions between the graphene layers and the metal ionic species.

REFERENCES

1. Druley, R.M. and Ordway, G.L., The Toxic Substances Control Act, The Bureau of National Affairs, Washington, D.C., 1977.
2. Hueper, W.C. and Payne, W.W., *Am. Soc. Clin. Pathol.*, 39, 475, 1963.
3. *National Cancer Inst. Report on the Carcinogenesis Bioassay of Chloroform*, Carcinogenesis Program, Division of Cancer Cause and Prevention, Bethesda, MD, March 1, 1976.
4. Ongreth, H.J., Spath, P., Crook, J., and Greenberg, E., *J. Am. Water Works Assoc.*, 65, 495, 1973.
5. Suffet, I.H. and McGuire, M.J., in *Activated Carbon Adsorption*, Ann Arbor Science Publishers, Ann Arbor MI, Vol. II, 1981.
6. Cheremisnoff, P.N. and Ellerbusch, F., in *Carbon Adsorption Handbook*, Ann Arbor Science Publishers, Ann Arbor MI, 1978.
7. Nigam, I.N., Snoeyink, V.L., Suidan, M.T., Lee, C.H., and Richard, Y., *J. Am. Water Works Assoc.*, 82, 65, 1990.
8. Hutchins, R.A., *Chem. Engg.*, 87, 101, 1980.
9. Derbyshire, F., Jagtoyen, M., Andrews, M., Rao, A., Martin-Gullon, I., and Grulke, E.A., in *Chemistry and Physics of Carbon,* L.R. Radovic, Ed., Vol. 27, Marcel Dekker, New York, 2001.
10. Mahajan, O.P., Morino-Castilla, C., and Walker, P.L., Jr., *Sep. Sci. and Technol.*, 15, 1773, 1980.
11. Millet, P.C., J. *NEWWA*, 1995, p. 141–48.
12. Goldfarb, A.S., Vegel, G.A., and Lundquist, G.E., Waste Management, 14, 145, 1994.
13. Bansal, R.C., Aggarwal, D., Goyal, M., and Kaistha, B.C., *Indian J. Chem. Technol.*, 9, 290, 2002.
14. Puri, B.R., Bhardwaj, S.S., and Gupta, U., *J. Indian Chem. Soc.,* 53, 1095, 1976.
15. Puri, B.R., in *Chemistry and Physics of Carbon*, Vol. 6, P.L. Walker, Jr., Ed. Marcel Dekker, New York, 1970, p.191.
16. Boehm, H.P., in *Advances in Catalysis*, Vol. 16, Academic Press, New York, 1966, p. 179.
17. Bansal, R.C., Donnet, J.B., and Stoeckli, F., in *Active Carbon*, Marcel Dekker, New York, 1988.
18. Donnet, J.B., Bansal, R.C., and Wang, M.J., in *Carbon Black*, Marcel Dekker, New York, 1993.
19. Mattson, J.S., Mark, H.B., Jr., in *Activated Carbon*, Marcel Dekker, New York, 1971.
20. Jankowska, H., Swiatkowski, A., and Choma, J., in *Active Carbon*, Ellis Horwood, 1991.
21. Kinoshita, K., in *Carbon*, John Wiley and Sons, 1988.
22. Smisek, M. and Cerny, S., in *Active Carbon*, Elsevier Publishing Company, 1970.

23. Linstedt. K.D., Houck, C.P., and O'Conner, J.T., *J. Water Poll. Control Fed.*, 43, 1507, 1971.
24. Cheremisinoff, P.N. and Habib, Y.H., *Water Sew. Works*, Aug. 1972, p. 46.
25. Argaman, Y. and Weddle, C.L., *Alchem. Symposium Series*, 70, 400, 1973.
26. Netzer, A., Wilkinson, P., and Berzedits, S., *Water Res.,* 8, 813, 1974.
27. Faust, S.D. and Aly, O.M., *Chemistry of Water Treatment,* Chelsia, M.L., Ed., Ann Arbor Press, 1988.
28. Huang, C.P., in *Carbon Adsorption Handbook*, Cheremisinoff, P.N. and Ellerbusch, F., Eds., Ann Arbor Science Publishers, Ann Arbor MI, 1978, p. 281–329.
29. Peters, R.W. and Ku, Y., *Reactive Polymers*, 5, 93, 1987.
30. Camara, S., Wang, Z., Ozeki, S., and Kaneko, K., *Colloid Interface Science*, 162, 520, 1994.
31. Bagreev, A.B. and Tarasenko, Y.A., Russian, *J. Phys. Chem.*, 67, 1111, 1993.
32. Bao, M.L., Griffini, O., Santiauni, D., Barbieri, K., Burrini, D., and Pantani, F., *Water Res.,* 33, 2959, 1999.
33. Gonce, E. and Voudries, E., *Water Res.,* 28, 1059, 1994.
34. Adams, M.D., *Miner Eng.,* 7, 1165, 1994.
35. Ilic, M.R., Jovanic, P.B., Radosevic, P.B., and Rajakovic, L.V., *Sep. Sci. Technol.*, 30, 2707, 1995.
36. Rajakovic, L.V. and Ristic, M.D., *Carbon,* 34, 769, 1996.
37. Siddiqi, M., Zhai, W., Amy, G., and Mysore, C., *Water Res.,* 30, 1651, 1996.
38. Seron, A., Benaddi, H., Beguin, F., Frackoviak, E., Bretelle, J.L., Thiry, M.C., Bendosz, T.J., Jagiello, J., and Schwaz, J.A., *Carbon,* 34, 481, 1996.
39. Frackowiak, E., *Fuel*, 77, 571, 1998.
40. Green-Pedersen, H. and Korshin, G., *Environ. Sci. Technol.*, 33, 2633, 1999.
41. Onganer, Y. and Temur, C., *J. Colloid Interface Sci.*, 205, 241, 1998.
42. Pakula, M., Biniak, S., and Swiatkowski, A., *Langmuir*, 14, 3082, 1998.
43. Qadeer, R., Hanif, J., Saleem, M., and Afzal, M., *Collect Czech. Chem. Common*, 57, 1, 1992.
44. Qadeer, R. and Hanif, J., *Carbon,* 33, 215, 1995.
45. Saito, L., *Kogai Shigen Kenkyusho Iho*, 5, 57, 1976, in Japanese C.A. 086100601346.
46. Kaya, M. and Akyol, F., *Kim Kim Muhendisligi*, Sem. 8th, 3, 213, 1992.
47. Naganuma, K., Okazaki, M., Yonekayashi, K., Kyuma, K., Vijarnsorn, P., and Abu Baker, G.Z., *Soil Science: Plant Nutr.*, Tokyo, 39, 455, 1993.
48. Rivera-Utrilla, J. and Ferro-Garcia, M.A., *Adsorp. Sci. Technol.*, 3, 293, 1986.
49. Corapcioglu, M.D. and Huang, C.P., *Water Res.*, 21, 1031, 1987.
50. Petrov, N., Budinova, T., and Khavesov, I., *Carbon*, 30, 135, 1992.
51. Usmani, T.H., Ahmed, T.W., and Ahmad, S.Z., Pak. *J. Sci. Ind. Res.*, 34, 26, 1991.
52. Andreev, I. Yu., Minko, I.L., and Kakakevich, I. Yu., *Z. Prikl. Khim.*, Leningrad, 64, 1276, 1991.
53. Mathur, A., Khare, S.K., and Rupainwar, D.C., *J. Indian Poll. Control*, 5, 52, 1989.
54. Swart, H., Cronje, I.J., Dekker, J., and Cloctc, T.E., *Gas, Oil, Coal, Environ. Biotechnol.*, 3 pap. IGTs, 3rd Int. Symposium, p. 315, 1990, published 1991.
55. Abdul Shafy, H., Ek-Gomal, I.M., Abdel, S.M.F., and Abo-el-Wafa, U., *Environ. Protection Eng.*, 15, 63, 1990.
56. Tewari, D.P., Promod, K., Mishra, A.K., Singh, R.P., and Srivastav, R.P.S., *Indian J. Environ. Health*, 31, 120, 1989.
57. Jankowska, H., Choma, J., Burakiewiez-Mortka, W., and Swiatkowski, A., *Prezem. Chem.*, 6, 39, 1987.
58. Yoshida, H., Kamegawa, K., and Arita, S., *Nippon Kagaku Kaishi*, 3, 387, 1977.

59. Huang, C.P. and Wu, M.H., *Water Res.*, 11, 673, 1977.

60. Bautista-Teledo, I., Rivera-Utrilla, J., Ferro-Garcia, M.A., and Moreno-Castilla, C., *Carbon*, 32, 93, 1994.

61. Jayson, G.G., Sangster, J.A., Thompson, G., and Wilkinson, M.C., *Carbon*, 31, 487, 1993.

62. Gu., L. and Sanghai, H., *Kexue*, 31, 1989.

63. Cici, M. and Kales, E., Pak. *J. Sci. Ind. Res.*, 33, 347, 1990.

64. Abe, T., *Japan*, Kokai 740717 in Japanese, C.A. 082804751213, 1974.

65. Tagashira, Y., Takagi, H., Ingaki, K., and Minoura, H., *Japan*, Kokai 750802 in Japanese, C.A. 08410065057c, 1975.

66. Huang, C.P. and Wu, M.H., *J. Water Pollution Control Federation*, 47, 2437, 1975.

67. Singh D. and Rawat, S., *Indian J. Chem. Technol.*, 1, 266, 1994.

68. Viraraghavan, J. and Rao, G.A.K., *Proc. 45th Industrial Waste Conference*, Purdue Univ., Lafayette IN, Lewis Publishers, 1991.

69. Huang, C.P. and Ostovic, F., *J. Environment, Engg.*, 104, 863, 1978.

70. Dobrowolski, R., Jaroniec, M., and Kosmulski, M., *Carbon*, 24, 15, 1986.

71. Liu, L., Xic, Q., Liu, Y., Li, C., Wang, L., and Liu, F., *Huanjing*, Kexue Xeubao, 10, 112, 1990.

72. Vishwakarma, P.S., *Pertanika*, 12, 357, 1989.

73. Dimitrov, S.V., *Water Res.*, 30, 228, 1996.

74. Shakir, K., Benyamin, K., and Aziz, M., *J. Radioanal. Nuclear Chem.*, 173, 141, 1993.

75. Shakir, K., Flex. H., and Benyamin, K., *J. Radioanal, Nuclear Chem.*, 173, 303, 1993.

76. Moore, R.H., U.S. DOI Report No. 651, 1971.

77. Nelson, F., Phillips, H.O., and Kraus, K.A., *29th Purdue Industrial Waste Conference*, Lafayette IN, 1974.

78. Kahn. M.A. and Khattak, J.I., *Carbon*, 30, 957, 1992.

79. Low, K.S., Lee, C.K., and Wong, S.L., *Environ. Technol.*, 16, 877, 1995.

80. Mu., G. and Yang, C., *Wuli Huaxue Xeubar*, 11, 157, 1995.

81. Bencheikh, L.M., *Environ. Technol. Letters*, 10, 101, 1989.

82. Kunz, R.G., Gianneli, J.F., and Stensel, H.D., *J. Water Pollut. Control. Fed.*, 48, 762, 1976.

83. Koustyuchenko, P.I., Taskovskaya, F.A., Konouchnk, T.L., Kovalenko, T.J., and Glushenkova, Z.L., in *Adsorption and Adsorbents*, D.N.Strazhesko, Ed., Wiley, New York, 1973, p. 37.

84. Riaz, Q., Zaved, H., Mohammad, H., and Mohammad, A., *Collect Czech. Chem. Common*, 57, 2065, 1992.

85. Riaz, Q., Zaved, H., Mohammad, H., and Mohammad, A., *Colloid Polymer Sci.*, 271, 83, 1983.

86. Gurwagen, J.J. and Pienarr. A.T., *Polyhedron*, 8, 71, 1989.

87. Tarsenko, Yu. A., Bangareev, A.A., Dudarenko, V.V., Mardanenko, V.K., and Soldovnikov, Yu, I., *Ukr. Chim. Zh.*, 55, 233, 1989.

88. Qadeer, R. and Hanif, J., *J. Chem. Soc. Pakistan*, 15, 227, 1993.

89. Qadeer, R. and Hanif, J., *Radiochim. Acta*, 65, 259, 1994.

90. Mu, G., Yang, Z., and Yang, C., *Linchan Huaxue Yu Gongye*, 14, 61, 1994.

91. Fu, R., Zang, H., Lu, Y., Lai, S.Y., Chan, W.H., and NG, C.F., *Carbon*, 33, 657, 1985.

92. Oppold, W.A., *44th Water Pollution Control Fed. Conf.*, San Francisco, CA, October 5, 1971.

93. Thiem, L., Badorek, D., and O'Conner, J.T., *J. Am. Water Works Assoc.*, 68, 447, 1997.

94. Homenick, M.J., Jr., and Schnoor, J.L., *J. Am. Soc. Environ. Eng. Div.*, 100, 1249, 1974.

95. Yoshida, H., Kamegawa, K., and Arita, S., *Nippon Kogaku Kaishi*, 5, 808, 1976.
96. Ohtsuki, S., Miyanohara, I., and Mizul, N., *Japan Kokai*, 74, 5, 1974.
97. Sinha, R.K. and Walker, P.L., Jr., *Carbon,* 10, 754, 1972.
98. Ammons, R.D., Daungharty, N.A., and Smith, J.M., *Ind. Eng. Chem. Fundamentals*, 16, 263, 1977.
99. Kanoko, K., *Carbon,* 26, 903, 1988.
100. Saleem, M., Afzal, M., Qadeer, R., and Hanif. J., *Sep. Sci. Technol.*, 27, 239, 1992.
101. Abbasi, W.A. and Streat, M., *Sep. Sci. Technol.*, 29, 1217, 1994.
102. Qadeer, R. and Saleem, M., *Adsorp. Sci. Technol.*, 15, 373, 1997.
103. Park, G.L., Park, S., and Woo, S., *Sep. Sci. Technol.*, 34, 833, 1999.
104. Dun, J.W., Gularia, E., and Ng, Kys., *Appl. Catalysis*, 15, 247, 1985.
105. El Bayoumy, S. and El Kolaly, M., *Radioanal. Chem.*, 68, 7, 1982.
106. Mohapatra, S.P., Siebel, M., and Alaerts, G., *J. Environ. Sci., Health,* A. 28, 615, 1993.
107. Khalfaoui, B., Meniai, A.H., and Borja, R., *J. Chem. Technol. Biotechnol.*, 65, 153, 1993.
108. Periasamy, K. and Namasivayam, C., *Chemsorphere*, 32, 769, 1996.
109. Wilkins, E. and Yang, Q., *J. Environ. Sci. Health*, A. 31, 2111, 1996.
110. Karthikeyan, A.G., Elliott, H.A., and Cannon, F.S., *Environ. Sci. Technol.*, 31, 2721, 1997.
111. Gupta, V.K., *Ind. Eng. Chem. Res.,* 37, 192, 1998.
112. Deorkar, N.V. and Tavlarides, L., *Environ. Prog.*, 17, 120, 1998.
113. Faust, S.D. and Aly, O.M., in *Chemistry of Water Treatment*, Chelsea, M.I., Ed., Ann Arbor Press, Ann Arbor, MI, 1998.
114. Rodda, D.P., Johnson, B.B., and Wills, J.D., *J. Colloid Interface Sci.*, 161, 57, 1993.
115. Budinova, T.K., Gergova, K.M., Petrov, N.V., and Minkova, V.N., *J. Chem. Technol. Biotechnol.*, 60, 177, 1994.
116. Mangalardi, T., Paolinic, A.E., and Pellorca, F., *Russian Chim.*, 42, 2570, 1990.
117. Tarkovskaya, I.A., Tomaskevskaya, A.N., Goba, U.E., Nagorskaya, O.E., Shurupova, L.A., and Povazhnyi, B.S., *Ukr. Khim. Ali*, 57, 480, 1991.
118. Allen, S.J. and Brown, P.A., *J. Chem. Technol. Biotechnol.*, 62, 17, 1995.
119. Allen, S.J., Whitton, L.J., Murray, M., Dugan, D., and Brown, P. *J. Chem. Technol. Biotechnol.*, 68, 442, 1997.
120. Johns, M.M., Marshall, W., and Toles, C., *J. Chem. Technol. Biotechnol.*, 68, 442, 1998.
121. Biniak, S., Pakula, M., and Swiatkowski, A., *Chemistry and Physics of Carbon*, Vol. 27, L.R. Radovic, Ed., Marcel Dekker, New York, 2001, p. 125.
122. Boehm, H.P., *Carbon*, 32, 759, 1994.
123. Solar, J.M., Leon, Y., Leon, C.A., Osseo-Asare, K., and Radovic, L.R., *Carbon*, 28, 369, 1990.
124. Leon, Y., Leon, C.A., and Radovic, L.R., in *Chemistry and Physics of Carbon*, P. Thrower, Ed., Vol. 24, Marcel Dekker, New York, 1994.
125. Radovic, L.R. and Rodriguez-Reinoso, F., in *Chemistry and Physics of Carbon*, P. Thrower, Ed., Vol. 25, Marcel Dekker, New York, 1997.
126. Ferro-Garcia, M.A., Rivera-Utrilla, J., Rodriguez-Gordillo, J., and Bautista-Toledo, I., *Carbon*, 26, 363, 1998.
127. Kahn, M.A. and Khattak, Y.I., *J. Chem. Soc. Pak.*, 12, 213, 1990.
128. Goyal, M., Rattan, V.K., Aggarwal, D., and Bansal, R.C., Colloids and Surfaces, A. Physico-Chem. and Engg. Aspects, 190, 229, 2001.
129. Goyal, M., Ph.D. thesis submitted to Panjab University, Chandigarh, India, 1997.
130. Chang, C. and Ku, Y., *Sep. Sci. Technol.*, 33, 483, 1995.

131. Chang, C. and Ku, Y., *J. Chin. Inst. Eng.*, 20, 651, 1997.

132. Bhattacharya, D. and Cheng, C.Y.R., *Environ. Prog.*, 6, 110, 1987.

133. Narayana, N. and Krishnaiah, A., *Indian J. Environ. Protection*, 9, 30, 1989.

134. Narayana, N. and Krishnaiah, A., *Indian J. Environ. Health*, 31, 304, 1989.

135. Kannan, N. and Vanangamudi, A., *Indian J. Environ. Protection*, 11, 241, 1991.

136. Ouki, S.K. and Newfeld, R.D., *Hazard Ind. Waste*, 21, 146, 1989.

137. Cici, M. and Kale, E., *Pak. J. Sci. Ind. Res.*, 33, 347, 1990.

138. Dikshit, U.P., Aggarwal, I.C., and Shukla, N.P., *Asian Environ.*, 12, 1990.

139. Golub, D. and Oren, Y., *J. Appl. Electro. Chem.*, 19, 311, 1989.

140. Farmer, J.C., Bahowick, S.M., Harrar, J.E, Fix, D.V., Martinelli, R.F., Vu, A.K., and Carroll, K.L., *Energy Fuels*, 11, 337, 1997.

141. Lalvani, S.B., Wiltowski, T., Hubner, A., Weston, A., and Mandich, N., *Carbon*, 36, 1219, 1998.

142. Tereshkova, S.G., Russian, *J. Phys. Chem.*, 70, 1020, 1996.

143. Huang, C.P. and Wu, M.H., *Water Res.*, 11, 673, 1977.

144. Perez-Candela, M., Martin-Martinez, J.M., and Terregrosa-Macia, R., *Water Res.*, 29, 2174, 1995.

145. Kim, J.L., *Diss. Abstr. Int.* B 37, 3566, 1977, C.A. 08620145537R.

146. Nagasaki, Y. and Tereda, A., *Japan Kokai*, 75, 721, 1975.

147. Huang, C.P. and Bowers, A.R., *10th Intern. Conf. Water Pollution Res.*, 1977.

148. Roerma, R.E., Alsema, G.L., and Anlhonissen, J.H., Belg, *Ned. Tijdscur, Opperblakte Tech. Met. Ser.*, 19, 53, 1975.

149. Miyagawa, T., Ikeda, S., and Koyama, K., *Japan Kokai*, 76, 417, 1976 C.A. 08520148621D.

150. Leyva-Ramos, R., Juarez-Martinez, A., and Guerrero-Coronado, R.M., *Water Sci. Technol.*, 30, 191, 1994.

151. Leyva-Ramos, R., Fuentes-Rubio, L., Guerrero-Coronado, R.M., and Mendoza-Barron, J., *J. Chem. Technol. Biotechnol.*, 62, 64, 1975.

152. Bello, G., Cid, R., Garcia, R., and Arriagada, R., *J. Chem. Technol. Biotechnol.*, 74, 904, 1999.

153. Park, S.J., Park, B.J., and Ryu, S.K., *Carbon*, 37, 1223, 1999.

154. Bautista-Toledo, I., Rivera-Utrilla, J., Ferro-Garcia, M.A., and Moreno-Castilla, C., *Carbon*, 32, 93, 1994.

155. Moreno-Castilla, C., Ferro-Garcia, M.A., Rivera-Utrilla, J., and Joly, J.P., *Energy Fuels*, 8, 1233, 1994.

156. Puri, B.R. and Satija, B.R., *J. Indian Chem. Soc.*, 45, 298, 1968.

157. Jayson, G.G., Lowless, T.A., and Farihurst, D., *J. Coll. Interface Sci.*, 86, 397, 1982.

158. Park, S.J., Jung, W.F., and Jung, C.H., *Carbon '02' Intern. Conf. on Carbon*, Beijung, 2002, Paper PI 86 D092.

159. Aggarwal, D., Goyal, M., and Bansal, R.C., *Carbon*, 37, 1989, 1999.

160. Aggarwal, D., Ph.D. thesis, Panjab Univ., Chandigarh, India, 1997.

161. Nightmighle, E.R., *J. Phys. Chem.*, 63, 1381, 1959.

162. Bansal, R.C., Bhatia, N., and Dhami, T.L., *Carbon*, 16, 65, 1978.

163. Bansal, R.C., Aggarwal, D., Goyal, M., and Kaistha, B.C., *Indian J. Chem. Technol.*, 9, 290, 2002.

164. Huang, C.P. and Blankenship, D.W., *Water Res.* 18, 37, 1984.

165. Joensun, D.H., *Science*, 172, 1027, 1971.

166. O'Gorman, J.V., Suhr. N.H., and Walker, P.L. Jr., *Appl. Spectrosco*, 26, 44, 1972.

167. Lopez-Gonzalez, J.D., Moreno-Castilla, C., Guerrero-Ruiz, A., and Rodriguez-Reinoso, F., *J. Chem. Technol. Biotechnol* ., 32, 575, 1982.
168. Adams, M.D., *Hydrometallurgy*, 26, 201, 1991.
169. Jayson, G.G., Sangster, J.A., Thompson, G., and Wilkinson, M.C., *Carbon*, 25, 523, 1987.
170. *Handbook of Chemistry and Physics*, 59th Edition, CRC Press, Boca Raton FL, 1978, p. F213.
171. Gomez-Serrano, V., Macias-Garcia, A., Espinosa-Mansilla, A., and Valenzuela-Calahorro, C., *Water Res.*, 32, 1, 1998.
172. Ma, X., Subramanian, K.S., Chakrabarti, C.L., Guo, R., Cheng. J., Lu, Y., and Pickering, W.F., *J. Environ Sci. Health*, A27, 1389, 1992.
173. Namasivayam, C. and Periasami, K., *Water Res.*, 27, 1663, 1993.
174. Namasivayam, C. and Periasami, K., *Carbon*, 37, 79, 1999.
175. Sasaki, A., and Kimura, Y., *Nippon Kagaku Kaishi*, 12, 880, 1997 C.A. 52603e 1998.
176. Peraniemi, S., Hannonen, S., Mustalahti, H., and Ahlagren, N., Fresenuis J. *Anal. Chem.*, 349, 510, 1994.
177. Kanako, K., *Carbon*, 26, 903, 1988.
178. Shirakashi, T., Tanaka, K., Tamura, T., and Yoshihara, S., *Nippon Kagater Kaishi*, 1999, p. 137.
179. Carrot, P., Ribero-Carrot, M., and Nabias, J., *Carbon*, 36, 11, 1998.
180. Rangel-Mendez, J.R. and Streat M., *Carbon '01 Intern. Conf. on Carbon*, Lexington, KY, 2001.
181. Huang, C.P., Smith, E.H., in *Chemistry in Water Reuse*, W.J. Cooper, Ed., Ann Arbor Science Publishers, Ann Arbor MI, 1981, p. 355.
182. Polovina, M., Surbek, A., Lausevic, M., and Kaludjerovic B., *J. Serb. Chem. Soc.*, 60, 43, 1995.
183. Jevtitch, M.M. and Bhattacharya, D., *Chem. Eng. Commun.*, 23, 191, 1983.
184. Reed, B.E. and Nonavinakere, S.K., *Sep. Sci. Technol.*, 27, 1985, 1992.
185. Rubin, A.J. and Mercer, D.L., in *Adsorption of Inorganics at Solid-Liquid Interfaces*, M.A. Anderson and A.J. Rubin, Eds., Ann Arbor Science Publishers, Ann Arbor MI, 1981, p. 295.
186. Periasamy, K. and Nanasivayam, C., *Ind. Eng. Chem. Res.*, 33, 317, 1994.
187. Ferro-Garcia, M.A., Rivera-Utrilla, J., Rodriguez-Gordillo, J., and Bautista-Toledo, I., *Carbon*, 26, 363, 1988.
188. Kragten, J., *Atlas of Metal-Ligand Equilibria in Aqueous Solution* , Ellis Horwood, New York, 1978.
189. Marcias-Garcia, A., Valenzuela-Calahorro, C., Espinosa-Mansilla, A., and Gomez-Serrano, V., *Anal. Quim.*, 91, 547, 1995.
190. Brennsteiner, B.A., Zondlo, J.W., Stiller, A.H., Stansberry, P.G., Tian, D., and Xu, Y., *Energy Fuels*, 11, 348, 1997.
191. Jia, Y. and Thomas, K., *Langmuir*, 16, 1114, 2000.
192. Berman, E., in *Toxic Metals and Their Analysis*, Heyden and Son, London and Philadelphia, 1980.
193. Tewari, P.A. and Lee, W., *J. Colloid Interface Sci.* 52, 77, 1975.
194. Gray, M.J. and Malati, M.A., *J. Chem. Technol. Biotechnol.*, 29, 135, 1979.
195. Huang, C.P. and Lin, Y.T., in *Adsorption from Aqueous Solutions*, P.H. Tewari, Ed., Plenum Press, New York, 1981.
196. Netzer, A. and Hughes, D.E., *Water Res.*, 18, 927, 1984.

197. Huang, C.P., Tsang, M.W., and Hsieh, J.S., in *Separation of Heavy Metals and Trace Contaminant*, R.W. Peters and B.M. Kim, Eds., American Instt. of Chemical Engineers, New York, 1985, p. 85.

198. Rivera-Utrilla, J., Ferro-Garcia, M.A., Mata-Arjona, A., Gonzalez-Gomez, C.J., *Chem. Technol. Biotechnol.*, 34a, 243, 1984.

199. Rivera-Utrilla, J. and Ferro-Garcia, M.A., *Carbon*, 25, 645, 1987.

200. Baes, C.F., Jr., and Hermer, R.E., in *The Hydrolysis of Cations*, Wiley Interscience Publications, New York, 1976.

201. Bansal, R.C., Goyal, M., and Agarwal, D., *23rd Bienn. Conf. on Carbon,* Penn. State University, Univ. Park PA, July 13–17, 1997.

202. Aggarwal, D., Ph.D. thesis, Panjab Univ., Chandigarh, India, 1997.

203. Takiyama, L.R. and Huang, C.P., *23rd Bienn. Conf. on Carbon,* Penn State University, Univ. Park PA, July 13–17, 1997.

204. Kim, H.S., Ryu, S.K., Park, K.K., Lee, J.B., and Jung, C.H., *Carbon '02' Intern. Conf. on Carbon,* Beijung, 2002, Paper PI 91 DO 98.

205. Paajanon, A., Lehte, J., Sautpakka, T., and Morneau, J.P., *Sep. Sci. Technol.*, 32, 813, 1997.

206. Tekker, M., Sallabar, O., and Imamoglu, M., *J., Environ. Sci. Health*, A 32, 2077, 1997.

207. Seco, A., Marzal, P., Gabaldon, C., and Ferror, J., *J. Chem. Technol. Biotechnol.*, 68, 23, 1997.

208. Goyal, M., Rattan, V.K., and Bansal, R.C., *Indian J. Chem. Technol.*, 6, 305, 1999.

209. Ferro-Garcia, M.A., Rivera-Ultrilla, J., and Bautista-Toledo, I., *Carbon*, 28, 545, 1990.

210. Cheng, J., Subramanian, K., Chakrabarti, C., Guo, R., Ma, X., Lu, Y., and Pickering, W.F., *J. Environ. Science Health*, A. 28, 51, 1993.

211. Taylor, R.M., and Kuennen, R., *Environ. Prog.*, 13, 65, 1994.

212. Burges, J., in *Metal Ions in Solution*, Ellis Horwood Ltd., New York, 1978.

213. Greenwood, N.N. and Farnshow, A., in *Chemistry of the Elements*, Pergamon Press, Oxford, 1984.

214. Bailar, J.C., Jr., Emeleus, H.J., Nyholm, Sir Ronald, and Trotman Dickerson, A.F., in *Comprehensive Inorganic Chemistry*, Oxford, 1973.

215. Sohail A. and Qadeer, R., *Adsorp. Sci. Technol.*, 15, 815, 1997.

216. Yang, J., Zhangfing, Q., Songying, C., and Shaoyi, P., *Huanging Huaxue*, 16, 423, 1997.

217. Reed, B.E., *Sep. Sci. Technol.*, 30, 101, 1995.

218. Carriere, P., Mohaghegh, S., Gaskari, R., and Reed, B.E., *Sep. Sci. Technol.*, 31, 965, 1996.

219. Rinkus, K., Reed, B.E., and Lin, W., *Sep. Sci. Technol.*, 32, 2367, 1997.

220. Reed, B.E., Jamil, M., and Thomas, B., *Water Environ. Res.*, 68, 877, 1996.

221. Reed, B.E., Robertson, A.M., and Jamil, M., *Hazard Ind. Waste*, 26, 250, 1994.

222. *World Health Organization International Standards for Drinking Water*, WHO, Geneva, 3rd edition, 1971.

223. Council Directive 80/778/EEC, *Official Journal of the European Communities*, Aug. 30, 1980, L229 23, p. 11.

224. Mishra, S.P. and Chaudhary, G.R., *J. Chem. Technol. Biotechnol.*, 59, 359, 1994.

225. Marzel, P., Seco, A., Gabaldon, C., and Ferror, J., *J. Chem. Technol. Biotechnol.*, 66, 279, 1996.

226. Tai, M.H. and Streat, M., *Chem. Eng. Res. Event, Two-day Symp.*, 1998, p. 578, C.A. 100456d 1298, 1998.

227. Karounou, E. and Saha, B., *Carbon 01 Intern. Conf. on Carbon,* Lexington, KY, 2001, Paper 25.6.

228. Ferguson, J.F. and Gavis, J., *Water Res.,* 6, 1259, 1972.

229. Pierce, M.Z. and Moore, C.B., *Environ. Sci. Technol.,* 14, 214, 1980.

230. Gulledge, J.H. and OConnor, J.T., *J. Am. Water Works Assoc.,* 65, 548, 1973.

231. Rosenblum E. and Clifford D., U.S. EPA report, *Cooperative Agreement,* 1984 CR 807939.

232. Sen, A.K. and De, A.K., *Indian J. Technol.,* 25, 259, 1987.

233. Wasay, S.A., Haron, M.J., and Tokunaga, S., *Water Environ. Res.,* 68, 295, 1996.

234. Gupta, S.K. and Chen. K.Y., *J. Water Pollut. Control Fed.,* 50, 493, 1978.

235. Kamegawa, K., Yoshida, H., and Arita, S., *Nippon Kagaku Kaishi,* 1979, p. 1365.

236. Huang, C.P. and Fu, P.L.K., *J. Water Pollut. Control Fed.,* 56, 233, 1984.

237. Rajakovic, L.V., *Sep. Sci. Technol.,* 27, 1423, 1992.

238. Huang, C.P. and Van, L.M., *J. Water Pollut. Control Fed.,* 61, 1596, 1989.

239. Raji, C. and Anirudhan, T.S., *J. Sci. Ind. Res.,* India, 57, 10, 1998.

240. Lorenzen, L., Van Deventer, J., and Landi, W., *Miner. Eng.,* 8, 557, 1995.

241. Sen, A.K. and De, A.K., *Indian J. Technol.,* 25, 259, 1987.

242. Nightingale, E.R., *J. Phys. Chem.,* 63, 1381, 1959.

243. Biniak, S., Swiatkowski A., and Pakula, M., in *Chemistry and Physics of Carbon,* L.R. Radovic, Ed., Vol. 27, Marcel Dekker, New York, 2001.

244. Radovic, L.R., Morino-Castilla, C., and Rivera-Utrilla, J., in *Chemistry and Physics of Carbon,* L.R. Radovic, Ed., Vol. 27, Marcel Dekker, New York, 2001.

7 Activated Carbon Adsorption and Environment: Adsorptive Removal of Organics from Water

Drinking water has been found to contain small amounts of a large number of synthetic organic compounds such as phenols, pesticides, herbicides, aliphatic and aromatic hydrocarbons, and their halogen derivatives, dyes, surfactants, organic sulfur compounds, ethers, amines, and nitro compounds. More than 1500 different organic compounds are suspected to be present in drinking waters. These organic compounds are derived from industrial and municipal waste water, rural and urban runoff, natural decomposition of animal and vegetable matter, and from agricultural practices. In addition, several halogenated compounds such as trihalomethanes and chlorophenols are produced during chlorination practices for disinfection of drinking water. The leaching of solid industrial wastes is also a source of several organic chemical compounds in water.

Several chemical and biological methods have been used for the removal of organic compounds from waste water, but they have achieved limited success. This is due to the fact that the amount and variety of chemicals in waste water is ever increasing due to the development of chemical, pharmaceutical, and other industries. Thus, a considerable effort has been directed to develop more efficient and effective technologies for their removal. Activated carbons, because of their high surface area, highly microporous structure, and a high reactivity of their surface, have been considered to be potential adsorbents for the removal of organic compounds from waste water.

The more important parameters that determine the adsorptive removal of organic compounds from water are the nature and the molecular dimension of the organic compound, the porous and the chemical structure of the carbon surface, and the pH of the aqueous solution. This chapter will present a review of the work carried out on the adsorption of different groups of organic compounds on different activated carbons. The parameters that determine and influence the adsorption will be examined, and the adsorption mechanisms based on the adsorption data will be suggested.

7.1 ACTIVATED CARBON ADSORPTION
OF HALOGENATED ORGANIC COMPOUNDS

The presence of light halogenated hydrocarbons in general and of trihalomethanes (THM) in particular in drinking waters was first reported in 1974.[1] Since then, chloroform and other THM have been observed in all water supplies where chlorine is used as a disinfecting agent. These halogenated compounds are toxic, carcinogenic, and mutagenic to human beings when ingested over extended periods of time.[2-4] The U.S. Congress thus passed the Safe Drinking Water Act in 1974, because earlier studies in certain parts of U.S.A. showed the presence of several carcinogenic organic compounds and their possible correlation between the quality of water supply and increased cancer mortality. The legislation was to ensure that public water supply systems conform to some minimum national standards with respect to the control of organic chemical pollutants in drinking waters. The U.S. Environmental Protection Agency (EPA) then carried out a national Monitoring Survey in 1974 to 1975 in several areas and found that trihalomethanes such as chloroform, brome dichloromethane, dibromo chloro ethane, bromoform, and dichloroidomethane were the major halogenated contaminants in drinking waters. Their concentration in drinking waters could be as high as 784 µg/L.[5] According to the National Organics Reconnaissance Study (NORS) of EPA,[6] these light halogenated hydrocarbons and THM$_S$ are formed during the disinfection of domestic water supply by chlorine, which reacts with certain naturally occurring organic materials, such as humic and fulvic acids produced by the degradation of vegetable matter.

The current regulations in the U.S.A. stipulate that water supply systems serving a population of more than 10,000 should maintain a total THM concentration below 100 µg/L, based on yearly average. However, the U.S. EPA has proposed a two-stage Water Disinfection or Disinfection By-products rule, according to which the maximum contaminant level (MCL) should be lowered from 100 to 80 µg/L for total THMs, and an MCL of 60 µg/L for a total of five helioacetic acids: monochloro, dichloro, trichloro, monobromo, and dibromo acetic acids in the first stage. In the second stage, the MCLs are proposed to be further decreased to 40 and 30 µg/L for THMs and haloacetic acids, respectively. The stage 1 of the rule that was promulgated on December 16, 1998 has to be complied with by the water utilities between the years 2001 and 2002, depending on their size.[4]

Youssefi and Faust[7] investigated the formation of these light halogenated hydrocarbons by the chlorination of naturally occurring organic compounds in water such as humic acid, tannic acid, D-glucose, gallic acid, and vanillic acid. These workers found that appreciable amounts of chloroform ($CHCl_3$) were formed when 10 mg/L of these precursors were contacted with 10 mg/L of chlorine, and the amount of $CHCl_3$ formed increased with the time of contact. The reaction was quite fast, and about 90% of the ultimate yield of $CHCl_3$ was formed in the first two hours (Figure 7.1) of the reaction time in the absence of residual chlorine. The amount of the organic compound used was less than 2%.

These workers[7] determined the adsorption of chloroform, bromoform, bromodichloromethane, dibromochloromethane, and carbon tetrachloride on a Nuchar granulated activated carbon from aqueous solutions. The pH of the solution was kept at

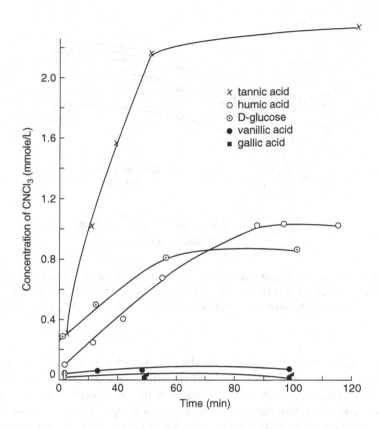

FIGURE 7.1 Formation of chloroform by the chlorination of water containing tannic acid, humic acid, D-glucose vanillic acid, and gallic acid (From Youssefi M. and Faust, S.D., in *Activated Carbon Adsorption*, Vol. 1, I.H. Suffet and M.J. McGuire, Eds., Ann Arbor Science Publishers, Ann Arbor MI, 1981, p. 133. With permission.)

7 by using a phosphate buffer, and the time of contact with the carbon was 24 hr. It was observed that these contaminants were adsorbed at different rates. For example, the percentage adsorption varied from 53.5% for chloroform to 81.0% for bromoform in the first hour (Figure 7.2). The adsorption followed the Freundlich isotherm equation and gave linear log x/m versus log C plots. The breakthrough studies using adsorption columns showed that the breakthrough point followed the order $CHCl_3 > CHCl_2Br > CHClBr_2 > CHBr_3$. But there was a point on the breakthrough curve at which the sum of the concentrations of all the halomethanes was equal to 100 mg/L, which was the required standard for total THM in drinking waters. This break point was attained quite early, indicating the high potential of using activated carbon for the control of THM in drinking water.

Ishizaki et al.[8] studied the adsorption of chloroform from aqueous solutions in the concentration range 10 to 200 μg/L on as-received Filtrasorb-200 and after removing the carbon-oxygen surface groups by evacuation at 1000°C. Both carbon samples adsorbed appreciable amounts of chloroform, but the outgassed sample showed a stronger affinity and a large capacity compared to the as-received carbon

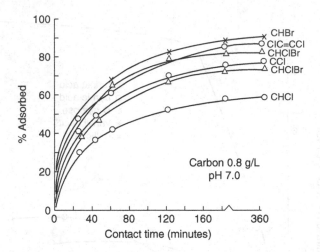

FIGURE 7.2 Removal of light halogenated hydrocarbon mixture by granular activated carbon. (From Youssefi, M. and Faust, S.D., in *Activated Carbon Adsorption*, Vol. 1, I.H. Suffet, and M.J. McGuire, Eds., Ann Arbor Science Publishers, Ann Arbor MI, 1981, p. 133. With permission.)

sample (Figure 7.3). The adsorption data obeyed the Langmuir equation. The Langmuir equation constants calculated from the linearized Langmuir plots gave adsorption energies, which were different in the as-received and the degassed samples. The adsorption energy on the degassed sample was about 2 to 10 times the adsorption energy on the as-received carbon sample. Because the degassing of the carbon did not appreciably change the pore-size distribution of the carbon, the difference in the adsorption energy was attributed to the presence of carbon-oxygen surface groups in case of the as-received carbon sample.

When the linear Langmuir plots were extrapolated for the calculation of monolayer capacities, the two linear plots crossed each other at higher equilibrium concentrations (Figure 7.4). This indicated that the adsorption capacity of the as-received sample was higher at equilibrium concentrations greater than 270 µg/l. Thus, these workers concluded that the adsorption of chloroform from aqueous solution was a function of the equilibrium concentration of the aqueous solution and the nature of the carbon surface. Nakamura et al.[9] investigated the removal of chloroform from water using activated carbons and observed that the amount adsorbed did not depend upon the surface area or the pore volume of the carbon and the pH of the aqueous solution. The important factor in the adsorption of chloroform was the hydrophobicity of the carbon surface.

Rey et al.[10] removed chlorinated aromatic hydrocarbon from a pharmaceutical plant by stripping the chlorinated hydrocarbons with air and then adsorbing in activated carbon beds. The removal of chlorinated volatile organic compounds such as trichloroethane, cis, and trans dichloroethane from ground water contaminated with these nonaqueous liquids was carried out by Yu et al.[11] using activated carbon fibers. All the halogenated compounds were adsorbed rapidly by the activated carbon

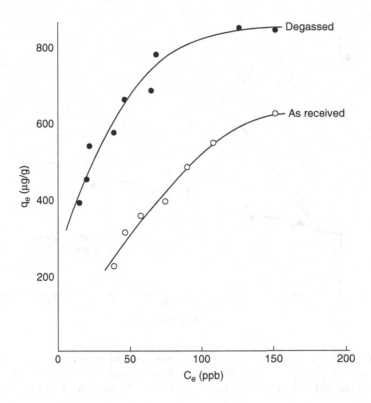

FIGURE 7.3 Adsorption isotherms of chloroform from aqueous solution on carbon before and after outgassing. (Adapted from Ishizaki, C., Marti, I., and Ruiz, M., in *Advances in Chemistry,* No. 202, I.H. Suffet and M.J. McGuire, Eds., *Amer. Chem. Soc.*, Washington, D.C., 1983, p. 95. With permission.)

fiber. The adsorption isotherms of trihalomethanes (THM) on a sample of granulated activated carbon (GAC) and a sample of activated carbon fiber (ACF) from aqueous solutions were determined by Li et al.[12] The adsorption data agreed well with the Freundlich isotherm equation both for GAC and ACF. The adsorption capacity of ACF was twice as large as that of GAC and increased with the hydrophobicity of the carbon surface, and by the substitution of Br atoms for Cl atoms. When the adsorptive removal was carried out in a fixed bed reactor, the adsorption of THM was many times larger on ACF than on GAC under similar operating conditions. The running time of the ACF column was twofold longer than that of the GAC column. The ACF mass transfer zones were smaller than those of GAC zones and decreased when chlorine in caloroform was replaced by bromine atoms. In a later publication, these workers[13] observed that the adsorption of trihalomethanes on activated carbon fiber was in principle but not completely monolayer adsorption. The adsorption capacity of ACF for the four THMs increased as the hydrophobicity of the THM increased. It also increased by the replacement of chlorine atoms by bromine atoms. Uchida et al.[14] studied the adsorption characteristics of trihalomethanes on an activated carbon fiber from their solution in chlorinated water and

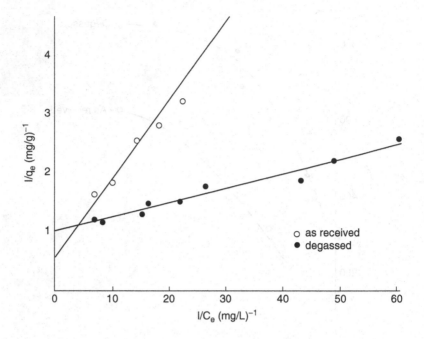

FIGURE 7.4 Langmuir plots of adsorption of chloroform from water on carbon before and after outgassing. (From Ishizaki, C., Marti, I., and Ruiz, M., in *Advances in Chemistry*, No. 202, I.H. Suffet and M.J. McGuire, Eds., *Amer. Chem. Soc.,* Washington, D.C., 1983, p. 95. With permission.)

found that the amount of THMs adsorbed was not determined by the carbon fiber surface properties but by the polarity of the trihalomethanes. The magnitude of adsorption was in the order bromoform > dibromochloromethane > bromodichloromethane > chloroform. This indicated that the greater the polarity, the lower the adsorption of trihalomethanes by the activated carbon fiber.

Karanfil and Kilduff[15] studied the adsorption of two synthetic organic contaminants, trichloroethylene (TCE) and trichlorobenzene (TCB), on coal-based and wood-based granulated activated carbons. The activated carbon surface was modified by liquid-phase oxidation with nitric acid and by degassing in an inert atmosphere. The activated carbons were characterized by elemental analysis, surface area and pore-size distribution, and acid-base adsorption techniques. The adsorption isotherms were determined by equilibrating a known weight of the carbon sample with different concentrations of TCE and TCB solutions, and analyzing the solution by gas chromatography.

Acid-washed wood-based activated carbon (WVBAW) showed significantly lower adsorption than the coal-based acid-washed activated carbon (F 400 AW) sample on both a mass and surface area basis over the entire concentration range. This indicated that the uptake of both TCE and TCB is not determined by the surface area alone. However, the size of the pores may also play an important role, because the coal-based activated carbon was found to have a considerably larger amount of

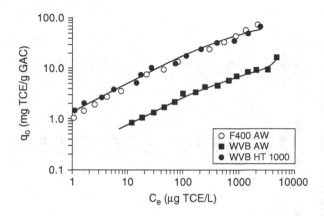

FIGURE 7.5 Adsorption of Trichloroethylene y as-received and degassed wood based activated carbon. (Adapted from Karanfil, T. and Kilduff, J.E., *Environ. Sci. Technol.*, 33, 3217, 1999. With permission.)

micropore surface area than the wood-based carbon. However, when the adsorption isotherms were determined on both the carbon samples after degassing them at 1000°C in nitrogen, there was no effect of degassing on the adsorption of TCE and TCB on the coal-based (F 400) activated carbon, although the adsorption on the wood-based degassed carbon increased nearly tenfold for both TCE and TCB. The increase in adsorption capacity was significant because the degassed wood-based carbon (degassed WVB) showed almost similar adsorption capacity as the coal-based acid washed sample (Figure 7.5).

Because the micropore surface area in both the coal-based and wood-based carbons did not change significantly on heat treatment, the results clearly indicated that the chemical structure of the carbon surface played a very important role in the adsorption compared with the pore structure. This received further support from the adsorption isotherms on the carbon samples after oxidation with nitric acid and subsequent degassing at 650°C in nitrogen gas. Although the oxidation of the activated carbons decreased the adsorption of both TCE and TCB, subsequent degassing at 650°C enhanced the adsorption. These results, when analyzed with respect to the surface characteristics of the activated carbons (Table 7.1), suggested that the adsorption of these hydrophobic compounds (TCE and TCB) was influenced strongly by the surface acidity of the activated carbon surface. These workers were of the view that the acidic surface groups, which were present on the edges of the layer planes, were likely to form water clusters on the carbon surface as well as at the entry of the micropores. Although these edges constitute the main adsorbing surface, these clusters may prevent the adsorption of the low molecular weight hydrophobic TCE and TCB molecules. Because the concentration of the acidic surface groups on the carbon surface increased on oxidation and decreased on degassing, there was a corresponding decrease or increase in the adsorption of these compounds. This was consistent with the adsorption of organic compounds such as phenols[16–18] and 2-methylisoborneol.[19] The oxidation of the heat-treated carbons significantly

TABLE 7.1
Surface Area and Carbon-oxygen Surface Groups on Different As-received and Modified Activated Carbons

Carbon Type	Surface Area (M²/G)	Strong Carboxylic Groups[a] (µEq/M²)	Week Carboxylic and Lactonic[b] (µEq/M²)	Hydroxyl and Carbonyl Groups[c] (µEq/M²)	Total Acidic Groups[d] (µEq/M²)	Total Basic Groups (µEq/M²)
As-received Carbons						
WPLL	293	0.44	0.10	0.31	0.85	1.26
FS100	750	0.23	0.04	0.20	0.47	0.61
BPL	1200	0.11	0.04	0.08	0.23	0.57
F400[g]	945	0.09	0.04	0.11	0.24	0.40
Surface Modified Carbons						
F400[g] AW	1040	0.00	0.00	0.14	0.15	0.34
F400 HT 1000	1000	0.00	0.00	0.00	0.00	0.38
F400 OX 2/70	1041	0.22	0.18	0.54	0.94	0.22
F400 OX 9/70	1030	0.61	0.35	0.76	1.71	0.11
F400 OX 2/70, HT650	1029	0.00	0.00	0.26	0.26	0.29
F400 OX 9/70, HT650	1065	0.05	0.17	0.33	0.55	0.24
WVB AW	1745	0.13	0.12	0.33	0.57	0.18
WVB HT 1000	1169	0.00	0.08	0.21	0.28	0.24
WVB OX 2/50	1270	0.19	0.20	0.36	0.75	0.17
WVB OX 2/70	1295	0.35	0.27	0.52	1.14	0.13

Adapted from Karanfil, T. and Kilduff, J.E., *Environ. Sci. Technol.*, 33, 3217, 1999.

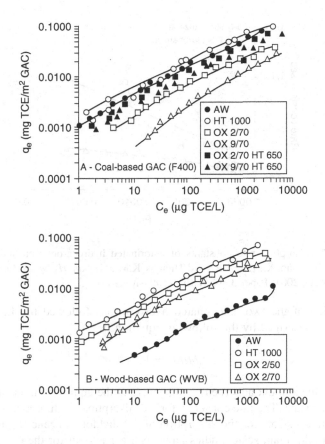

FIGURE 7.6 Adsorption of Trichloroethylene on coal-based (a) and wood-based (b) carbons before and after heat treatment (HT) and oxidation (OX). (Adapted from Karanfil, T. and Kilduff, J.E., *Environ. Sci. Technol.*, 33, 3217, 1999. With permission.)

increased the total surface acidity or the polarity while it decreased the total basicity. Thus, the adsorption of both TCE and TCB was decreased on both the coal-based and wood-based activated carbons. Degassing of the oxidized carbons at 650°C selectively removed the strongly acidic functional groups from the surface, thereby enhancing the adsorption of TCE and TCB.

Xiao et al.[20] studied the adsorption of chlorobenzene and 1,3-dichlorobenzene at 313 to 453 K on a wood-based activated carbon, which was prepared by the activation of a char in CO_2 at 1073 K. The adsorption isotherms for both the halogenated hydrocarbons are Type I of the BET classification (Figure 7.6). The amount adsorbed decreased systematically with the increasing temperature. The total volume of the halogenated hydrocarbon adsorbed was equal to the total pore volume of the carbon obtained from nitrogen adsorption at 77 K. The isosteric heats of adsorption of chlorobenzene calculated using the van Hoff isochore was 37 KJ/mol at an adsorption of 1 mmol/g and compared well with the heat of vaporization 35.19 KJ/mol. The adsorption appeared to be physical in nature involving micropores. The kinetics

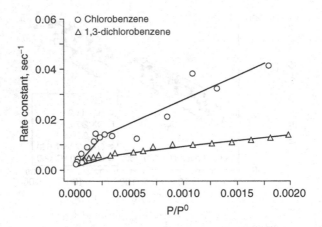

FIGURE 7.7 Adsorption rate constants of chlorinated hydrocarbons on activated carbon. (After Xiao, B., Zhao, X., Yavuz, R., and Thomas, K.M., *Carbon 01, Intern. Conf. on Carbon*, Lexington, K.Y., 2001, Paper 35.6. With permission.)

of adsorption of the two halogenated hydrocarbons followed the Linear Driving Force model described by the following equation.[21,22]

$$M/Me = 1 - e^{kt}$$

where M is the amount adsorbed at time t, Me is the equilibrium adsorption, and k is the rate constant. The rate constants for the adsorption of chlorobenzene are faster than the rate constants for the adsorption of 1.3 dichlorobenzene (Figure 7.7).

Several pilot plant scale studies have been carried out for the removal of these halogenated organic compounds using different activated carbons. For example, Love[23] studied the influence of carbon type, bed depth, hydraulic loading, and contact time under pilot scale conditions and observed that GAC was capable of removing 90% or more of chloroform for about 3 weeks, after which the chloroform level increased and the influent and the effluent concentrations became equal after the ninth week. Furthermore, it was also observed that the brominated THM were reduced significantly and for longer periods of time than chloroform. O'Connor et al.[24] also made pilot scale studies of THM removal using GAC and found that a 3-feet-deep column of GAC was capable of removing more than 90% of influent chloroform for 9 weeks of continuous operation. Smith et al.[25] studied mass transfer properties for a fixed bed GAC for removing THM from water using four different brands of GAC, and used breakthrough curves to calculate parameters that determine mass transfer properties of GAC columns. It was found that the mass transfer zone was shallow, with a height of about 0.21 to 0.27 meter, which indicated a rapid interaction between the chloroform and the carbon. The GAC column exhaust rate was about 0.5 to 1.0 cm per day. Mullins et al.[26] evaluated several brand commercial GACs for the removal of THM from drinking water using a bench scale vertical column operating in continuous down mode. Seven brands of activated carbons were studied in a 2.5 cm diameter column, using water that contained varying amounts

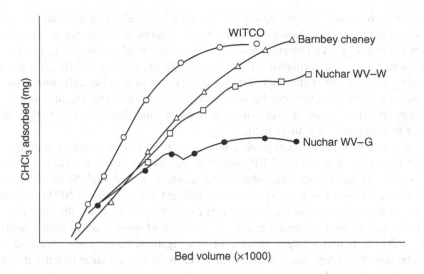

FIGURE 7.8 Adsorption isotherms of chloroform by different activated carbons. (After Mullins, R.L., Jr, Jogorski, J.S., Hubbs, S.A., and Allgerei, G.D., in *Activated Carbon Adsorption*, Vol. I, I.H. Suffet and M.J. McGuire, Eds., Ann Arbor Science Publishers, Ann Arbor MI, 1981, p. 273. With permission.)

of THM ranging from 50 to 200 µg/L. The influent and effluent of each column was examined for residual chlorine, pH, temperature, total organic carbon, and THM concentration. The effluent was found to contain 0.02 to 0.22 µg/L of the THM. Each brand of GAC was capable of removing appreciable amounts of THM, although the amounts removed varied with the GAC sample (Figure 7.8). The rate of removal of THM was quite large, but the ultimate adsorption capacity of the carbon was very low, between 0.05 and 0.12% by weight of the carbon for chloroform, and between 0.02 and 0.04% for bromodichloromethane. Thus, although the removal of THMs could be carried out rapidly and efficiently, the carbon needed more frequent regeneration because of its lower adsorption capacity.

7.2 ACTIVATED CARBON ADSORPTION OF NATURAL ORGANIC MATTER (NOM)

Naturally occurring organic materials (NOM), a heterogenous mixture of complex organic compounds that include humic acids, lipids, proteins, hydrophilic acids, carboxylic acids, amino acids, polysaccharides, and hydrocarbons,[27,28] are invariably present in surface and ground waters, in dissolved or colloidal forms. These dissolved organic materials (DOM) are very difficult to remove by conventional treatments. The DOM can act as substrates for the growth of bacteria in the distribution system. Consequently, it becomes essential to disinfect water before domestic supply. Chlorination of the water supply is the most common water supply treatment. The chlorination of water results in the formation of disinfection by products consisting of several halogenated and other organic compounds, which are formed by reactions involving chemical oxidants, NOM, and bromides. Trihalomethanes and haloacetic acids are the two major classes of organic disinfection by products that are commonly found in water given a chlorine treatment. Thus, the simultaneous control of microbial activity and the formation of disinfection by products is a difficult and challenging problem. As mentioned earlier, the presence of these disinfection products in potable water is undesirable because they are toxic, carcinogenic, and mutagenic to humans when ingested over longer periods of time. Besides, the NOM impart color to water, which is unaesthetic.

Because NOM is present in all natural waters and is the primary precursor to DBP formation, most of the DBP control strategies have been directed at removal of NOM from water before it is subjected to disinfection. Although NOM is a mixture of organic compounds, it is expected that different components of NOM will show different reactivity toward the disinfectant to produce DBP. Thus, the control technologies are generally directed toward the removal of more reactive components of the NOM. Although a larger number of investigations have been concerned with determining the composition of NOM in water using fractionation of natural waters using adsorption, chromatography, and employing synthetic resins. The fractions separated by adsorption chromatography have been analyzed using elemental analysis and several spectroscopic techniques such as GC/MS, C^{13}-NMR and IR/FTIR.[4] Although these techniques provide insight into the composition of NOM, they are only semiquantitative. Besides, they require large quantities of NOM for analysis and do not have practical values for the water treatment plants, because composition of NOM can change during the treatment.

Granulated activated carbons have been considered potential adsorbents for the removal of both regulated synthetic organic materials and dissolved naturally occurring organic materials (NOM) and, consequently, a large number of studies have been reported. Karanfil et al.[29] investigated the role of carbon surface chemistry and pore structure on the adsorption of 4 model dissolved organic material (DOM) isolates and 4 surface water natural organic materials (NOM) by using 10 carbons prepared by modifying the surfaces of a coal-based and a wood-based carbon, and by using 7 different as-received granulated activated carbons. The modification of the carbon surface was carried out by oxidation with nitric acid for different intervals of time and by degassing the oxidized sample at 650°C. The carbon samples were washed with acid and then heat treated in nitrogen at 1000°C. These treatments did not significantly change the surface area or the pore volume of the carbon but appreciably changed the concentration of the carbon-oxygen surface functionality of the carbon surface. The four model dissolved organic materials (DOMs) used in this study included polymaleic acid (PMA), a natural solid humic acid (HA), and a fulvic acid (FA) extracted from Laurentian soil and Aldrich humic acid (AHA). The aqueous solutions of these compounds were equilibrated with a known weight of the activated carbon for one month under oxic conditions, and the solution was analyzed using a uv-spectrophotometer. The adsorption isotherms of PMA and FA by the untreated and acid-washed carbons were similar. This was attributed to the similar molecular size of the two macromolecules and the presence of similar amount of carboxylic type acidity in their structure. However, the adsorption of AHA was significantly lower as compared to PMA and FA. This was attributed partly to the larger size of AHA, so that it cannot have access to many of the micropores, and partly to the smaller interactions with the positively-charged GAC surface due to the lower acidity of AHA molecule. However, when the GAC surface was oxidized with HNO_3, the adsorption of all the DOM samples decreased considerably as shown by a significant fall in the Freundlich affinity parameter. When the oxidized carbon samples were degassed by heat treatment at 650°C, the adsorption of all the dissolved organic materials increased considerably.

The oxidation of the carbon samples with nitric acid enhanced the amount of carbon-oxygen acidic surface groups that make the carbon surface hydrophilic and negatively charged. The increase in hydrophilicity of the carbon surface resulted in the formation of water clusters around the pore entrances, while the negative charge of the carbon surface reduced the attractive interactions between the carbon surface and the DOM molecules. Both these factors tend to reduce the adsorption of these compounds on oxidation. When the acidic surface groups were eliminated on degassing of the carbons, the surface became hydrophobic and less negatively charged, thereby increasing the adsorption of these compounds. Thus, these workers suggested that the adsorption capacity of a granulated activated carbon for naturally occurring materials (NOM) was determined by two parameters: the pore size and the surface acidity of the carbon. The former determines the availability of a pore for adsorption, and the latter determines the attractive interactions between the carbon surface and the NOM molecule.

Newcomb[30] studied the adsorption of natural organic matter (NOM) on activated carbon to understand the effect of carbon surface charge and pore-size distribution, and NOM charge and molecular weight on the adsorption. The effect of pH and the ionic strength of the aqueous solution were also examined. It was found that the adsorption occurred by a pore-filling mechanism in the absence of strong electrostatic effects. At neutral pH, the NOM has a significant negative charge, which affects the adsorption. When the activated carbon surface had a low positive surface charge, the adsorption took place by direct surface-NOM electrostatic attractive interactions. However, when the carbon surface was hydrophobic, then the adsorption possibly involved interactions between the NOM and the aromatic rings on the carbon surface.

Several investigators have shown that the addition of several cations such as calcium, magnesium, and sodium[31–34] helped the adsorptive removal of natural organic materials from water at certain pH values.

Vidic and Suidan,[35] Warta et al.[36] and Cerminara et al.[37] on the other hand, observed that the dissolved oxygen in water also enhanced the adsorption of NOM, which was attributed to some sort of oxidative polymerization or aggregation of NOM under oxic conditions. Lee et al.[38] while studying the adsorption of humic acids on as-received and chemically modified activated carbon fibers and carbon blacks found that pore-size distribution in an activated carbon was an important factor in determining the adsorption (Figure 7.9). These workers applied the Freundlich adsorption isotherm to their data and observed that the Freundlich adsorption constant increased with increase in the pore volume of the carbon. Kilduff and coworkers[39–41] also observed that the adsorption of humic acids, which have a large molecular diameter, was determined by the pore-size distribution in carbons, because some of the micropores were not accessible to large humic acid molecules. Because the natural

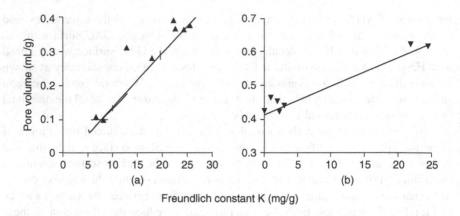

FIGURE 7.9 Effect of porosity on the removal of humic substances by a series of carbon adsorbents: (a) humic sustances fractionated at pH = 2.8; (b) humic substances fractionated at pH = 6.8. (Adapted from Lee, M.C., Snoeyink, V.L., and Crittenden, J.C., *J. Am. Water Works Assoc.*, Aug. 1981, p. 440. With permission.)

organic materials are polymeric substances, different activated carbons will show different adsorption capacities toward different size fractions of the NOMs, depending upon the pore-size distribution of carbons. Many NOM fractions have molecular diameters of about 3 nm so that only activated carbons with a large proportion of mesoporous will be more suitable for their adsorptive removal.

Thus, it is apparent that the adsorption of natural organic materials and dissolved organic materials present in surface and ground waters is determined both by the porous and the chemical structure of the carbon surface.

Utrera-Hidalgo et al.[42] studied the adsorption of gallic acid from aqueous solutions by activated carbons prepared from almond shells, olive stones, and lignite. The adsorption isotherms (Figure 7.10) obtained using static method obeyed the Langmuir adsorption equation, and the adsorption increased with the degree of activation of the carbon. In the case of dynamic adsorption, some characteristics of the carbon bed such as breakthrough volume, adsorption capacity at different breakthrough volumes, and height of the mass transfer zone were calculated from the breakthrough curves. The adsorption capacity under dynamic conditions was very well related to the pore volume accessible to water molecules. However, the height of the transfer zone was dependent upon the raw material and the history of the formation of the activated carbon. The presence of inorganic electrolytes in the gallic acid solution enhanced the adsorption capacity of the carbon bed. According to these workers, the anion of the electrolyte, being smaller in size than the gallic acid molecule, was first adsorbed on the carbon surface and increased the negative surface charge of the activated carbon. This enhanced the adsorption of gallic acid, because the carboxylic acid group in gallic acid could enhance the donor-acceptor interaction by acting as an electron withdrawing group as in the case of phenol adsorption, discussed elsewhere in the book.

Weber and coworkers[31] investigated the adsorption of humic acid from tap water and deionized water in the presence of Ca^{2+}, Mg^{2+}, Fe^{3+}, and OCl^- ions, which were added to water in different concentrations. The presence of each ion in water enhanced

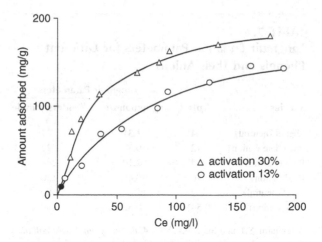

FIGURE 7.10 Adsorption isotherms of gallic acid on two activated carbons (After Uterera-Hidalgo, E., Moreno-Castilla, C., Rivera-Utrilla, J., Ferro-Garcia, M.A., and Carrasco-Morina, F., *Carbon*, 30, 107, 1992. With permission.)

the adsorption of humic acid, but the effect was the largest in the case of Mg^{2+} and smallest in the case of Fe^{3+} cation. This enhanced adsorption was attributed to the modification of the humic acid through coordinate covalent bonding, or by substitutive oxidation, to make humic acid macromolecules capable of attaching to oxygen functional group sites on the carbon surface. The interaction between the humic acid macromolecules and the carbon surface oxygen groups may also be enhanced by the coordination of the Ca^{2+}, Mg^{2+} metal ions with oxygen of the surface functional groups.

7.3 ACTIVATED CARBON ADSORPTION OF PHENOLIC COMPOUNDS

Phenol and its derivatives such as methyl phenols, ethyl phenols, and dimethyl phenols constitute a group of pollutants that are invariably present in the effluents from industries engaged in the manufacture of a variety of chemical compounds such as plastics, dyes and plants used for coal gasification, and petrochemical units. Schmidt et al.[43] carried out mass spectrometric analysis of the methylene chloride extract of coal gasification plant effluents and found that about 60 to 80% of the total organic content of the effluent consisted of phenolic components, primarily phenol, methyl phenol, dimethyl phenols, and ethyl phenols. Some of these phenolic compounds also originate from natural sources by the decay of vegetation and are released into surface and ground waters. Many of these compounds are carcinogenic, even when present in low concentrations. The presence of phenols in drinking water also produces foul-smelling chlorophenols during disinfection of water by chlorination. Thus, investigations relating to the removal of phenols from water have engaged the attention of a large number of investigators. Activated carbons, because of their large area and a high degree of surface reactivity, have a high efficiency for the removal of phenolic compounds from water.

TABLE 7.2
Langmuir Equation Parameters for Different Phenols and Their Anions

Species	pH	Langmuir Parameters	
		x_m mmoles/g	k_m liters/mmole
Phenol (neutral)	4	1.31	77
Phenolate (anion)	12	0.37	22
PNP (neutral)	4	2.10	240
PNP (anion)	10	0.46	130
DNP (neutral)	2	2.5	400
DNP (anion)	9.5–10	0.52	190

After Jaim, S.J. and Snoeyink, V.L., *45th Ann. Conf. Water Pollution Control Federation*, Georgia, Oct. 8–13, 1972.

Jaim and Snoeynik[44] studied the removal of several phenols and their anions on activated carbon Filtrasorb-400 from aqueous solutions and observed that the anions showed a much lower adsorption affinity for activated carbon than the neutral molecules (Table 7.2). Knadarov & Verteshev[45] found that the phenol was adsorbed by activated carbons in a semicontinuous manner. The adsorption capacity of the carbon was dependent upon the nature of the carbon, the history of its formation, and the nature of the ash constituents present as impurities. Mironova and coworkers[46] were able to remove about 80% of phenol present in the waste gases of the semi coking process by adsorption on a sample of coke activated at 800 to 850°C. Rodriguez et al.[47] achieved success in the removal of phenol from industrial effluents in the concentration range 80 to 1000 ppm using activated carbons. Bethel[48] and Radeke et al.[49] could almost completely recover phenols from aqueous effluents of resin manufacturing units by adsorption on active carbons. The phenol was eluted from the carbon surface using hydrocarbon solvents and then recovered by distillation. In an experiment, an effluent containing 4.87% phenol, 2.86% formaldehyde, and 0.49% furfuraldehyde was passed through a bed of 3.5-ton activated carbon. The phenol was stripped off the carbon surface by toluene at 100°C and obtained almost in pure form by distillation. The rejected effluent contained only 2 ppm of phenol.

Kitagara[50] used two commercial granular activated carbons for the adsorption of phenol, p.nitrophenol, and 2,4 dichlorophenol from aqueous solutions, and found that the adsorption data could be explained by the Freundlich isotherm equation. The adsorption at a given concentration decreased with increase in the temperature of adsorption, although the rate of adsorption increased with increase in the adsorption temperature. Scharifov[51] derived a mathematical model for the adsorption of phenols by activated carbons from aqueous solutions and obtained an equation for the static adsorption isotherm, which could help in the calculation of adsorption of phenol at any concentration. Chakravorti and Weber[52] used batch and fixed-bed systems for the removal of phenol from aqueous solutions by activated carbons. The pore-diffusion model and a homogenous solid model were used to explain the results.

Juang[53] studied the adsorption of 8 phenolic compounds from aqueous solutions on activated carbon fibers in the concentrations range 40 to 500 mg/L at 303 K. Chlorinated phenols were found to be adsorbed in larger amounts compared to methyl substituted phenols. Several two- or three-parameter isotherm equations were tested. Among the two parameter equations, the Langmuir equation gave the poorest overall fit, while the Freundlich equation was better although it also showed deviations at lower concentrations. Among the three-parameter equations, the Jossens equation[54] gave the most satisfactory fit over the entire range of concentrations.

Puri and coworkers[17,55] studied the adsorption of phenol and p.nitrophenol from aqueous solutions on a number of activated carbons and carbon blacks at low and moderate concentrations, and found that the adsorption was partly reversible and partly irreversible. At moderate concentrations, there was a small irreversible adsorption when phenol concentration was 0.12 M in the case of carbons associated with greater than 1.5% oxygen, which they attributed to the complexation of π electrons of the benzene nucleus with the carbonyl groups present on the carbon surface. The irreversible adsorbed amount, however, was only 3 to 4% of the total adsorption. The adsorption isotherms of reversibly adsorbed phenol for different activated carbons, and carbon blacks (Figure 7.11) were almost similar. In the case of carbon blacks, the isotherms showed a well-defined plateau followed by a distinct rise, indicating completion of the monolayer and starting of a second layer. However, in the case of activated carbons, there was no indication for the commencement of the second layer, although the formation of the monolayer was completed at about the same concentration of the phenol solution as in the case of carbon blacks.

The adsorption isotherms for p.nitrophenol from aqueous solutions both on carbon blacks and activated carbons also showed a plateau indicating completion of a monolayer. Assuming parallel orientation of the phenol molecule, these workers calculated the specific surface areas of various carbons using the plateau of the adsorption isotherms and by calculating V_m values from the linear BET plots of x (i.e., c/co) against $\frac{x}{v(1-x)}$, where v is the amount of phenol adsorbed (mg/g). These surface-area values were found to be in agreement with the BET (N_2) surface areas (Table 7.3), in spite of large variations in the nature of the carbons, their oxygen contents, and surface areas, which shows that reversible adsorption of phenol from water at moderate concentrations can be used as a convenient method for the determination of the surface area of carbons. Weber and Moris[56] also observed two plateaus in their adsorption isotherms of phenol on activated carbons. The upper plateau was observed at a concentration three orders of magnitude higher than the lower plateau (Figure 7.12).

Seidi and Kriska[57] and Drozhalina and Bulgakova[58] also observed that the adsorption of phenol from aqueous solutions was reversible and could be used to determine the surface area of activated carbons. Brand and coworkers[59] used the adsorption of phenol as a measure of surface area of alumina, but the adsorption in this case was determined from solutions of phenol in heptane. Jankowska et al.[60] Gruszek et al.[61] Koltesev et al.[62] and Badnar and Nagi[63] found that the adsorption of phenol from aqueous solutions was determined by the pore-size distribution of the activated carbon. The carbons with finer micropores had a better adsorption capacity, which could be as much as 34% by weight at an initial phenol concentration of 1 g/L.

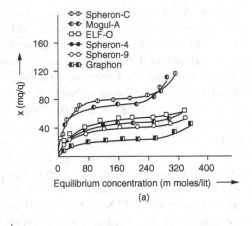

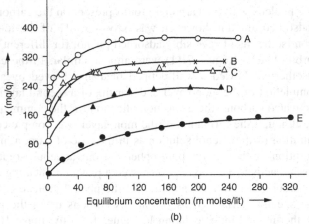

FIGURE 7.11 Adsorption isotherms of phenol from aqueous solutions of moderate concentration on (a) carbon blacks and (b) activated carbons. (After Puri, B.R., Bhardwaj, S.S., and Gupta, W.J., *Indian Chem. Soc.*, 53, 1095, 1976. With permission.)

Moreno-Castilla[64] investigated the adsorption behavior of mono-substituted phenols of different solubilities on activated carbons obtained from an original and demineralized bituminous coal. The adsorption capacity of the activated carbon was found to depend on the surface area and the porosity of the carbon, the solubility of the phenolic compound, and the hydrophobicity of the substituent. The relative affinity of the phenolic compound toward the carbon surface was related to the donor-acceptor complexes formed between the basic sites on the carbon surface and the organic ring of the phenol. The adsorption was also influenced by the pH of the solution. Marsh and Campbell[65] observed that the adsorption of p. nitrophenol at low concentrations on polyfurfuryl alcohol and polyvinylidene chloride activated carbons was extremely sensitive to the microporous structure of these carbons and showed adsorption isotherms similar to those of nitrogen or carbon dioxide. The adsorption of p.nitrophenol was so strong at low c/c_o values that the adsorption

TABLE 7.3
Specific Surface Areas of Different Carbons from Phenol Adsorption Isotherms and BET Plots

Sample	Surface Area (m²g) from		
	BET Plots of Reversible Adsorption of Phenol	Plateau of the Reversible Adsorption Isotherm of Phenol	Conventional BET (N_2) Method
Carbon Blacks			
Spheron-6	114.9	120.5	120.0
Spheron-9	141.2	144.9	146.0
Spheron-4	149.5	153.4	152.7
Spheron-C	254.1	253.0	253.7
Philblack-A	49.2	57.6	45.8
Philblack-I	110.9	110.0	116.8
Philblack-E	133.7	134.9	135.1
Mogul	305.7	306.7	308.0
Mogul-A	229.8	232.8	228.4
Elf-0	169.9	173.0	171.0
Graphon	72.7	76.6	76.0
Activated Carbons			
I	1284.7	1317.7	1308.0
II	1034.3	1065.0	1089.0
III	1043.0	1048.5	1093.0
1V	982.5	1021.0	987.0
V	475.7	453.0	468.0
Degassed Carbon Blacks			
Spheron-C 600° degassed	315.7	319.3	320.0
Mogul-600° degassed	315.7	313.0	321.0
Mogul-A 600° degassed	279.0	286.4	—
Elf-0 600° degassed	192.8	193.1	189.0

After Puri, B.R., Bhardwaj, S.S., and Gupta, W.J., *Indian Chem. Soc.*, 53, 1095, 1976.

could be used to compare the microporosities of the carbons. The surface areas calculated were smaller than the nitrogen surface areas. This was attributed to differences in the efficiencies in the packing of micropore and transitional pore volumes and to the effect of the solvent during building up of the layers in transitional porosity of the phenol molecules. Singh,[66] while studying the adsorption of phenol on two carbon blacks with different surface acidities, observed that the adsorption isotherms were of Type II. The molecular diameters of phenol calculated from the knee of the isotherm were found to be 3.3 nm and 2.8 nm for the two carbon blacks. This indicated that the orientation of the phenol molecule in the adsorbed state on the carbon surface was in edge-on position and not flat as suggested by Puri,[55] so that the phenolic groups were directed toward the liquid phase. Economy and Lin,[67] working with carbon

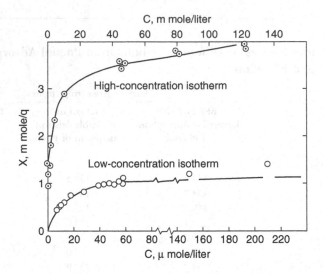

FIGURE 7.12 Isotherms for phenol on Columbia LC carbon. (After Weber, W.J., Jr., and Moris, S.C., J., *San. Eng. Div. Am. Soc. Civil Eng.*, 90, 79, 1964. With permission.)

fibers, and Dondi et al.[68] working with activated carbons, and several other workers,[69–72] however, did not find a linear relationship between the amount of phenol adsorbed and the surface area. Aytekin,[73] Chaplin,[74] and Kiselev and Krasilinkov[75] observed that the adsorption isotherms of phenol from aqueous solutions were stepwise, suggesting the possibility of rearrangement of phenol molecules in the adsorbed phase and their interaction with carbon active sites.

Singer and Yen[76] carried out a series of batch experiments to evaluate the effect of alkyl substitutents on the adsorption of phenols by activated carbon, with particular attention directed at the position, the length, and the number of the alkyl substitutents, and compared the competitive adsorptive behavior of phenol and alkyl phenols in two component mixtures. The adsorption data obeyed the Langmuir adsorption equation. The adsorption isotherms for different phenols (Figure 7.13) showed that the alkyl substituted phenols were adsorbed more strongly than phenol itself, and the adsorption increased with increase in the length of the alkyl chain. The adsorption was not affected by the position of the alkyl group but was increased significantly by an increase in the number of substitutents. The increase or decrease in adsorption was attributed to the variation in the solubility of the phenol. The substitution of an alkyl group made phenol less polar, resulting in a decrease in its solubility and increase in its adsorption. An ideal solution adsorption model was tried to explain the results obtained from a two-component adsorption system but deviations were obtained, probably because of the heterogeneous nature of the carbon surface.

In addition to surface area and the microporosity of the activated carbons, the adsorption of phenolic compounds has also been found to be influenced by the presence of carbon-oxygen groups on carbon surfaces and the pH of the carbon-solution suspension. Urano et al.[77] found that the adsorption of phenolic compounds

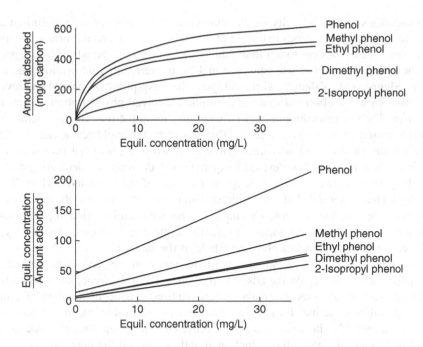

FIGURE 7.13 Adsorption isotherms of different phenols on activated carbon. (From Singer, P.C. and Yen, C. Yu, in *Activated Carbon Adsorption*, I.H. Suffet and M.J. McGuire, Eds., Ann Arbor Science Publishers, Ann Arbor MI, 1980, Vol. I, p. 167. With permission.)

on carbon is partly irreversible and involved high adsorption energies, indicating a strong bonding between the surface oxygen groups such as quinones and ketones present on the carbon surface. Because these groups are electron donors, the adsorption may involve π electron combinations on the carbon surface. Radeke et al.[78] and Hanmin and Yigun[79] observed that the adsorption of phenol depended upon the acidity of the carbon surface. The adsorption increased on degassing of the carbon, which reduced the surface acidity without any appreciable change in the surface characteristic of the carbon, such as surface area and microporous character. Mahajan et al.[80] and Keizo et al.[81] observed that the phenol uptake by porous carbons decreased sharply on surface oxidation, and it increased when the chemisorbed oxygen was removed on heat treatment in nitrogen. The phenol adsorption also increased considerably on activation of the carbon with CO_2 or water vapor. Coughlin et al.[82] and Graham[83] studied the adsorption at lower concentrations of phenol and reported that the presence of surface oxide groups on the carbon surface reduces the adsorption. However, Clauss et al.[84] while working at relatively higher concentrations, did not observe any such effect. The negative effect of the surface oxide groups was attributed to the depletion of π electron bond of graphite-like layers as a result of lowering of van der Walls forces of attraction. Peel and Benedek[85] reviewed the adsorption isotherms of phenol from aqueous solutions on an activated carbon and found that the adsorption variations could not be explained by variations in the carbon surface properties or the variations in the experimental conditions. They attributed a part of the variations to the difficulty in the attainment of the adsorption equilibrium and another part to the variations in the pH of the carbon surface and its surface chemical structure. These workers found that although about 60% of the adsorption capacity of the carbon for phenol was obtained in 4 hr, the remaining adsorption was very slow. Magne and Walker,[86] while studying the adsorption of phenol on several activated carbons, observed that the adsorption involved physisorption and chemisorption. The chemisorption increased with increase in the time and temperature of the adsorption, indicating that a part of the physisorbed phenol became chemisorbed. The chemisorbed phenol was irreversibly adsorbed and could not be removed as such. On heat treatment at elevated temperatures, it decomposes depositing carbon, resulting in a decrease in the adsorption capacity of the activated carbon. These workers also observed that the chemisorption was inhibited by the presence of carbon-oxygen surface groups. This indicated that the chemisorption of phenol takes place on oxygen-free sites located primarily at the edges of the graphene layers, whereas physisorption takes place on whole of the surface.

Coughlin et al.[82,87,88] investigated the influence of surface oxides present on the activated carbon surface on the adsorption of phenol, nitrobenzene, and sodium benzene sulfate from aqueous. The carbons were oxidized in aqueous solutions of ammonium persulphate, reduced with zinc amalgam in acid solutions, and degassed in vacuum at 890°C. The oxidation of the activated carbon significantly reduced the adsorption capacity while the reduction slightly enhanced the adsorption capacity. Outgassing of the carbon restored the original adsorption capacity of the carbon. Mattson et al.[89] observed several differences in the adsorption of phenol and its derivatives, which indicated that the interaction between the carbon surface and the phenolic compound was through the pi (π) electrons of the benzene ring. These workers suggested that the aromatic compound adsorption on carbons involved a donor-acceptor complex interaction in which the carbon surface acts as the electron donor and the aromatic ring of the solute as the acceptor. The presence of nitro group in phenol enhanced the donor-acceptor interaction by acting as electron withdrawing group. These results were confirmed by Epstein et al.[90] in his studies on the adsorption of p.nitrophenol on isotropic pyrolytic carbons, which were associated with quinonic and hydroquinonic groups. These results also indicated that the adsorption of p.nitrophenol was favored when the carbon surface was associated with quinonic groups.

Vidic and coworkers[91–93] and Abuzad and Nakhla[94] observed that the adsorption capacity of granular activated carbons for several phenolic compounds was highly influenced by the presence of molecular oxygen. The adsorption capacity of a carbon for 2-methyl phenol was enhanced by about 200% in the presence of oxygen. The adsorption of phenol was reversible in the absence of molecular oxygen and could be recovered almost completely by solvent extraction, but only 10 to 30% of the adsorbed phenolic compound was recoverable when the adsorption had been carried out in the presence of oxygen gas. This was attributed to oxidative coupling of the phenolic compounds on the carbon surface under oxic conditions (in the presence of oxygen). The compounds formed under oxic conditions were very difficult to remove from the carbon surface. A considerable amount of oxygen was consumed in this process and the amount consumed was linearly proportional to the amount of irreversibly adsorbed phenolic compound. Grant and King[95] also observed that the adsorption of phenol on activated carbons was partly reversible and partly irreversible. However, they were of the view that the carbon-oxygen surface groups are not primarily involved in the irreversible adsorption, but this was caused by oxidative coupling of phenol in the presence of molecular oxygen.

Leng and Pinto[96] studied the adsorption of phenol and o-cresol on activated carbons in the presence of dissolved oxygen and observed no relationship between adsorption and BET surface area under either oxic or anoxic conditions. The adsorption of these phenols was much higher in the presence of molecular oxygen. When the activated carbon surface was associated with chemisorbed oxygen, the adsorption of both the phenols decreased both under oxic and anoxic conditions. The adsorption of o-cresol was larger than that of phenol. The authors were of the view that the adsorption involved neither the solubility factor nor the surface chemistry of the carbon, but it involved oxidative coupling of the phenolic compounds in the presence of molecular oxygen, resulting in surface polymerization. Tessmer et al.[97] however, observed that the presence of acidic surface oxygen groups on the carbon surface reduced the adsorption capacity of the carbon for phenolic compounds under oxic conditions by reducing oxidative coupling reactions.

Teng and Hsieh[98,99] studied the adsorption of phenol on partially oxidized bituminous coal char and observed that, although the surface coverage of the carbon with low burn-off was consistently about 100%, the surface coverage was only about 40% for a carbon with 60% burn-off, because the latter carbon was associated with a high concentration of the carbon-oxygen surface groups. They attributed this change in surface coverage to the change in the particle size of the carbon with burn-off. Leyva-Ramos et al.[100] on the other hand attributed it to the change in the available surface as the particle size was reduced. Nevskaia et al.[101] also observed a decrease in surface coverage with increase in surface oxidation of the carbon.

Oda et al.[102] determined the adsorption isotherms of phenol on six commercially available activated carbons and their heat-treated products containing varying amounts of surface acidic groups. The carbons with high surface acidity adsorbed smaller amounts of phenol. The adsorption increased when the carbon samples were heat-treated at 800°C. The heat treatment reduced the total surface acidity without any appreciable change in the physical properties of the carbons. This indicated that the presence of acidic surface groups reduces the adsorption of phenol. These workers also determined the selectivity of the carbons for phenol and benzoic acid from solutions of equimolar binary components. In this case, the amount of benzoic acid adsorbed increased with increase in the equilibrium concentration while that of phenol remained more or less unchanged up to a certain concentration and decreased abruptly at higher concentrations. The absolute quantities adsorbed varied with the type of the activated carbon, but the adsorption of benzoic acid was larger in the case of all the carbons. The amounts of benzoic acid adsorbed from binary solutions were 10 to 20% less than those from the single component solutions. But the amount of phenol adsorbed decreased as the increase in the amounts of benzoic acid adsorbed increased. In the case of the heat-treated carbons, the total amount adsorbed (phenol + benzoic acid) increased, but there was a larger increase in the adsorption of phenol. This was attributed to an increase in the solubility of benzoic acid in water in the presence of phenol. It was also observed that preadsorbed benzoic acid could not be easily removed by phenol, but the reverse could be done easily.

Puri and coworkers[103] studied the adsorption of phenol by sugar charcoal and carbon blacks from dilute aqueous solutions in the concentration range up to ~20 m moles per liter. The extent of adsorption was found to be maximum in the case of graphon, which was almost free of any chemisorbed oxygen and minimum in the case of sugar charcoal, which contained maximum amount of oxygen. This indicated that the presence of oxygen on the carbon surface tends to decrease the adsorption of phenol. The surface overage by phenol in the case of different carbons varied between 8 and 40%. While studying the adsorption of phenol on sugar charcoal after degassing at 600 and 1000°C, these workers observed that the presence of chemisorbed oxygen as CO_2-complex reduced the adsorption of phenol, while the presence of the CO-complex enhanced the adsorption. This was attributed to the interaction of quinone groups of the CO-complex with π electrons of the benzene ring in phenol. Thus, the view of the earlier investigators that the combined oxygen as a whole has an adverse effect on the adsorption of phenol from aqueous solutions was not substantiated by these studies. The adsorption by oxygen-free carbons was influenced largely by porosity and surface area of the carbon.

Bansal and coworkers[104,105] investigated the influence of carbon-oxygen surface groups on the adsorption of phenol from aqueous solutions in the concentration range of 20 to 1000 mg/L by activated carbons, using two samples of granulated activated carbons (GAC) and two samples of activated carbon fibers (ACF). The activated carbons had different surface areas and were associated with different amounts of the carbon-oxygen surface groups. The adsorption isotherms were Type I of the BET classification (Figure 7.14) and obeyed the Langmuir adsorption

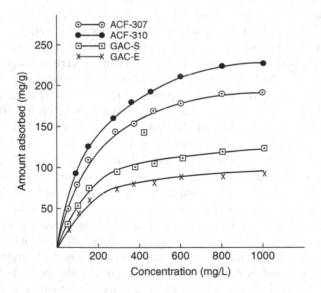

FIGURE 7.14 Adsorption isotherms of phenol on different as-received activated carbons. (After Bansal, R.C., Aggarwal, D., Goyal, M., and Kaistha, B.C., *Indian J. Chem. Technol.*, 9, 290, 2002. With permission.)

TABLE 7.4
Surface Areas and Amounts of Oxygen Evolved on Degassing
As-received Activated Carbons at 950°C

Sample	Surface Area (M²/G)	Oxygen Evolved (g/100 g) As			
		CO_2	CO	H_2O	Total
ACF-307	190	1.00	5.30	1.30	7.60
ACF-310	1184	1.90	4.20	1.40	7.50
GAC-S	1256	2.10	1.05	1.24	4.39
GAC-E	1190	2.13	1.66	1.33	5.12

After Bansal, R.C., Aggarwal, D., Goyal, M., and Kaistha, B.C., *Indian J. Chem. Technol.*, 9, 290, 2002.

equation. The adsorption could not be explained on the basis of surface area of the carbons as ACF-307, which had a smaller surface area (910 m²/g) and adsorbed much larger amounts of phenol compared with GAC-E, which had a larger surface area (1190 m²/g). The adsorption was influenced by the presence of carbon-oxygen surface groups. A comparison of the oxygen present on the four activated carbon samples (Table 7.4) showed that although the GACs are associated with larger amounts of oxygen groups evolved as CO_2, the fibrous activated carbons (ACFs) contain larger amounts of oxygen evolved as CO. The former groups have been postulated as carboxylic or lactonic groups[106–108] and render the carbon surface polar and hydrophilic in character. It appears that these surface groups enhanced the preference of the carbon surface for the adsorption of water than that of phenol from aqueous solutions.[109–112] The carbon-oxygen surface groups that evolve as CO have been postulated as quinones. These are nonacidic and enhance the preference of the carbon surface for phenol.

These workers then enhanced the amounts of these surface groups by oxidation of the carbons with nitric acid, ammonium persulphate, and hydrogen peroxide, and decreased their amounts by degassing the activated carbons at gradually increasing temperature of 400°, 650°, and 950°. The oxidation of the carbon decreased the adsorption of phenol (Figure 7.15), the extent of decrease depending on the nature of the oxidative treatment. The decrease in adsorption was maximum in the case of the oxidation with nitric acid, which incidentally created the maximum amounts of acidic surface groups (Table 7.5). This indicated that the presence of acidic surface groups decreased the adsorption of phenol. The adsorption of phenol increased when the activated carbons were degassed at gradually increasing temperatures. However, the increase in adsorption was maximum in the case of the sample degassed at 650°C (Figure 7.16). This has been attributed to the fact that the 650°C-degassed sample had lost most of its acidic groups and the only remaining surface groups were the quinonic groups.

Similar results were obtained by Goyal[113,114] on the adsorption of p.nitrophenol from aqueous solutions on the same activated carbons as used by Bansal et al.[104,105]

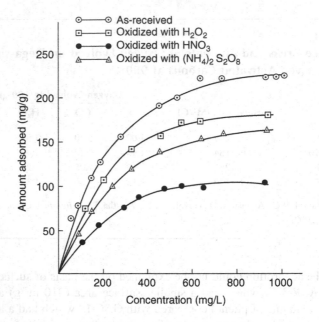

FIGURE 7.15 Adsorption isotherms of phenol on ACF-310 before and after oxidation. (After Bansal, R.C., Aggarwal, D., Goyal, M., and Kaistha, B.C., *Indian J. Chem. Technol.*, 9, 290, 2002. With permission.)

TABLE 7.5
Oxygen Evolved on Degassing Different Oxidized Activated Carbon Samples at 950°C

Sample	Oxygen Evolved (g/100 g) As			
	CO_2	CO	H_2O	Total
ACF-307				
As-received	1.00	5.30	1.30	7.60
Oxidized with-				
HNO_3	12.90	7.47	2.40	22.77
$(NH_4)_2S_2O_8$	5.40	7.51	4.91	17.82
H_2O_2	2.55	7.42	2.10	12.07
ACF-310				
As-received	1.90	4.20	1.40	7.50
Oxidized with-				
HNO_3	11.96	7.20	2.20	21.36
$(NH_4)_2S_2O_8$	4.70	6.30	2.10	13.10
H_2O_2	2.1	6.20	2.10	10.40

After Bansal, R.C., Aggarwal, D., Goyal, M., and Kaistha, B.C., *Indian J. Chem. Technol.*, 9, 290, 2002.

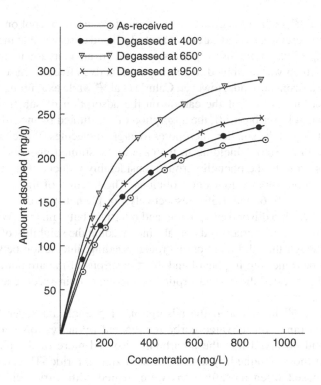

FIGURE 7.16 Adsorption isotherms of phenol on ACF-310 before and after degassing. (After Bansal, R.C., Aggarwal, D., Goyal, M., and Kaistha, B.C., *Indian J. Chem. Technol.*, 9, 290, 2002. With permission.)

The adsorption of p.nitrophenol decreased on oxidation and increased on degassing of the oxidized activated carbons. Thus, the results on the adsorption of phenol and p.nitrophenol showed clearly that the oxidation of carbons with nitric acid, which preferably creates carboxylic or lactonic groups and suppresses the adsorption of both phenol and p.nitrophenol. These acidic groups, which are bulky, block entrances to some of the narrow micropores and result in a decrease in adsorption. In addition, the water molecules cluster around the polar acidic surface groups,[115–123] again reducing the adsorption capacity. When these acidic surface groups were removed by degassing, the adsorption of both the phenols was increased, the extent of increase depending on the amount of acidic groups removed. The amount adsorbed was at a maximum when the activated carbon surface had lost most of its acidic surface groups and when carbonyl groups (quinonic groups) were the dominating surface groups, as in the case of 650°C-degassed samples. In other words, the presence of acidic surface groups inhibits the adsorption, but the presence of nonacidic surface groups favors the adsorption of both the phenols. This increase in adsorption was attributed to the binding of the phenols with the partial positive change on the quinonic carbon atoms. The p.nitrophenol could not be recovered completely, even on prolonged evacuation at room temperature. This was attributed to the complexing of p.nitrophenol at the quinonic sites within the micropores.

Yong et al.[124] studied the effect of substitutent groups in phenol on their adsorption on a commercial activated carbon at neutral pH and observed that the substitution of a methyl, ethyl, or methoxy groups resulted in an increase in the reversible adsorption, which was attributed to a stronger inductive effect of these substitutents and forming a stronger bond. However, Caturla et al.[125] while examining the influence of the degree of burn-off of the carbon on the adsorption of substituted phenols, observed that, at lower burn-off, the adsorption of substituted phenols decreased due to inaccessibility of the smaller micropores to larger molecules. Wheeler and Levy[126] also observed that steric shielding in the case of substituted phenols, having large alkyl groups ortho to the phenolic group, considerably reduced their adsorption by activated carbons from cyclohexane solutions. The fraction of the surface covered was about 17, 12, 9, 6, and 1.5%, respectively, for phenol, p.tertiary butyl phenol, o.dimethyl phenol, o.diisopropyl phenol, and o.tertiary butyl phenol. While explaining their results, these workers did not take into account the solubility of the different phenols, although they did mention hydrogen bonding, interaction between π electrons of the benzene ring in phenol and the π electrons of the aromatic structure in carbons and suggested that the adsorption occurs on certain specific active sites on the carbon surface.

Deng et al.[127] investigated the adsorption of phenol and several substituted phenols on a range of activated carbons prepared by activation with potassium carbonate and observed that these carbons adsorbed more of the phenolic compounds than those obtained by activation with zinc chloride. The adsorption was further increased when these carbons were treated with nitric acid followed by treatment with sodium hydroxide. This was attributed to changes in surface area, porosity, and the hydrophobicity of the carbon surface. The adsorption was more in the case of nitro and chlorophenols as compared to cresol or dimethyl phenol. Because the nitro and chlorol groups were electron withdrawing groups, they decreased the electron density of the benzene ring, which increased the interaction of the adsorbate with the carbon surface and thus increased the adsorption of these phenolic compounds.

Arafat[128] studied the effect of adding KCl to aqueous solutions of phenol, benzene, and toluene on their adsorption by activated carbons associated with varying amounts of surface oxides and observed that the salt effect was different in different cases. The observed influences in different cases were interpreted in terms of the charge neutralization between the carbon surface and the adsorbate molecules and on the basis of the adsorption of water. The adsorption of water on the carbon surface was found to be crucial, especially in those carbons, which contained larger amounts of surface oxygen groups.

Laszlo et al.[129] studied the adsorption of phenol and 2.3,4 trichlorophenol from dilute aqueous solutions on a granular activated carbon prepared from PAN by a two-step physical activated process. The adsorption isotherms were Type I of the BET classification and followed the Langmuir adsorption equation. The adsorption capacity and the adsorption constant K values, obtained using the Langmuir equation (Table 7.6), were found to depend on the pH of the solution. The results were discussed in terms of the acid-base character of the carbon surface and the acidic character of the two phenols. The effect of pH is more significant in the

TABLE 7.6
Langmuir Adsorption Parameters for Phenol and Trichlorophenol on a PAN Based Activated Carbons

Adsorption Parameter	Phenol			2,3,4-Trichlorophenol		
	pH = 3	Unbuffered	pH = 11	pH = 3	Unbuffered	pH = 11
n_m [mmol/g]	0.78	0.78	0.72	0.99	0.96	0.63
K [dm^3/mmol]	0.66	7.54	5.00	26.01	131.57	11.18
Surface area available for one adsorbate molecule, nm^{2a}	0.77	0.77	0.83	0.61	0.63	0.95
Cross sectional area nm^2/molecule [8]		0.30–0.52			0.63–0.72	

After Laszlo et al (129).

case of adsorption of 2,3,4 trichlorophenol than phenol. Three possible interactions suggested were:

- Dispersion effect between the aromatic ring and π electrons of the graphite structure
- Electron donor-acceptor interaction between the aromatic ring and the basic surface sites
- Electrostatic attraction and repulsion when ions are present

At acidic pH = 3, the carbon-oxygen surface groups as well as the phenols are present in the unionized form being either neutral or positively charged. This results in a weak dispersive interaction between the carbon surface and the phenol molecule. The enhanced interaction at this pH in the case of 2,3,4 trichlorophenol was attributed to the electron withdrawing effect of the three chlorine substitutents. At pH 11, the phenols dissociate to give negatively charged phenolate ions and the carbon surface functional groups are partially ionized giving the surface a negative charge. This results in electrostatic repulsive interactions between the carbon surface and the phenolate ions thus decreasing the adsorption of phenols. However, some dispersive interactions may still take place. Furthermore, the undissociated basic surface groups may form donor-acceptor complexes. In the case of adsorption from unbuffered solutions, the pH of the solution will depend on its concentration so that both phenols and the carbon surface groups coexist in their ionized and unionized forms depending on their pK values. Under these conditions all the three types of interactions stated above will determine the adsorption of phenols. Because the dissociation of phenol is negligible in this case, the ionic interactions may be excluded. But in the case of 2,3,4 trichlorophenol, which has a low pK value, there will be some dissociation resulting in attractive interactions between the positively charged carbon surface and the negatively charged phenolate ions.

7.4 ADSORPTION OF NITRO AND AMINO COMPOUNDS

Toxic and hazardous nitro compounds are often present in process waste water from explosive manufacturing plants and armament industries. Their removal from this waste water before discharge into water streams is an important aspect of getting good quality drinking water. Consequently, considerable research work has been directed toward the removal of nitro compounds from water. Aggarwal et al.[130,131] investigated the adsorption of nitrobenzene from aqueous solutions in the concentration range 20 to 200 ppm at pH 7 on five different commercial-grade activated carbons prepared from coconut shell and pine wood. These activated carbons had surface areas varying between 650 and 1300 m^2/g. The adsorption isotherms were Type I of the BET classification (Figure 7.17) and followed the Freundlich adsorption equation. The amount adsorbed increased with increase in surface area of the carbon, being maximum in the case of AC-70, which has the largest surface (1300 m^2/g), and minimum in the case of AC-30, which has the smallest surface area (650 m^2/g). However, no direct relationship could be obtained between the amount adsorbed and the surface area, which indicated that the surface area is not the only factor that determines the adsorption of nitrobenzene on carbon. In order to modify the surface chemistry of the carbons, three of these carbons were oxidized with nitric acid, hydrogen peroxide, and ammonium persulphate solutions. The oxidation of the activated carbon considerably decreased the adsorption of nitrobenzene (Figure 7.18).

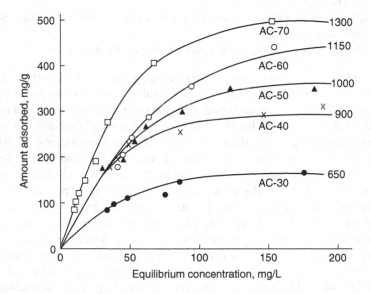

FIGURE 7.17 Adsorption isotherms of nitrobenzene on different activated carbons with different surface areas. (After Aggarwal, P., Kapoor, S.K., Kapoor, J.C., Bhalla, A.K., and Bansal, R.C., *Indian J. Chem. Technol.*, 3, 187, 1996, and Aggarwal, P., Misra, S., Kapoor, S.K., and Bansal, R.C., in *Recent Trends in Carbon*, O.P. Bahl, Ed., Shipra Publ., Delhi, 1996, p. 203. With permission.)

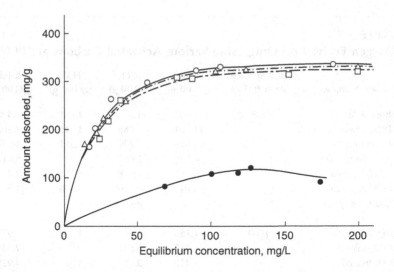

FIGURE 7.18 Adsorption isotherms of nitrobenzene on activated carbon (1000 mg/g) before and after surface oxidation. (O = Original; Δ = (NH$_4$)$_2$S$_2$O$_8$-treated; ❑ = H$_2$O$_2$-treated; ● = HNO$_3$ treated) (After Aggarwal, P., Kapoor, S.K., Kapoor, J.C., Bhalla, A.K., and Bansal, R.C., *Indian J. Chem. Technol*., 3, 187, 1996, and Aggarwal, P., Misra, S., Kapoor, S.K., and Bansal, R.C., in *Recent Trends in Carbon,* O.P. Bahl, Ed., Shipra Publ., Delhi, 1996, p. 203. With permission.)

However, the decrease in adsorption was only slight in case of the oxidations with hydrogen peroxide and ammonium persulphate, whereas the decrease was appreciably large in the case of oxidation with nitric acid. The nitric acid oxidized sample adsorbed about 1/3 of the amount adsorbed by the as-received sample. This was attributed to the fact that the oxidation with nitric acid was a stronger oxidative treatment and resulted in the formation of appreciable amounts of carbon oxygen surface groups, the acidic groups evolved as CO$_2$ on evacuation, and nonacidic evolved as CO (Table 7.7). In order to examine the effect of these surface groups on the adsorption of nitrobenzene, the activated carbons were degassed at 400°, 600°, and 1000°C in vacuum and the adsorption predetermined. The adsorption isotherms on the degassed AC-50 (surface area 1000 m^2/g) (Figure 7.19) show that the adsorption increased on degassing, the extent of increase depending upon the temperature of degassing of the carbon. The increase in adsorption was maximum in the case of 600°-degassed carbon while the 1000°-degassed carbon showed a lower adsorption of nitrobenzene.

The large decrease in adsorption of nitrobenzene in the case of the oxidation with nitric acid, which predominantly produces acid surface groups on the carbon surface and a maximum increase in adsorption in case of the carbon samples degassed at 600°C, a treatment that results in the elimination of a large proportion of the acidic surface groups indicated that the adsorption of nitrobenzene was suppressed by the presence of acidic carbon-oxygen surface groups. Miyahara and Okazaki,[132] while studying the concentration dependence of surface diffusivity of nitrobenzene and benzonitile in liquid-phase adsorption on an activated carbon also

TABLE 7.7
Oxygen Evolved on Outgassing Various Activated Carbons at 1000°C

Carbon Sample	BET Surface Area (m²/g)	CO_2 (g/100 g)	CO (g/100 g)	H_2O (g/100 g)	Total (g/100 g)
Original AC-30	650	1.033	1.402	1.061	3.496
HNO_3-treated		11.201	7.086	1.815	20.102
H_2O_2-treated		1.018	1.829	1.211	4.058
$(NH_4)_2S_2O_8$-treated		4.727	2.067	1.824	8.618
400°C degassed		0.788	1.068	0.552	2.408
600°C degassed		0.041	0.926	0.576	1.543
1000°C degassed			traces		
Original AC-50	1000	1.155	1.495	1.323	3.973
HNO_3-treated		8.713	6.841	2.092	17.646
H_2O_2-treated		1.415	2.042	1.471	4.928
$(NH_4)_2S_2O_8$-treated		2.356	2.921	5.411	10.688
400°C degassed		0.567	1.132	1.554	3.253
600°C degassed		0.631	1.097	1.134	2.862
1000°C degassed			traces		
Original AC-70	1300	1.439	1.161	1.361	3.961
HNO_3-treated		12.414	7.601	2.741	22.756
H_2O_2-treated		1.658	2.731	1.752	6.141
$(NH_4)_2S_2O_8$-treated		2.175	2.471	3.012	7.658
400°C degassed		1.123	1.167	1.268	3.558
600°C degassed		0.489	0.761	0.446	1.716
1000°C degassed			traces		

After Aggarwal, P., Kapoor, S.K., Kapoor, J.C., Bhalla, A.K., and Bansal, R.C., *Indian J. Chem. Technol.*, 3, 187, 1996. With permission.

obtained Type I isotherms, which followed the Freundlich adsorption equations. Haghsereshi et al.[133] while studying the adsorption of nitrobenzene from aqueous solutions on activated carbons, observed that the data obeyed the Langmuir adsorption equation reasonably well. The surface covered depended upon the packing of the adsorbate molecules on the carbon surface. In the case of nitrobenzene, these workers suggested that the adsorbed nitrobenzene probably had perpendicular orientation, although all the adsorbed molecules might not have exactly perpendicular orientation. Different molecules may be adsorbed with different till angles.

Jain and Bryce[134] investigated the feasibility of the removal of dissolved explosives from waste water using an adsorption-oxidation process. The materials studied were trinitrotoluene (TNT), hexahydro trinitro triazine (RDX), and octahydro tetranitro tetrazocine (HMX). The process involved contacting the activated carbon with the ammunition waste water containing these materials. The activated carbon adsorbed the organic compound. Simultaneously, an oxidant was introduced into the

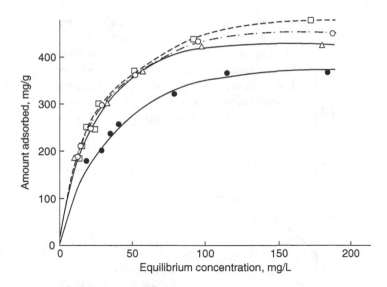

FIGURE 7.19 Adsorption isotherms of nitrobenzene on activated carbon (1000 mg/g) before and after outgassing at different temperatures. (● = Original; Δ = 400°C degassed; ❑ = 600°C degassed; O = 1000°C degassed) (After Aggarwal, P., Kapoor, S.K., Kapoor, J.C., Bhalla, A.K., and Bansal, R.C., *Indian J. Chem. Technol.*, 3, 187, 1996, and Aggarwal, P., Misra, S., Kapoor, S.K., and Bansal, R.C., in *Recent Trends in Carbon*, O.P. Bahl, Ed., Shipra Publ., Delhi, 1996, p. 203. With permission.)

system, which oxidized the adsorbed material and regenerated the activated carbon, which could be reused. In their experiments Jain and Bryce used a granulated activated carbon as the adsorbent and ozone as the oxidant. The concentration of the components in the waste water was determined using gas chromatography. The adsorption-oxidation and the adsorption data obtained from solutions of TNT, RDX, and HMX (Figure 7.20, Figure 7.21, and Figure 7.22) clearly indicate that the former

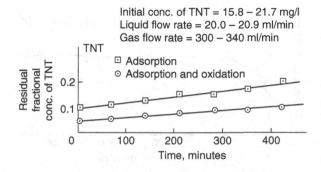

FIGURE 7.20 Adsorption-oxidation and adsorption of TNT. (After Jain, K. and Bryce, A. J., in *Carbon Adsorption Handbook*, P.N. Chemreisinoff and F. Ellerbusch, Eds., Ann Arbor Science Publishers, Ann Arbor MI, Chapter 17, p. 661. With permission.)

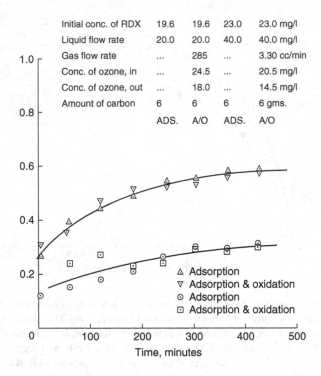

FIGURE 7.21 Adsorption-oxidation and adsorption of RDX. (After Jain, K. and Bryce, A. J., in *Carbon Adsorption Handbook*, P.N. Chemreisinoff and F. Ellerbusch, Eds., Ann Arbor Science Publishers, Ann Arbor MI, Chapter 17, p. 661. With permission.)

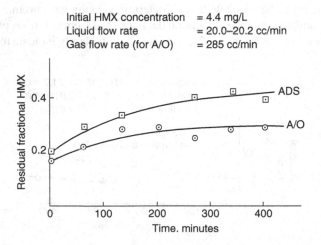

FIGURE 7.22 Adsorption and adsorption-oxidation of HMX. (After Jain, K. and Bryce, A. J., in *Carbon Adsorption Handbook*, P.N. Chemreisinoff and F. Ellerbusch, Eds., Ann Arbor Science Publishers, Ann Arbor MI, Chapter 17, p. 661. With permission.)

process exhibited a significantly more pronounced tendency to reach equilibrium, as compared to the latter process (adsorption alone). Furthermore, the removal of these explosive materials is much better with the adsorption-oxidation process than with the adsorption alone. This indicates that the performance was better when adsorption-oxidation process was used. The authors, however, were of the view that there was need for further work to evaluate the life of the carbon adsorbent and the oxidant utilization efficiency, because these parameters will determine the economic viability of the adsorption-oxidation process. They also suggested that the process performance may also be enhanced by using a catalyst.

Aggarwal et al.[135] carried out adsorption isotherms of styphnic acid from aqueous solutions in the concentration range 20 to 700 mg/L on five different commercial activated carbons having surface areas varying between 650 and 1300 m^2/g. The adsorption was Type II of the BET classification with well-defined plateau. The adsorption depended on surface area, but the fraction of the surface occupied by styphnic acid at monolaryer coverage increased with surface area and was maximum in the case of the carbon sample having a surface area of 1000 m^2/g. The adsorption showed a small decrease at higher surface areas (Figure 7.23). This may be due to the presence of some fine micropores in the higher surface area carbons, which could not accommodate the large styphnic acid molecules. The adsorption increased appreciably on oxidation of the activated carbons with hydrogen peroxide and ammonium persulphate (Figure 7.24). There was a linear relationship between the amount of styphnic acid adsorbed and the total oxygen content of the carbon surface (Figure 7.25). This increase in adsorption on oxidation, which enhanced the amount of carbon-oxygen surface groups on the carbon sample, was explained on the basis of the unique structure of styphnic acid, which contains nitro groups in ortho position

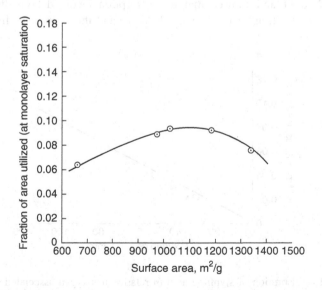

FIGURE 7.23 Fraction of surface area utilized for monolayer adsorption of styphnic acid on activated carbons. (After Aggarwal, P., Misra, S., Kapoor, S.K., Bhalla, A.K., and Bansal, R.C., *Indian J. Chem. Technol.*, 4, 42, 1997. With permission.)

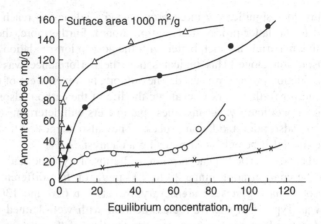

FIGURE 7.24 Adsorption isotherms of styphinic acid on activated carbons before and after oxidation. (O = as-received, x = HNO$_3$ treated; • = H$_2$O$_2$ treated; Δ = (NH$_4$)$_2$S$_2$O$_8$ treated) (After Aggarwal, P., Misra, S., Kapoor, S.K., Bhalla, A.K., and Bansal, R.C., *Indian J. Chem. Technol.*, 4, 42, 1997. With permission.)

to two phenolic groups. Such an arrangement favors the formation of intermolecular hydrogen bonds. The remaining nitro groups in styphnic acid may interact with the carboxylic and phenolic groups of the carbon surface and form hydrogen bonds.

Radovic et al.[136] studied the adsorption of aniline and nitrobenzene on activated carbons after modifying their surface by oxidation, degassing, and after treatment with ammonia gas. The adsorption studies were carried out at controlled pH in the range pH 2 to 11 and showed that the adsorption involved both dispersive and electrostatic interactions between the adsorbate and the adsorbent. In the case of

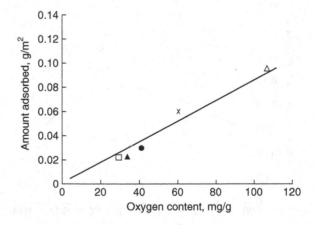

FIGURE 7.25 Adsorption of stypfinic acid in relation to oxygen associated with activated carbons. (• = as-received, x = H$_2$O$_2$ treated; Δ = (NH$_4$)$_2$S$_2$O$_8$ treated; ▲ = 400°C degassed; □ = 600°C degassed) (After Aggarwal, P., Misra, S., Kapoor, S.K., Bhalla, A.K., and Bansal, R.C., *Indian J. Chem. Technol.*, 4, 42, 1997. With permission.)

aniline, the adsorption was at a maximum in the case of oxidized carbons at a pH near the carbon zero point charge (ZPC). This indicated that the adsorption of aniline involved the electrostatic interactions between the aniline cations and the negatively-charged surface as well as the dispersive interactions between the aniline molecules and the graphene layers. In the case of nitrobenzene, which does not dissociate, the adsorption was found to be maximum in the case of degassed carbon samples, particularly at a pH approximately equal to pH ZPC. This indicates that the adsorption mainly involves dispersive interactions between nitrobenzene molecules and the graphene layers. Because the adsorption studies were carried out near the zero point charge pH (pH$_{ZPC}$), the repulsive interactions between charged surface groups and the uncharged nitrobenzene molecules were expected to be minimum.

Nitrosoamines are dangerous health hazards discharged into waste water by certain chemical process industries. Their removal has, therefore, been examined by several workers. Bornehoff[137] investigated the removal of several nitroasoamines from water using activated carbons and observed that when 1 L of dosed water was filtered through 23 g of an activated carbon at 40 to 50 L/Kg/hr, the removal was more than 99% in the case of dimethyl, diethyl, and diphenyl nitrosoamines. Fochtman and Dobbs[138] used Darco and Filtrasorb-activated carbons for the removal of several carcinogenic nitrosoamines and hydrocarbons such as naphthalene benzidine from water and observed that appreciable amounts of these substances could be adsorbed. The extent of adsorption was dependent on the nature, molecular weight, and the chemical structure of the compound (Table 7.8). The adsorption data obeyed the Freundlich isotherm equation. About 20 to 30 mg carbon was sufficient to reduce the concentration of these organic compounds from 1.0 to 0.1 mg/L, which is the safe limit for drinking water. Kipling,[139] from a study of the adsorption of n.butylamine on activated carbons from aqueous solutions, concluded that the process in every case took place through undissociated molecules and was essentially physical in nature and independent of the physical or electrochemical nature of the carbon surface. Anderson and Emmett,[140] on the other hand, while studying the adsorption of methylamine by a carbon black, showed that the adsorption was influenced by the nature of the surface and that the adsorption in excess of a monolayer in the case of oxygen containing carbon blacks could be attributed to the polarity of the molecules and their ability to form hydrogen bonds with the carbon-oxygen surface groups. Puri et al.[141] studied the adsorption of n-butyl, dimethyl, diethyl, and triethyl amines from their aqueous solutions on sugar charcoals before and after degassing the charcoals at different temperatures. The amounts of n.butylamine adsorbed by each charcoal was almost equal to the surface acidity of the charcoal as determined by base neutralization capacity. The adsorption amounts decreased on degassing in proportion to the elimination of the acidic carbon-oxygen surface groups. At degassing temperatures above 750°C, the carbon lost its capacity to adsorb n.butylamine, because these samples no more contained any acidic surface groups. Thus, these workers are of the view that the adsorption of amines by activated carbons was primarily a neutralization reaction between the acidic groups on the carbon surface and the basic NH$_2$ groups of the amine. These workers also observed that the adsorption of dimethyl, diethyl, and triethyl amines were appreciably lower than the adsorption of n.butylamine and that adsorption values decreased with increase in the length and the number of the carbon chains. This decrease in adsorption

TABLE 7.8
Adsorption of Different Carcinogenic Compounds on Activated Carbons

Compounds	Molecular Weight	Equilibrium Concentration Range (mg/L)	Adsorption Capacity	Carbon Dose to Reduce 1.0 mg/L to 0.1 mg/L (mg/L)
Naphthalene	128			
Darco		0.002–9.3	6.3	29
Filtrasorb		0.06–5.3	16.8	19
1,1-Diphenylhydrazine	184			
Darco		0.2–1.5	9.4	18
Filtrasorb		0.4–9.1	14.9	10
Naphthyl amine	143			
Darco		0.2–1.5	7.7	27
Filtrasorb		0.1–1.5	16.6	10
4,4′-Mathylene-bis (2-chloroaniline)	264			
Darco		0.08–1.5	12.1	27
Filtrasorb		0.05–1.2	24.2	15
Benzidine	184			
Darco		0.2–10	8.4	19
Filtrasorb		0.003–8.2	17.2	10
Dimethylnitrosamine	74			
Darco		6.5–9.5	6×10^5	$>10^5$
Filtrasorb		6.5–9.0	6×10^7	$>10^7$
3,3′-Dichlorobenzidine	253.1			
Darco		0.8–2.5	12.6	13
Filtrasorb		0.6–2.5	24.1	5

After Fochtman, E.G. and Dobbs, R.A., in *Activated Carbon Adsorption*, I.H. Suffet and M.J. McGuire, Eds., Ann Arbor Science. Publishers, Ann Arbor MI, 1980, Vol. I, p. 157. Reproduced with permission from Ann Arbor Science Publishers.

was attributed to steric effects making some of acidic sites unavailable to amine molecules.

The adsorption of cationic polymers such as nonylphenol diethylamine, NP triethylamine, NP tetraethyl pentamine, Octyl phenol (OP) diethylene triamine, and OP tetraethylene pentamine from aqueous solutions on an activated carbon showed a two-stage adsorption.[142] At low adsorptions the molecules were postulated to have flat orientation on the surface of the carbon. A reorientation of the molecules occurred in the second stage as the adsorption increased, until at saturation the molecules were almost perpendicular to the surface. The variation of heats of adsorption with surface coverage showed two inflections (Figure 7.26).

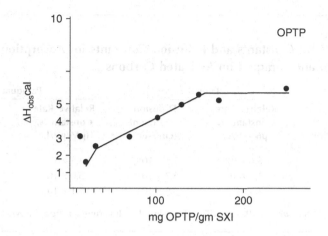

FIGURE 7.26 Heat of adsorption as a function of amount of octylphenol tetraethylene pentamine (OPTP) on Norit-activated carbon. (After Capelle, A., in *Activated Carbon, A Fascinating Material*, A. Capelle and F. Devooys, Eds., Norit N.V. Netherlands, 1983, p. 191. With permission.)

7.5 ADSORPTION OF PESTICIDES

The production and use of pesticides, which include insecticides, algicides, repellants, herbicides, fungicides, and other agrochemicals has become an important factor in increasing the production of food and cotton, and to have increased freedom from disease and obnoxious plant and animal life. During the last 50 years, the number of these pesticides available, the variety of their applications, and the volume of active ingredients used and their formulated products has seen a tremendous growth. It is estimated that more than 1500 chemical pesticides have been produced and as many as 8000 individual formulations are available for specific applications.[143] This has caused many undesirable effects, such as disturbing the ecosystem and producing environmental contamination. These harmful and hazardous materials are leached into surface and ground waters. The potable water thus has to be made free of these materials, because they may accumulate in the food chain and many have short- and long-term effects on human and animal life. Consequently, many adsorption studies using activated carbons have been carried out.

Prakash[144] studied the adsorption of two cationic pesticides, diquat and paraquat, from aqueous solutions on a lignite-based Darco activated carbon and two coconut carbons of different origins, and activated with steam to different degrees of activation. The BET (N_2) surface areas of these carbons varied from 600 to 1200 m²/g. The rate of adsorption when plotted against the square root of time indicated that the rate limiting step for the adsorptive removal of these two pesticides in agitated nonflow systems was one of interparticle transport of the solute into the pores of the activated carbons. The relative rates of adsorption could be characterized by the diffusion constant, D, which varied inversely with the nitrogen BET surface area of an activated carbon for each adsorbate (Table 7.9). The higher values of diffusion

TABLE 7.9
**Relative Rate Constants and Diffusion Constants for Adsorption
of Diquat and Paraquat on Activated Carbons**

	Diquat		Paraquat	
Carbon Identification	Relative Rate Constant, k, $(\mu M/g)^2/hr$	Diffusion Coefficient (cm^2/sec)	Relative Rate Constant, k, $(\mu M/g)^2/hr$	Diffusion Coefficient (cm^2/sec)
Darco	8.6×10^3	7.5×10^{-10}	3.3×10^3	11.5×10^{-10}
Coconut-A	13.1×10^3	5.7×10^{-10}	5.6×10^3	7.3×10^{-10}
Coconut-B	17.5×10^3	3.9×10^{-10}	7.1×10^3	5.1×10^{-10}

After Prakash, S., *Carbon*, 19, 483, 1974. Reproduced with permission from Elsevier.

constant for paraquat were attributed to the different molecular dimensions of the two pesticides.

The adsorption isotherms for both pesticides were Type I of the BET classification (Figure 7.27) and fitted very well to the Langmuir equation. The amount of the two pesticides adsorbed on different activated carbons varied between ~18 and 36% for diquat, and between ~6 and 14% for paraquat, depending upon the surface area of the activated carbon (Table 7.10). These workers also carried out calculations for thermodynamic quantities, such as differential heat of adsorption ΔH and activation energy E, which indicated that the rate of removal of these pesticides by activated carbons is an endothermic process, which agreed with the suggested interparticle transport rate control mechanism. However, the equilibrium

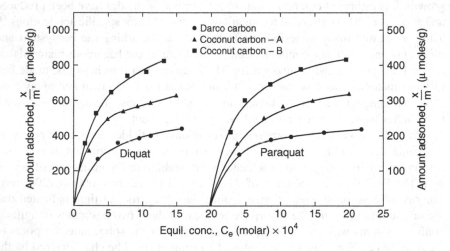

FIGURE 7.27 Diquat and paraquat adsorption isotherms for activated carbons. (After Prakash, S., *Carbon*, 19, 483, 1974. With permission.)

TABLE 7.10
Langmuir Parameters for Diquat and Paraquat Adsorption on Activated Carbons

Carbon Identification	Diquat			Paraquat		
	Langmuir Parameters		% Adsorption Capacity by Wt. of Carbon	Langmuir Parameters		% Adsorption Capacity by Wt. of Carbon
	X_m (μM/g)	b (l/μM) $\times 10^3$		X_m (μM/g)	b (l/μM) $\times 10^3$	
Darco.	500	3.7	18.1	250	2.8	6.4
Coconut-A	710	4.7	25.7	410	2.0	10.5
Coconut-B	990	6.9	35.8	550	2.2	14.1

After Prakash, S., *Carbon,* 19, 483, 1974. Reproduced with permission from Elsevier.

adsorption was an exothermic reaction, as is generally the case. The adsorption of diquat and paraquat was also determined from equimolecular mixtures of the two components in the aqueous solutions. The adsorption data indicated (Table 7.11) that the activated carbons show a clear preference for diquat. This was attributed to the adsorption of the two pesticides on different sites and to the differences in the steric hindrance, due to their different molecular dimensions.

Klaus[145] studied the removal of atrazine from industrial waste water, using a powdered activated carbon (PAC) in the adsorber. The adsorber also contained some fibrous material along with PAC. The calculations of the flow concentration at different operational conditions were made on the basis of the estimation of break-down lines. The effluent from the carbon bed contained atrazine well below the

TABLE 7.11
Competitive Adsorption of Diquat and Paraquat by Activated Carbons

Carbon Identification	Amt. Adsorbate Added		Amount Adsorbed			Adsorption Ratio Diquat
	Diquat	Paraquat	Diquat	Paraquat	Total	$\dfrac{\text{Diquat}}{\text{Diquat} + \text{Paraquat}}$
	(millimoles/100 g)					
Darco	200	—	40	—	40	
	—	200	—	20	20	
	100	100	16	8	24	0.67
Coconut-A	200	—	70	—	70	
	—	200	—	30	30	0.64
	100	100	36	20	56	
Coconut-B	200	—	84	—	84	
	—	200	—	42	42	
	100	100	50	30	80	0.63

After Prakash, S., *Carbon,* 19, 483, 1974. Reproduced with permission from Elsevier.

limits required by German legislation (i.e., below 0.1 μg/L). The powdered activated carbon was able to treat six times more water for equivalent weight of the granulated activated carbon. Ralph et al.[146] treated drinking water containing atrazine using PAC embedded in a layer of permeable synthetic collector with a high inner porosity. The removal of atrazine was better in the case of PAC than the granular activated carbon and was not affected by the carbon fouling in the PAC adsorber. Ayele et al.[147] investigated the influence of three surfactants the anionic, sodium dodecyl sulphate (SDS), cationic, hexadecyl trimethyl ammonium bromide (HTAB), and nonionic, 2 dodecyl oxy penta ethanoxy ethanol (DE 6) on the adsorption of atrazine from water by powdered activated carbon (PAC). The adsorption of atrazine at pH 5.5 was 230 mg/g. The adsorption decreased in the presence of the surfactants independent of the nature of the surfactant (cationic, anionic, or nonionic). The Langmuir adsorption capacity and the kinetic constants decreased in the presence of the surfactants. The decrease was more in the case of cationic surfactants HTAB, which was attributed to steric hindrance. The adsorption of molecular form of atrazine was not affected by the pH of the solution. However, when the pH of the solution in the presence of anionic surfactant SDS was changed from 3.5 to 10 the adsorption of atrazine increased while that of the surfactant decreased. In the presence of HTAB in the solution, the adsorption of atrazine decreased in the pH range 3.5 to 10. In this pH range the adsorption of HTAB increased. The presence of nonionic surfactant DE6 in the solution had no effect on the adsorption of atrazine. The low-dose ozonation and filtration of the Rhine river water[148] through granulated carbon bed enhanced the removal of atrazine, whereas addition of acetate to water did not make any difference in the removal of atrazine by carbon.

Kauras and coworkers[149,150] studied the adsorption of a fungicide Dodine and an insecticide Lindane on PAC from water in the presence of several coagulants. The coagulants used were $FeCl_3$ and basic polyaluminum chlorosulphate $[Al(OH)_x \, Cl_6 \, (SO_4)_n]$. Poly-acrylomie was used a coagulant aid (polyelectrolyle). The efficiency of Dodine removal was determined with respect to the PAC dose, the pH of the solution, and the type and dose of the coagulant and the polyelectrolyte. It was found that at an initial concentration of Dodine as 250 μg/L in the pH range 5 to 8, a dose of 100 μg/L of PAC could achieve more than 98% removal of Dodine, whereas a lower removal between 91 and 93% was obtained using half a dose of PAC under the same conditions. However, when 10 to 100 mg/L of the coagulant $FeCl_3$ was simultaneously added with the PAC, the removal efficiency was enhanced to more than 98%, even with half the dose of PAC.

In the case of Lindane, the PAC adsorption efficiency was not appreciably influenced by solution pH, but PAC doses greater than 20 mg/L were necessary to reduce Lindane from initial concentration of 10 mg/L down to 0.1 μg/L within 1-hr contact time. The efficiency of Lindane removal, however, decreased with the addition of the coagulants and, therefore, there was need for the carbon dose to be doubled to attain the required residual concentration of 0.1 μg/L.

Matsui et al.[151] examined the adsorptive removal efficiency of several pesticides using a GAC adsorber preloaded with background organic matter. It was found that the pesticides with higher water solubility were weaker adsorbates and showed lower removal efficiency and smaller adsorption capacity. This tendency was prominent when humic substances coexisted as background organic matter. The decrease in adsorption capacity of the carbon for weaker adsorbates was greater than for the stronger adsorbates. The results were explained by the Langmuir ideal adsorption theory.

Khappe et al.[152] used GAC filters as a primary treatment or in a polishing step for the removal of pesticides from spring runoff water. The carbon filters became preloaded with background organic matter (natural organic matter) over a period of use and needed to be regenerated. These workers developed a technique to determine the remaining life of a GAC filter for pesticide removal and to make predictions regarding the life of the fresh GAC adsorber.

Ashley et al.[153] studied the adsorption of S.triazines and their metabolites on granulated and powdered activated carbons at pH 7 and 9. The equilibrium adsorption data followed the Freundlich adsorption equation. The adsorption of S.triazine, and its metabolites diethyl atrazine and diisopropyl atrazines, was different on GAC and PAC and at different pH values.

Becker and Wilson,[143] in a comprehensive survey of the removal of pesticides from water using different methods including activated carbon adsorption, concluded that the activated carbon adsorption had the largest potential (Table 7.12). The activated carbon beds were capable of removing greater than 99% of most of the pesticides from water.

TABLE 7.12
Cumulative Pestiside Removal at 10 ppb Load from Water

	Pesticide Removed, %					
Process	DDT	Lindane	Parathion	Dieldrin	2, 4, 5-T Ester	Endrin
Chlorination–5 ppm chlorine (Batch test–90-minute contact)	<10	<10	75[a]	<10	<10	<10
Alum Coagulation and Sand Filtration	98	<10	80[b]	55	65°	35
Carbon						
Slurry-Slow mixing						
5 ppm (2×10^{-3} lb/lb carbon)		30	>99	75	80	80
10 ppm (1×10^{-3} lb/lb carbon)		55	>99	85	90	90
20 ppm (5×10^{-4} lb/lb carbon)		80	>99	92	95	94
Bed–0.5 gpm/ft³ of carbon	>99	>99	>99	>99	>99	>99

[a]Oxidized to paraoxon, which is more toxic than parathion.

[b]A value of 20% is reported.

[c]A value of 63% is reported.

After Becker, D.L. and Wilson, S.C., in *Carbon Adsorption Handbook*, P.N. Cheremisinoff and F. Ellerbusch, Eds., Ann Arbor Science. Publ., Ann Arbor MI, Vol. I, Chapter 5, p. 167. Reproduced with permission from Ann Arbor Science Publishers.

It is apparent from the perusal of the literature that the investigations concerning the adsorptive removal of different pesticides from water have generally been directed toward determining the efficiencies of powdered and granulated activated carbons for their removal. None of the studies have discussed the effect of such parameters as the surface area, pore-size distribution, or the chemistry of the carbon surface on the adsorption and its mechanism. Only in one paper Prakash[144] indicated that the adsorption of diquat and paraquat depended on the surface area of the carbon. Thus, there is need to study the influence of these parameters on the adsorption of pesticides.

7.6 ADSORPTION OF DYES

Industrial effluents from dyes, textile, and pulp and paper industries are highly colored due to the presence of residual dyes. These dyes cause microtoxicity to aquatic life and slow down self-purification of streams by reducing light penetration. The color in water also generates public resentment. Therefore, stringent standards are being fixed by the regulating agencies for the removal of dyes before the effluent is discharged into rivers and lakes. The methods that have generally been used for the removal of dyes from waste water are flocculation and coagulation using metallic compounds. However, these methods introduce metallic impurities and produce a large quantity of sludge, the disposal of the sludge being another environmental problem. The sludge-free treatments are, therefore, gaining importance. Activated carbon adsorption is one such method that has a potential for the removal of dyes from waste water. Consequently, a considerably amount of research has been carried out in this direction.

The activated carbon adsorption of cationic and anionic dyes has been studied from two points of view. Their removal from waste water and the characterization of the activated carbon surface for their surface area, microporous character, and polarity. For example, Graham[83] studied the adsorption of two dyes of opposite character but of approximately the same molecular dimensions (methylene blue and metanil yellow) from aqueous solutions on a number of activated carbons, and measured separately the effect of pore size and the surface acidic groups. Graphon, a graphitized carbon black that essentially has a uniform and homogenous surface, was used as a standard model substance. The activated carbon surface available for adsorption of methylene blue and metanil yellow was 100% of the BET surface area in the case of graphon that is almost free of any associated oxygen. However, in the case of activated carbons only a fraction of the BET surface was available (Table 7.13) for adsorption of both the dyes. This was attributed to the presence of acidic carbon-oxygen surface groups on the surface of activated carbons, which modified the adsorption characteristics of the carbons. Furthermore, the adsorption of the cationic dye (methylene blue) increased, while that of anionic dye (metanil yellow) decreased with increase in the acidity of the carbon surface. The adsorption of metanil yellow decreased linearly with increase in the acidity of the carbon surface (Figure 7.28), which indicated that the presence of acidic groups on the carbon surface tends to reduce the adsorption of anionic dyes roughly in proportional to the concentration of these groups. Similar results were obtained by Puri[55] and Arora[154] in their studies on the adsorption of methylene blue and Rhodamine B on

TABLE 7.13
Adsorption of Methylene Blue and Metanil Yellow in Relation to Surface Area and Surface Acidity

Carbon	Total Surface Area (m²/g)	Methylene Blue Accessible Area (m²/g)	Methylene Blue Accessible Area (%) of Total Surface	Metanil Yellow Accessible Area (m²/g)	Metanil Yellow Accessible Area (%) of Total Surface	Ratio of Accessible Areas of Metanil Yellow and Methylene Blue	Acidity of Carbon Surface (meq/100 m)
Graphon	83.9	83.9	100	83.9	100	1.00	0
1	1120	757	68	721	64	0.95	0.12
2	1130	836	74	629	56	0.76	0.45
3	1300	912	70	880	68	0.97	0.06
4	1300	602	46	329	25	0.55	0.94
4	1300	950	73	721	55	0.76	0.80
5	600	430	72	323	54	0.75	0.70
6	580	374	64	294	51	0.79	0.60

After Graham, D., *J. Phys. Chem.*, 59, 896, 1955. With permission.

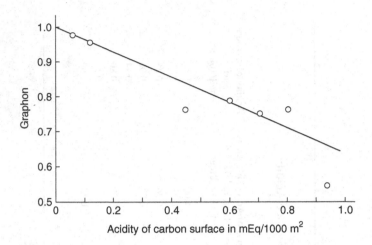

FIGURE 7.28 Influence of surface acidity of a carbon surface on its adsorption for metanil yellow. (After Graham, D., *J. Phys. Chem.*, 59, 896, 1955. With permission.)

graphon and a number of commercial activated carbons. The adsorption isotherms were Type I, which indicated completion of the monolayer. However, when surface areas were calculated using $119\,A^{\circ 2}$ and $144\,A^{\circ 2}$ as the molecular areas of methylene blue and metanil yellow, respectively, the values agreed with the BET area for graphon and differed appreciably for active carbons. It was suggested that some of the micropores were not accessible to larger dye molecules.

Sastri[155,156] examined the adsorption of methylene blue on several commercial activated carbons and explained their results on the basis of the carbon surface charge as a function of pH of the solution. According to Sastri, the charge on the carbon surface is negative, due to the ionization of carboxylic surface groups, and this resulted in higher adsorption of the cationic methylene blue molecules. However, when the pH of the solution was lowered, the carbon surface tended to attain a positive charge, which resulted in a decrease in the adsorption of the dye. A pH between 3 and 7.5 was found to be the best for maximum adsorption of methylene blue on different carbons. Perineau et al.,[157] while studying the adsorption of acidic and basic dyes, however, found that the adsorption capacity was maximum at pH2 for acidic dyes and a less acidic pH (>5) showed maximum adsorption for the basic dyes. Dai and coworkers[158–160] also investigated the effect of pH on the adsorption of anionic dyes (phenol red, carmine, and titan yellow) on activated carbons and observed that the zeta potential of the carbon surface determined the adsorption. When the pH of the dye solution was greater than the zero point charge (ZPC), the electrostatic repulsion between the anionic dye and the negatively-charged carbon surface reduced the adsorption of the dye. Gao et al.[161] studied the adsorption of anionic and nonionic compounds using several activated carbons by modifying their surface by oxidation with nitric acid or reduction with hydrogen. They observed that the adsorption capacity of anionic and nonionic compounds decreased with an increase in surface acidity of the carbon, while the adsorption of cationic compounds decreased with decrease in surface acidity. Listiskaya et al.[162] observed that the influence of pH on the adsorption of anionic dyes on carbon adsorbents depended on the molecular form of the dye and its tendency to form miscelles in the solution. When the dye did not form miscelles in the solution, the change in pH did not affect the adsorption of the dye.

Al-Degs and coworkers[163,164] investigated the removal efficiency of several activated carbons using three highly reactive anionic dyes in the textile industry: Ramazol reactive golden, yellow, and Ramazol black B. These workers observed that all three dyes were adsorbed in considerable amounts, but Ramazol yellow showed the largest adsorption when F-400 activated carbon was used. However, the amount adsorbed was not related to surface area or the surface acidity of the carbon surface. The differences in adsorption were attributed to the carbon surface charge leading to electrostatic interactions. Chen and Lu[165] and Chen and Wu[166] studied the adsorption rates of methylene blue and eriochrome blue-black R, and crystal-violet dyes on sisal-based activated carbon fibers (ACF) in single and binary systems. The adsorption rates were different on different ACFs and much faster in the case of methylene blue. All the carbon fibers showed a high adsorption capacity for all the dyes. The variation in adsorption amounts and rates were attributed to the varying surface properties (e.g., surface area, pore size, and surface functional groups) of the carbon fibers and to the geometric dimensions of the dye molecules. These workers suggested that the adsorption rates of different dyes could be used as parameters for the pore structure of activated carbon fibers. Honas and Bakir[167] observed that the adsorption of methylene blue on a commercial activated carbon followed the Langmuir isotherm equation. The adsorption capacity of the carbon was higher in the alkaline solution than in the acid solution with an ultimate adsorption capacity of 367.6 mg/g. There was no significant effect of temperature on the adsorption. Khalil and Girgis,[168] while studying the adsorption of methylene blue on a sugar baggase-activated carbon from aqueous solutions in the concentration range 160 to 960 mg/L, also observed that the adsorption data followed the Langmuir isotherm equation better than the Freundlich isotherm equation. The surface area accessible to methylene blue decreased with an increase in the degree of activation of the carbon, which indicated that the activation of the carbon develops fine microporosity. Some of these micropores are not accessible to larger dye molecules. Thus, the efficiency of a given activated carbon depended on the pore-size distribution rather than the surface area of the carbon.

Mittal and Venkovachar[169] studied the adsorption of Rhodamine B (basic violet 10) and Sendolan Rhodine (acid Red 1) dyes on sulphonated coal from aqueous solutions and found that the coal exhibited moderate removal potential for both the dyes. The data followed the Freundlich isotherm equation, which showed that the uptake of the basic dye was higher than the acid dye. Desorption studies with water, H_2SO_4, CH_3COOH, and $HCOOH$ indicated that the adsorption of Rhodamine B was a chemical rather than a physical or ion exchange, whereas adsorption of Sendolan Rhodine was physical with the absence of ion exchange process. Viraraghavan and Mihial[170] observed that the adsorption of a basic and an acidic dye on peat followed Langmuir and Freundlich adsorption isotherms. The adsorption of the basic dye was better at 99% compared to the 48% of the acidic dye. The equilibrium time for the adsorption of the basic dye was also smaller. The effect of powdered activated carbon particle size on the removal of color of a disperse dye was studied by Lin.[171,172] Lin observed that the multiplayer BET isotherm was a better fit than the monolayer Langmuir or Freundlich models. Lin also carried out the adsorption of the disperse dye (Red 60) on several adsorbents (e.g., molecular sieves, activated alumina, and granular and powdered activated carbons). The powdered activated carbons, activated alumina, and molecular sieves were found to be better than the granular activated carbons.

The removal of two basic dyes (Soframine Red T and Deorlene yellow) by activated carbons from the effluents of textile, tanning, and food industries was studied by Mckay and Al Duri.[173] The monolayer capacities for the two dyes were 390 and 1240 mg/g, respectively. Kinetic studies in batch vessels showed that initial rate of adsorption increased with agitation. The adsorption data indicated that intra-particle diffusion played a role in the removal of the dye. In later investigations[174–177] these workers studied the adsorption of three basic dyes: Basic Red 22, Basic yellow 21, and Basic Blue 69 on Filtrasorb-400, using single-, double-, and triple-component systems. The kinetic experiments were carried out at different dye concentrations and using different amounts of carbon. The prediction of the experimental equilibrium data was carried out by using the ideal adsorption theory and several other isothermic equations. When using an agitated batch adsorption technique, these workers calculated an intraparticle diffusion rate parameter, which was related to dye concentration and carbon particle size. The single-component batch mathematical model was extended to describe the mutlicomponent adsorption data. Dusant et al.[178] studied the adsorption of anionic and cationic dyes from waste water on peat and charcoal using a dynamic system. The breakthrough curves showed that while peat columns were able to remove sufficient amounts of basic red dye, the pure charcoal showed smaller adsorption values.

Sasaki et al.[179] studied the adsorption of acid dye (Acid Red 88) and direct dye (Direct Yellow 11) on activated carbon fibers (Al-ACF, Ti-ACF, and Y-ACF) obtained from pitch containing metal complexes. The different carbon fibers had different isoelectric points that varied between pH 4.4 and 6.6. The adsorption of acid Red 88 on the three ACFs was lower than that of the metal-free ACF in the pH range 2 to 12. The adsorption of Direct yellow 11 on Y-ACF, which contained a large microporosity, was higher than that of the other ACFs in the acidic pH range. However, this adsorption of Direct yellow 11 decreased with increasing pH above the isoelectric point of pH 6.7. Thus, according to these workers, the adsorption of dyes of ACFs depended on the pore-size distribution in ACF and on the electrostatic interactions between the dye molecules and the carbon surface. Miau and Dai[180–181] examined the effect of pH of the solution on the adsorption on anionic dye Titan yellow by an activated carbon. The adsorption of the dye varied appreciably by a change in the zeta potential of the activated carbon. Juang and Swei,[182] In their studies on an anionic and a cationic dye by an activated carbon, suggested that the higher adsorption of the basic dye was due to the electrostatic interaction between the electron acceptor groups on the carbon surface and the positive charge on the basic dye molecule. However, this explanation does not appear to the very convincing. Nandi and Walker[183] investigated the adsorption of two acidic and one basic dye by coals, charcoals, and activated carbons, and observed that the fraction of the

BET surface area occupied by the different dyes depended on the nature of the carbon surface.

Guzel[184] studied the effect of carbon-surface acidity on the methylene blue and metanil yellow by activated carbons at 298 K. The adsorption of methylene blue was much larger compared to the adsorption of metanil yellow. It has been suggested that the molecules of methylene blue (a cationic dye) are attracted by the carbon surface, but those of metanil yellow (an anionic dye) are repelled by the carbon surface. Tamai[185] examined the adsorption of several acid, basic, and direct dyes on a highly mesoporous activated carbon fiber Y-ACF obtained from pitch containing yttrium acetylacetonate in terms of the size of the dye molecule and pore size and surface charge of the activated carbon fiber. The results were compared with that on the microporous activated carbon fiber ACF-20. The small-sized acid and basic dyes were adsorbed in large amounts both on the Y-ACF and ACF-20. The adsorption of direct dyes, which had large sizes in one or two dimensions of the molecular structure, were much larger on the mesoporous Y-ACF than those on the highly microporous ACF-20. Furthermore, the adsorption of direct dyes decreased with increasing pH in the alkaline range as the molecular dimensions of the direct dye increased. Thus, the adsorption of direct dyes decreased in the order Direct yellow 50 > Direct black 19 > Direct yellow 11.

This order was the same as that of one large dimension of the dye molecules: 2.39 nm for Direct yellow 50, 30.4 nm for direct black 19, and 31.2 for Direct yellow 11. These workers suggested that pore-size distribution plays an important role in the adsorption of Direct dyes with one or two large dimensions. Because Y-ACF was mesoporous and its average pore diameters were larger than the sizes of these dye molecules, they were adsorbed in larger amounts on Y-ACF. The dye molecules diffused directly into the mesopores connected with the external surface. The high adsorption capacity of the carbon for basic dyes was attributed to the negative charge on the activated carbon fiber surface and to the relatively smaller size of the dye molecules. In other words, the electrostatic attractive interactions between the carbon surface and the dye molecule and the diffusion of the dye molecule into meso- and micropores resulted in increased adsorption in the case of the basic dyes. Kasaoka et al.[186] also observed that in the adsorption of relatively small-sized dyes on activated carbons, which have mainly micropores, the short axis size of the dye molecule plays an important role in adsorption by microporous carbons.

Goyal[114, 212] studied the adsorption isotherms of methylene blue on two samples of granulated and two samples of fibrous activated carbons from aqueous solutions in the concentration range 20 to 3000 mg/L. The adsorption isotherms (Figure 7.29) are generally Type I of the BET classification in the case of fibrous activated carbons (ACF-307 and ACF-310), and they tend to become Type II in the case of granulated activated carbons (GAC-E and GAS-S). The adsorption data conformed to the Langmuir adsorption equation. The linear Langmuir plots (Figure 7.30) were used to calculate the monolayer capacity and the surface area occupied by methylene blue using molecular area of methylene blue as 119 $A^{\circ 2}$. Although almost all of the BET surface area was available for the adsorption in granulated activated carbons, only ~50% of the BET area was available in the case of fibrous activated carbons (Table 7.14). This indicated that about 50% of the BET area in the case of fibrous

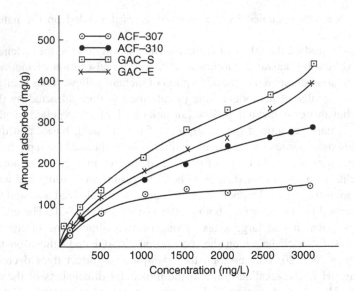

FIGURE 7.29 Adsorption isotherms of methylene blue on as-received activated carbons. (After Goyal, M., Ph.D. thesis submitted to Panjab University, Chandigarh, India, 1997. With permission.)

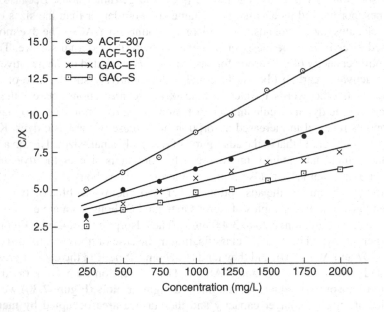

FIGURE 7.30 Linear Langmuir plots for as-received activated carbons. (After Goyal, M., Ph.D. thesis submitted to Panjab University, Chandigarh, India, 1997. With permission.)

TABLE 7.14
Fraction of BET Surface Area Occupied by Methylene Blue
for Different Activated Carbons

Sample Identification	BET(N$_2$) Surface Area (m^2/g)	Monolayer Capacity x$_m$ (from Linear Langmuir Plots (mg/g)	Surface Area Covered by Methylene Blue (m^2/g)	% BET Area Covered
ACF-307	910	200	451	49.5
ACF-310	1184	300	676	57.1
GAC-E	1190	454	1123	94.3
GAC-S	1256	556	1253	99.7

After Goyal, M., Ph.D. thesis submitted to Panjab University, Chandigarh, India, 1997. With permission.

activated carbons was present in the form of fine micropores, which were inaccessible to larger methylene blue molecules. The adsorption of the dye was found to increase on oxidation of the activated carbons with nitric acid and oxygen, and the increase in adsorption was larger in the case of oxidation with nitric acid (Figure 7.31). This increase in adsorption on oxidation has been attributed to the formation of carbon-oxygen surface groups. Because the nitric acid is a stronger oxidative

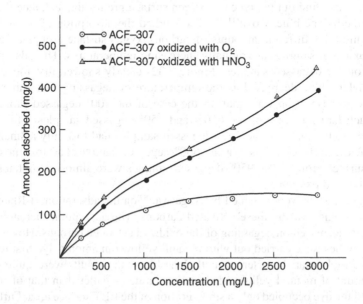

FIGURE 7.31 Adsorption isotherms of methylene blue on ACF-307 before and after oxidation. (After Goyal, M., Ph.D. thesis submitted to Panjab University, Chandigarh, India, 1997. With permission.)

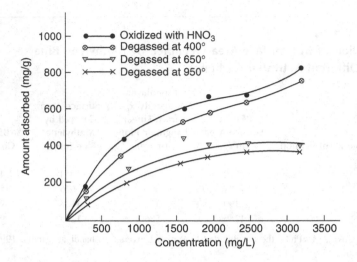

FIGURE 7.32 Adsorption isotherms of methylene blue on oxidized GAC-E before and after degassing at different temperatures. (After Goyal, M., Ph.D. thesis submitted to Panjab University, Chandigarh, India, 1997. With permission.)

treatment than oxygen gas, it resulted in the formation of larger amounts of carbon-oxygen surface groups and, therefore, caused a larger increase in the adsorption of methylene blue.

In order to find out the type of oxygen surface groups that enhance the adsorption of methylene blue, Goyal[114,212] also studied the adsorption of methylene blue after eliminating different amounts of carbon-oxygen surface groups by degassing of the oxidized samples at gradually increasing temperatures. The adsorption isotherms on the degassed samples (Figure 7.32) clearly showed that the adsorption of methylene blue decreased as the temperature of degassing was enhanced. The decrease in adsorption was small in the case of the 400°-degassed sample, but it was much larger in case of the 650°- and 950°-degassed samples. This has been attributed to the fact that the 400°-degassed samples had lost only a small amount of the acidic surface groups, but the 650°-degassed samples had lost most of their acidic surface groups. The 950°-degassed samples were almost completely free of any associated oxygen.

Similar results were obtained by Aggarwal[187] on the adsorption of Rhodamine B (also a cationic dye) by these activated carbons. The adsorption increased on oxidation but decreased on degassing of the oxidized activated carbons. However, when similar studies were carried out with metanil yellow (an anionic dye that has almost the same molecular diameter as methylene blue), the results were quite different. The amount of metanil yellow adsorbed was much smaller than that of methylene blue, and the dye occupied only a small portion of the BET surface area. Furthermore, the adsorption of metanil yellow decreased on oxidation of the carbons with nitric acid or oxygen, and the decrease was much larger in case of the nitric

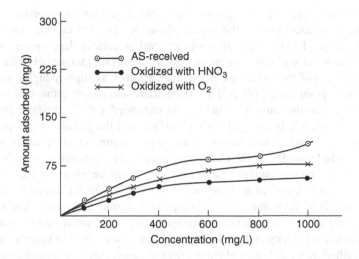

FIGURE 7.33 Adsorption isotherms of metanil yellow on ACF-307 before and after oxidation. (After Aggarwal, D., Ph.D. thesis, Panjab University, Chandigarh, India, 1997. With permission.)

acid oxidized samples (Figure 7.33). However, the adsorption increased when the oxidized samples were degassed at 400°, 650°, and 950° (Figure 7.34). The increase in adsorption increased with increase in the temperature of degassing.

The adsorption results have been explained on the basis of electrostatic attractive and repulsive interactions. The oxidized activated carbons were associated with

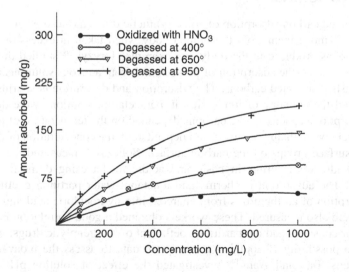

FIGURE 7.34 Adsorption isotherms of metanil yellow on oxidized ACF-307 before and after degassing. (After Aggarwal, D., Ph.D. thesis, Panjab University, Chandigarh, India, 1997. With permission.)

acidic carbon-oxygen surface groups that ionized in solution to produce H^+ ions, which were directed toward the liquid phase, leaving the carbon surface with a negative charge. The negative charge of the carbon surface depends on the amount of the acidic-surface groups present on the carbon surface and the pH of the solution. Because the pH of the oxidized carbon suspensions was slightly higher (pH 5 to 6) than the zero point charge pH (pH = 2 to 3), there was preponderance of negatively-charged sites on the carbon surface. This enhanced the electrostatic interactions between the negatively-charged carbon surface and the positively-charged cations in the case of methylene blue and Rhodamine B, resulting in an increased adsorption on the oxidized samples. In the case of the anionic dye, metanil yellow, the negative charge on the carbon surface enhanced the repulsive electrostatic interactions between the carbon surface and the anions of the dye, which reduced the adsorption of Metanil yellow. When these acidic groups were eliminated from the carbon surface by degassing at gradually increasing temperatures, the carbon surface became less and less negatively charged, thereby decreasing the adsorption of cationic methylene blue and Rhodamine B dyes, and enhancing the adsorption of anionic Metanil yellow dye. A small increase in the adsorption of both the cationic and anionic dyes on the 950°-degassed carbon samples was attributed to widening of some of the micropores so that these were now accessible to larger molecules of the dyes.

Thus, the adsorption of cationic and anionic dyes by activated carbons from aqueous solutions depends largely on the chemistry of the carbon surface and to a smaller extent on the porous structure of the carbon surface.

7.7 ACTIVATED CARBON ADSORPTION OF DRUGS AND TOXINS

The activated carbon adsorption of some synthetic drugs has also been studied with a view to remove them from the human body when taken in excess or to remove certain toxins produced in the body. For example, Teryzyk[188] studied the influence of temperature on the adsorption of paracetamol from aqueous solutions at pH 7 on three deashed activated carbons. The adsorption and desorption isotherms at 300 to 320 K and the enthalpy of immersion in paracetamol solutions were determined. The adsorption capacity of the carbons depended on the temperature of adsorption, and the relative enthalpy of adsorption depended on the concentration of the carbon-oxygen surface groups on the carbon surface. Sobezak[189] measured the adsorption of six β-adrenolytic drugs in basic and acidic media using domestic medicinal charcoal. The adsorption isotherms and n.octanol-water partition coefficients and the desorption of all the drugs from charcoal into hydrochloric acid and phosphate buffer were also measured. These workers obtained a good correlation between the Langmuir constants and the partition coefficient of β-adrenolytic drugs, which indicated the possibility of applying the sorption data to assess the bioavailability of these drugs. Bao and Wang[190] investigated the effect of solution pH, adsorption temperature, and isotonic regulating agents on the adsorption of Norfloxacin (NFR), Ofloxacin (OFX), and Ciprofloxacin (FX) by activated carbons. The saturation adsorption amounts of these drugs were found to be 374, 290, and 320 mg/g for

NFX, OFX, and FX, respectively. However, the equilibrium value increased with increasing pH and temperature. Dutta et al.[191] studied the adsorption of semisynthetic cephalosporins such as 7-amino cephalosporanic acid (7 ACA), cephalexin, and cephadroxil from aqueous solutions using an activated carbon. The adsorption capacity of the carbon depended on the pH of the solution. 7 ACA exhibited relatively low adsorption capacity compared to cephalexin and cephadroxil, which had about the same adsorption capacities. The adsorption data for all three drugs fitted the Langmuir adsorption equation. The adsorption rate studied showed that the external film diffusion controls the adsorption in the initial stages, and particle diffusion controls adsorption in the later stages. Bredberry and Allister[192] observed that multiple-dose activated carbon adsorption therapy in animals results in a significant elimination of drugs. These workers have suggested that this therapy may be successfully used in case of patients who have ingested life-threatening amounts of phenobarbital (phenobarbitone), carbamazepine, theophylline, quinine, dapsone, or salicylate, to eliminate these poisons.

Active carbons are widely used for the removal of toxins from blood.[193–195] Creatinine is one such substance that is adsorbed by activated carbons. Smith et al.[196] observed that when an activated carbon was placed in contact with a creatinine solution in air, the concentration of the solution continued to decrease considerably over a number of days. The decrease is concentration was considerably smaller when granulated carbons were used, and greater when powdered carbons were used (Table 7.15). In the case of the controlled sample (i.e., without any carbon), the

TABLE 7.15
Catalytic Oxidation of Creatinine and Creatine by Powdered Charcoal[a]

Time (hr)	Creatinine Oxidized Using 0.1 g Charcoal at 83°C (mmol/g)		Time	Creatine Oxidized Using 0.5 g Charcoal at 37°C (mmol/g)
	Control	With Charcoal		
0	0	0	0	0
2.5	1.4	5.1		
21.5	10.5	24.4	1 hr, 5 min	0
69.5	11.4	30.3		
75	12.1	32.2	18 hr, 25 min	0.59
120	11.5	32.5		
144	11.7	32.6	24 hr, 40 min	1.12
168	11.5	32.6		
216	11.5	32.7		

[a]Solution taken = 500 ml 10^{-3} M.

Source: Smith, E.M., Affrossman, S., and Courtney, J.M., *Carbon*, 17, 149, 1979. Reproduced with permission from Elsevier.

concentration of the solution decreased for 24 hr and then remained more-or-less constant. The decrease in the concentration of the control was attributed to the conversion of small amounts of creatinine into creatine, for which the equilibrium was established in one day.[197] The decrease in the concentration of the solution in the presence powdered carbon was the result of the oxidation of the creatinine into creatine, for which the equilibrium was also established in one day.[197] However, no oxidation of creatinine was observed when the reaction was carried out in an inert atmosphere. The kinetic measurements with granulated carbons in the low concentration range in inert atmosphere and in the presence of oxygen showed that the concentration of creatinine in the inert atmosphere decreased only very slightly after the first rapid fall, whereas it showed a steep fall in the presence of oxygen, ultimately reaching a zero concentration after about 45 hr (Figure 7.35). The initial rapid fall in the concentration in the nitrogen atmosphere, which was similar to the fall in the presence of oxygen, was attributed to the adsorption of creatinine on the carbon surface because there was no fall in concentration in the case of the control. This further continued fall in the concentration of creatinine in the presence of oxygen has been attributed to the catalytic oxidation of the creatinine by the carbon. Thus, according to these workers, the creatinine first adsorbs on the surface of the carbon, followed by its oxidation. Furthermore, the rate of oxidation of creatinine was 1.6 times higher than the rate of oxidation from a solution containing equimolecular quantity of creatine, because creatine then competes with creatinine for the carbon-active sites.

Rybolt et al.[198] studied the adsorption of a biomedically active aromatic compound acetaminophen, which is the active constituent of Tylenol, and an aliphatic compound N-acetylcysteine, which is an antidote for acetaminophen overdose on an activated carbon from aqueous solutions at gastric pH 1.2 and intestinal pH 7.0. The adsorption data fitted the Langmuir isotherm equation for both the adsorbates.

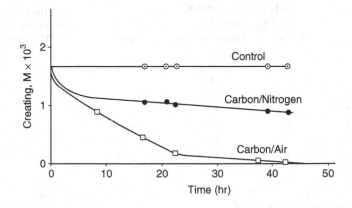

FIGURE 7.35 Kinetics of catalytic oxidation of creatinine in inert atmosphere and in the presence of air. (After Smith, E.M., Affrossman, S., and Courtney, J.M., *Carbon,* 17, 149, 1979. With permission.)

The authors are of the view that the adsorption of acetaminophen involved van der Walls interactions and the adsorbed molecule had a flat configuration. The adsorption of N-acetylcysteine, however, showed different configurations of the molecules depending on the pH of the solution. Furthermore, the adsorption of N-acetylcysteine was more at pH = 1.2 than at pH = 7.

7.8 ADSORPTION OF MISCELLANEOUS ORGANIC COMPOUNDS

In addition to the group of organic compounds discussed above, the activated carbon adsorption has been used for the removal of a large number of other organic compounds such as amines, organic acids, organic sulfur compounds, polycyclic aromatic hydrocarbons, surfactants, cationic polymers, aromatic hydrocarbons, aldehydes, and amino acids. The next few pages discuss important work that has been carried out to remove the organic compounds using activated carbons.

Tamamushi[199] and Allali et al.[200] Smolin[201] and Chobamu et al.[202] studied the adsorption isotherms of ionic surfactants such as dedecyl ammonium chloride, dodecyl pyridenium bromide, and sodium dedecyl sulphate from their aqueous solutions on activated carbons and obtained S-shaped isotherms with two maxima. The equilibrium concentration for the second maxima coincided with the critical miscelle concentration (CMC) of the surfactants. The adsorption involved two steps: monomolecular adsorption of the Langmuir type and multilayer adsorption of the BET type. The polar groups of the surfactants were attracted by the polar groups on the carbon surface by electrostatic attractive forces and formed oriented monomolecular layers and then multimolecular layers between nonpolar chains of the surfactant molecules. The adsorption data clearly showed the influence of adsorbate-adsorbate and adsorbate-adsorbent interaction forces during adsorption from solution. Gao et al.[203] studied the adsorption of a specialty surfactant Darex II from aqueous solutions on several chars after modifying their surface by air oxidation at 400°C for 10 hr or by ozonization at room temperature (2 Vol. % ozone in air). The adsorption capacity of the chars increased with increase of surface area, but a linear relationship was obtained when the amount adsorbed was plotted against the surface area contained in pores larger than 2 nm (Figure 7.36(a) and Figure 7.36(b)). The adsorption capacity of the chars for the surfactant also depended on the source raw material and the history of preparation of the carbons. The adsorption was decreased considerably on oxidation of the chars in air or ozone. This was attributed to the formation of polar surface groups on the carbon surface. Zouboulis et al.[204] studied the adsorption of anionic and cationic surfactants at different solution pHs. The removal of cationic surfactants was very good at higher pH values while intermediate pH values showed good results for the removal of anionic surfactants. This was attributed to electrostatic interactions between the carbon surface and the surfactant molecules and ions.

Bornehoff[137] examined the adsorption of a number of polycyclic aromatic hydrocarbons (PAH) using a variety of activated carbons under laboratory and field

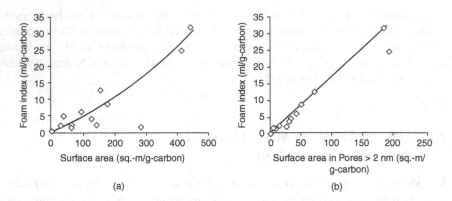

FIGURE 7.36 (a) Surfactant adsorptivity vs. surface area of various solid fuel char. (After Gao, Y., Farber, M., Chem. X., Suuberg, E.M., and Hurt, R.H., *Carbon 02 Intern. Conf. on Carbon*, Beijung, 2002, Paper 29.3. With permission.) (b) Surfactant adsorptivity vs. the sum of mesopore and macropore area for various solid fuel chars. (After Gao, Y., Farber, M., Chem. X., Suuberg, E.M., and Hurt, R.H., *Carbon '02 Intern. Conf. on Carbon*, Beijung, 2002, Paper 29.3. With permission.)

conditions and observed that activated carbon were good adsorbents for the removal of PAH from water. These workers also determined the presence of different PAH$_s$ on the surface of different carbons by extracting 25 g of an activated carbon with 300 ml of benzene in a soxhlet extraction for 8 hr. Ten different PAH were found to be present in the extract, the amount varying between 1 and 227 mg/kg, with a mean value of 63.4 mg/kg. However, in spite of the presence of PAH, the adsorption capacity of the carbons for PAH from water did not decrease. Furthermore, it was observed in laboratory experiments that the amounts of these PAHs leached with water from most of the activated carbons were small (on the order of 1 mg/Kg), so that the use of activated carbons did not result in any detectable contamination of water.

The adsorption experiments carried out using 10 and 100 mg of Norit active carbon with contact time varying between 30 min and 45 hr for 8 different PAHs showed (Table 7.16 and Table 7.17) that the removal was 80 to 98% in the first 30 min of contact time. Furthermore, the PAHs with lower molecular weights were adsorbed to a larger extent than the higher molecular weight PAHs. Similarly, when the experiments were carried out by using PAHs of comparable molecular weights but differing in their structural configurations, the adsorption was found to depend on the pore-size distribution of the activated carbons. The carbons showed a certain degree of molecular sieving behavior toward larger molecules.

When the removal of PAH was carried out on actual plant scale, only 10% of the PAH present in water could be removed by the carbon with an effluent concentration of 10 mg/L of the hydrocarbon. However, the removal efficiency was enhanced when the effluent was pretreated with chlorine or ozone.

TABLE 7.16
Adsorption of Polycyclic Aromatic Hydrocarbons
by Norit-Activated Carbon

PAH	Amount of Carbon (mg/L)	Amount Adsorbed, Contact Time of 0.5−45 hr (%)				
		0.5 hr	4 hr	22 hr	28 hr	45 hr
Fluoranthene	100	98.8	99.2	99.5	99.2	99.4
	10	81.8	92.8	98.4	98.7	99.2
Benzo[a]anthracene	100	97.2	99.0	97.7	99.3	98.8
	10	79.3	84.5	97.6	98.4	99.1
Benzo[b]fluoranthene	100	80.9	97.8	99.1	99.0	98.3
	10	99.8	66.1	95.0	97.3	98.9
Benzo[a]pyrene	100	80.7	96.9	99.5	98.8	98.6
	10	68.6	73.2	96.2	98.1	99.0
Benzo[g,h,i]perylene	100	57.6	81.8	97.7	97.3	98.7
	10	37.8	49.0	93.7	96.6	98.1
Benzo[j]fluoranthene	100	85.4	96.8	99.4	99.0	98.4
	10	59.1	74.9	96.1	98.0	98.6
Pyrene	100		98.4		99.6	98.5
	10		93.6		99.3	99.5

After Bornehoff, J., in *Activated Carbon Adsorption*, I.H. Suffet and M.J. Mcguire, Eds., Ann Arbor Science Publishers, Ann Arbor MI, 1980, Vol. 1, p. 145. Reproduced with permission from Ann Arbor Science Publishers.

TABLE 7.17
Adsorption of Polycyclic Aromatic Hydrocarbons by Activated Carbon

PAH	Molecular Weight	Amount of Active Carbon Added (mg/L)	Amount Adsorbed in 30 min (%)
Anthracene	178	100	98
		10	89
Benzo(a)anthracene	228	100	83.7
		10	67.5
Dibenzopyrene	302	100	70
		10	51.3

After Bornehoff, J., in *Activated Carbon Adsorption*, I.H. Suffet and M.J. Mcguire, Eds., Ann Arbor Science Publishers, Ann Arbor MI, 1980, Vol. 1, p. 145. Reproduced with permission from Ann Arbor Science Publishers.

The adsorptive removal of benzene from aqueous solutions containing 0.5 to 500 ppm of benzene was investigated by Garcia-Garcia et al.[205] using a pine-waste char activated with varying amounts of potassium hydroxide. The kinetics of benzene adsorption showed that the rate of benzene adsorption progressively increased as the KOH-to-char ratio was enhanced. This was attributed to the formation of highly microporous activated carbons containing mesopores. The equilibrium studies of benzene adsorption with contact time of 48 hr at two concentrations, 200 and 500 ppm. of benzene, showed that the adsorption capacity of the carbons was also dependent on the development of microporosity. At equilibrium an adsorption capacity of benzene as high as 370 mg/g could be achieved with some of these carbons.

Kamegawa et al.[206] prepared water-soluble Nanographite (WSNG) from carbon black and studied the adsorption of several organic compounds from acidic and basic solutions. The adsorption of 2-naphthol was much larger on WSNG than on the graphitized carbon black (GCB), both from acid and neutral solutions. The maximum uptake of 2-naphthol at the saturated concentration was 1.28 m mol/g and was independent of the pH of the solution. This amount adsorbed indicated that each WSNG molecule, on average, adsorbed 1.5 molecules of 2-naphthol, and GCB showed an adsorption of 0.2 m mol/g. The adsorption of 2-napthol indicated an apparent surface area of ~500 m^2/g for WSNG. The adsorption of 9-phenanthrol on WSNG was 1.8 m mol/g, which indicated that each WSNG molecule adsorbed two molecules of 9-phenanthrol, the apparent surface area amounting to 820 m^2/g. These surface areas correspond to many active carbon surface areas. Thus, these workers have considered WSNG as *Water Soluble Activated Carbon*. These workers also determined the adsorption isotherms of naphthalene derivatives on WSNG and observed that with the exception of ethoxy group, the electron-donating groups enhanced the adsorbate-adsorbent interaction, and the electron-attracting groups such as -COOH, -CN, -NO$_2$ suppressed the interaction, as in the case of activated carbons. These workers also observed that WSNG could adsorb an equal number of aromatic acid molecules, regardless of their molecular size and, in some cases, adsorbed a larger number of larger-sized acid molecules. This unique adsorptive behavior of WSNG has been attributed to the existence of molecular-size hydrophobic adsorption sites on each WSNG molecule.

The removal of 2-methyl isoborneol (MIB), the primary cause of disagreeable odors in surface waters, was studied by Nawack and Cannon[207] using lignite-based activated carbon after heat treatment from natural waters. The heat treatment of the activated carbon resulted in increasing the adsorption of MIB from 60 to 75%. The physical and chemical characteristics of the different carbons suggested that changes in pore-size distribution and the relative amounts of carbon-oxygen surface groups were important factors that were responsible for the enhanced adsorption of MIB. Ishizaki and Cookson[208] selected four typical organic compounds (i.e., n.butyl mercaptan, butyl disulfide, decane, and p-hydroxy benzaldehyde) for adsorption studies on granulated carbons from aqueous solutions. The mercaptan and the disulfide imparted an unpleasant odor to water, decane represented the nonpolar paraffins, and p-hydroxy benzaldehyde represented degradation products of large organic molecules identified as natural-color-producing compounds in water. The adsorption isotherms for all the adsorbates obeyed the Langmuir adsorption equation.

The adsorption capacity of the carbon for butyl sulfide increased on outgassing the carbon at 900°C. When the degassed carbon was oxidized in dry oxygen in the temperature range 200 to 300°C, the adsorption capacity remained unchanged on oxidation at 200°C but increased at 300°C oxidation temperature. This increase in adsorption of butyl sulfide was attributed to an increase in surface rather than to surface functionality because the values, when normalized with respect to surface area, showed no change. Furthermore, this explanation was supported by the fact that there was no change in the adsorption capacity of the carbon for butyl sulfide when the surface functionality of the carbon was enhanced by oxidation with ammonium persulphate. The adsorption of the mercaptan and the disulfide also decreased on methylation of the carbon. This could be attributed to a decrease of surface functionality due to methylation but these authors attributed it to a decrease in surface area as a result of pore blocking. In the case of p-hydroxy benzaldehyde, the adsorption on a per-unit area basis decreased on oxidation, which indicated that surface oxygen groups played a role in adsorption.

The kinetic studies showed that the role of adsorption of butyl disulfide remained unchanged with degassing temperature up to 700°C but decreased when the outgassing temperature was raised above 900°C. Similarly, the rate of adsorption remained unchanged on oxidation but decreased when the oxidation was followed by outgassing at 900°C (Table 7.18). This was attributed to the presence of metallic impurities such as copper and iron, which were removed on outgassing at 900°C and which enhanced the adsorption rate. However, the presence of these metallic impurities did not influence the rate of adsorption of decane, although the presence of surface oxides suppressed its adsorption. The rate of adsorption of decane increased when the surface oxygen groups were removed, and decreased when the surface oxygen groups were formed on oxidation. This was suggested to be due to the nonpolar nature of the decane molecule, which will be preferred by a nonpolar carbon surface.

TABLE 7.18

Rates of Adsorption of Butyl Disulfide, Decane, and p-Hydroxybenzaldehyde on Untreated Carbons

	Rate Constant (L/mmol-min $\times 10^3$)		
Carbon	Butyl Disulfide	Decane	p-Hydroxy-benzaldehyde
Untreated	78.4	3.80	12.6
Outgassed at 900°C	6.65	16.6	11.5
Outgassed at 900°C and oxidized at 100°C	2.21	—	—
Outgassed at 900°C and oxidized at 200°C	0.95	—	—
Outgassed at 900°C and oxidized at 300–500°C	0.29	1.18	—
Chemically oxidized	0.37	—	11.1

After Ishizaki, C. and Cookson, J.T., in *Chemistry of Water Supply Treatment, and Distribution*, A.J. Rubin, Ed. Ann Arbor Science. Publ., Ann Arbor MI, 1980, p. 201. Reproduced with permission from Ann Arbor Science Publishers.

In the case of p-hydroxy benzaldehyde, neither polar carbon surface nor oxidation of the carbon surface caused any change in the rate of adsorption. This indicated that pore diffusion rather than the presence of oxygen surface groups or metallic impurities was the limiting step in the adsorption.

Usmani and Wahab[209] determined the adsorption of lower aliphatic and some organic unsaturated fatty acids from their aqueous solutions on activated carbons prepared from wood, rice husk, and lignite coal. The adsorption depended on the porosity and the chemical structure of the carbon surface. It was also found that for a given acid the adsorption depended on the polarity of the carboxylic group, the length of the side chain, and the nature of the type of bonding. The adsorption followed both the Langmuir and the Freundlich adsorption equations.

The adsorption of higher n-alkanols from their dilute solutions in n.heptane[210] exhibited a pronounced step indicating a strong cooperative adsorption mechanism leading to a close-packed monolayer of alkanol moelecules oriented with their chain axis parallel to the graphite basal planes. This unusual adsorption behavior in dilute nonaqueous solutions was attributed to an order-disorder transition of the pore alkanols at the liquid-carbon interface. Urano et al.[18] examined the adsorption of 17 organic compounds, including m-hydroxy benzaldehyde on granulated activated carbons from aqueous solutions, and observed that the adsorption of m.hydroxy benzaldehyde was reversible. This was attributed to the presence of a phenolic hydroxyl group in m.hydroxy benzaldehyde, which does not act as a strong electron donor.

7.9 MECHANISM OF ADSORPTION OF ORGANICS BY ACTIVATED CARBONS

To propose a simple and coherent mechanism for the adorption of organic compounds from aqueous solutions by activated carbons is not an easy task, as in the case of adsorption of metal cations. The adsorption of cations mainly involves electrostatic attractive or repulsive interactions between the metal ionic species in the solution and the carbon-oxygen surface groups on the carbon surface. The type of these interactions depend on the pH of the solution in contact with the carbon surface, which also determines the carbon surface charge. The dispersive interactions between the ionic species in the solution and graphene layers and the surface area of the carbon played a smaller role in the adsorption of metal cations. In the adsorption of organic compounds, however, the situation is completely different. The organic compounds present in water can be polar as well as nonpolar so that not only electrostatic interactions but also dispersive interactions will play on important role. In addition, the hydrogen bonding will be an important consideration in certain cases. The size of the organic molecules also has a very wide variation so that the porous structure of the activated carbon, which includes the existence of mesopores and micropores, shall also be an important consideration for the adsorption of organic compounds, because a certain proportion of the microporosity may not be accessible to very large organic molecules. Furthermore, the pH of the solution of organic compounds in contact with the carbon has seldom been mentioned.

This makes it difficult to asses the role of the carbon surface charge on the adsorption of organics from aqueous solutions. Thus, the adsorption of organics from aqueous solutions by activated carbons has been interpreted differently by different workers. A very comprehensive review of the adsorption of organics and inorganics from aqueous solutions citing more than 770 research papers has been published by Radovic et al.[211] In this review the role of these different types of interactions has been critically examined. Thus, we shall only include some of the more important postulations regarding the mechanisms of adsorption of organic compounds.

In the case of the adsorption of chlorinated hydrocarbons, most of the adsorption data has been explained by the Langmuir and Freundlich adsorption isotherm equations. There are conflicting views regarding the parameters that determine the adsorption. For example, Ishizaki et al.[8] while studying the adsorption of chloroform on an as-received activated carbon and after heat treatment at 1000°C, observed that the degassed sample had a stronger affinity for chloroform than the as-received sample. Because the degassing of the carbon did not appreciably change the surface area or the pore-size distribution of the carbon, the interaction between the chloroform and the carbon surface played an important role. Nakamura et al.[9] in a similar study of the adsorption of chloroform suggested that the adsorption did not depend upon the surface area, the pore volume, or the pH of the solution but depended on the hydrophobicity of the carbon surface. In the case of trichloroethylene (TCE) and trichlorobenzene (TCB), the adsorption depended on the surface area and the size of pores on the carbon surface, and it was strongly influenced by the presence of acidic surface groups on the carbon surface,[15] but the adsorption of chlorobenze and dichlorobenzene was determined by surface area and the microporosity of the carbon.[20]

The adsorption of natural organic matter such as humic acid and lipids that have large molecular dimensions but also have polar groups in their molecule, has been attributed to the pore-size distribution in carbons and to the nature of the carbon surface charge.[29,30] The former determines the availability of the pore for adsorption, and the latter determines the electrostatic interactions between the carbon surface and the adsorbate molecule. Many natural organic materials have molecular diameters of about 3 nm so that only activated carbons with a large proportion of mesopores will be more suitable for their adsorption.

The adsorptive removal of phenols from aqueous solutions by activate carbons is the most studied group of organic compounds. However, there is still no single coherent mechanism suggested for their adsorption. The adsorption data has been interpreted differently. Mattson et al.[109] and Moreno-Castilla[64] attribute adsorption of phenols to charge transfer interactions between the adsorbate and the surface carbonyl groups or between the adsorbate and the fused-ring system of the basal planes. The carbon-oxygen groups on the carbon surface act as electron donors and the aromatic ring of the solute is the acceptor. Coughlin et al.[82] and Graham[83] are of the view that the presence of carbon-oxygen surface functional groups also influence the adsorption of phenols through the formation of a bond with the nonacidic oxygen surface groups. Gonzalez-Martin et al.[122] Marsh and Campbell,[65] and Puri et al.[55] observed that the adsorption at low and moderate concentrations was dependent on the surface area and the microporous structure of the carbon. Bansal and coworkers[104] and Goyal,[113] on the other hand, feel that the adsorption of phenols involves interaction between the π electrons of the benzene ring in phenols and the partial positive charge on the carbonyl group. Furthermore, although it has been observed by a large number of these workers that the presence of acidic surface groups suppresses the adsorption of phenols, the electrostatic repulsive interactions between the negatively-charged carbon surface and the negatively-charged phenolate ions in the solution phase have rarely been mentioned as a reason for this suppression. Phenol is a weak acid and can ionize to produce negatively-charged phenolate ions.

The adsorption of several aliphatic amines has been attributed by Puri et al.[141] to a simple acid-base neutralization reaction between the acidic groups on the carbon surface and the basic group of the amine, although Anderson and Emmett[140] attributed the adsorption of methyl amine partly to acid surfaces groups and partly to hydrogen bonding. Radovic et al.[136] on the other hand, is of the view that the adsorption of aniline by carbon involves both electrostatic and dispersive interactions depending upon the pH of the solution. Although at low pH values of the solution close to ZPC, where the carbon surface has a negative charge, the adsorption involves attractive electrostatic interactions between the negatively-charged carbon surface and the aniline cations at higher pH values, where aniline predominantly exists as unionized neutral molecules, the adsorption involves dispersive interactions between the carbon surface and the electrons in the graphene layers. Similarly, in the case of nitrobenzene, which exists in the molecular state, the adsorption mainly involves dispersive interactions because the electrostatic interactions are at a minimum.

The adsorption of dyes, surfactants, and several other polar compounds has indicated that the adsorption is determined by the electrostatic interactions between the carbon surface and the adsorbate cation or the anions. Thus, the mechanism of adsorption of organic compounds involves one or several of the three interactions:

- Dispersive interactions between the aromatic ring and the π electrons of the graphenes
- Electron donor-acceptor interactions between the aromatic ring and the nonacidic surface sites
- Electrostatic attractive or repulsive interactions between the charged carbon surface and the ions in the solution.

In addition, the surface area and the microporosity also play an important role in the adsorption of nonpolar and nonionizable compounds. The molecular dimensions of the organic compound shall become important for larger molecules when some of the micropores may become inaccessible.

REFERENCES

1. Rook, J.J., *J. Water Treatment Exam.*, 23, 234, 1974.
2. Glaze, W.H., Andelman, J.B., Bull, R.J., Conolly, R.B., Hertz, C.D., Hood, R.D., and Pegram, R.A., *J. AWWA*, 3, 53, 1993.
3. Minear, R.A. and Amy, G.L., in *Disinfection By-products in Water Treatment, The Chemistry of Their Formation and Control*, Lewis Publishers, Boca Raton, FL, 1996.

4. Kitis, M., Karanfil, T., Kilduff, J.E., and Wigton A., *Water Sci. Technol.*, 43, 9, 2001.
5. Cotruvo, J.A. and Wu, C., in *Activated Carbon Adsorption*, Vol. I, I.H. Suffet and M.J. McGuire, Eds., Ann Arbor Science Publishers, Ann Arbor MI, 1980, p. 1.
6. Druley, R.M. and Ordway, G.L., *The Toxic Substance Control Act*, Washington, D.C.: Bureau of National Affairs, 1977.
7. Youssefi, M. and Faust, S.D., in *Activated Carbon Adsorption*, Vol. 1, I.H. Suffet and M.J. McGuire, Eds., Ann Arbor Science Publishers, Ann Arbor MI, 1981, p. 133.
8. Ishizaki, C., Marti, I., and Ruiz, M., in *Advances in Chemistry*, No. 202, I.H. Suffet and M.J. McGuire, Eds., *Amer. Chem. Soc.*, Washington, D.C., 1983, p. 95.
9. Nakamura, T., Kawasaki, M., Tanada, S., Kawatak, K., and Imaki, M., *Toxicol. Environ. Chem.*, 57, 187, 1996.
10. Rey, F. and Oles, V., *Textile Veredhing*, 32, 252, 1997.
11. Yu, J. and Chou, S., *Proc. Contam. Site Rem. Conf.*, 1999, p. 671.
12. Li, J.W., Yu, Z., and Gao, M., *Water, Air, Soil Pollutant*, 97, 367, 1997.
13. Li, J.W., Yu, Z., Gao, M., Cai, Z., and Chao, F., *J. Environ. Sci.*, 8, 167, 1996.
14. Uchida, M., Nakamura, T., Kawasaki, N., and Tanada, S., *Bull. Environ. Contam. Toxicol.*, 59, 6, 935, 1997.
15. Karanfil, T. and Kilduff, J.E., *Environ. Sci. Technol.*, 33, 3217, 1999.
16. Bansel, R.C., Aggarwal, D., Goyal, M., and Kaistha, B.C., *Indian J. Chem. Technol.*, 9, 290, 2002.
17. Puri, B.R., Bhardwaj, S.S., and Gupta, W.J., *Indian Chem. Soc.*, 53, 1095, 1976.
18. Urano, K., Kano, N., and Tabota, T., *Bull. Chem. Soc. Japan*, 57, 2307, 1984.
19. Pendelton, P., Wong, S.H., Shumann, R., Levay, G., Denoyel, R., and Rouquerol, J., *Carbon*, 35, 1141, 1997.
20. Xiao, B., Zhao, X., Yavuz, R., and Thomas, K.M., *Carbon 01, Intern. Conf. on Carbon*, Lexington, KY, 2001, Paper 35.6.
21. Chagar, H.K., Ndaje, F.E., Sykesand, M.L., and Thomas, K.M., *Carbon*, 33, 1405, 1995.
22. Fletcher, A.J. and Thomas, K.M., *Langmuir*, 15, 6908, 1999.
23. Love, O.T., Jr., *Appendix 3 to Interim Treatment Guide for the Control of Chloroform and Other Trihalomethanes*, EPA, Cincinnati, OH, 1976.
24. O'Connor, J.R., Badorek, D., and Popaliski, J.R., *Proc. 97th American Water Works Association Conf.*, Anaheim CA, 1977.
25. Smith, J.D., Zogorski, J.S., Wilding, D.A., and Arbuckle, A.N., *Proc. 97th American Water Works Association Conf.*, Anaheim, CA, 1977.
26. Mullins, R.L., Jr, Jogorski, J.S., Hubbs, S.A., and Allgerei, G.D., in *Activated Carbon Adsorption*, Vol. I, I.H. Suffet and M.J. McGuire, Eds., Ann Arbor Science Publishers, Ann Arbor MI, 1981, p. 273.
27. Hayer, M.H.B., McCarthy, P., Malcolin, R.L., and Swift, R.H., in *Humic Substances II: In Search of Structure*, H.B. Michel, P. McCarthy, R.L. Malcoln, and R.H. Swift, Eds., John Wiley and Sons, New York 1989, Chapter 24, p. 689.
28. Thurman, E.M. in *Organic Geochemistry of Natural Waters*, Martinus Nijoff, Dr. Junk Publishers, Dordracht, The Netherlands, 1985.
29. Karanfil, T., Kitis, M., Kilduff, J.E., and Wigton, A., *Environ. Sci. Technol.*, 33, 3225, 1999.
30. Newcomb, G., *Water Sci. Technol.*, 40, 191, 1999.
31. Weber, W.J., Jr., Pirbazari, M., Long, J.B., and Barton, D.A., in *Activated Carbon Adsorption*, I.H. Suffet and M.J. McGuire, Eds., Ann Arbor Science Publishers, Ann Arbor, MI, Vol. I, 1980, p. 317.
32. Randke, S.J. and Japson, C.P., *J. Am. Water Works Assoc.*, Feb. 1982, p 84.
33. Weber, W.J., Jr., Voice, T.C., and Jodellah, A., *J. Am. Water Works Assoc.*, Dec. 1983, p. 612.

34. Lafrance, P. and Mazet, M., *J. Am. Water Works Assoc.*, April 1989, p. 155.
35. Vidic, R.D. and Suidan, M.T., *Water Environ. Res.*, 67, 892, 1995.
36. Warta, C.L., Papadimas, S.P., Sorial, G.A., Suidan, M.T., and Speth, T.F., *Water Res.*, 29, 562, 1995.
37. Cerminara, P.J., Sorial, G.A., Papadimas, S.P., Suidan, M.T., Moteleb, M.A., and Speth, T.F., *Water Res.*, 29, 409, 1995.
38. Lee, M.C., Snoeyink, V.L., and Crittenden, J.C., *J. Am. Water Works Assoc.*, Aug. 1981, p. 440.
39. Karanfil, T., Schlautman, M.A., Kilduff, J.E., and Weber, W.J., Jr., *Environ. Sci. Technol.*, 30, 2187, 1996.
40. Kilduff, J.E., Karanfil, T., Chin, Y.P., and Weber, W.J., Jr., *Environ. Sci. Technol.*, 30, 1336, 1996.
41. Kilduff, J.E., Karanfil, T., and Weber, W.J., Jr., *Environ. Sci. Technol.*, 30, 1344, 1996.
42. Uterera-Hidalgo, E., Moreno-Castilla, C., Rivera-Utrilla, J., Ferro-Garcia, M.A., and Carrasco-Morina, F., *Carbon*, 30, 107, 1992.
43. Schmidt, C.E., Sherkey, A.G., and Friedel, R.A., in *Mass Spectrometric Analysis of Product Water from Coal Classification*, U.S. Bureau of Mines, Pittsburgh Energy Research Center Technical Progress Report No. 86, 1974.
44. Jaim, S.J. and Snoeyink, V.L., *45th Ann. Conf. Water Pollution Control Federation*, Georgia, Oct. 8–13, 1972.
45. Knadarov, E.I. and Verteshev, M.S., *Vodosnabzh Sanit. Tekh.*, 11, 1, 1972.
46. Mironova, N.A., Zilberman, A.G., Lichagina, V.I., Kreich, Z.A., and Eppel, S.A., *Koks Khim.*, 11, 34, 1979.
47. Rodriguez, A.F., Costa, C.A., and Almeida, F., *Proc. Water Industries Conf.*, Brighton, U.K., 1981, p. 420.
48. Bethel, J.B., Brit. Patent 1537835, Jan. 10, 1979, Appl. 75/20 May 15, 1975.
49. Radeke, K.H., Jung, R., Seidel, A., and Marutovskii, R.M., *Chem. Tech.*, 42, 335, 1990.
50. Kitagara, H., Kugai, 10, 11, 1975.
51. Scharifov, R.R., Schultz, E.V., and Boudaresa, N.I., *Azerb. Khim. Zh.*, 3, 400, 1970.
52. Chakravorti, R.K. and Weber, T.W., *AIChE. Symp. Ser.*, 71, 392, 1975.
53. Juang, R.S., Wu, F.C., and Tsang, R.I., *J. Chem. Engg. Data*, 41, 487, 1996.
54. Jossens, L., Prausnitz, J.M., Fits, W., Schlunder, E.U., and Meyers, A.L., *Chem. Eng. Sci.*, 33, 1106, 1978.
55. Puri, B.R., in *Activated Carbon Adsorption*, I.H. Suffet and M.J. McGuire, Eds., Ann Arbor Science Publishers, Ann Arbor MI, 1980, Vol. I, p. 353.
56. Weber, W.J., Jr., and Moris, S.C., J., *San. Eng. Div. Am. Soc. Civil Eng.*, 90, 79, 1964.
57. Seidi, J. and Kriska, F., *Angew Macromol. Chem.*, 28, 87, 1973.
58. Drozhalina, N.E and Bulgakova. N.O., *Zh. Prikl Khim.*, 47, 298, 1971.
59. Brand, P., Koenig, A., and Lachman, R., *Freiberg Forchungsh.*, A 16, 125, 1979.
60. Jankowska, H, Swiatowkski, A., Radeke, K.H., and Seidel, A., *Chem. Stosow.*, 33, 585, 1989.
61. Gruszek, A.J., Partyka, S., and Raguel, P., *20th Bienn. Conf. on Carbon*, Univ. of California, Santa Barbara, CA, June 23–28, 1998, Ext. Abstracts, p. 58.
62. Koltesev, A.V., Korolev, Yu G., and Syskov, A.I., Deposited Doc., Viniti, 5, 1627, 1978.
63. Badnar, J. and Nagi, L.G., *Period. Polytech. Chem. Eng.*, 20, 187, 1976.
64. Moreno-Castilla, C., Rivera-Utrilla, J., Lopez-Ramon, M.V., and Corresco-Marin, F., *Carbon*, 33, 845, 1965.
65. Marsh H. and Campbell, H.G., *Carbon*, 9, 489, 1971.

66. Singh, D.D., *Indian J. Chem.*, 9, 1369, 1971.
67. Economy, J. and Lin, R.Y., *Appl. Polymer Symp.*, 29, 199, 1976.
68. Dondi, F., Betti, A., Blo, G., and Bighi, C., *Annali Chim.*, 68, 293, 1978.
69. Askawa, T. and Ogino, K., *J. Colloid Interf. Sci.*, 102, 348, 1984.
70. Barton, S.S., Evans, M.J.B., and MacDonald, J.A.F., *Polish J. Chem.*, 71, 651, 1996.
71. Kilduff, J.E. and King, C.J., *Ind. Eng. Chem. Res.*, 36, 1603, 1997.
72. Le Cloirec, P., Brasquet, C., and Subrenat, E., *Energy Fuels,* 11, 331, 1997.
73. Aytekin, C., *Spectrosc. Lett.*, 25, 653, 1991.
74. Chaplin, J., *Phys. Colloid Chem.*, 36, 909, 1932.
75. Kiselev, A.V. and Krasilinkov, D.K., *Dokalady Akad. Nauk, USSR,* 86, 111, 1952.
76. Singer, P.C. and Yen, C. Yu, in *Activated Carbon Adsorption*, I.H. Suffet and M.J. McGuire, Eds., Ann Arbor Science Publishers, Ann Arbor MI, 1980, Vol. I, p. 167.
77. Urano, K., Kano, H., and Tobato, T., *Bull Chem. Soc. Japan*, 57, 2307, 1984.
78. Radeke, K.H., Seidel, A., Spitzer, P., Jung, R., Jankowska, H., and Noffe, S., *Chem. Tech.*, 41, 32, 1989.
79. Hanmin, Z. and Yigun, L., *Shvi Chuli Jishu*, 13, 342, 1987.
80. Mahajan, O.P., Castilla, M.C., and Walker, P.L., Jr., *J. Sep. Sci. Technol.*, 15, 1733, 1980.
81. Keizo, O., Hiroyuki, T., Kiyoski, Y., and Hiroshi, T., *Nippon Kagaku Kaishi*, 3, 321, 1981.
82. Coughlin, R.W., Ezra, F. S., and Tan, R.N., *J. Colloid Interf. Sci.*, 28, 386, 1968.
83. Graham, D., *J. Phys. Chem.*, 59, 896, 1955.
84. Clauss, A., Boehm, H.P., and Hofman, U., *Z. Anorg. Chem.*, 35, 290, 1957.
85. Peel, R.G. and Benedek, A., *Environ. Sci. Technol.*, 14, 66, 1980.
86. Magne, P. and Walker, P.L., Jr., *Carbon,* 24, 101, 1986.
87. Coughlin, R.W. and Tan, R.N., *Chem. Eng. Prog. Symp. Sci.*, 64, 207, 1968.
88. Coughlin, R.W., Ezra, F.S., and Tan, R.N., *J. Colloid Interf. Sci.*, 28, 386, 1968.
89. Mattson, J.S., Mark, H.B., Jr., Mulbin, M.D., Weber, W.J., Jr., and Crittenden, J.C., *J. Colloid Interf. Sci.*, 31, 116, 1969.
90. Epstein, B.D., Dolle-Mole, E., and Mattson, J.S., *Carbon,* 9, 609, 1971.
91. Vidic, R.D., Suidan, M.T., and Brenner, R.C., *Environ, Sci. Technol.*, 27, 2069, 1993.
92. Vidic, R.D., Suidan, M.T., Traegnen, D.K., and Nakhla, G.S., *Water Res.*, 24, 1187, 1990.
93. Vidic, R.D. and Suidan, M.T., *Environ. Sci. Technol.*, 25, 1612, 1991.
94. Abuzad, N.S. and Nakhla, G.S., *Environ. Sci. Technol.*, 28, 216, 1994.
95. Grant, T.M. and King, C.J., *Ind. Eng. Chem. Res.*, 29, 264, 1990.
96. Leng, C.C. and Pinto, N.G., *Carbon,* 35, 1375, 1997.
97. Tessmer, C.H., Uranowski, L.J., and Vidic, R.D., *Environ. Sci. Technol.*, 31, 1872, 1972.
98. Teng, H. and Hsieh, C.T., *Ind. Eng. Chem*, 37, 3618, 1998.
99. Teng, H. and Hsieh, C.T., *J. Chem. Technol. Biotechnol.*, 74, 123, 1999.
100. Leyva-Ramos, R., Suto-Zuniga, J., Mendoza-Barron, J., and Guerrero-Coronada, R., *Adsorp. Sci. Technol.*, 17, 533, 1999.
101. Nevskaia, D.M., Sanlianes, A., Munov, V., and Guerrero-Ruiz, A., *Carbon,* 37, 1065, 1999.
102. Oda, H., Kishida, M., and Yokokawa, C., *Carbon,* 19, 243, 1981.
103. Puri, B.R., Bhardwaj, S.S., Kumar, V., and Mahajan, O.P., *J. Indian Chem. Soc.*, 52, 26, 1975.
104. Bansal, R.C., Aggarwal, D., Goyal, M., and Kaistha, B.C., *Indian J. Chem. Technol.*, 9, 290, 2002.

105. Aggarwal, D., Goyal, M., and Bansal, R.C., in *Recent Trends in Carbon*, O.P. Bahl, Ed., Shipra Publishers, Delhi, 1996, p. 244.
106. Bansal, R.C., Bhatia, N., and Dhami, T.L., *Carbon*, 16, 65, 1978.
107. Puri, B.R. and Bansal., R.C., *Carbon*, 1, 457, 1964.
108. Aggarwal, D., Goyal, M., and Bansal, R.C., *Carbon*, 37, 1989, 1999.
109. Mattson, J.S., Mark, H.B., Jr., Mikin, M.D., Weker, W.J., Jr., and Crittenden, J.C., *J. Colloid Interface Sci.*, 31, 116, 1969.
110. Epstein, B.D., Dalle-Molle, E., and Mattson, J.S., *Carbon*, 9, 609, 1971.
111. Boehm, H.P., in *Advances in Catalysis*, Vol. 16, Academic Press, New York, 1966, p. 179.
112. Garten, V.A. and Weiss, D.E., *Rev. Pure Appl. Chemistry*, 1957, p. 769.
113. Goyal, M., *Carbon Science*, 2003 Accepted.
114. Goyal, M., Ph.D. thesis submitted to Panjab University, Chandigarh, India, 1997.
115. Pierce, C. and Smith, R.N., *J. Phys. Colloid Chem.*, 54, 784, 1950.
116. Pierce, C., Smith, R.N., Wiley, J.K., and Corder, H.J., *J. Am. Chem. Soc.*, 73, 455, 195.
117. Dubinin, M.M., Zaverina, E.D., and Serpinski, V.V., *J. Chem. Soc.*, 1955, p. 1760.
118. Dubinin, M.M., *Carbon*, 18, 355, 1980.
119. Puri, B.R., Murari, K., and Singh, D.D., *J. Phys. Chem.*, 65, 37, 1961.
120. Puri, B.R., *Carbon*, 4, 391, 1966.
121. Bansal, R.C., Dhami, T.L., and Prakash, S., *Carbon*, 16, 389, 1978.
122. Gonzalez-Martin, M.I., Valenzuela, C.C., and Gomez-Serrano, V., *Langmuir*, 7, 1269, 1991.
123. Gomez-Serrano, V., Valenzuela, C.C., and Gonzalez-Martin, M.I., *Appl. Surface Sci.*, 74, 337, 1994.
124. Yong, D.R., Keinath, T.M., Poznanska, K., and Ziang, Z.P., *Environ. Sci. Technol.*, 19, 690, 1985.
125. Caturla, F., Martin-Martinez, J.M., Molina-Sabio, M., Rudriguez-Reinoso, F., and Torregrosa, R.J., *Colloid Interface Sci.*, 124, 528, 1988.
126. Wheeler, O.H. and Levy, E.M., *Can. J. Chem.*, 37, 1235, 1959.
127. Deng, X., Yue, Y., and Gao, Z., *J. Colloid Interface Sci.*, 192, 475, 1997.
128. Arafat, H.A., *Langmuir*, 15, 5997, 1999.
129. Laszlo, K., Tombacz, E., Josipovits, K., and Kerepesi, P., *Carbon 01 Intern. Conference on Carbon*, Lexington, KY, 2001, Paper 25.5.
130. Aggarwal, P., Kapoor, S.K., Kapoor, J.C., Bhalla, A.K., and Bansal, R.C., *Indian J. Chem. Technol.*, 3, 187, 1996.
131. Aggarwal, P., Misra, S., Kapoor, S.K., and Bansal, R.C., in *Recent Trends in Carbon*, O.P. Bahl, Ed., Shipra Publ., Delhi, 1996, p. 203.
132. Miyahara, M. and Okazaki, M., *J. Chem. Engg. Japan*, 25, 408, 1992.
133. Haghsereshi, F., Nouri, S., and Lu, G.Q., *Carbon 01 Intern. Conf. on Carbon*, Lexington, KY, 2001, Paper 25.3.
134. Jain, K. and Bryce, A. J., in *Carbon Adsorption Handbook*, P.N. Chemreisinoff and F. Ellerbusch, Eds., Ann Arbor Science Publ., Ann Arbor MI, Chapter 17, p. 661.
135. Aggarwal, P., Misra, S., Kapoor, S.K., Bhalla, A.K., and Bansal, R.C., *Indian J. Chem. Technol.*, 4, 42, 1997.
136. Radovic, L.R., Silva, I.F., Ume, J.I., Menendez, J.E., Leon y Leon, C.A., and Scaroni, A.W., *Carbon*, 35, 1339, 1997.
137. Bornehoff, J., in *Activated Carbon Adsorption*, I.H. Suffet and M.J. Mcguire, Eds., Ann Arbor Science Publishers, Ann Arbor MI, 1980, Vol. 1, p. 145.
138. Fochtman, E.G. and Dobbs, R.A., in *Activated Carbon Adsorption*, I.H. Suffet and M.J. McGuire, Eds., Ann Arbor Science. Publishers, Ann Arbor MI, 1980, Vol. I, p. 157.

139. Kipling, J.J., *J. Chem. Soc.*, 1483, 1948.
140. Anderson, R.B. and Emmett, P.H., *J. Phys. Chem.*, 56, 756, 1952.
141. Puri, B.R., Talwar, C., and Sandle, N.K., *J. Indian Chem. Soc.*, 41, 581, 1964.
142. Capelle, A., in *Activated Carbon: A Fascinating Material*, A. Capelle and F. Devooys, Eds., Norit N.V. Netherlands, 1983, p. 191.
143. Becker, D.L. and Wilson, S.C., in *Carbon Adsorption Handbook*, P.N. Cheremisinoff and F. Ellerbusch, Eds., Ann Arbor Science. Publ., Ann Arbor MI, Vol. I, Chapter 5, p. 167.
144. Prakash, S., *Carbon,* 19, 483, 1974.
145. Klaus, E., GWF, *Gas Wassefach Wasser/Abuasser,* 136, S17, 1995.
146. Ralph, H., Kochler, J., and Gimbel, R.M., *Vom Wasser*, 86, 321, 1996.
147. Ayele, J., Matri, A., and Mazet, M., *Rev. Sci.*, Eau, 8, 355, 1995.
148. Orlindini, E., Siebel, M.A., Graveland, A., and Schippers, G.C., *Water Supply*, 14, 99, 1996.
149. Kauras, A., Zouboulis, A., Samara, C., and Kouimtzis, T., *Chemisphere*, 30, 2307, 1995.
150. Kauras, A., Zouboulis, A., and Samara, C., *Environ. Pollut.*, 103, 193, 1999.
151. Matsui, Y., Kamer, T., Yuasa, A., and Tambo, N., *Water Supply*, 14, 31, 1996.
152. Khappe, D.R.V., Snoeyink, V.L., and Prades, M.J., *Proc. Ann. Conf., Am. Water Works Assoc.*, 1995, p. 754.
153. Ashley, J.K., and Dvorak, B.I., *Proc. Natl. Conf. Environ. Eng.*, 1998, p. 338.
154. Arora, V.M., Ph.D. thesis, Panjab University, Chandigarh, India, 1977.
155. Sastri, M.V.C., *Indian Instt. Sci. Quart. J.*, 5, 145, 1942.
156. Sastri, M.V.C., *Indian Instt. Sci. Quart. J.*, 5, 162, 1942.
157. Perineau, F., Malinier, J., and Gaset, A., *J. Chem. Technol. Biotechnol.*, 32 749 1982.
158. Dai, M., *J. Colloid Interface Sci.*, 164, 223, 1994.
159. Dai, M., *J. Colloid Interface Sci.*, 198, 6, 1998.
160. Dai, M., He, L., and Lin, T., *Yinguong Huaxue*, 12, 62, 1995.
161. Gao, S., Tanada, S., Abe, I., Kitigara, M., and Matsubara, Y., *Tanso*, 163, 138, 1994.
162. Listiskaya, I.G., Lazarev, L.P., Khavalov, V.V., and Makovetskaya, G.D., *Zh. Prikil Khim.*, 65, 2575, 1942.
163. Al-Degs, Y., Khraisheh, M., Allen, S., and Ahmed, M., *Adv. Environ. Res.*, 3, 132, 1999.
164. Al-Degs, Y., and Khraisheh, M., *Water Res.*, 34, 927, 2000.
165. Chen, S., Lu, G., and Jiao, L., *Huan Yu Xifu*, 16, 267, 2000.
166. Chen, S. and Wu, C., *Hecheng Xianwci Gongue*, 21, 22, 1998.
167. Honas, A. and Bakir, I., *J. Chim. Phys-Chim. Biol.*, 96, 479, 1999.
168. Khalil, L.B. and Girgis, B.S., *Bull. Nat. Res. Catalysis*, Egypt, 22, 349, 1997.
169. Mittal, A.U. and Venkovachar, C., *Indian J. Environ. Health*, 31, 105, 1989.
170. Viraraghavan, T. and Mihial, D.J., *Fresenius Environ. Bull.*, 4, 346, 1995.
171. Lin, S.H., *J. Chem. Technol., Biotechnol.*, 57, 387, 1997.
172. Lin, S.H., *J. Chem. Technol., Biotechnol.*, 58, 159, 1997.
173. McKay, G. and Al Duri, B., *Coulorage*, 35, 24, 1998.
174. McKay, G. and Al Duri, B., *Ind. Eng. Chem*, 30, 385, 1991.
175. McKay, G. and Al Duri, B., *Environ. Prot. Eng.*, 14, 15, 1989.
176. McKay, G. and Al Duri, B., *Colour Ann.*, p. 23, 1990.
177. Al Duri, B. and McKay, G., *Carbon*, 29, 191, 1991.
178. Dusant, O., Masmier, D.P., and Seapand, B., *Trib. Eau*, 44, 15, 1991.
179. Sasaki, M., Tamai, H., Yoshida, T., and Yasuda, H., *Tanso*, 183, 151, 1998.
180. Miau, R. and Dai, M., Fuzhou Daxue Xeubao, *Ziran Kexueban*, 24, 83, 1996.

181. Miau, R. and Dai, M., *Wuli Huaxue Xeuban*, 12, 173, 1996.
182. Juang, R.S. and Swei, W.L., *Sep. Sci. Technol.*, 31, 2143, 1997.
183. Nandi, S.P. and Walker, P.L., Jr., *Fuel*, 50, 345, 1971.
184. Guzel, F., *Sep. Sci. Technol.*, 31, 283, 1996.
185. Tamai, H., Yoshida, T., Sasaki, M., and Yasuda, H., *Carbon,* 37, 983, 1999.
186. Kasaoka, S., Sabata, Y., Tanaka, E., and Naitoh, R., *Nihon Kagaku Kaishi*, 1987, p. 2260.
187. Aggarwal, D., Ph.D. thesis, Panjab University, Chandigarh, India, 1997.
188. Teryzyk, A.P., *Adsorp. Sci. Technol.*, 17, 441, 1999.
189. Sobezak, H., *Acta Pol. Pharm.*, 56, 187, 1999.
190. Bao, J. and Wang, W., *Zhonggue Yiyuan Yuaxue Zazki*, 19, 409, 1999.
191. Dutta, M., Barriah, R., Dutta, N.N., and Ghosh, C., *Colloids Surf.*, A. 127, 25, 1997.
192. Bredberry, S.M. and Allister, V.J., *J. Toxicol. Clin. Toxicol.*, 33, 407, 1995.
193. Volans, G.M., Vole, J.A., Crème, P., Widdop, B., and Goulding, R., in *Artificial Organs*, R.M. Kendi et al., Eds., McMillan, London, 1977, p. 178.
194. Chang, T.M.S., Can, *J. Physiol. Pharmacol.*, 47, 1043, 1969.
195. Yatzidis, H., *BOC Eur. Dial Transpt. Assoc.*, 83, 1, 1964.
196. Smith, E.M., Affrossman, S., and Courtney, J.M., *Carbon,* 17, 149, 1979.
197. Edgar, G. and Shiver, H.E., *J. Am. Chem. Soc.*, 47, 1179, 1925.
198. Rybolt, T.R., Burrell, D.E., Shults, J.M., and Kelly, A.K., *J. Chem. Educ.*, 65, 1009, 1988.
199. Tamamushi, B., in *Adsorption from Solutions*, R.H. Ottewill, C.H. Rochester, and A.L. Smith, Eds., Academic Press, New York, 1983.
200. Allali, H.M., Dusart, O., and Mazet, M., *Water Res.*, 24, 699, 1990.
201. Smolin, S.K., Tinoshenko, M.N., and Klimenko, N.A., *Khim. Tekhnol. Vody.*, 13, 495, 1991.
202. Chobamu, M.M., Kesdivarenko, M.A., and Ropot, V.M., *Izu Akad. Nauk. Mold SSR Biol. Ikhim Nauk.*, 3, 49, 1990.
203. Gao, Y., Farber, M., Chem. X., Suuberg, E.M., and Hurt, R.H., *Carbon '02, Intern. Conf. on Carbon*, Beijung, 2002, Paper 29.3.
204. Zouboulis, A.I., Lazaridis, N.K., and Zamboulis, D., *Sep. Sci. Technol.*, 29, 385, 1994.
205. Garcia-Garcia, A., Gregorio, A., Franco, C., Pinto F., Baavida, D., and Gulyurthi I., *Carbon 01 Intern. Conf. on Carbon*, Lexington, KY, 2001.
206. Kamegawa, K., Nishikubo, K., Kodama, M., Adachi, Y., Imamura, T., and Yoshida, H., *Carbon '02, Intern. Conf. on Carbon*, Beijung, 2002, paper180, D683.
207. Nawack, K.O., Cannon, F.S., and Mazyck, D.W., *Carbon 01 Intern. Conf. on Carbon*, Lexington, KY, 2002, Paper 25.1.
208. Ishizaki, C. and Cookson, J.T., in *Chemistry of Water Supply Treatment, and Distribution*, A.J. Rubin, Ed. Ann Arbor Science. Publ., Ann Arbor MI, 1980, p. 201.
209. Usmani, T. and Wahab, A.T., *J. Chem. Soc.,* Pakistan, 20, 1, 1998.
210. Furdenegg, G.H., Koch, C., and Liphard, M., in *Adsorption from Solutions*, R.H. Ottewill, C.H. Rochester, and A.L. Smith, Eds., Academic Press, New York, 1983, p. 79.
211. Radovic, L.R., Moreno-Castilla, C., and Rivera-Utrilla, J., in *Chemistry and Physics of Carbon*, L.R. Radovic, Ed. Marcel Dekker, New York, Vol. 27, p. 227.
212. Goyal, M., Singh, S., and Bansal, R.C., *Carbon Science*, 5, 170, 2004.

8 Activated Carbon Adsorption and Environment: Removal of Hazardous Gases and Vapors

Rapid development of chemical, pharmaceutical, and other process industries, and the ever-increasing number of automobiles on the roads, together with emissions from the combustion of fossil fuels in power plants and similar industrial processes, is releasing large amounts of hazardous gases and vapors into the air. This puts considerable pressure on the availability of clean air. These effluents contain small amounts of a large number of volatile organic compounds (VOC). These flue gases also contain small amounts of oxides of sulfur and oxides of nitrogen, the most important being SO_2 and NO. These gases combine with VOC to produce smog and sometimes also produce acid or smog clouds. One such smog cloud was seen over London in 1950, and it caused several thousand human deaths due to respiratory problems.

Consequently, there has been a strong push to reduce these gases and particulate matter from domestic and industrial combustion processes. Because the regulations for pollution control are becoming more and more stringent, a lot of research effort has been directed to remove these gases and vapors from flue gases. The effort has been directed into two directions: usage of clean fuels that discharge only very small amounts of these gases and vapors, and the treatment of flue gases. Several methods have been tried, such as wet- or dry-scrubbing for the removal of SO_2, and the introduction of low nitrogen-oxide burners. Catalytic converters are also used in automobiles for the conversion of more harmful gases into harmless (or less harmful) constituents. But these conventional methods have their associated disadvantages. Consequently, a considerable amount of work has been carried out for removing these gases and vapors by adsorption. Activated carbons are unique adsorbents that have a great potential for the treatment of fuel and indoor gases before discharging into the air or recycling.

8.1 REMOVAL OF VOLATILE ORGANIC COMPOUNDS AT LOW CONCENTRATIONS

Low-concentration VOCs are present in the effluent gases from the vent stacks from flaxiographic printing, from paint booths in automotive assembly plants, and from bakeries. The major VOCs are acetates and alcohols, and volume flows are typically

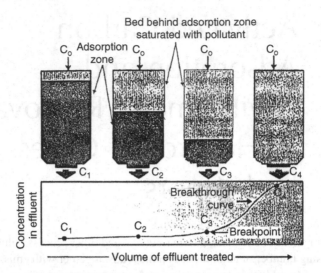

FIGURE 8.1 Progression of adsorption front through the adsorption bed. (After Derbyshire, F., Jagtoyen, M., Andrews, R., Rao, A., Martin-Gullon, I., and Grulke, E.A., in *Chemistry and Physics of Carbon*, L.R. Radovic, Ed., Marcel Dekker, New York, 2001, Vol. 27, p. 1. With permission.)

in the 10,000 to 14,000 kg/hr range. Bakeries alone release about 3 kg ethanol/1000 kg of processed dough.[1] Styrene, a common monomer that is used in a variety of process industries including the manufacture of fiberglass-reinforced products such as recreational and sports vehicles and car and truck body parts, is another important VOC. The consumption of styrene was about 4.0 billion kg in 1989,[2] and the reported styrene emissions were about 15 million kg/yr in 1990.[1] For the removal of low-concentration VOCs, there is need to develop new technologies because the airflow is high, so the contact time of the polluted air with the activated carbon adsorbent has to be large, which results in a high pressure drop in the carbon bed. Furthermore, the adsorption zone, which is also called *mass transfer zone* (MTZ) and is the "region of the bed between the activated carbon which is already saturated and the point where the gas phase concentration of the adsorbate VOC has the maximum acceptable limit in the effluent stream" (Figure 8.1) has to be large. The length of the adsorption region determines the adsorption efficiency of the activated carbon bed. When granulated activated carbons are used, the rate of adsorption is low, because it is determined by interparticulate diffusion process.

Because the removal of VOC from the effluents is gaining importance due to stringent clean air regulations, attempts are being made to develop newer technologies. In one recent technology used for the removal of styrene emissions from plastic molding industry, the VOCs are concentrated in a rotary concentrator made of activated carbon or a substrate impregnated with activated carbon. The concentrated VOCs are then destroyed by thermal or catalytic oxidation. In the rotary concentrator (Figure 8.2) the wheel rotates slowly with a speed of 1 to 3 rotations per hr so that about 90% of its face is exposed to the incoming air stream, and the remainder about 10% is under

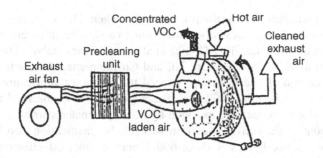

FIGURE 8.2 Schematic of rotary concentration for VOC recovery. (After Derbyshire, F., Jagtoyen, M., Andrews, R., Rao, A., Martin-Gullon, I., and Grulke, E.A., in *Chemistry and Physics of Carbon*, L.R. Radovic, Ed., Marcel Dekker, New York, 2001, Vol. 27, p. 1. With permission.)

regeneration where hot air flows in the counter direction to the VOC laden air. This rotary adsorber can increase the concentration of VOC by a factor of 100.[2] The process can also be used for emissions from automotive paint booths, where modular rotary concentrators are used and the VOCs are captured and destroyed by thermal oxidizers.

Jagtoyen and Derbyshire[3] prepared rigid activated carbon fiber (ACF) composites by incorporation of ACF in forms, such as felt paper and rigid monoliths, and examined their potential for the removal of low-concentration VOCs. The VOCs were easily accessible to the activated carbon fiber surface and presented little resistance to the flow because of the open internal structure of the composite. These workers also compared the ability of the same weights of GAC and an ACF composite for the removal of butane with a concentration of 20 ppm in a flow of nitrogen carrier gas. The uptake of butane at breakthrough was twice on the composite as on the GAC. Furthermore, the ACF composite was not susceptible to the attrition problems associated with packed granular beds during adsorption desorption cycles.

8.2 REMOVAL OF OXIDES OF NITROGEN FROM FLUE GASES

In addition to SO_2, flue gases from the combustion of fossil fuels in power plants and similar industries also contain other noxious gases such as oxides of nitrogen, NO_x (e.g., NO, NO_2, or N_2O). Typical flue gas concentrations of nitrogen oxides range of 400 to 1500 ppm. Furthermore, the low-level emissions from automobiles are a large source of air pollution, especially in cities where millions of vehicles are on the roads. Japan and several other countries have made stringent regulations for bringing down these noxious emissions in the flue gases below a certain limit.

Activated carbon when used can work as a reducing agent, as a catalyst, and as an adsorbent. When used as a catalyst, the activated carbons are impregnated with metals that can reduce the temperature required for the reduction of the oxides of nitrogen to nitrogen. The reduction in temperature could be 600 to 700°C for an uncatalyzed reaction to as low as 300°C for a catalyzed reaction. Copper is one of the metals which, when impregnated on the surface of carbon, can significantly enhance the reaction rate[4] in the presence of oxygen. However, in the absence of oxygen the reduction process does not occur to a significant extent, regardless of the presence of the metal. Illan-Gomez et al.[5] studied the catalytic effect of several metals such as K, Ca, Cr, Fe, Co, Ni, and Cu impregnated on activated carbon, and observed that all these metals reduced the reaction temperature although K, Co, and Fe were the most effective. These workers[6] also observed that nitrogen oxides can also be reduced to nitrogen by nonimpregnated carbons, but an appreciable amount of the activated carbon was lost by gasification into CO and CO_2 because the reaction begins at about 600°C and becomes effective only at 700°C. They also found that the presence of oxygen considerably enhanced the process, but this also increased the rate of gasification of the carbon and, consequently, the loss of carbon.

The activated carbon can also catalyze the removal of nitrogen oxides by reaction with reducing gases such as ammonia, CO, and hydrogen, although ammonia has generally been used. Because NO is the major component of nitrogen oxides in the flue gases, its removal by reaction with ammonia on the carbon surface has been largely studied, and several selective catalytic reduction (SCR) catalysts have been developed that are typically metal-impregnated active carbons. The reaction can be carried out in the temperature range 100 to 150°C[7] in the presence or absence of oxygen.

$$6NO + 4NH_3 \xrightarrow{\text{active carbon}} 5N_2 + 6H_2O$$

$$4NO + 4NH_3 + O_2 \xrightarrow{\text{active carbon}} 4N_2 + 6H_2O$$

Several workers have used activated carbon fibers (ACF) as catalyst supports for the preparation of SCR catalyst for the reduction of NO. Kaneko et al.[8] studied the adsorption of NO on ACF impregnated with copper and iron salts, and observed that the presence of these salts considerably increased the amount of NO adsorption. Yoshikawa et al.[9] impregnated coal tar pitch-based activated carbon fibers with Co, Fe, and Mn salts by a pore volume impregnation technique and studied the SCR activity of these catalysts for NO reduction. Mn_2O_3 impregnated ACF after heat treatment showed higher activity for low-temperature SCR compared to other transition metal oxides such as Fe_2O_3 and cobalt oxide formed on the ACF surface. When compared to similarly impregnated granulated activated carbons, the ACF supports were found to be more effective. The ACF supports also showed a low degree of oxidation and negligible formation of nitrous oxide during the catalytic reaction. Several other studies[10–12] also showed that the activity of manganese oxide supported on carbons was higher.

Westwood et al.[13] prepared SCR catalyst for NO reduction by impregnating a surface oxidized activated carbon fiber from aqueous solutions of ferric nitrate, nickel nitrate, or copper nitrate followed by calcinations at 300°C under nitrogen for 2 hr. The catalytic activity of these metal loaded ACFs was determined in a plug-flow microreactor using line FTIR analysis of the feed and flue gases containing 800 ppm

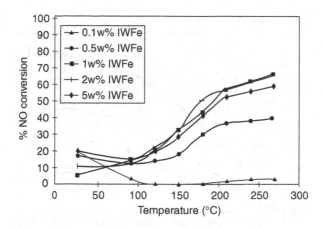

FIGURE 8.3 Effect of variation in metal loading on SCR activity of iron impregnated ACF catalysts. (After Westwood, A.V.K., Modill, D., Hampastsoumian, E., and Jones, J.M., *Carbon 02, Intern. Conf. on Carbon,* Beijung, 2002, Paper 39.2. With permission.)

NO, 800 ppm NH_3, and 0.6 volume percent oxygen with helium as the carrier gas. The temperature range studied was 25 to 270°C at a flow rate of 100 ml/min. Unoxidized ACF appeared to be a little more active catalyst than the surface oxidized ACF probably due to screening of some of active carbon sites, by oxygen containing surface functional groups in the latter case. In the case of loading by iron, the SCR activity at temperatures up to 250°C reached a plateau at a loading of 1% by weight (Figure 8.3). This has been attributed to the filling of available oxygenated sites on the carbon surface by the well dispersed iron particles which offered maximum catalytic activity. When the loading was increased, the catalytic activity went down due to

- Blocking of the ACF pores by the impregnated metal so that the reactant gases cannot enter the pores
- Agglomeration of the metal species resulting in a loss of surface area available to the reactants
- Sintering of the metal particles chemisorbed at the oxygenated sites created during the oxidation of the ACF

In the presence of oxygen the SCR activity mechanism involved adsorption of NO at the metal sites followed by oxidation of adsorbed NO to NO_2. The adsorbed NO_2 was then reduced to nitrogen gas and water by ammonia and the products were desorbed.

The conversion of NO into NO_2 by ammonia reached a maximum at temperatures up to 180°C (Figure 8.4). At higher temperatures the conversion of NO into N_2O began to increase, which indicated that the SCR of NO was dominant at temperatures below ~200°C, although at higher temperatures oxidation of NH_3 to N_2O was the preferable reaction.

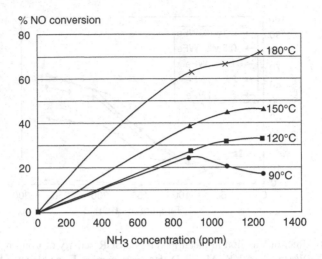

FIGURE 8.4 Influence of ammonia concentration on NO conversion over wt.% iron-impregnated ACF catalyst (NO = 800 ppm, O_2 = 6 vol%). (After Westwood, A.V.K., Modill, D., Hampastsoumian, E., and Jones, J.M., *Carbon '02, Intern. Conf. on Carbon,* Beijing, 2002, Paper 39.2. With permission.)

$$4NO + 4NH_3 + O_2 \xrightarrow{<200°C} 4N_2 + 6H_2O$$

$$4NH_3 + 4NO + 3O_2 \xrightarrow{>200°C} 4N_2O + 6H_2O$$

or

$$2NH_3 + 2O_2 + O_2 \xrightarrow{>200°C} N_2O + 3H_2O$$

These workers also carried out kinetic studies of SCR activity and determined the activation energies over the 1% iron-impregnated ACF catalyst and estimated the reaction orders with respect to ammonia and oxygen. The kinetic data suggested that the presence or oxygen considerably enhances the NO reduction due to the formation of NO_2, which was far more readily reduced to N_2 than NO. At higher temperatures the rate of conversion was high so that diffusional limitations started influencing the kinetics.

Inagaki and coworkers[14] studied the decomposition of NO on an activated carbon loaded with different amounts of palladium (Pd) metal by passing NO containing air (0.8 ppm NO) through the sample bed at a flow rate of 3 l/min. They observed that the performance of the active carbon for NO decomposition was considerably enhanced, although there was no appreciable difference between the performance of the two activated carbon samples loaded with different amounts of palladium (Figure 8.5).

Juntgen[15] and Juntgen and Kuhl[17] observed that the catalytic activity on the carbon surface was higher when it was associated with basic nitrogen surface groups.

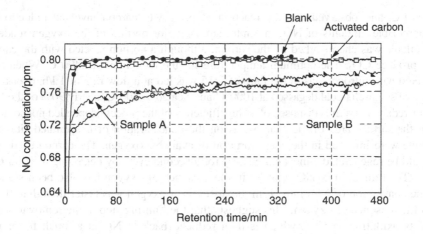

FIGURE 8.5 Performance of palladium loaded activated carbons for NO decomposition. (After Inagaki, M., Okuni, T., Tanaika, O., Yoshikawa, T., and Takahashi, K., *Carbon 02 Intern. Conf. on Carbon,* Beijung, 2002, Paper 35.2. With permission.)

Ahmed et al.[16] on the other hand, observed that the activity was considerably enhanced on oxidation of the carbon surface, which produced acidic carbon-oxygen surface groups such as carboxyls and lactones. The catalytic activity, however, did not depend upon the surface area. It was found that the presence of gas-phase oxygen increased the catalytic activity by a factor of about 50 (Figure 8.6). The presence of even small amounts of water vapor, on the other hand, considerably inhibited the activity. The decrease in activity was to the extent of about 25% with a moisture content of 5% as compared to the activity in the dry air.[17]

Puri et al.[18] passed a mixture of nitrogen dioxide (NO_2) and nitrogen gas at 300 to 500°C temperature over a bed of sugar or coconut-shell charcoal degassed at

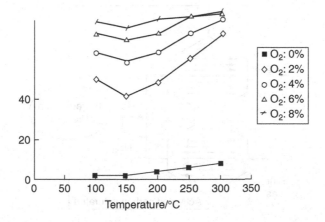

FIGURE 8.6 Influence of oxygen concentration and temperature on NO conversion over activated carbon. (After Juntgen, H., *Fuel,* 65, 1436, 1986. With permission.)

1000°C, and observed that the reaction of NO_2 with carbon involved reduction of appreciable amounts of NO_2 into nitrogen. A major portion of the oxygen rendered available was chemisorbed by the carbon, although a portion reacted with the carbon to produce CO_2. The optimum temperature for the reduction was 400°C. When NO_2 mixed with nitrogen gas in the ratio of 1:15 was led at a flow rate of 1 l/h. over a 5 g bed of sugar charcoal degassed at 1000°C and heated to 400°C, nearly 90% of the NO_2 was reduced and thus disposed off. The efficiency of the process depended not so much on the surface area of the charcoal as on the concentration of unsaturated sites that alone were involved in the chemisorption of available oxygen. The carbon used once could be regenerated almost completely for a second cycle by degassing at 1000°C.

The reduction of NO_2 into N_2 in the presence of oxygen has also been studied. The conversion of NO_2 into N_2 in the absence of oxygen was faster than that of NO in the presence of oxygen. This suggests that the limiting step in the removal of NO is its oxidation to NO_2, which is then reduced back to NO at a much faster rate because it is more reactive.[2]

Processes have been developed by which both SO_2 and oxides of nitrogen (NO_x) are removed from the flue gases together, using activated carbons. One such plant, Mitsui process, is an extension of the Burgbau-Forschung process used for the removal of SO_2. In the Mitsui process the granulated carbon or the activated coke is used in the adsorber, which is divided into two vertical sections (Figure 8.7). The active carbon flows downward through both sections and meets the flue gases moving in the countercurrent direction. Most of the SO_2 is removed in the lower section of the absorber, although NO_x is reacted in the upper section where ammonia is injected. The oxides of nitrogen are reduced by reaction with ammonia. If some SO_2 is remaining in the flue gases in the upper section of the absorber, it reacts with ammonia to produce ammonium salts, which may result in the loss of ammonia and may sometimes cause blocking of the absorber. Therefore, NO_x removal is generally

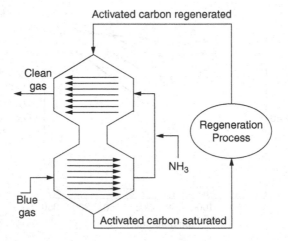

FIGURE 8.7 Schematic of Mitsui process for removal of SO_2 and NO. (Adapted from Tsuzi, K. and Shiraishi, I., *Fuel*, 76, 549, 1997. With permission.)

carried out after removing most of the SO_2, which occurs in the lower portion of the adsorber.

The activated carbon is removed from the bottom of the adsorber and regenerated by heat treatment, where some gasification of the active carbon also takes place, increasing its surface area. The Mitsui process for combined removal of SO_2 and NO_x is carried out at temperatures of 100 to 200°C and does not require a higher heat treatment as in the case of processes employing metal-oxide catalysts. Because the removal of NO_x requires a comparatively higher temperature than the removal of SO_2, two separate temperatures can be easily maintained in the two sections of the adsorber. The process is capable of removing 99% SO_2 and more than 70% NO_x.[2] The Mitsui process, however, has high operating costs and requires a high investment because of the high pressure drop through the bed, high resistance to mass transfer within the particles, mass loss by attrition, and large amounts of processed active carbon. But several modifications in the process have been made in Germany and Japan to make it more cost effective.

8.3 REMOVAL OF SULFUR DIOXIDE FROM FLUE GASES

Sulfur dioxide is the major air pollutant obtained from the combustion of fossil fuels. The 1950 acid and smog clouds over London, which took the life of several thousand persons due to respiratory ailments, acted as a catalyst in making stringent clean air regulations. Consequently, a lot of research has gone into the removal of harmful gases from air. Activated carbons have been found to be very promising adsorbents for SO_2 removal because they are highly efficient and can be regenerated. Furthermore, the more efficient temperature for the adsorptive removal of SO_2 by activated carbons is about the same as the temperature of the fuel gas (60 to 150°C). Although a number of research groups are currently engaged in the development of activated carbon desulfurizers, the factors that play a dominant role in the removal of SO_2 are not yet clearly understood. Somel workers,[19-25] while working on the removal of SO_2 by several carbonaceous materials, found that the chemical nature of the carbon surface plays an important role and influences the adsorption and desorption of SO_2, although the porous structure of the carbon surface is of less importance. These workers observed that an increase in the basicity of the activated carbon surface enhanced the adsorption capacity of the carbon for SO_2. Several other workers,[26-31] on the other hand, are of the view that the adsorptive removal of SO_2 by activated carbons is mainly determined by the surface area, specifically when oxygen and water are present, the role of chemical nature of the carbon surface being of little consequence.

When activated carbon is used for the removal of SO_2 from flue gases, three different processes take place:

- Physical adsorption
- Chemisorption of the gas, resulting in the formation of carbon-sulfur surface compounds
- Catalytic oxidation of the gas on the carbon surface, resulting in the deposition of elemental surface on the carbon surface

The physically adsorbed gas can be removal by degassing in vacuum or in an inert gas, although the elemental sulfur that is present in the form of very stable carbon-sulfur surface groups can only be recovered by heat treatment in hydrogen gas at 500°C. The work carried out by different workers on the desulfurization of fuel gases using activated carbons is discussed in more detail in Chapter 5 of this book.

8.4 EVAPORATED LOSS CONTROL DEVICE (ELCD)

With a continuous increase in the number of automobiles in big urban agglomerates, the availability of clean air has become a big casualty. These automobiles produce two types of emissions: those from the by-products of combustion such as CO, CO_2, oxides of nitrogen, sulfur and carbon particles, and those from the evaporation of the fuel itself. Consequently, attempts have been made to reduce these polluting emissions by improving upon the working of internal combustion engines, including automobile carburetors and fuel injection system. As a result of this work, the modern automobiles are fitted with more efficient systems, producing lesser amounts of polluting emissions. Furthermore, the exhaust from modern, more efficient internal combustion engines is further cleaned by using catalytic converters to reduce the more harmful constituents of the exhaust. The interest has thus, been focused upon uncontrolled emissions due to the evaporation of gasoline or diesel oil. This has been remedied by installing activated carbon canisters on motor vehicles to prevent emissions of volatile petroleum constituents. These activated carbon canisters are also called *evaporative loss control devices* (ELCDs).[2] With the more stringent clean air regulations in developed countries, the installation of ELCDs is becoming mandatory. The activated carbon canister is located between the fuel tank and the engine. The evaporation emissions occur as a result of the rise in the daytime temperature as well as during filling of the fuel tank, when the gasoline vapors are displaced by liquid gasoline. The activated carbon canister is regenerated by a bypass flow of combustion air when the engine is running. The vapor-saturated air is then passed through the inlet manifold so that the desorbed gasoline forms a part of the engine fuel mixture. This provides a small but significant benefit toward increasing fuel efficiency and controlling air pollution.

8.5 PROTECTION OF UPPER RESPIRATORY TRACT
IN HAZARDOUS ENVIRONMENT

The hazards of chemical warfare were realized for the first time during World War I, when the German army released chlorine gas on April 22, 1915, on the Allied forces. The Allies immediately started an intensive effort to develop and introduce new and more effective substances for chemical warfare and quickly developed means of protection against inhaling the choking gas. The first protection against chlorine was provided through adsorption by filtration through a fabric soaked in a solution of sodium hyposulphite and sodium carbonate, and respirators were prepared by impregnating this solution on pumice or diatomaceous earth. However, the continued threat of chemical warfare and the development of highly advanced chemical and biological weapons, the methods of protection had to be improved. This resulted in a more sophisticated design of adsorbents and respirators. Active carbon adsorption became indispensable in respirators (gas masks) when low reactivity war gases, which could be removed only by physical adsorption, were introduced. At the present moment, a wide range of hazardous gases are available that can be removed using activated carbon respirators, both by physical adsorption (e.g., nerve gases) and by chemical adsorption (e.g., cynogen chloride, phosgene, hydrogen cyanide, phosphine, and arsine).

Activated carbons used in respirators have to conform to stringent requirements and standards. They should be uniform in size to avoid channeling and thereby shortening the life of the respirator. Furthermore, the activated carbon alone can provide sufficient protection only against substances that are easily adsorbed at ordinary temperature. To be able to undertake protection through chemisorption or through a surface reaction, the activated carbon is impregnated with suitable mixtures of metal compounds, which include zinc, manganese, sodium, silver, chromium, and sometimes organic compounds such as pyridines and aromatic amines. The impregnated carbons provide protection against toxic substances through

- Adsorption, as in the case of nerve gases and chloropicrin
- Chemical reaction with the impregnant, such as neutralization, hydrolysis, and complex formation, as in the case of hydrogen cyanide, phosgene
- Catalytic reaction such as oxidation, as in the case of arsine

The protection of the respirator against the toxic substance depends upon the nature of the activated carbon, which includes the porous and chemical structure of the carbon surface, the nature of the impregnant, as well as on the method of impregnation. However, the composition of the impregnant is usually kept a secret and is seldom published. The reactions of some of the toxic gases with impregnated activated carbon are shown in Table 8.1.[32,33] The activated carbons generally preferred are those that have a sufficient transitional porosity, which helps in the fixation of the impregnant and has smaller water adsorption capacity. Although some toxic substances show a better reaction in the presence of moisture, usually the presence of moisture reduces the protection effectiveness of the impregnant. For protection against carbon monoxide, special filters containing a desicant, and a catalyst that is usually a mixture of cupric oxide and manganese dioxide is used. The protection is provided by the catalytic oxidation of CO into CO_2.

Activated carbons impregnated with Cu^{2+}, Ag^+, Cr^{6+}, NH_4^+, and CO_3^{2-} have been found to be efficient adsorbates for arsine, hydrogen cyanide, cynogen chloride, chloroform, and phosgene. Barnir and Aharoni[34] compared the adsorption of cynogen chloride on an active carbon before and after impregnation with Cu^{2+}, Cr^{6+}, Ag^+, and NH_4^+ in a given ratio and found that, although the adsorption of cynogen chloride was reversible in the case of active carbon, it was irreversible on the impregnated sample. When the cynogen chloride-loaded carbon was heated, cynogen chloride was the main gas evolved from the active carbon. On the other hand, CO_2 was mainly desorbed from the loaded impregnated carbon. It was postulated that the cynogen chloride chemisorbed on the impregnant surface reacted with water adsorbed on the carbon surface or linked with the impregnant material, producing CO_2 and ammonium chloride:

$$CNCl + 2H_2O \rightarrow CO_2 + NH_4Cl$$

TABLE 8.1
Reactions Taking Place on the Surface of Active Carbon Acting as a Catalyst

Impregnant Component	Effective Form of Impregnant	Function of Impregnant	Stoichiometric Equation
Copper	Cu_2O	Reactant	$Cu_2O + 2HCN \rightarrow 2CuCN + H_2O$
	CuO	Reactant	$CuO + COCl_2 \rightarrow CuCl_2 + CO_2$
Zinc	Na_2ZnO_2	Catalyst	$2AsH_3 + 3O_2 \rightarrow As_2O_3 + 3H_2O$
	ZnO	Reactant	$ZnO + 2HCN \rightarrow Zn(CN)_2 + H_2O$
	ZnO	Reactant	$ZnO + COCl_2 \rightarrow ZnCl_2 + CO_2$
Silver	Ag, Ag_2O	Catalyst	$2AsH_3 + 3O_2 \rightarrow As_2O_3 + 3H_2O$
Copper	$CuSO_4 \cdot 5H_2O$	Reactant	$CuSO_4 \cdot 5H_2O + 4NH_3 \rightarrow$ $\rightarrow [Cu(NH_3)_4]SO_4 + 5H_2O$
Chromium	$CuCO_4 \cdot NH_3 \cdot 5H_2O$	Catalyst	$ClCN + 2H_2O \rightarrow CO_2 + NH_4Cl$
	$CuCrO_4$	Catalyst	$ClCN + 2H_2O \rightarrow CO_2 + NH_4Cl$
Pyridine	C_5H_5N	Reactant	$C_5H_5N + ClCN + H_2O \rightarrow CHOCH =$ $= CH–CH = CHNHCN + HCl$

Adapted from Smisek, M. and Cerny, S., in *Active Carbon*, Elsevier Publ. Co., Amsterdam, 1970; and Jankowska, H., Swiatkowski, A., and Choma, J., in *Active Carbon*, Ellis Howard, England, 1991. With permission.

This process regenerates the surface of the impregnant, where more cynogen chloride can be chemisorbed. Reucroft and Chion[35,36] compared the adsorption behavior of BPL activated carbon with ASC whetlerite (prepared by impregnating BPL active carbon with Cu^{2+}, CrO_4^{2-}, and Ag^+) and ASB carbons (prepared by impregnating BPL active carbon with Cu^{2+} and BO_3^{3-} in different mole ratios) for phosgene, chloroform, cynogen chloride, and hydrogen cyanide. The adsorption of chloroform and phosgene increased by about 17% on the ASB impregnated carbons, although the adsorption of cynogen chloride increased by about 20% on ASC whetlerite. The impregnated carbons showed both chemisorption and physisorption, but chemisorption was more pronounced in the case of phosgene, cynogen chloride, and hydrogen cyanide on ASC and ASB carbons compared to BPL-activated carbon. Both ASC and ASB carbons retained appreciable amounts of the three adsorbates after evacuation at 150°C, the amount retained being more in ASC whetlerites.

Singh and coworkers[37] prepared NBC canisters using a coconut–shell activated carbon impregnated with 6.0% Cu (II), 2.5% Cr (VI), 0.2% Ag (I), 2.0% sodium hydroxide, and 2.5% pyridine. The canisters were studied for their adsorptive protection against phosgene ($COCl_2$) under wet (50% relative humidity) and dry conditions (Figure 8.8).

Phosgene mixed with dry or wet air was passed through the canister at a flow rate of 30 lpm. The phosgene breakthrough was monitored through $COCl_2$ detection paper by observing color change from off-white to brick red. The protection against phosgene was found to depend upon the compactness of the carbon in the canister. Higher degree of compactness produced protection for a longer period of time. The protection was

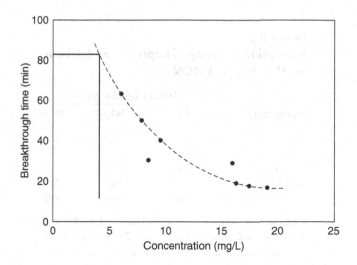

FIGURE 8.8 Effect of Phosgene concentration on breakthrough time. (After Singh, B., Saxena, A., Sharda, D., Yadav, S.S., Chintamani, D., Pandey, D., and Sekhar, K., *Proc. Indo. Carbon 2003 Conf.*, G.N. Mathur, V.S. Tripathi, T.L. Dhami, and O.P. Bahl, Eds., *Indian Carbon Soc.* 2003, p. 33. With permission.)

also longer under wet conditions than under dry conditions. The phosgene breakthrough curves indicated that at 4.0 mg/l phosgene concentration the canister could provide protection for 84 min under wet and only 37 min under dry conditions at 30°C and 30 lmp airflow. $COCl_2$ breakthrough studies also showed that the protection was mainly provided by (a) physical adsorption of $COCl_2$ by the impregnated activated carbon as indicated by the increased protection time with increased compactness of the carbon and (b) by the hydrolysis of $COCl_2$. In addition, a chemical reaction between the impregnant and phosgene could also take place. These three processes have been represented as

$$>C + COCl_2 \longrightarrow >C-COCl_2 \qquad \text{Adsorption}$$

Carbon surface

$$>C-COCl_2 + H_2O \longrightarrow >C + CO_2 + HCl \quad \text{Hydrolysis}$$

$$>C-CuO + HCl \longrightarrow >C + CuCl_2 + H_2O$$

Impregnant   Carbon surface

Capon et al.[38] studied the adsorption of pollutants such as SO_2, NO_2, H_2S, HCN, and CNCl on an activated charcoal cloth after impregnating with oxidizing agents such as $KMnO_4$, $Na_2Cr_2O_7$, and ClO_2 with an organic tertiary amine triethylenediamine (TEDA) and $AgNO_3$ from aqueous solutions. The presence of copper and oxidizing agents as impregnants enhanced the adsorption capacity of all the pollutant gases (Table 8.2), although the presence of organic amine (TEDA) enhanced considerably

TABLE 8.2
Adsorption Capacity of Impregnated Carbons for H_2S, SO_2 and HCN

Impregnant	Weight Uptake ($\mu g/cm^2$)		
	H_2S	SO_2	HCN
None	50	220	120
$AgNO_3$	210	—	—
$Na_2Cr_2O_7$	310	480	120
Cu (5%)	360	470	320
Cu (5%) + $Na_2Cr_2O_7$	760	960	890

After Capon, A., Alves, V.R., Smith, M.E., and White, M.P., *15th Bienn. Conf. on Carbon*, 1981, Reprints, p. 232. With permission.

the adsorption of CNCl (Table 8.3). The adsorption of phophine (PH_3)on an activated carbon cloth before and after impregnation with $AgNO_3$ and $Cu(NO_3)_2 \cdot 3H_2O$ was studied by Hall et al.[39] The techniques used included adsorption by gravimetry and surface characterization using XPS, SEM, ED x-ray analysis, and IR spectroscopy. Both the raw and impregnated samples of carbon cloth retained appreciable amounts of PH_3, which was bonded strongly. The amount of PH_3 retained was five times greater on the impregnated carbon cloth when impregnated with $AgNO_3$ and three time larger when it was impregnated with $Cu(NO_3) \cdot 3H_2O$. Wilson and Whetzel[40] have reported that active carbons impregnated with silver and copper compounds are particularly useful for the removal of chlorine, phosgene, arsine, and hydrogen cyanide from air. The addition of zinc oxide improves the activity of these impregnated carbons for phosgene and diphosgene. Jankowska et al.[33] and Choma et al.[41]

TABLE 8.3
Adsorption Capacity of Impregnated Carbons for NO_2 and CNCl

Impregnant	Weight Uptake of NO_2 ($\mu g/cm^2$)	Impregnant	Weight Uptake of CNCl ($\mu g/cm^2$)
None	10	None	60
$KMnO_4$	30	Cu (5%)	160
Cu (14%)	110	Cu (5%) +$Na_2Cr_2O_7$ + pyridine	320
Cu (14%) + $KMnO_4$	160	Cu (5%) + $Na_2Cr_2O_7$ + TEDA (1%)	860
Cu (14%) + ClO_2	520	Cu (5%) + $Na_2Cr_2O_7$ + TEDA (3%)	1530

After Capon, A., Alves, V.R., Smith, M.E., and White, M.P., *15th Bienn. Conf. on Carbon*, 1981, Reprints, p. 232. With permission.

have reported the preparation of an activated carbon impregnated in a solution of tetrammine copper (II) carbonate (cuperammonium carbonate), ammonium chromate, and silver salts and called it *chromium copper adsorbent*. One typical composition of the solution per 100 kg of the activated carbon is basic copper carbonate 6 kg, ammonium carbonate 5 kg, liquor ammonia (25%) 10 liters, potassium dichloromate 3 kg, silver nitrate 170 g and water 59.5 liters. It has been suggested that this chromium copper adsorbent has a mixture of copper chromates of approximate composition $CuCrO_4 \cdot NH_3 \cdot H_2O$ and $CuCrO_4 \cdot 2Cu \cdot 2H_2O$ on its surface.

The chromium copper adsorbent exhibited good protection against hydrogen cyanide, cynogen chloride, and arsine after heat treatment at 170 to 190°C. However, the protection was better against cynogen chloride when the proportion of chromate was more on the adsorbent, and better against arsine when it contained more of copper oxide.

The adsorptive removal of chloropicrin ($CCl_3 NO_2$) vapors on activated carbon fibers (ACF), metal impregnated activated carbon (ASC-AC), and activated carbon-deposited cloths was studied by Choi et al.[42] The chloropicrin vapor saturated air obtained by bubbling dry air through liquid chloropicrin was diluted by mixing with dry air and then passed through the adsorbent packed column. A constant flow rate was maintained by using a calibrated flow meter. The influent and elluent chloropicrin concentration was measured using a gas chromatograph. The breakthrough point time was longest for ACF than for the other adsorbents (Figure 8.9), although the amount of the ACF was the smallest in the column packing (0.54 g). The breakthrough curve for ACF showed a very steep rise just after the break point, which indicated that the mass transport zone (MTZ) of ACF was small relative to the packing height and that the most of the adsorption capacity of the ACF was utilized

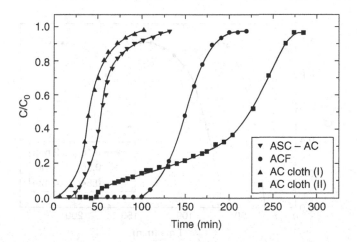

FIGURE 8.9 Breakthrough curves of chloropicrin on different activated carbons. (ACF = 0.54 g, ASC-AC = 3.0 g, AC-Cloth = 1.03 g, AC-Cloth (II) = 1.8 g, Flow rate = 0.4 L/min, Input concentration = 7.0 g/L). (After Choi, S.R., Lim, J.S., Baeg, S.J., Kim, M.J., Kim, K.H., and Ryu, S.K., in *Carbon 02 Intern. Carbon Conf.* Beijung, 2002, Paper PI 63D055. With permission.)

TABLE 8.4
Adsorption Capacity of Different Activated Carbons for Chloropicrin

Carbon Adsorbent	ACF	ASC-AC	AC-deposited Cloth
Adsorption Amount at Break Point	0.194	0.013	0.021
Theoretical Maximum Capacity	0.264	0.034	0.079

After Choi, S.R., Lim, J.S., Baeg, S.J., Kim, M.J., Kim, K.H., and Ryu, S.K., in *Carbon 02 Intern. Carbon Conf.*, Beijung, 2002, Paper PI 63D055. With permission.

for the adsorption of chloropicrin. A narrow MTZ is desirable to make efficient use of the adsorbent and to reduce the energy costs during regeneration.

The breakthrough curves for ASC-AC and AC-deposited carbon cloth (I) were also steep after the breakpoint, but the break point times were short in these cases. In the case of AC-deposited cloth (II), the breakthrough was typically extended, showing a less adsorption efficiency than ACF. The adsorption amounts at the break point and theoretical maximum adsorption capacities of the adsorbents (Table 8.4) showed that the removal efficiencies of the different carbon adsorbents, as calculated by the ratio of the amount adsorbed at the break point, and the maximum adsorption capacity, were 67, 38, 24, and 20%, respectively, for ACF, ASC-AC, and AC-Cloth (I) and AC-Cloth (II), which indicated that ACF was a far superior adsorbent for the removal of chloropicrin vapors. These workers also studied the influence of packing height and the concentration of chloropicrin in the influent stream an the adsorption amount at the break point. The amount of chloropicrin adsorbed increased with increase in the packing height (Figure 8.10) but decreased with the influent concentration (Figure 8.11).

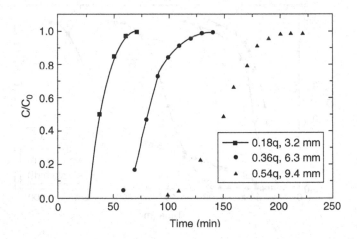

FIGURE 8.10 Breakthrough curve of chloropicrin on ACF at different packing heights. (Flow rate = 0.4 L/min, Input = 7.0 mg/L). (After Choi, S.R., Lim, J.S., Baeg, S.J., Kim, M.J., Kim, K.H., and Ryu, S.K., in *Carbon 02 Intern. Carbon Conf.*, Beijung, 2002, Paper PI 63D055. With permission.)

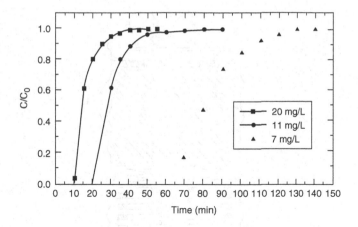

FIGURE 8.11 Breakthrough curves of chloropicron ACF at different influent concentrations. (ACF = 0.36 g, Flow rate = 0.4 L/min). (After Choi, S.R., Lim, J.S., Baeg, S.J., Kim, M.J., Kim, K.H., and Ryu, S.K., in *Carbon 02 Intern. Carbon Conf.*, Beijung, 2002, Paper PI 63D055. With permission.)

Impregnation with certain organic compounds such as pyridines has also been found to improve the adsorption of certain hazardous compounds. Baker and Poziomek[43] modified adsorption properties of coal-based activated carbons for cynogen chloride by impregnation with pyridine, 4-vinylpytridine, 4-amino pyridine, 4-cynopyridine and 4-n.propylpyidine. The carbons were impregnated with volatile impregnants in a rotary evaporator, and the amount impregnated was determined by the increase in the weight of the carbon. The nonvolatile impregnants were adsorbed on the carbon surface from alcohol solutions. The reactivity of the impregnated activated carbons increased with increase in the amount of the impregnant. The reactivity, however, varied from one pyridine to another, but not necessarily in the order of their nucleophilicities or basicities (Table 8.5). The activated carbon impregnated with 4-vinylpyridine showed higher

TABLE 8.5
Influence of Impregnation of Carbon on the Adsorption of Cynogen Chloride

Impregnant	Amount Impregnated (mmol/100 g)	Saturation Adsorption Capacity for CNCl₃ (mmol/100 g)	Ratio of Additional CNCl Adsorbed to Impregnant (mmol)
None	0	85	0
4-Aminopyridine	212	272	0.882
Pyridine	253	207	0.482
4-Vinylpyridine	190	200	0.605

After Baker, J.A. and Poziomek, E.J., *Carbon*, 12, 45, 1974. Reproduced with permission from Elsevier.

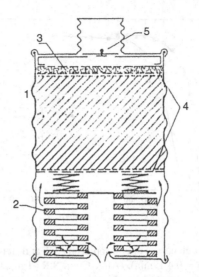

FIGURE 8.12 A respirator filter. (1. activated carbon layer, 2. smoke filter, 3. cotton plug, 4. wire mesh, 5. breathing valve) (After Choi, S.R., Lim, J.S., Baeg, S.J., Kim, M.J., Kim, K.H., and Ryu, S.K., in *Carbon '02, Intern. Carbon Conf.,* Beijung, 2002, Paper PI 63D055. With permission.)

adsorption of cynogen chloride (0.605 mmol/g) compared to pyridine impregnated carbon (0.482 mmol/g), although the former was less basic than the latter. Furthermore, the 4-vinyl pyridine-impregnated carbon failed to result in any weight loss on heating at 150°C for many hours, and not more than about 66% of the adsorbed 4-vinyl pyridine could be removed by soxhlet extraction with ethanol for 6 hr.

The respirator filters for protection against toxic gases generally consist of an antismoke screen for trapping biological aerosols, radioactive dust, and irritating arsine smokes and a layer of activated carbon for the removal of toxic gases and vapors. The activated carbon is held between two sheets of wire gauge and is held in position by a spring (Figure 8.12). The filter is made from thin corrugated steel sheet. Air enters the filter at the bottom through a stopper, which is kept closed during storage. It has a rubber breathing valve that closes when breathing out to prevent moist air enter the canister.

With the continued threat of chemical warfare, and with the development of more sophisticated chemical and biological weapons, there is need to develop better activated carbon adsorbents for use in respirators. Furthermore, considering the threat to use nerve gases, complete body suits have been developed for the protection of the whole body. Because the nerve gases are generally physically adsorbed, the activated carbons used should have a high surface area and a high pore volume. To provide high surface area, and to allow rapid and effective adsorption, the activated carbons used are in the form of small spheres, which are fixed on a fabric with the help of suitable adhesives.

These filters with modified activated carbons are also being used in underground shelters and as inlet filters in armed vehicles and other forms of military transport. Several chemical and process industries where hazardous and toxic gases and vapors can endanger the life of workers in emergencies are also making use of these activated carbons in respirators for personal protection of the individuals, in air treatment filters, and in protective clothing. In industrial protective filters, the adsorbent carbons are generally impregnated with potash for acidic pollutants, with zinc salts for protection against ammonia and H_2S, with copper salts for HCN, with potassium bicarbonate against SO_2, and with iodine or sulfur compounds for protection against mercury vapors.

8.6 ACTIVATED CARBON ADSORPTION OF MERCURY VAPORS

Mercury, along with several other heavy metals, is always present in small amounts in certain fossil fuels, particularly in coal and waste materials, and it is released into the atmosphere as a part of the flue gases on combustion of these materials. Mercury vapors cannot be easily recovered from the flue gas stream because of its highly volatile nature, although several procedures are being developed to capture the mercury vapor. Sinha and Walker[44] studied the removal of mercury vapor from contaminated gas streams at 150°C using a sulfurized Saran charcoal. The charcoal was sulfurized by the oxidation of hydrogen sulfide on the carbon surface at 140°C in a fluidized bed, when a sulfur loading between 1.0 and 11.8% by weight was obtained. The mercury-contaminated gas stream was passed through a bed of the sulfur-loaded carbon at 150°C for a given period of time. The rate of mercury build-up in the effluent stream after passing through the bed was extremely low (Figure 8.13), indicating that the sulfurized carbon was a good adsorbent for the removal of mercury vapors. This was attributed to the interaction between the sulfur present on the carbon surface and the mercury vapors, due to the formation of less volatile mercuric sulfide.[45] Because the temperature of the stream of flue gases containing mercury vapors is usually high, it

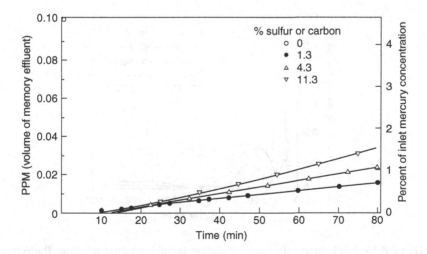

FIGURE 8.13 Effect of sulphur content on carbon on the removal of mercury vapors from air. (After Sinha, R.K. and Walker, P.L., Jr., *Carbon*, 10, 754, 1972. With permission.)

is often impractical to cool these large volumes of the gas for an efficient removal of mercury vapors.

8.7 REMOVAL OF ORGANOSULFUR COMPOUNDS

Methyl mercaptan is one organosulfur compound that is present in the flue gases of certain process industries. When present even in small amounts, they have an obnoxious odor and toxicity. Bashkova et al.[46] studied the adsorption of methyl mercaptan (CH_3SH) under wet, dry, and oxidizing conditions, using several activated carbons of different origins, and examined the influence of carbon surface on the removal and oxidation of the compound, and identified the reaction products. They found that the activated carbons were highly efficient for the removal of methyl mercaptan, although different carbons adsorbed different amounts. The adsorption of CH_3SH on the carbon surface decreased the pH of the carbon slightly, which was attributed to the formation of small amounts of sulfur containing acid along with disulfides. The species present on the carbon surface were analyzed using DTG curves in nitrogen (Figure 8.14). The peak at temperatures lower than 100°C represents the desorption of weakly adsorbed species as water, although the peak between 100 and 300°C might represent desorption of disulfides.[47] The disulfide, CH_3SSCH_3, is formed by the decomposition of methyl mercaptan, and its formation is enhanced in the presence of water. The disulfide when formed is more strongly adsorbed on the surface of the activated carbon than water, and it is possible that the disulfide may replace water from the carbon surface. The adsorption of methyl mercaptan has also been found to be determined by the size of the pores and the pore volume of the carbon.

Suzuki et al.[48] studied the adsorption of dimethyl sulfide (DMS) from city gas in a dynamic adsorption system using Pitch-based and PAN-based activated carbon

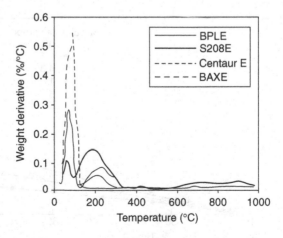

FIGURE 8.14 DTG curves of exhausted carbon sample in nitrogen. (After Bashkova, S., Bagreev, A., and Bandosz, T.J., in *Carbon 02, Intern. Carbon Conf.,* Beijung, 2002, Paper 21.2. With permission.)

fibers and two samples of activated carbons before and after treatment of the carbons with nitric acid and sulfuric acid. The city gas containing DMS was passed through a bed of carbon using nitrogen as a carrier gas. The influent and effluent concentration of DMS was measured using a gas chromatograph. The breakthrough profiles showed that among the carbon fibers, the PAN-based ACF showed better adsorption. The oxidation of the carbon enhanced the adsorption capacity depending upon the degree of oxidation of the carbon.

8.8 ADSORPTIVE REMOVAL OF MISCELLANEOUS VAPORS AND GASES

Formaldehyde and other aldehydes are main pollutants in indoor air. Formaldehyde is mainly obtained from some decorated materials, paint, polymerizing plate, bindings of some furniture, and chemical fiber carpets. It is also a constituent of cigarette smoke. In newly decorated and painted rooms, the concentration of formaldehyde sometimes is quite high. Being toxic, a high concentration of formaldehyde in indoor air can cause symptoms such as headache, nausea, coryza, pharyngitis, and emphysema, and may lead to lung cancer.[49,50] It is thus essential to remove aldehyde vapors from indoor air. Activated carbon adsorbents are the most promising adsorbents when adsorption of traces of gases and vapors is concerned.

Domingo-Garcia et al.[51] studied the adsorption of formaldehyde on activated carbons and observed that the adsorption involved strong forces with an isosteric heat of adsorption of the order of 15 to 33 kJ per mol. The amount adsorbed increased with increase in the surface area of the activated carbon. Rong and coworkers[49,52] investigated the adsorption of formaldehyde on cellulose-based activated carbon fibers using static and dynamic adsorption systems. The ACF was oxidized with air at different temperatures between 300 and 500°C and for different intervals of time and by varying the oxygen flow rate. The ACF was also impregnated with p.aminobenzoic acid (PABA). The adsorption of formaldehyde increased with increase in oxidation up to 420°C and decreased at higher temperature. The adsorption was also influenced by the time of oxidation and the air flow rate. The increase in adsorption on oxidation was attributed to the formation of polar carbon-oxygen surface groups. The adsorption also increased with increase in the surface area of ACF.

The static and dynamic adsorption capacity of the ACF increased on impregnation with PABA. The breakthrough time for formaldehyde and the amount adsorbed also increased on impregnation. The removal of formaldehyde involved both physisorption and chemisorption. The chemisorption was suggested to be due to the introduction of the amino groups in the impregnant.

El-Sayed and Bandosz[53] used three activated carbon samples of different origin, namely BPL from Calgon and MVP from Norit, both prepared from bituminous coal, and BAX from Westvaco, made from wood, using chemical activation with phosphoric acid for the adsorption of acetaldelyde. These carbons were washed in a soxhlet apparatus to remove water-soluble impurities and then oxidized with nitric acid. The adsorption of acetaldehyde was determined by inverse gas chromatography at infinite dilution and finite concentration. The heats of acetaldehyde adsorption at zero coverage were calculated from the retention volumes. The heats of adsorption were also calculated from isotherms derived from chromatographic peaks. The results indicated that the heat of adsorption depended on the pore structure and the chemical structure of the activated carbon surface. The interaction of acetaldehyde was the strongest in case of the activated carbons, which contained smaller pores and were associated with larger amounts of carbon-oxygen surface groups. These workers suggested that the smaller pores enhanced the dispersive interaction, although the presence of carbon-oxygen surface groups enhanced the hydrogen bonding between the acetaldehyde molecule and the carbon-oxygen surface groups.

Salame and Bandosz[54] investigated the adsorption of diethyl ether on two samples of activated carbons obtained from wood before and after oxidation. The adsorption was found to depend on the surface area of the carbon and had no relationship with the amount of acidic carbon-oxygen surface groups. It was observed that at very low surface coverages and at low pressures, the adsorption occurred in narrow micropores, although at higher surface coverages the adsorption also occurred in wider pores.

The adsorption of halogenated hydrocarbons has also been studied by a number of workers. The adsorption of carbon tetrachloride vapors known as the CTC index is used for the characterization of activated carbons and for quality control. Puri et al.[55] studied the adsorption of carbon tetrachloride vapors on five different commercial grade carbon blacks and six different activated carbons. The adsorption, in each case, was reversible, the adsorption and desorption isotherms being completely superimposed on each other (Figure 8.15 and Figure 8.16). The surface areas of the

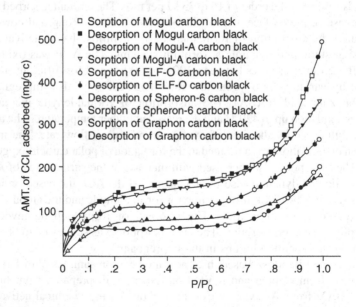

FIGURE 8.15 Adsorption and Desorption isotherms of CCl_4 at 35°C on different carbon blacks. Adsorption = open; desorption = closed. (After Puri, B.R., Arora, V.M., and Verma, S.K., *J. Indian Chem. Soc.*, 56, 802, 1979. With permission.)

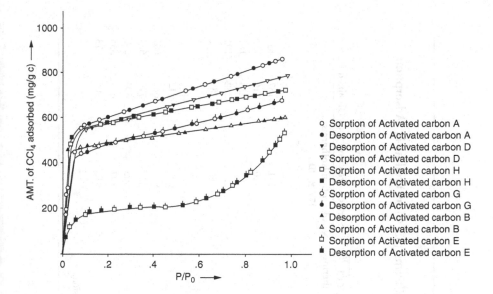

FIGURE 8.16 Adsorption and desorption isotherms of CCl_4 at 35°C on activated carbons, Adsorption = open; desorption = closed. (After Puri, B.R., Arora, V.M., and Verma, S.K., *J. Indian Chem. Soc.*, 56, 802, 1979. With permission.)

carbon blacks and the activated carbons (Table 8.6) were calculated using the B point method from the CCl_4 adsorption isotherms, where the monolayer formation is considered to be complete and by using the BET equation for the adsorption of CCl_4 and by nitrogen adsorption isotherms at 77 K. For calculation of surface area, the molecular area of the CCl_4 molecule was taken as 36.6 $A^{°2}$.

It was found that, although for carbon blacks the surface areas calculated from CCl_4 adsorption agreed fairly well with the $BET(N_2)$ surface areas, the values were smaller than the BET surface areas in the case of activated carbons. This indicated that a large fraction of the BET surface was not accessible to larger CCL_4 molecules. The fraction of the surface that remained in accessible to CCl_4 vapors varied between 5 to 40% in different activated carbons. These workers concluded that, although the adsorption of CCl_4 in the case of carbon black is significant in giving an estimate of surface area, its relevance in the case of highly porous activated carbons was limited to approximation of relative proportions of ultra fine micropores in the porous network of the system.

Yun and coworkers[56] studied the adsorption equilibrium of dichloromethane, trichlomethane, and trichloroethylene on activated carbons using static volumetric technique. The adsorption isotherms were measured for the pure vapors in the temperature range 283 to 683 K and at different pressures. The Dubinin-Redushkevich equation was applied to experimental adsorption isotherms to determine thermodynamic properties such as heat of adsorption and the Henry constant. The value of isosteric heat of adsorption was found to vary with surface loading. Li et al.[57]

TABLE 8.6
Specific Surface Areas of Carbons as Obtained from CCl₄ Adsorption Isotherms (35°C) as Well as N₂ Adsorption Isotherms (77 K)

Carbon	Oxygen Content (mg/g)	Monolayer Values (X_m) from (mg/g)		Surface Area (m³/g) as Obtained from		
		The 'Point of Inflection' or 'Knee' of the CCl₄ Adsorption Isotherms (i)	BET Plots of CCl₄ Adsorption Isotherms (ii)	N₂ Adsorption Isotherms (77 K) BET	(i)	(ii)
Carbon Black						
Elf-O	40.4	115.0	94.7	171	167	137
Spheron-6	30.8	80.0	75.8	120	116	110
Mogul	78.3	175.0	125.2	301	254	181
Mogul-A	71.1	160.5	131.8	221	232	191
Graphon	nil	60.0	56.4	86	87	82
Active Carbon						
A	10.7	560	489.9	1093	812	710
B	9.4	470	414.1	987	682	601
D	11.5	535	471.7	821	776	684
E	8.3	210	166.6	505	304	242
G	12.9	444	382.9	897	644	555
H	7.2	555	464.1	912	805	673

After Puri, B.R., Arora, V.M., and Verma, S.K., *J. Indian Chem. Soc.*, 56, 802, 1979. With permission.

investigated the mechanism of adsorption of trihalomethanes on an activated carbon fiber (ACF). The adsorption was exothermic and monomolecular and increased when the trihalomethanes became more and more hydrophobic by the substitution of chlorine atoms by bromine atoms.

Reid and Thomas[58] studied the adsorption characteristics of a series of planar (ethylene, benzene, and pyridine) and tetrahedral (methane, chloromethane, dichloromethane, chloroform, and carbon tetrachloride) molecules on a carbon molecular sieve used for air separation over a range of temperatures. The size-exclusion characteristics of planar and tetrahedral molecules indicated that the selective porosity behaved as though the pores were spherical in shape. The partial screening of probe molecules into the microporous structure allowed the quantification of selective and nonselective microporosity and mesoporosity. Hazourli and Bemnecaze[59] studied the influence of applied potential and number of cycles of adsorption and electrosorption of chloroform on three different granulated activated carbons. A negative potential increased the adsorption, although the positive potential had the opposite effect. The repetition of potential cycles resulted in an increase in retention of chloroform.

Choi et al.[60] studied the adsorption of CCl_4 vapor on an activated carbon fiber (ACF), a metal impregnated activated carbon ASC-AC, and an activated carbon deposited cloth (AC-Cloth). Carbon tetrachloride vapor-saturated air was passed through activated carbon packed column at 0.6 L/min flow rate. The vapor concentration in the influent and effluent stream was measured by gas chromatography. The breakthrough curves were very steep after the break point (0.005 mg/l). The ACF showed higher adsorption capacity for CCl_4 (Figure 8.17) compared with the other carbon adsorbents. This was attributed to the presence of a uniformly developed micropore structure of ACF in addition to its larger surface area. The adsorption capacity for CCl_4 on ACF at the break point increased with the increase of packing height, influent concentration, and lower flow rate. Desorption studies of CCl_4

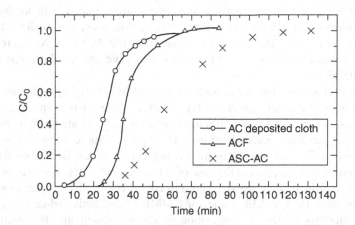

FIGURE 8.17 Breakthrough curves of CCl_4 on different activated carbons. (After Choi, S.R., Kim, B.S., Lim, J.S., Lee, S.K., and Ryu, S.K., in *Carbon 02 Intern. Carbon Conf.,* Beijing, 2002, Paper 9.1. With permission.)

showed that it could be desorbed almost completely in 30 min at 121°C and 1.2 atm in an autoclave. This indicated that CCl_4 was adsorbed physically on ACF so that it can be used for the removal of CCl_4 vapors and can be regenerated for further use.

Huang and coworkers[61] studied the adsorption of methyl ethyl ketone and benzene vapors on activated carbon fibers (ACF) prepared by activation in CO_2 at different temperatures and for different intervals of time. The adsorption of both vapors increased with increased surface area of the ACF sample.

The adsorption of benzene vapors on activated carbons has been studied by a number of investigators. Dubinin and coworkers used benzene as a standard vapor for the derivation of their theories of physical adsorption. These workers were of the view that the adsorption of benzene involved purely dispersion interactions. However, Puri and coworkers,[62] in their studies on the adsorption of benzene on sugar charcoals and carbon blacks, and Bansal and Dhami,[63] on polymer carbons observed that the presence of acidic carbon-oxygen groups on the carbon surface, which is known to impart polar character to the carbon surface, suppressed the adsorption of benzene. When the acidic groups were removed almost completely and the carbon surface predominantly had nonacidic quinonic groups, the adsorption of benzene increased. The 600°-degassed carbon samples that were associated with relatively larger amounts of quinonic groups showed higher adsorption of benzene than those degassed at higher temperatures. It was further found that the additional adsorption of benzene at each relative pressure > 0.3 amounted roughly to one mole of benzene per mole of quinonic oxygen (Table 8.7). This indicated the probability of interaction of π electrons of the benzene ring with the partial positive charge on the carbonyl carbon atom. Thus, for the removal of benzene vapors that are now considered carcinogenic, heat-treated carbons free of acidic surface oxygen groups be preferred.

The adsorptive removal of xenon (Xe) using activated carbons is important from the radiation protection point of view as well as its recovery for being of practical importance. Shuixia et al.[64] studied the adsorption of Xe on viscose based and pitch-based ACFs. The activated carbon fibers were impregnated with methylene blue (Mb), p.nitrophenol (PNP) and sodium chloride solutions, and oxidized in $KMnO_4$ and HNO_3 solutions. The adsorption capacity of the ACFs for Xe decreased with each impregnation and by oxidation with $KMnO_4$ but increased on oxidative treatment with HNO_3.

Saxena et al.[65] studied the adsorption of dimethyl methyl phosphonate (DMMP) as a stimulant for Sarin, which is a highly toxic warfare gas agent. The adsorption was carried out on an activated carbon, a whetlerite and activated carbon impregnated with copper hexafluorocacetylacetonate (I) copper trifluoro acetyl-acetonate and a copper oxime. The adsorption isotherms are Type I of the BET classification and show maximum adsorption in the case of activated carbon (I). The amount adsorbed was 68.5 weight percent on (I). The adsorption involved both physisorption and chemisorption. TGA and IR studies of DMMP loaded activated carbons, and mass spectric analysis of the decomposition products showed that the decomposition products were methyl methyl phosponic acid and methylphosphonic acid.

Perfluoroisobutylene (PFIB) is a very hazardous compound that is produced when fluoro-organic materials are produced or processed. PFIB causes pulmonary edema and thus activated carbon canisters have been prepared for protection against

TABLE 8.7
Increased Benzene Adsorption in Relation to Quinonic Oxygen in the Case of 600°C-Outgassed Carbon Blacks

Relative Vapor Pressure	Increase In Benzene Adsorption on Outgassing at 600°C (mmol/g)	Increased Benzene Adsorption / Quinonic Oxygen
Mogul (quinonic oxygen = 0.658 mmol/g)		
0.30	0.564	0.86
0.40	0.577	0.88
0.50	0.596	0.89
0.60	0.615	0.94
0.80	0.621	0.94
ELF-0 (quinonic oxygen = 0.239 mmol/g)		
0.30	0.269	1.16
0.40	0.263	1.10
0.50	0.282	1.18
0.60	0.307	1.28
Spheron-4 (quinonic oxygen = 0.149 mmol/g)		
0.30	0.141	0.95
0.40	0.147	0.98
0.50	0.157	1.05
0.60	0.166	1.11
Spheron-6 (quinonic oxygen = 0.0843 mmol/g)		
0.30	0.0772	0.92
0.40	0.0859	1.02
0.50	0.0897	1.06
0.60	0.0987	1.16

After Puri, B.R., Kaistha, B.C., Vardhan, Y., and Mahajan, O.P., *Carbon*, 11, 329, 1973. Reproduced with permission from Elsevier.

PFIB. Zhang et al.[66] studied the parameters that determine the protective performance of activated carbons against PFIB. The adsorption capacity of an activated for PFIB was also determined after impregnation with Cu, Ag, and Cr, and after chlorination. The adsorption capacity was determined from the breakthrough curves. The results indicated that the micropore volume of the activated carbon was an important factor for the adsorption of PFIB. The activated carbon acted both as an adsorbent and as a catalyst. Activated carbons containing carbon-oxygen surface groups that enhanced the adsorption of water vapor decreased the adsorption only slightly. However, water present on the carbon surface could initiate catalytic reactions, producing a nontoxic compound by degradation of PFIB.

REFERENCES

1. Gupta, A., *Proc. Emerging Solutions to VOC Air Toxics Control*, San Diego, CA, Feb. 26–28, 1997, p. 286.
2. Derbyshire, F., Jagtoyen, M., Andrews, R., Rao, A., Martin-Gullon, I., and Grulke, E.A., in *Chemistry and Physics of Carbon*, L.R. Radiovic, Ed., Marcel Dekker, New York, 2001, Vol. 27, p. 1.
3. Jagtoyen, M. and Derbyshire, F., *Proc. Emerging Solutions to VOC Air Toxics Control*, Florida, March 4–6, 1998.
4. Yamashita, H., Tomito, A., Yamada, H., Kyotani, T., and Radovic, L.R., *Energy Fuels*, 7, 85, 1993.
5. Illan-Gomez, M.J., Linares-Solano, A., Radovic, L.R., and Solanas-Martinez, de Lecea, C., *Energy Fuels*, 9, 97, 1995.
6. Illan-Gemez, M.J., Linares-Solano, A., Solanas-Martinez de Lecea, C., and Calo, J.M., *Fuel, Energy Fuels*, 7, 146, 1993.
7. Tsuzi, K. and Shiraishi, I., *Fuel*, 76, 549, 1997.
8. Kaneko, K., Wang, Z., Suzuki, T., and Ozeki, A., *J. Colloid Interface Science*,142, 489, 1991.
9. Yoshikawa, M., Yasutake, A., and Machida, I., *Appl. Catalysis A General*, 173, 239, 1998.
10. Yoshikawa, M. and Kakazu, T., Japanese Patent No. JP 10225641 Az, Aug. 25, 1998.
11. Grzybek, T., Klinik, J., Roger, M., and Pap. H., *J. Chem. Soc. Faraday Trans.*,94, 2843, 1998.
12. Grzybek, T., Pasal, J., and Pap, H., *Phys. Chem. Chem. Phys.*, 1, 341, 1999.
13. Westwood, A.V.K., Modill, D., Hampastsoumian, E., and Jones, J.M., *Carbon '02, Intern. Conf. on Carbon,* Beijung, 2002, Paper 39.2.
14. Inagaki, M., Okuni, T., Tanaika, O., Yoshikawa, T., and Takahashi, K., *Carbon '02, Intern. Conf. on Carbon,* Beijung, 2002, Paper 35.2.
15. Juntgen, H., *Fuel*, 65, 1436, 1986.
16. Ahmed, A.N., Baldwan, R., Derbyshire, F., McEnany, B., and Stencil, J., *Fuel*, 72, 287, 1993.
17. Juntgen, H. and Kuhl, H., in *Chemistry and Physics of Carbon*, P.A. Thrower, Ed., Marcel Dekker, New York, 1989, Vol. 22.
18. Puri, B.R., Bansal, R.C., and Bhardwaj, S.S., *Indian J. Chem.*, 11, 1168, 1973.
19. Davini, P., *Fuel*, 68, 145, 1969.
20. Davini, P., *Carbon*, 28, 565, 1990.
21. Davini, P., *Carbon*, 29, 321, 1991.
22. Rodriguez-Reinoso, F., Molina-Sabio, M., and Munecas, M.A., *J. Phys. Chem.*, 96, 2707, 1992.
23. Tartarelli, R., Davini, P., Morelli, F., and Corsi, P., *Atmos. Environ.*, 12, 289, 1978.
24. Muniz, J., Horrero, J.E., and Fuertes, A.B., *Appl. Catal. B.*, 18, 171, 1998.
25. Puri, B.R., Mahajan, O.P., and Bhardwaj, S.S., *Indian J. Chem.*, 11, 1170, 1973.
26. Rubio, B. and Izquierdo, M.T., *Carbon*, 35, 1005, 1997.
27. Rubio, B. and Izquierdo, M.T., *Carbon*, 36, 263, 1998.
28. Rubio, B. and Izquierdo, M.T., *Fuel*, 77, 631, 1998.
29. Moreno-Castilla, C., Corrasco-Marin, F., Utrera-Hidalgo, E., and Rivera-Utrilla, J., *Langmuir*, 9, 1378, 1993.
30. Liu, Q., Li, C., and Li, Y., *Carbon*, 41, 2217, 2003.
31. Liu, Q., Guan, J.S., Li, J., and Li, C., *Carbon*, 41, 2225, 2003.

32. Smisek, M. and Cerny, S., in *Active Carbon*, Elsevier Publ. Co., Amesterdem, 1970.
33. Jankowska, H., Swiatkowski, A., and Choma, J., in *Active Carbon*, Ellis Howard, England, 1991.
34. Barnir, J. and Aharoni, C., *Carbon*, 13, 363, 1975.
35. Chion, C.T. and Reocroft, P.J., *Carbon*, 15, 49, 1977.
36. Reocroft, P.J. and Chion, C.T., *Carbon*, 15, 285, 1977.
37. Singh, B., Saxena, A., Sharda, D., Yadav, S.S., Chintamani, D., Pandey, D., and Sekhar, K., *Proc. Indo. Carbon 2003 Conf.*, G.N. Mathur, V.S. Tripathi, T.L. Dhami, and O.P. Bahl, Eds., *Indian Carbon Soc.*, 2003, p. 33.
38. Capon, A., Alves, V.R., Smith, M.E., and White, *M.P., 15th Bienn. Conf. on Carbon*, 1981, Reprints, p. 232.
39. Hall, P.G., Gittins, P.M., Wuin, J.M., and Reherston, J., *Carbon*, 23, 353, 1985.
40. Wilson and Whetzel, U.S. Patent 1519470, 1924 Quoted in Ref. 33.
41. Choma, J., Horak, J., and Rozmarijanewiez, M., *Biul WAT*, 32, 75, 1983.
42. Choi, S.R., Lim, J.S., Baeg, S.J., Kim, M.J., Kim, K.H., and Ryu, S.K., in *Carbon '02, Intern. Carbon Conf.*, Beijing, 2002, Paper PI 63D055.
43. Baker, J.A. and Poziomek, E.J., *Carbon*, 12, 45, 1974.
44. Sinha, R.K. and Walker, P.L., Jr., *Carbon*, 10, 754, 1972.
45. Waker, P.L., Jr., *Carbon*, 10, 369, 1972.
46. Bashkova, S., Bagreev, A., and Bandosz, T.J., in *Carbon '02, Intern. Carbon Conf.*, Beijung, 2002, Paper 21.2.
47. Nuzzo, R.G., Zegarski, R.G., and Bubois, L.H., *J. Am. Chem. Soc.*, 109, 773, 1987.
48. Suzuki, T., Nishimura, I., Satogawa, S., Korai, Y., and Mochida, I., in *Carbon '02, Intern. Carbon Conf.*, Beijing, 2002, Paper PI 73 D072.
49. Rong, H. and Zheng, J., *Carbon 01 Intern. Carbon Conf.*, Lexington, KY, 2001, Paper 22.5.
50. Figueiredo, J.L., Pereira, M.F.R., Freitas, M.M.A., and Orfao, J.J.M., *Carbon*, 37, 1379, 1999.
51. Domingo-Garcia, M., Fernandez-Morales, F.J., Lopez-Garzon, F.J., Mereno-Castilla, C., and Perez-Mendoza, M., *Langmuir*, 15, 3226, 1999.
52. Rong, H., Ryu, Z., and Zheng, J., *Carbon '02, Intern. Carbon Conf.*, Beijing, 2002, Paper PI 94 D107.
53. EL-Sayed, Y. and Bandosz, T.J., in *Carbon '02, Intern, Carbon Conf.*, Beijing, 2002, Paper 33.2.
54. Salame, I.I. and Bandosz, T.J., in *Carbon 01 Intern. Carbon Conf.*, Lexington, KY, 2001.
55. Puri, B.R., Arora, V.M., and Verma, S.K., *J. Indian Chem. Soc.*, 56, 802, 1979.
56. Yun, J.H., Choi, D.K., and Kim, S.H., *Ind. Eng., Chem.*, 37, 1422, 1998.
57. Li, J., Yu, Z., Gao, M., Chao, F., and Cai, X., *J. Environ. Sci.*, 8, 167, 1996.
58. Reid, C.R. and Thomas, K.M., *J. Chem. Phys.*, 105, 1061, 2001.
59. Hazourli, S. and Bemnecaze, G., *Environ. Technol.*, P. 1275, 1996.
60. Choi, S.R., Kim, B.S., Lim, J.S., Lee, S.K., and Ryu, S.K., in *Carbon '02, Intern. Carbon Conf.*, Beijing, 2002, Paper 9.1.
61. Huang, Z.H., Kang, F., Yang, J.B. and Liang, K.M., *Carbon '02, Intern. Carbon Conf.*, Beijung, 2002, Paper 9.3.
62. Puri, B.R., Kaistha, B.C., Vardhan, Y., and Mahajan, O.P., *Carbon*, 11, 329, 1973.
63. Bansal, R.C. and Dhami, T.L., in *Proc. First Indian Conf. Carbon*, Dec. 1982, Indian Carbon Soc. Publ., 1984.

64. Shuixia, C., Ying, L., and Hanmin, Z., *Carbon '02, Intern. Carbon Conf.,* Beijung, 2002.
65. Saxena, A., Singh, B., Dubey, V., Banerjee, S., Semwal, R.P., and Panday, D., in *Proc. Indo. Carbon 2003*, G.N. Mathur, V.S. Tripathi, T.L. Dhami, and O.P. Bahl, Eds., Shipra Publishers, 2003, p. 44.
66. Zhang, Z., Luan, Z., and Cheng, D., in *Carbon '02, Intern. Carbon Conf.,* Beijung 2002, Paper PI 77 DO78.

Author Index

Numbers before the parentheses indicate that the author's work is referred to on this page in the text. The numbers in parentheses are the reference numbers, and the numbers that follow the parentheses gives the page on which the complete reference is listed.

A

Abbasi, W.A. 303 (101) 367
Abdel, S.M.F. 300 (55) 365
Abdi, M.A. 205 (41) 239; 229 (86) 241
Abdul Shafy, H. 300 (55) 365
Abe, I. 149 (10) 196; 418 (161) 441; 300 (64) 366
Abedinzadegan, M. 206 (45) 239
Abo-el-Wafa, U 300 (55) 365
Abu Baker, G.Z. 300 (47) 365
Abuzad, N.S. 394 (94) 439
Adachi, Y. 432 (206) 442
Adams, L.S. 201 (14) 238
Adams, M.D. 257 (44) 293; 257 (49) 293;
    257 (51) 293; 257 (52) 293;
    257 (53) 293; 300 (34) 365;
    331 (168) 369; 257 (50) 293
Addoun, A. 223 (81) 240; 58 (205) 65;
    136 (99) 143
Addoun, F. 58 (205) 65
Aemer, T.A. 286 (144) 295
Afanansev, A.D. 151 (38) 197
Afzal, M. 303 (100) 367; 300 (43) 365
Aggarwal, D. 8, 44 (32) 61; 44 (139) 63; 44 (141)
    64; 8, 44 (31) 61; 119, 120 (80) 143; 81 (9)
    141; 151, 152, 172 (39) 197; 151 (41) 197;
    157 (52) 197; 254, 259 (34) 292; 254, 259
    (33) 292; 321, 363 (159) 368; 321, 363 (160)
    368; 343 (201) 370; 343 (202) 370; 298 (13)
    364; 321, 363 (163) 368; 307, 321, 363 (128)
    367; 396, 397 (105) 440; 397 (108) 440; 424
    (187) 442; 379 (16) 437; 396, 397, 435 (104)
    439; 157 (54) 197; 316 (138) 368
Aggarwal, P. 402 (130) 440; 402 (131) 440; 407
    (135) 440
Aggarwal, V.K. 8 (18) 61
Aghahabazadeh, A. 229 (86) 241
Aharoni, C. 59 (209) 65; 453 (34) 471
Ahlagren, N. 333 (176) 369
Ahmad, S.Z., 300, 353 (51) 365
Ahmadpour, A. 206 (45) 239; 205 (41) 239; 229
    (86) 241

Ahmed, A.N. 449 (16) 470
Ahmed, M. 419 (163) 441
Ahmed, T.W. 300, 353 (51) 365
Akyol, F. 300, 341 (46) 365
Al-Baharani, K.S. 149 (8) 196
Al-Degs, Y. 419 (163) 441; 419 (164) 441
Al Duri, B. 420 (173) 441; 420 (174) 441; 420
    (175) 441; 420 (176) 441; 420 (177) 441
Alaerts, G. 304 (106) 367
Alcaniz-Monge, J. 206 (43) 239
Alchem, 299 (25) 365
Allali, H.M. 430 (200) 442
Allardice, D.J. 47 (159) 64
Allen, A.W. 253, 260 (23) 292
Allen, S. 419 (163) 441
Allen, S.J. 305 (119) 367; 305 (118) 367
Allgerei, G.D. 382 (26) 437
Allister, V.J. 426 (192) 442
Allmand, A.J. 126 (71) 142; 126 (72) 142
Almeida, F. 388 (47) 438
Alsema, G.L. 317 (148) 368
Alves, V.R. 59 (212) 65; 455 (38) 471
Aly, O.M. 300 (27) 365; 304,
    353 (113) 367
Amamkwah, K.A.G. 291 (177) 296
Ammons, R.D. 303, 333 (98) 367
Amy, G. 300 (37) 365
Amy, G.L 374 (3) 436
Andelman, J.B. 374 (2) 436
Anderson, R.B. 17 (67) 62; 409,
    435 (140) 441
Anderson, T.N. 255 (40) 292
Ando, J. 270 (95) 294
Andrade, J.D. 280 (125) 295
Andreeva, I. Yu. 300, 304, 316, 353 (52) 365
Andrews, M. 298 (9) 364
Andrews, R. 269 (92) 294; 263, 269, 280 (90)
    294; 290 (167) 296; 444, 445, 450, 451, 452
    (2) 470; 252, 260 (18) 292
Anirudhan, T.S. 356 (239) 371
Anjia, W. 204 (34) 239
Anlhonissen, J.H. 317 (148) 368

473

Subject Index

Printed in the United States
by Baker & Taylor Publisher Services